APPLICATIONS OF CALORIMETRY IN A WIDE CONTEXT – DIFFERENTIAL SCANNING CALORIMETRY, ISOTHERMAL TITRATION CALORIMETRY AND MICROCALORIMETRY

Edited by **Amal Ali Elkordy**

Applications of Calorimetry in a Wide Context – Differential Scanning Calorimetry, Isothermal Titration Calorimetry and Microcalorimetry
http://dx.doi.org/10.5772/2898
Edited by Amal Ali Elkordy

Contributors

Adriana Gregorova, Safia Alleg, Saida Souilah, Joan Joseph Suñol, P.V. Dhanaraj, N.P. Rajesh, Jose C. Martinez, Javier Murciano-Calles, Eva S. Cobos, Manuel Iglesias-Bexiga, Irene Luque, Javier Ruiz-Sanz, Diana Romanini, Mauricio Javier Braia, María Cecilia Porfiri, Ruel E. McKnight, Stefka G. Taneva, Sonia Bañuelos, María A. Urbaneja, Amal A. Elkordy, Robert T. Forbes, Brian W. Barry, Laura T. Rodriguez Furlán, Javier Lecot, Antonio Pérez Padilla, Mercedes E. Campderrós, Noemi E. Zaritzky, Pratima Parashar, Luis Alberto Alcazar-Vara, Eduardo Buenrostro-Gonzalez, W. Steinmann, S. Walter, M. Beckers, G. Seide, T. Gries, Eliane Lopes Rosado, Vanessa Chaia Kaippert, Roberta Santiago de Brito, R. F. B. Gonçalves, J. A. F. F. Rocco, K. Iha, Kazu-masa Yamada, Daniel Plano, Juan Antonio Palop, Carmen Sanmartín, Jindřich Leitner, David Sedmidubský, Květoslav Růžička, Pavel Svoboda, Eric A. Smith, Phoebe K. Dea, M.D.A. Saldaña, S.I. Martínez-Monteagudo

Published by InTech

Janeza Trdine 9, 51000 Rijeka, Croatia

Copyright © 2013 InTech

Notice

Publishing Process Manager Marina Jozipovic
Technical Editor InTech Prepress, Novi Sad
Cover InTech Design team

Additional hard copies can be obtained from orders@intechopen.com

Applications of Calorimetry in a Wide Context – Differential Scanning Calorimetry, Isothermal Titration Calorimetry and Microcalorimetry, Edited by Amal Ali Elkordy
p. cm.
ISBN 978-953-51-0947-1

For EU product safety concerns:
IN TECH d.o.o., Prolaz Marije Krucifikse Kozulić 3, 51000 Rijeka, Croatia,
info@intechopen.com or visit our website at intechopen.com.

Meet the editor

Dr. Amal Ali Elkordy is a Senior Lecturer in Pharmaceutics in the Department of Pharmacy, Health and Well-being, Faculty of Applied Sciences, University of Sunderland, UK. Her area of research interest is the stabilisation of protein formulations (using spray drying and crystallisation technology) and their delivery via oral and pulmonary routes. Her work in this field has been recognised by the award of two prizes at the British Pharmaceutical Conference in 2002 and in 2004. Her more recent work is concerned with the enhancement of poorly water soluble drugs and gene therapeutics (awarded national recognition from the College of Mental Health Pharmacists, 2010). Dr. Elkordy has many publications in peer-review journals and she was invited speaker in a number of conferences.

Contents

Preface XI

Section 1 Application of Differential Scanning Calorimetry into Pharmaceuticals 1

Chapter 1 **Application of Differential Scanning Calorimetry to the Characterization of Biopolymers 3**
Adriana Gregorova

Chapter 2 **Thermal Stability of the Nanostructured Powder Mixtures Prepared by Mechanical Alloying 21**
Safia Alleg, Saida Souilah and Joan Joseph Suñol

Chapter 3 **Studies on Growth, Crystal Structure and Characterization of Novel Organic Nicotinium Trifluoroacetate Single Crystals 49**
P.V. Dhanaraj and N.P. Rajesh

Section 2 Application of Isothermal Titration Calorimetry for Analysis of Proteins and DNA 71

Chapter 4 **Isothermal Titration Calorimetry: Thermodynamic Analysis of the Binding Thermograms of Molecular Recognition Events by Using Equilibrium Models 73**
Jose C. Martinez, Javier Murciano-Calles, Eva S. Cobos, Manuel Iglesias-Bexiga, Irene Luque and Javier Ruiz-Sanz

Chapter 5 **Applications of Calorimetric Techniques in the Formation of Protein-Polyelectrolytes Complexes 105**
Diana Romanini, Mauricio Javier Braia and María Cecilia Porfiri

Chapter 6 **Insights into the Relative DNA Binding Affinity and Preferred Binding Mode of Homologous Compounds Using Isothermal Titration Calorimetry (ITC) 129**
Ruel E. McKnight

Chapter 7 **Thermodynamic Signatures of Macromolecular Complexes – Insights on the Stability and Interactions of Nucleoplasmin, a Nuclear Chaperone 153**
Stefka G. Taneva, Sonia Bañuelos and María A. Urbaneja

Section 3 Application of MicroCalorimetry to Study Protein Stability and Folding Reversibility 183

Chapter 8 **Determination of Folding Reversibility of Lysozyme Crystals Using Microcalorimetry 185**
Amal A. Elkordy, Robert T. Forbes and Brian W. Barry

Chapter 9 **Calorimetric Study of Inulin as Cryo- and Lyoprotector of Bovine Plasma Proteins 197**
Laura T. Rodriguez Furlán, Javier Lecot, Antonio Pérez Padilla, Mercedes E. Campderrós and Noemi E. Zaritzky

Section 4 Thermal Analysis of Phase Transitions of Polymers and Paraffinic Wax 219

Chapter 10 **Silver Particulate Films on Compatible Softened Polymer Composites 221**
Pratima Parashar

Chapter 11 **Liquid-Solid Phase Equilibria of Paraffinic Systems by DSC Measurements 253**
Luis Alberto Alcazar-Vara and Eduardo Buenrostro-Gonzalez

Chapter 12 **Thermal Analysis of Phase Transitions and Crystallization in Polymeric Fibers 277**
W. Steinmann, S. Walter, M. Beckers, G. Seide and T. Gries

Section 5 Indirect Calorimetry to Measure Energy Expenditure 307

Chapter 13 **Energy Expenditure Measured by Indirect Calorimetry in Obesity 309**
Eliane Lopes Rosado, Vanessa Chaia Kaippert and Roberta Santiago de Brito

Section 6 Applications of Calorimetry into Propellants, Alloys, Mixed Oxides and Lipids 323

Chapter 14 **Thermal Decomposition Kinetics of Aged Solid Propellant Based on Ammonium Perchlorate – AP/HTPB Binder 325**
R. F. B. Gonçalves, J. A. F. F. Rocco and K. Iha

Chapter 15 **Numerical Solutions for Structural Relaxation of Amorphous Alloys Studied by Activation Energy Spectrum Model 343**
Kazu-masa Yamada

Chapter 16 **Thermal Analysis of Sulfur and Selenium Compounds with Multiple Applications, Including Anticancer Drugs 365**
Daniel Plano, Juan Antonio Palop and Carmen Sanmartín

Chapter 17 **Calorimetric Determination of Heat Capacity, Entropy and Enthalpy of Mixed Oxides in the System $CaO–SrO–Bi_2O_3–Nb_2O_5–Ta_2O_5$ 385**
Jindřich Leitner, David Sedmidubský, Květoslav Růžička and Pavel Svoboda

Chapter 18 **Differential Scanning Calorimetry Studies of Phospholipid Membranes: The Interdigitated Gel Phase 407**
Eric A. Smith and Phoebe K. Dea

Chapter 19 **Oxidative Stability of Fats and Oils Measured by Differential Scanning Calorimetry for Food and Industrial Applications 445**
M.D.A. Saldaña and S.I. Martínez-Monteagudo

Preface

This book (carrying at the beginning the name of "Calorimetry") started when I received an invitation from the InTech Open Access Publisher to be the editor of the book for my experience and publications in the field of applications of calorimetry and biocalorimetry in the analysis of small and large drug molecules. I welcomed the invitation and I was enthusiastic to handle chapters submitted from colleagues all over the world with the aim of disseminating the high quality research in application of calorimetry for the benefits of scientists, students, academics and industry (pharmaceutical, biopharmaceutical and food industries).

Calorimetry is an analytical method which can thermodynamically characterise the phase transition by determining heat capacities, enthalpies and melting temperatures of substances including oils, lipids, biological macromolecules, small drug molecules and polymers. It was an honour to read submitted chapters, to write a chapter and to divide the book into sections. Accordingly, the name of the book was changed into "Applications of Calorimetry in a Wide Context - Differential Scanning Calorimetry, Isothermal Titration Calorimetry and Microcalorimetry" to reflect the content of the book.

Finally, without the support of many other expert colleagues, who helped in the review process, completion of this book would have been difficult. The editor would like to thank the following scientists who have helped in the peer-review process: Prof. Brian Barry, Bradford School of Pharmacy, University of Bradford, UK; Dr. Paul Carter, Department of Pharmacy, Health and Well-being, University of Sunderland, UK; Dr. Shu Cheng Chaw, Department of Pharmacy, Health and Well-being, University of Sunderland, UK; Dr. Eman Ali Elkordy, Faculty of Medicine, University of Tanta, Egypt; Prof. Gamal El Maghraby, Faculty of Pharmacy, University of Tanta, Egypt; Dr. Ebtessam Ahmed Essa, Faculty of Pharmacy, University of Umm Al Qura, Saudi Arabia; Prof. Robert Forbes, Bradford School of Pharmacy, University of Bradford, UK; Dr. Wendy Hulse, Formulation technical specialist 2, Ipsen, UK.

Dr. Amal Ali Elkordy,
Department of Pharmacy, Health and Well-being,
Faculty of Applied Sciences,
University of Sunderland,
Sunderland, United Kingdom

Section 1

Application of Differential Scanning Calorimetry into Pharmaceuticals

Chapter 1

Application of Differential Scanning Calorimetry to the Characterization of Biopolymers

Adriana Gregorova

Additional information is available at the end of the chapter

http://dx.doi.org/10.5772/53822

1. Introduction

Generally, polymers can be classified according to their thermal and mechanical properties into thermoplastics, thermosets and elastomers. Thermoplastics are amorphous or semi-crystalline polymers that soft or melt during heating and solidify during cooling. The heating/cooling/heating process can be repeated without perceptible changes in thermal and mechanical properties of thermoplastics. Thermosets during heating undergo chemical changes and this process is irreversible. Elastomers can be vulcanized (cross-linked under assistance of heat, light, or special chemicals like sulfur, peroxides) that makes them reversibly stretchable for small deformations but vulcanization is the irreversible process.

The resulted properties of polymer materials and mixtures depend on the chemical and physical properties of neat polymers, additives as well as the used processing methodology. Differential scanning calorimetry (DSC) is a physical characterization method used to study thermal behavior of neat polymers, copolymers, polymer blends and composites. Generally, the non-isothermal DSC is used for the identification of neat basic polymers as well as the determination of their purity and stability. Amorphous polymers exhibit a glass transition temperature and semi-crystalline polymers may possess the glass transition temperature, a crystallization temperature, a melting temperature with various crystallization and melting enthalpies. However, these properties alter by both a presence of additives and applied polymer processing methodologies. Basically, a small quantity of sample (up to 10 mg) in pan from various materials (e.g. aluminum pan) and empty pan (reference) are treated under a defined temperature program (various combinations of thermal scans-heating/cooling, and isothermal cycles), a pressure (stable) and an atmosphere (inert or reactive). Principally, sample and reference are maintained at the same temperature, while any transition occurred in the sample needs an energy supply, which is recorded by the DSC as a rate dQ/dt against a temperature or a time. The DSC is the thermal analysis mainly used to determine a first-order transition (melting) and a second order endothermic transition (glass transition). The sudden change in the specific heat value, C_p corresponds with the glass transition temperature as follows (Bower, 2002):

$$\frac{dQ}{dt} = mC_p \tag{1}$$

where m is the mass of the sample.

However, the determination of the glass transition of polymers with a high crystallinity content is limited. The first-order transitions such as the crystallization of a polymer during a heating (cold crystallization) or a cooling cycle (crystallization) and a melting of polymer crystals can be described by the following formula (Bower, 2002):

$$\frac{dQ}{dt} = \kappa\Delta T = \kappa\dot{T}(t - t_0) + \left.\frac{dQ}{dt}\right|_{t_0} \tag{2}$$

where κ is a thermal conductance between a sample holder and a sample, $\dot{T}$ is a temperature increase rate, and t_0 is the start of transition.

Figure 1 shows the example of thermal transitions occurring in the injection molded sample of poly(lactic acid) (PLA) such as the glass transition, the cold crystallization and the melting. PLA is a thermoplastic aliphatic semi-crystalline biodegradable polyester. The presented molded sample had been cooled very rapidly during the processing (injection molding), so as the consequence during the second heating cycle appeared the cold crystallization peak.

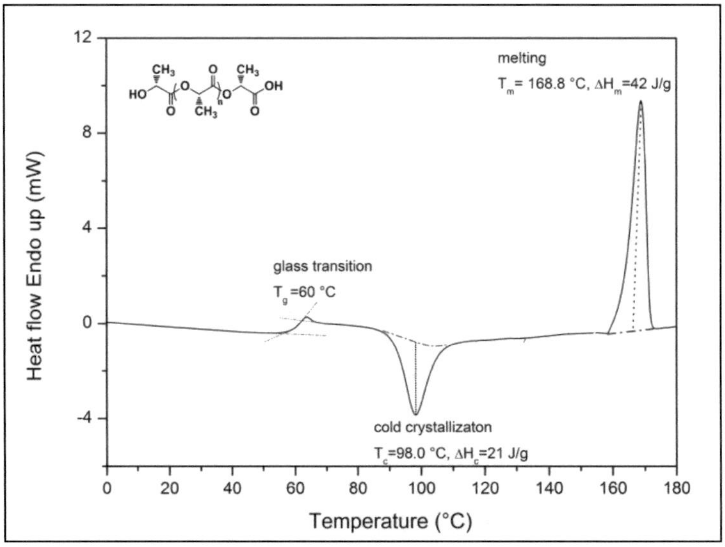

Figure 1. DSC thermogram of commercial poly(lactic acid) with $\overline{Mw}$ = 70 400 and PDI = 1.8 detected during 2nd heating cycle (0-180°C, 10°C/min, N_2 atmosphere)

There are two types of DSC systems: 1) heat-flux (sample and reference pans are in an identical furnace block) and 2) power compensation (sample and reference pans are in two separate furnace blocks). From the practical point of view, it is important to pay attention to issues influencing an accuracy of results as follows:

- an instrument calibration, baseline subtractions,
- a selection of working gas (N_2, He, O_2),
- a selection of pans (e.g. Al-, Pt-, Ni-, Cu-, Quartz-pans, hermetic or non-hermetic pans),
- a proper thermal contact between sample and pans,
- a temperature program (heating cycle usually should start about 50°C under and finish about 10-20°C above the expected measured transition temperature),
- a sufficient slow scanning rate (to avoid the neglecting of the requested thermal transition),
- a sufficient purity and source of sample (neat polymer, polymer blend, composite, before or after processing, kind of the processing).

The aim of this chapter is to show some examples of the practical use of the DSC within the investigation of an amorphous biopolymer – lignin and semi-crystalline biodegradable polymer – poly(lactic acid) as well as to discuss the dependence of the thermal thermal properties on the value of the molecular weight of polymer, the polymer processing methodology and the presence of additives in the polymer mixtures.

2. Effect of molecular weight on glass transition temperature

Amorphous and semicrystalline polymers undergo a phase change from a glassy to rubbery stage at a glass transition temperature (*Tg*).

At *Tg* the segmental mobility of molecular chains increases and a polymer is more elastic and flexible. The value of *Tg* is dependent on the various factors such as a molecular weight of polymer, a presence of moisture, a presence of the crystalline phase (in the case of semicrystalline polymers). The dependence of *Tg* on a number-average molecular weight is described by Flory-Fox equation:

$$T_g = T_g^{\infty} + \frac{K}{M_n} \tag{3}$$

where T_g^{∞} is a glass transition for polymer with the infinite number-average molecular weight, K is an empirical parameter related to the free volume in polymer and M_n is a number-average molecular weight of polymer.

2.1. Thermal properties of Kraft lignin extracted with organic solvents

In this sub-chapter, an example of the effect of various extraction solvents on molecular weight properties and thermal properties of Kraft lignin is shown.

Lignin is polydisperse amorphous natural polymer consisting of branched network phenylpropane units with phenolic, hydroxyl, methoxyl and carbonyl groups. Its molecular weight properties as well as functional groups depend on its genetic origin and used isolation method. Differential scanning calorimetry is the useful method to determine its glass transition temperature. The value of Tg depends on the molecular weight, the thermal treatment, the humidity content and the presence of various contaminants in lignin sample.

Generally, phenyl groups together with the cross-linking restrict the molecular motion of lignin as an amorphous polymer in contrast to propane chains. Moreover, the intermolecular hydrogen bonding decrease *Tg* in the contrast to the methoxyl groups (Hatakeyama & Hatakeyama, 2010). Lignin might be defined as a natural polymeric product produced by the enzymatic dehydrogenation polymerization of the primary methoxylated precursors such as *p*-coumaryl-, coniferyl- and sinapyl- alcohols (Figure 2).

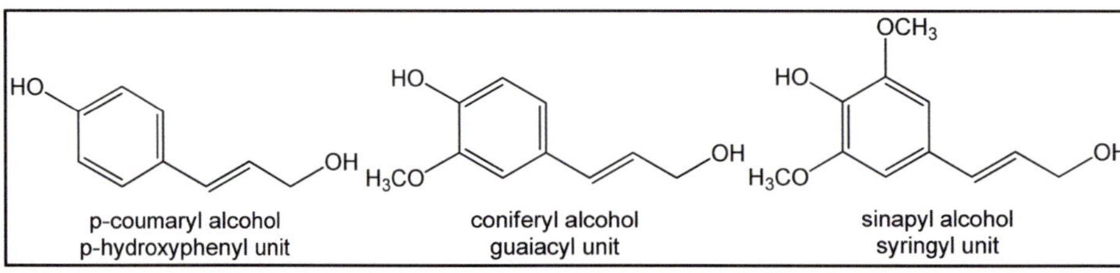

Figure 2. Lignin monomer building units

The structure of lignins depends on their natural origin and also on the external and internal conditions existing during lignin macromolecule synthesis and isolations. The large heterogeneity of lignin´s structures makes it difficult to determine the overall structure of lignin. High variability of substituents on phenyl propane unit together with auto-coupling reaction gives rise to different lignin´s structures depending on its origin and isolation method (Figure 3).

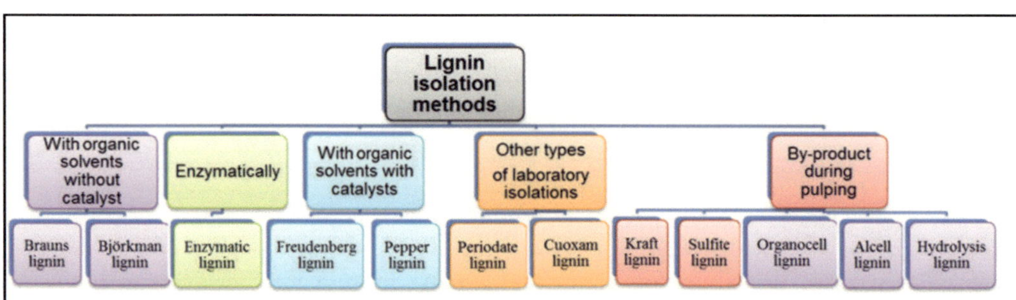

Figure 3. Lignin isolation methods

Kraft lignin used in this study was isolated from commercial spent pulping black liquor through the acidification with 98% sulphuric acid to pH=2 (Zellstoff Pöls AG, Austria). Precipitated, filtered, washed and dried Kraft lignin was extracted at the room temperature with organic solvents with Hildebrand solubility parameters in the range of 18.5-29.7 $MPa^{1/2}$ (see Table 1) and then again filtered and dried.

Solvent	Chemical formula	Hildebrand solubility parameter (MPa)$^{1/2}$	Polarity index
Dichlormethane	CH_2Cl_2	20.2	3.1
Tetrahydrofuran	C_4H_8O	18.5	4.0
Acetone	CH_3COCH_3	19.7	5.1
1,4-Dioxane	$C_4H_8O_2$	20.5	4.8
Methanol	CH_3OH	29.7	5.1

Table 1. Solvents used for Kraft lignin extraction

The determined thermal and molecular weight properties of Kraft lignins are shown in Table 2. The glass transition temperature (Tg) and the specific heat change (ΔC_p) were assessed by the differential scanning calorimetry (DSC) under the nitrogen flow, using the second heating cycle. Molecular weight properties were determined by a gel permeation chromatography (GPC) with the using of tetrahydrofuran as an eluent.

Sample	T_g (°C)	ΔC_p (Jg°C)	$\overline{M}_n$ (g/mol)	$\overline{M}_w$ (g/mol)	PDI
Kraft lignin_acetone	114	0.086	1030	1800	1.7
Kraft lignin_tetrahydrofuran	124	0.222	1170	3150	2.7
Kraft lignin_dichlormethane	59	0.260	750	940	1.3
Kraft lignin_methanol	105	0.368	910	1300	1.4
Kraf lignin_1,4-dioxane	120	0.367	1150	3070	2.7

Table 2. Thermal and molecular weight properties of Kraft lignins extracted in acetone, tetrahydrofuran, dichlormethane, methanol and 1,4-dioxane

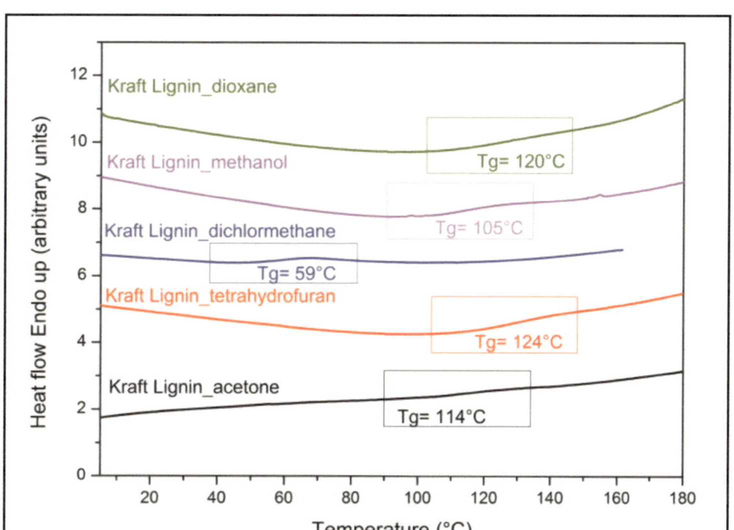

Figure 4. DSC thermograms of Kraft lignin extracted in acetone, tetrahydrofuran, dichlormethane, methanol and 1,4-dioxane detected during second heating scan (5-180°C, 10°C/min, N_2 atmosphere)

Figure 4 shows the thermograms of the individual Kraft lignins extracted with various organic solvents.

As can be seen from the results, the extraction as the last step used during the isolation process of Kraft lignin has a big effect on molecular as well as thermal properties of lignin.

2.2. Thermal properties of Poly(lactic acid) synthetized through azeotropic dehydration condensation

This sub-chapter shows the connection between PLA structure, its molecular weight properties and its thermal properties.

Poly(lactic acid) (PLA) is a biodegradable, thermoplastic, aliphatic polyester, which monomer can be derived from annually renewable resources. The glass transition temperature value is an important attribute that influences viscoelastic properties of PLA. The increase of the ambient temperature above Tg of PLA causes the sharp loss of its stiffness. The Tg values of PLA are influenced by its molecular weight, crystallinity, thermal history during processing, character of the side-chain groups and the presence of additives in the composition. The DSC analysis is one of the suitable methods to characterize the effect of the modification of PLA reactive side-chain groups on its thermal properties.

It is worth to mention that the melting temperature and the heat of fusion of polymers are influenced by thermal history applied during the polymer synthesis or processing. Therefore DSC results derived from 1st heating cycle give information concerning an actual state of polymer crystals and the application of cooling cycle erase the previous thermal history, e.g. annealing during processing. Some semi-crystalline polymers with the slow crystallization ability like poly(lactic acid) do not have time to crystallize during cooling and thus crystallize during 2nd heating cycle (cold crystallization) and consequently the melting peak may appear as double peak due to the content of different kinds of crystals. The melting behaviour of PLA is complex with regard to its multiple melting behaviour and polymorphism and has been intensively studied by several authors (Yasiniwa et al., 2004; Yasuniwa et al., 2006; Yasuniwa et al., 2007; Di Lorenzo, 2006).

PLA sample in the following example, marked as *PLA 0*, was synthetized by an azeotropic dehydration condensation in a refluxing boiling m-xylene from 80% L-lactic acid. During the azeotropic dehydration condensation samples *PLA_1-3* were modified by succinic anhydride in the concentration 0.7, 1.3 and 2.5 mol% (Gregorova et al., 2011a). Table 3 summarizes the nomenclature and molecular properties of non-modified PLA and PLA modified with various concentration of succinic anhydride.

Figure 5 shows DSC heating/cooling/heating thermogram of non-modified PLA with the molecular weight of 35 600 g/mol.

Generally, glass transition temperature is determined from the second heating cycle to provide T_g value independent on the thermal history during processing. The modification of PLA side-chain groups by succinic anhydride influenced not just molecular weight properties of PLA but also their thermal properties such as the glass transition temperature (*Tg*), the melting temperature (*Tm*) (in this case *Tm* was determined as the peak temperature of the melting peak) and the crystallinity (see Figure 6. and Table 4). As an adequate indicator of the crystallinity was chosen the specific heat of fusion, calculated as follows:

$$\sum \Delta H = (\Delta H_{m1} + \Delta H_{m2}) - \Delta H_c \quad (4)$$

where ΔH_{m1} and ΔH_{m2} are enthalpy values of the first and second melting peak, ΔH_c is the enthalpy of cold crystallization.

Sample	Concentration of succinic anhydride (mol%)	$\overline{M_n}$ (g/mol)	$\overline{M_w}$ (g/mol)	PDI
PLA_0	0	21400	35600	1.7
PLA_1	0.7	1950	3200	1.6
PLA_2	1.3	5600	9300	1.7
PLA_3	2.5	7000	13000	1.9

Table 3. Description of PLA samples and their molecular properties determined by GPC in chloroform

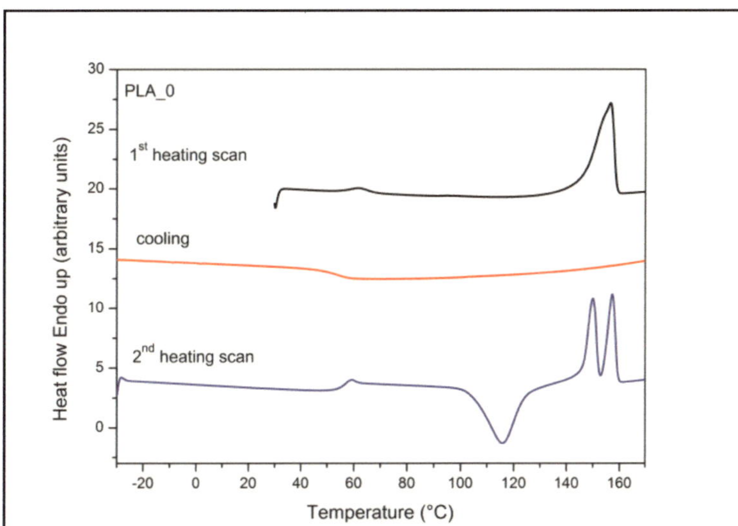

Figure 5. DSC thermogram of PLA_0 detected during heating/cooling/heating scan (30-170°C, 170-0°C, -30-170°C, 10°C/min, N_2 atmosphere)

By the comparison of the content of the crystalline phase determined from 1st heating and 2nd heating cycle, it can be seen that PLA samples during second heating cycle exhibit an amorphous character despite of the initially crystalline character determined from 1st heating scan. A thermal history is very important issue that influence the arrangement of amorphous/crystalline phase and consequently influence the physico-mechanical properties of poly(lactic acid).

Sample	1st heating cycle					2nd heating cycle							
	T_{m1} (°C)	ΔH_{m1} (J/g)	T_{m2} (°C)	ΔH_{m2} (J/g)	$\Sigma\Delta H$ (J/g)	T_g (°C)	T_c (°C)	ΔH_c (J/g)	T_{m1} (°C)	ΔH_{m1} (J/g)	T_{m2} (°C)	ΔH_{m2} (J/g)	$\Sigma\Delta H$ (J/g)
PLA_0	157	37.8	-	-	37.8	56	116	27.3	150	14.3	157	15.4	2.4
PLA_1	145	18.8	-	-	18.8	47	106	27.1	132	9.5	143	18.6	1.0
PLA_2	143	10.7	152	15.8	26,5	50	107	23.6	140	7.8	151	20.0	4.2
PLA_3	139	7.4	152	13.5	20,9	50	109	25.9	139	9.9	150	20.1	4.1

Table 4. Thermal properties of PLA synthetized through the azetropic dehydration condensation from 80% L-Lactic acid and modified by succinic anhydride

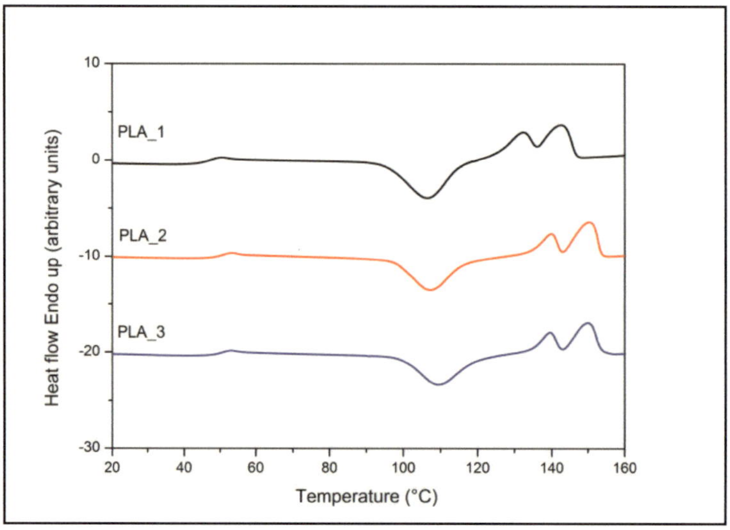

Figure 6. DSC thermograms of PLA samples with modified side chain groups and various molecular properties detected during second heating scan (-30-170°C, 10°C/min, N_2 atmosphere)

3. Effect of thermal treatment on thermal behavior of poly(lactic acid)

As was already discussed in the previous sub-chapter, PLA is the semi-crystalline polymer with the slow crystallization ability. Mechanical properties as well as gas barrier properties of PLA depend also on its gained crystallinity value. The resulting crystallinity of PLA can be modified by a thermal treatment (annealing) for some time at the crystallization temperature during the thermal processing of a sample. The change of a crystals size and a form during the annealing can be revealed by a X-Ray analysis but the change in the percentage of crystalline phase is detectable also by the DSC analysis. This section describes the progress of the PLA crystalline phase due to the applied annealing treatment. Moreover, the obtained DSC data are supported by a light microscopy study.

The followed data were obtained by the analysis of the thermal compression molded poly(lactic acid) synthetized by the azeotropic dehydration condenstation (PLA_3) (Figure 7).

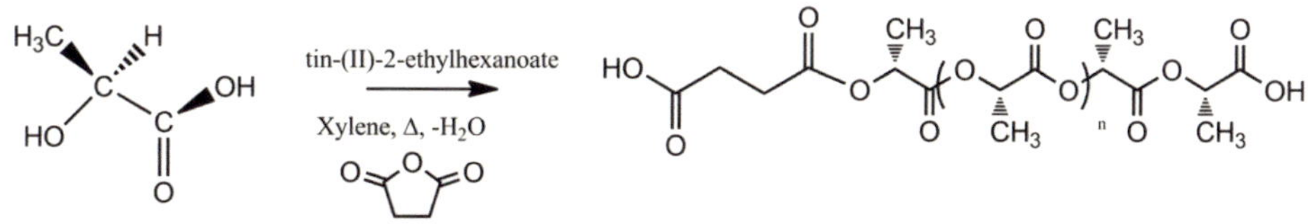

Figure 7. Structure of PLA_3 (PLA synthesized by the azeotropic dehydration condensation and modified by 2.5 mol% succinic anhydride)

The crystallinity value of PLA was modified during thermoprocessing by the thermal annealing at 110°C for 0, 5, 10, 15, 20, 30, 45, 60 and 120 min, respectively and afterwards cooled down to the room temperature. The samples are designated as PLA_3_110_X, where X indicates annealing time.

The clear effect of the thermal annealing on the PLA melting behavior is shown in Figure 8.

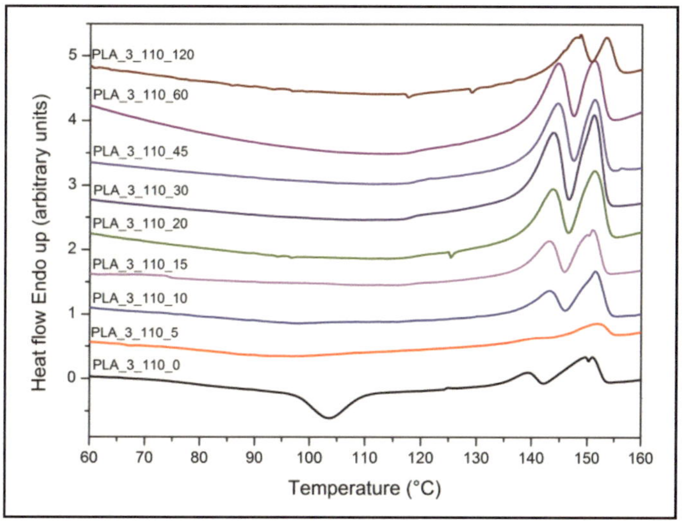

Figure 8. DSC thermograms of PLA_3 annealed at 110°C for 0-120 min (1st heating, 30-160°C, 10°C/min, N_2 atmosphere)

The change of the annealing time influenced the value of the specific melting enthalpy (ΣΔH) due to the enabling of a growth of crystals (Table 5).

The crystals morphology of PLA samples annealed at 110°C and various times were investigated by using of the light microscope with crossed polarizers (Figure 9). It can be seen that a shape and dimensions of the created crystals depend on the annealing time.

The DSC as well as the light microscopy analyses showed that the thermo-processed films without the annealing processing step have an amorphous character (Figure 9a), and on other side the application of the annealing processing step at 110°C during thermoforming instead of a quick direct cooling step (to the room temperature) promotes the growth of crystals. A kind, a size, a thickness, and a content of arisen crystals depend on the annealing temperature and time. DSC data displayed in Table 5 showed that the value of the specific

Sample	1st heating cycle								
	T_{c1} (°C)	ΔH_{c1} (J/g)	T_{m1} (°C)	ΔH_{m1} (J/g)	T_{m1}Peak height (mW)	T_{m2} (°C)	ΔH_{m2} (J/g)	T_{m2}Peak height (mW)	$\Sigma\Delta H$ (J/g)
PLA_3_110_0	104	21.8	139	4.6	0.14	151	20.5	0.39	3.3
PLA_3_110_5	-	-	-	-	-	152	12.6	1.18	12.6
PLA_3_110_10	-	-	143	6.1	0.26	151	15.4	0.67	21.5
PLA_3_110_15	-	-	143	20.1	0.51	151	13.5	0.65	33.6
PLA_3_110_20	-	-	143	19.9	0.76	151	15.9	1.0	35.8
PLA_3_110_30	-	-	144	20.8	1.1	151	15.6	1.4	36.4
PLA_3_110_45	-	-	144	14.5	0.87	151	13.0	1.03	27.5
PLA_3_110_60	-	-	144	15.2	0.87	151	11.8	0.88	27.0
PLA_3_110_120	-	-	149	10.3	0.44	154	5.6	0.44	15.9

Table 5. Thermal properties of PLA_3 films, annealed at 110°C for 0-120 min

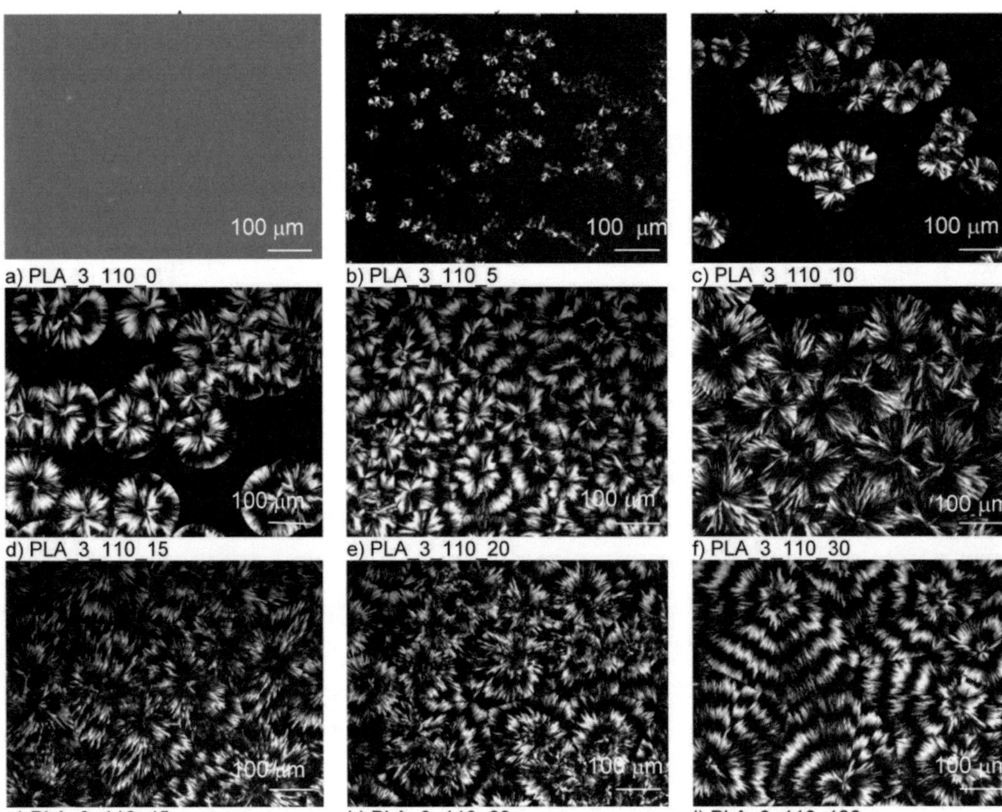

Figure 9. Polarized optical micrographs (magnification 400×) of crystals of polylactic acid modified with succinic anhydride (PLA_3) grown from the melt and annealed at 110°C for 5-120 min

heat of fusion markedly increased up to 15 min of the annealing time, but the extension of the annealing time up to 30 min increased $\Sigma\Delta H$ just slightly and further extension of the annealing time even decreased it. However, light micrographs of *PLA_3* (see Figure 9 b-i) show clear differences of the character of crystals, arisen from the samples annealed under and above 30 min. The application of the longer annealing time caused the creation of overgrowth crystals. The difference in the character of crystals can be also detected by the change of the height of the melting peak and by their shift to the higher temperatures. The value of $\Sigma\Delta H$ of PLA annealed for 120 min (*PLA_3_110_120*) is comparable to that of annealed just for 10 min, however the crystal morphology is markedly different. Furthermore, the change of the crystal morphology was indicated by the increase of the melting temperature (T_{m1} and T_{m2}) about 10 and 3°C, respectively. Also the optical micrograph displayed in Figure 9i showed the difference in the crystal morphology in a comparison to the previous samples annealed at the lower time. As a remark can be highlighted that the crystal morphology has an essential influence on resulting physico-mechanical properties of PLA materials.

4. Thermal stability of biopolymers determined by DSC

4.1. Effect of functional end groups on poly(lactic acid) stability

The intramolecular transesterification with the formation of cyclic oligomers and by-products like acrylic acid, carbon oxide and acetaldehyde is considered as one of the main mechanisms of the PLA thermal degradation. Above 200°C five reaction pathways have been found: intra-and intermolecular ester exchange, cis-elimination, radical and concerted nonradical reactions, radical reactions and Sn-catalyzed depolymerisation (Kopinke et al., 1996). It has been suggested that CH groups of the main chain and the character of functional end groups affect thermal and hydrolytic sensitivity of PLA (Lee et al., 2001; Ramkumar & Bhattacharya, 1998). In our previous work it was shown that thermal sensitivity of PLA might be improved by the modification of its functional end groups (Gregorova et al., 2011a). This sub-chapter shows that the DSC analysis can be used to determine the thermal stability of poly(lactic acid).

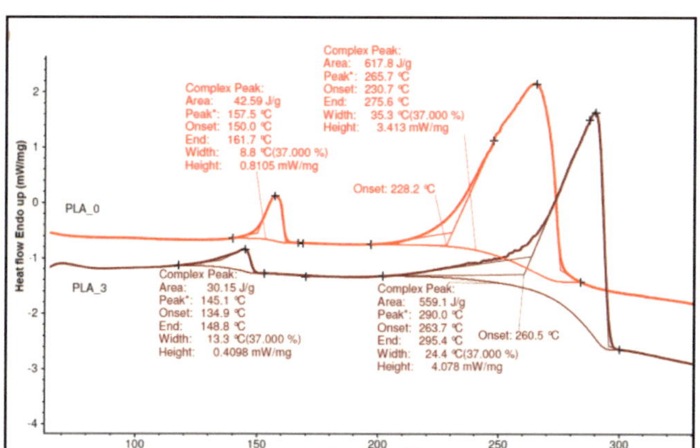

Figure 10. DSC curves of low molecular weight PLA synthetized by azeotropic dehydration condensation (PLA_0) and modified by 2.5 mol.% succinic anhydride (PLA_3), detected by 1st heating cycle from 30 to 350°C at heating rate of 10°C/min, in nitrogen flow.

The obtained DSC data, displayed in Figure 10, showed that the modification of low molecular weight PLA with succinic anhydride caused the decrease of its melting temperature and crystallinity. Furthermore, the detected values of the onset degradation temperature, the degradation temperature in peak and the enthalpy of degradation indicate the improvement of thermal stability, caused by the modification of hydroxyl functional end group by succinic anhydride.

4.2. Stabilizing effect of lignin used as filler for natural rubber

Natural rubber (NR) is highly unsaturated polymer exhibiting poor resistance to oxidation. For the inhibition of the degradation process during thermo-oxidation can be used stabilizers such as phenol and amine derived additives. NR for the production of vulcanized products is mixed with the number of the other compounding ingredients to obtain the desired properties of vulcanizates (e.g sulfur, accelerators, and filler). Lignin is biopolymer that can be used as an active filler for rubber. It was found that some lignins can play dual role in rubber compounds, influencing their mechanical properties as well as their stability [11].

The obtained data were obtained by using of vulcanizates based on natural rubber (NR) and filled with 0, 10, 20 and 30 phr of Björkman beech lignin (Mw= 2000, PDI= 1.2) (Kosikova et al., 2007). Samples are designated as NR_Lignin_X, where X presents concentration of lignin in phr (parts per hundred rubber).

Table 6 shows values of degradation temperature determined as the onset and the peak temperature in dependence on the lignin concentration in natural rubber vulcanizates. It can be seen that lignin used as filler exhibit also the stabilizing effect, while the best stabilizing effect was reached in the case of 20 phr presence of Björkman beech lignin.

Sample	T_{onset} (°C)	T_{peak} (°C)	ΔH (J/g)
NR_Lignin_0	184	326	886
NR_Lignin_10	183	349	833
NR_Lignin_20	301	368	363
NR_Lignin_30	296	364	318

Table 6. DSC data evaluated from 1st heating cycle analysis (30-500°C, 10°C/min, air atmosphere) of vulcanizates based on natural rubber (NR) and NR filled with Björkman beech lignin (Kosikova et al., 2007)

4.3. Stabilizing effect of lignin used as additive in polypropylene

It was already reported that the lignin in the certain circumstances can support the biodegradability of polymer samples (Kosikova et al., 1993a; Kosikova et al., 1993b; Mikulasova&Kosikova, 1999). On the other side lignin with the important functional groups and the low molecular weight with the narrow polydispersity can be used as the stabilizer for polypropylene (Gregorova et al., 2005a). This section shows that DSC is the sensitive method able to determine the stabilizing effect of lignin in polypropylene.

The polypropylene samples, stabilized with Björkman beech lignin (Mw= 2000, PDI= 1.2), used in this example were thermal processed with the injection molding (Gregorova et al., 2005a). Figure 11 shows the change of the onset oxidation temperature (T_{onset}) recorded for polypropylene stabilized with lignin. Generally, additives should be compatible with polymer matrix to keep physico-mechanical properties on the desired level; therefore it is necessary to know the lowest active concentration of the additive. It can be seen that the studied Björkman beech lignin increased T_{onset} about 15-30°C depending on the used concentration. On the base of the obtained mechanical properties of polypropylene/lignin composites, 2 wt% of Björkman beech lignin was determined as the optimal concentration to stabilize polypropylene. It was shown that the higher concentration of non-modified lignin deteriorated the mechanical properties of polypropylene (Gregorova et al., 2005a, Gregorova et al., 2005b).

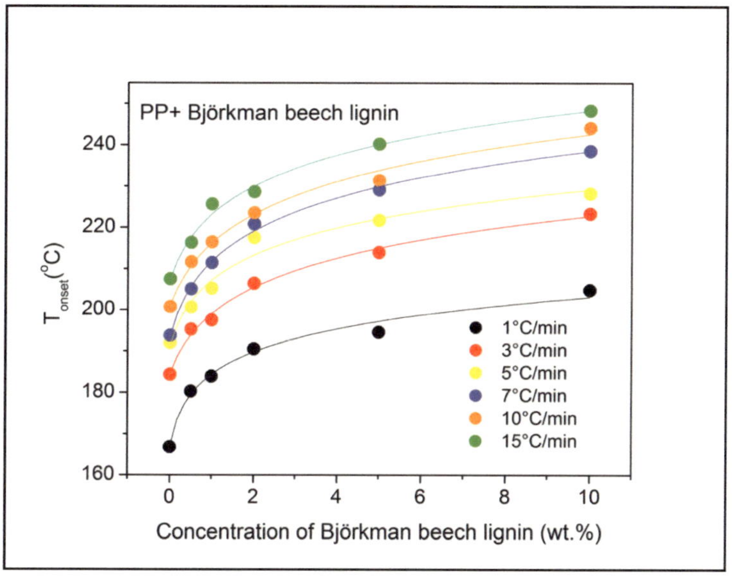

Figure 11. Thermal stability of polypropylene expressed as onset degradation temperature (T_{onset}) in dependence on lignin concentration, heating scan 30 to 500°C, heating rate of 1, 3, 5, 7, 10 and 15 °C/min, oxygen flow (Gregorova et al., 2005a).

5. Thermal properties of poly(lactic acid) composites

The incorporation of filler in PLA may change its crystallization behaviour and consequently its thermal properties. Some filler, such as wood flour or wood fibers, promote the transcrystallization and thus modify crystalline morphology of PLA (Mathew et al.; 2005 Pilla et al., 2008; Matthew et al., 2006; Hrabalova et al. 2010). This section describes the ability of hydrothermally pretreated beech flour to support a nucleation of PLA. Moreover, the effect of quick cooling and thermal annealing during thermal processing of PLA films is recorded.

The sample used in this section were thermoplastic processed compounds of commercial poly(lactic acid) (PLA 7000D, NatureWorks LLC, USA) plasticized with 10 vol% of polyethylene glycol 1500 and filled with 30 wt% of hydrothermally pretreated beech flour (Gregorova et al., 2011b). Composite films were prepared by thermal molding in press at 160°C, 10 MPa for 5 min and by modification of cooling process were prepared two morphologies: amorphous (quick cooling) and semi-crystalline (thermal annealing at 100°C for 45 min). The samples are designated as pPLA_X_100_Y, where X indicates filler (0-no filler, WF- hydrothermally pretreated beech flour) and Y shows annealing time.

The thermal behavior of quenched and annealed PLA composites, investigated by differential scanning calorimetry (heating cycle from 20 to 180°C, 10°C/min, 60 ml/L nitrogen flow) is summarized in Table 11 and shows that both filler incorporation of wood flour and thermal annealing influenced melting behavior and crystallinity of PLA composites. Specific melting enthalpy as an indicator of crystallinity degree of PLA in the composite was calculated as follows:

$$\Sigma\Delta H = \frac{(\Delta H_{m1} + \Delta H_{m2}) - \Delta H_c}{\upsilon} \quad (5)$$

where ΔH_{m1} and ΔH_{m2} are enthalpy values of the first and second melting peak, ΔH_c is the enthalpy of the cold crystallization and ν is volume fraction of PLA in the composite.

	1st heating cycle								2nd heating cycle
Sample	ΔH_c (J/g)	T_c (°)	ΔH_{m1} (J/g)	T_{m1} (°)	ΔH_{m2} (J/g)	T_{m2} (°)	ΣΔH (J/g)	$\Delta H_{m1}/\Delta H_{m2}$	T_g (°C)
pPLA_0_100_0	18.5	82	0.4	134	23.7	150	5.6	0.02	35
pPLA_0_100_45	-	-	11.8	142	14.2	152	26.0	0.83	35
pPLA_WF_100_0	14.2	95	1.7	140	12.9	151	0.4	0.13	43
pPLA_WF_100_45	-	-	6.1	148	8.4	155	14.4	0.73	45

Table 7. DSC thermal data of non-annealed and annealed PLA composites determined (Gregorova et al.; 2011b)

Samples that were after melting quickly cooled down to room temperature (quenched) exhibit cold crystallization and the double melting behavior that may be attributed to the melting of the original crystals and those of formed through the cold crystallization from the glassy state (Ling & Spruiell, 2006). The known slow crystallization ability of PLA and quick cooling process caused that quenched samples remained mostly amorphous that was proved by low value of specific enthalpy *ΣΔH*. Thermograms of annealed samples displayed a marked double melting peak showing high degree of crystallinity (Gregorova et al.;

2011b). The presence of filler marginally decreased specific enthalpy values of PLA. The presence of multiple melting peaks in thermograms of annealed samples can be explained by applied annealing that induce other crystal population, namely α' (initially created with grain like morphology) and α (during annealing created with spherulitic morphology) crystals (Zhang et al., 2008; Pan et al., 2008). Melting temperature for unannealed neat or filled PLA samples were recorded between 134-140°C for the first melting peak and 150-151°C for the second melting peak. The growth of crystals during annealing increased the values of temperature of both melting peaks depending on the mixture composition. The change in the value of the ratio of the first and the second melting peaks indicates the modification of size of the present crystals. The Tg value after an annealing treatment can be taken as an indicator for the occurred changes in an amorphous/crystal ratio but also in PLA/filler interaction level. The increase of an interfacial compatibility between wood filler and poly(lactic acid) can be detected by an shift of a glass transition to the higher temperature (Gregorova & Wimmer, 2012).

6. Conclusions

Differential scanning calorimetry is the method to characterize thermal behavior of polymeric materials on the base of the differences obtained in the heat flow between a sample and a reference under various temperature programs. In the addition to the quality and compositional analyses of polymers, DSC is applicable to the investigation of the thermal changes occurring in polymer systems during chemical reactions (e.g. polymerisation), oxidative degradation, vaporization, sublimation and desorptionThe selection of a proper temperature program is an important issue for the proper DSC analysis (e.g. a position and a shape of melting peak depend inherently on the nature of polymer and on the used heating scan rate). Thermal properties of biopolymers depend on many factors such as their natural origin, purity, composition, processing, thermal treatment, mechanical stressing, and aging. In this chapter, non-isothermal DSC was introduced as an method to investigate thermal properties of biopolymers, namely amorphous lignin and semi-crystalline poly(lactic acid). It can be concluded that DSC is one of the available methods to determine thermal properties of lignin with various molecular weight properties and composition.. Moreover, DSC can serve as a method to determine stabilizing effect of lignin used as an additive in polymer samples. Furthermore, DSC can be used as the quick method to measure melting behavior and the crystallinity of poly(lactic acid). The thermal history during polymer processing as well as the incorporation of some filler (e.g. wood flour) or additives can modify the crystallinity of PLA. The percentage of the crystallinity is one of the most important characteristics that influence its physico-mechanical behavior (stiffness, toughness, brittleness, barrier resistance, thermal stability and optical clarity). DSC is the valuable method for the investigation of thermal properties of biopolymers. However, it is necessary to use also the other additional physical and chemical testing methods to obtain complex data describing biopolymers, such as lignin and poly(lactic acid).

Author details

Adriana Gregorova
Graz University of Technology, Institute for Chemistry and Technology of Materials, Austria

7. References

Bower, B.I. (2002). *An Introduction to Polymer Physics.* Cambridge University Press, New York

Di Lorenzo, M.L. (2006). The Crystallization and Melting Processes of Poly(L-lactic acid). *Macromol. Symp.*, Vol.234, pp. 176-183

Gregorova, A.; Cibulkova, Z.; Kosikova, B.& Simon P. (2005a). Stabilization effect of lignin in polypropylene and recycled polypropylene. *Polymer Degradation and Stability*, Vol. 89, No. 3, pp. 553-558, ISSN 0141-3910

Gregorova, A.; Kosikova, B.& Osvald, A. (2005b). The study of lignin influence on properties of polypropylene composites. *Wood Research*, Vol. 50, No. 2, pp. 41-48, ISSN 1336-4561

Gregorova A.; Schalli M.& Stelzer F. (2011a). Functionalization of polylactic acid through azeotropic dehydrative condensation. *19th Annual Meeting of the BioEnvironmental Polymer Society BEPS, Book of Abstracts, PO-4*, ISBN 978-3-9502992-3-6, Vienna Austria, September 2011

Gregorova, A.; Sedlarik, V.; Pastorek, M.; Jachandra, H.& Stelzer, F. (2011b). Effect of compatibilizing agent on the properties of highly crystalline composites based on poly(lactic acid) and wood flour and/or mica. *Journal of Polymers and the Environment*, Vol. 19, No.2, pp. 372-381, ISSN 1566-2543

Gregorova, A& Wimmer R. (2012). Filler-Matrix Compatibility of Poly(lactic acid) Based Composites. In: Piemonte V., Editor. *Polylactic Acid: Synthesis, Properties and Applications*, Piemonte, V., pp. 97-119, Nova Science Publishers NY, ISBN 978-1-62100-348-9

Hatakeyama, H. & Hatakeyama T. (2010). Lignin Structure, Properties and Applications. In: *Biopolymers Lignin, Proteins, Bioactive Nanocomposites*, Abe A., Dusek K., Kobayashi S., pp. 11-12, Springer-Verlag Berlin Heidelberg

Hrabalova, M.; Gregorova, A.; Wimmer, R.; Sedlarik, V.; Machovsky, M.& Mundigler N. (2010). Effect of Wood Flour Loading and Thermal Annealing on Viscoelastic Properties of Poly(lactic acid) Composite Films. *Journal of Applied Polymer Science*, Vol. 118, No. 3, pp. 1534-1540, ISSN 1097-4628

Kopinke, F.D.; Remmler, M.; Mackenzie, K.; Möder, M.& Wachsen, O. (1996). Thermal decomposition of biodegradable polyesters-II. Poly(lactic acid). *Polymer Degradation and Stabilility*, Vol. 53, No. 3, pp. 329-342, ISSN 0141-3910

Kosikova, B.; Kacurakova, M.& Demianova V. (1993a). Photooxidation of the composite lignin/polypropylene films. *Chemical Papers*, Vol.47, pp. 132-136, ISSN 0366-6352

Kosikova, B.; Demianova, V.& Kacurakova, M. (1993b). Sulphur-free lignins as composites of polypropylene films. *Journal of Applied Polymer Science*, Vol. 47, No. 6, pp. 1065-1073, ISSN 1097-4628

Kosikova, B.; Gregorova, A.; Osvald, A.& Krajcovicova, J. (2007). Role of Lignin Filler in Stabilization of Natural Rubber-Based Composites. *Journal of Applied Polymer Science*, Vol. 103, No. 2, pp. 1226-1231, ISSN 1097-4628

Lee, S-H.; Kim, S.H.; Han, Y-K.& Kim Y.H. (2001). Synthesis and degradation of end-group-functionalized polylactide. *Journal of Polymer Science Part A: Polymer Chemistry*, Vol. 39, No. 7, pp. 973-985, ISSN 1099-0518

Ling, X.& Spruiell, J.E. (2006). Analysis of the complex thermal behaviour of poly(L-lactic acid) film. I. Samples crystallized from the glassy state. *Journal of Polymer Science Part B: Polymer Physics*, Vol. 44, No. 22, pp. 3200-3214, ISSN 1099-0488

Mathew, A.P.; Oksman, K.& Sain, M. (2005). Mechanical properties of biodegradable composites from poly lactic acid (PLA) and microcrystalline cellulose (MCC). *Journal of Applied Polymer Science*, Vol.97, No. 5, pp. 2014-2015, ISSN 1097-4628

Mathew, A.P.; Oksman, K.& Sain, M. (2006). The effect of morphology and chemical characteristics of cellulose reinforcements on the crystallinity of polylactic acid. *Journal of Applied Polymer Science*, Vol. 101, No. 1, pp. 300-310, ISSN 1097-4628

Mikulasova, M.& Kosikova, B. (1999). Biodegradability of lignin-polypropylene composite films. *Folia Microbiologica*, Vol. 44, pp. 669-672, ISSN 0015-5632

Pan, P.; Zhu, B.; Kai, W.; Dong, T.& Inoue, Y. (2008). Polymorphic Transition in Disordered Poly(L-lactide) Crystals Induced by Annealing at Elevated Temperatures. *Macromolecules*, Vol.41, No. 12, pp. 4296-4304, ISSN 1520-5835

Pilla, S.; Gong, S.; O'Neil, E.; Rowell, M.& Krzysik, A.M. (2008). Polylactide-pine wood flour composites. *Polymer Engineering Science*, Vol.48, No. 3, pp. 578-587, ISSN 1548-2634

Ramkumar, D.H.S.& Bhattacharya, M. (1998). Steady shear and dynamic properties of biodegradable polyesters. *Polymer Engineering Science*, Vol. 38, No. 9, pp. 1426-1435, ISSN 1548-2634

Yasuniwa, M.; Tsubakihara, S.; Sugimoto, Y.& Nakafuku, C. (2004). Thermal analysis of the double-melting behavior of poly(L-Lactic acid). *Journal of Polymer Science Part B: Polymer Physics*, Vol.42, No. 1, pp. 25-32, ISSN 1099-0488

Yasuniwa, M.; Tsubakihara, S.; Iura, K.; Ono, Y.; Dan, Y.& Takashashi K. (2006). Crystallization behavior of Poly(L-lactic acid). *Polymer*, Vol. 47, No. 21, pp. 7554-7563, ISSN 0032-3861

Yasuniwa, M.; Iura, K.& Dan Y. (2007). Melting behavior of poly(L-lactic acid): Effects of crystallization temperature and time. *Polymer*, Vol. 48, No. 18, pp. 5398-5407, ISSN 0032-3861

Zhang, J.; Tashiro, K.; Tsuji, H.& Domb, A.J. (2008). Disorder-to-Order Phase Transition and Multiple Melting Behavior of Poly(L-lactide) Investigated by Simultaneous Measurements of WAXD and DSC. *Macromolecules*, Vol.41, No. 4, pp. 1352-1357, ISSN 1520-5835

Chapter 2

Thermal Stability of the Nanostructured Powder Mixtures Prepared by Mechanical Alloying

Safia Alleg, Saida Souilah and Joan Joseph Suñol

Additional information is available at the end of the chapter

http://dx.doi.org/10.5772/54151

1. Introduction

Nanocrystalline materials present an attractive potential for technological applications and provide an excellent opportunity to study the nature of solid interfaces and to extend knowledge of the structure-property relationship in solid materials down to the nanometer regime. Nanocrystalline materials can be produced by various methods such as mechanical alloying, inert gas condensation, sol–gel process, electrodeposition, chemical vapour deposition, heat treatment of amorphous ribbons, high speed deformation, etc. Mechanical alloying is a non-equilibrium process resulting in solid state alloying beyond the equilibrium solubility limit. During the milling process, mixtures of elemental or prealloyed powders are subjected to heavy plastic deformation through high-energy collision from the balls. The processes of fracturing and cold welding, as well as their kinetics and predominance at any stage, depend mostly on the deformation characteristics of the starting powders. As a result of the induced heavy plastic deformation into the powder particles during the milling process, nanostructured materials are produced by the structural decomposition of coarser-grained structure. This leads to a continuous refinement of the internal structure of the powder particles to nanometer scales.

Solid-state processing is a way to obtain alloys in states far-from-equilibrium. The microstructural manifestations of the departures from equilibrium achieved by mechanical alloying can be classified as follows: (i) **augmented defect concentrations** such as vacancies, interstitials, dislocations, stacking faults, twin boundaries, grain boundaries as well as an increased level of chemical disorder in ordered solid solutions and compounds; (ii) **microstructural refinement** which involves finer scale distributions of different phases and of solutes; (iii) **extended solid solubility**; a stable crystalline phase may be found with solute levels beyond the solubility limit at ambient temperature, or beyond the equilibrium limit at any temperature; and (iv) **metastable phases** which may form during processing like crystalline, quasicrystalline and intermetallic compounds. Chemical reactions can proceed towards equilibrium in stages, and the intermediate stages can yield a metastable phase. In the solid state amorphization reaction, an amorphous alloy can be produced by the reaction of two solid metallic elements. Severe mechanical deformation can lead to metastable states. The deformation forces the production of disturbed configurations or brings different phases into intimate contact promoting solid-state reactions.

The alloying process can be carried out using different apparatus such as planetary mills, attrition mills, vibratory mills, shaker mills, etc. [1]. A broad range of alloys, solid solutions, intermetallics and composites have been prepared in the nanocrystalline, quasicrystalline or amorphous state [2-10]. A significant increase in solubility limit has been reported in many mechanically alloyed systems [11, 12]. Several studies of the alloy formation process during mechanical alloying have led to conflicting conclusions like the interdiffusion of elements, the interactions on interface boundaries and/or the diffusion of solute atoms in the host matrix. Indeed, the alloying process is complex and hence, involves optimization of several parameters to achieve the desired product such as type mill, raw material, milling intensity or milling speed, milling container, milling atmosphere, milling time, temperature of milling, ball-to-powder weight ratio, process control agent, etc. The formation of stable and/or metastable crystalline phases usually competes with the formation of the amorphous phase. For alloys with a negative heat of mixing, the phase formation has been explained by an interdiffusion reaction of the components occurring during the milling process [13]. Even though the number of phases reported to form in different alloy systems is unusually large [14], and property evaluations have been done in only some cases and applications have been explored, the number of investigations devoted to an understanding of the mechanism through which the alloy phase's form is very limited. This chapter summarizes the information available in this area. The obtained disordered structures by mechanical alloying are metastable and therefore, they will experience an ordering transition during heating resulting in exothermic and/or endothermic reactions. The thermal properties of materials are strongly related to the size of nanocrystals essentially when the radius of nanocrystals is smaller than 10 nm. Hence, an important task of thermal analyses is to find the size-dependent function of the thermodynamic amounts of nanocrystalline materials.

2. Thermodynamic stability

The state of a physical system evolves irreversibly towards a time-independent state in which no further macroscopic physical or chemical changes can be seen. This is the state of thermodynamic equilibrium characterized, for example, by a uniform temperature throughout the system but also by other futures. A non-equilibrium state can be defined as a state where irreversible processes drive the system towards the equilibrium state at different rates ranging from extremely fast to extremely slow. In this latter case, the isolated system may appear to have reached equilibrium. Such a system, which fulfils the characteristics of an equilibrium system but is not the true equilibrium state, is called a metastable state. Both stable and metastable states are in internal equilibrium since they can explore their complete phase space, and the thermodynamic properties are equally well defined for metastable states as for stable states. However, only the thermodynamically stable state is in global equilibrium; a metastable state has higher Gibbs energy than the true equilibrium state.

Thermodynamically, a system will be in stable equilibrium, under the given conditions of temperature and pressure, if it is at the lowest value of the Gibbs free energy:

$$G = H - TS \tag{1}$$

Where H is enthalpy, T absolute temperature and S entropy. According to equation (1), a system can be most stable either by increasing the entropy or decreasing the enthalpy or both. At low temperatures, solids are the most stable phases since they have the strongest atomic bonding (the lowest H), while at high temperatures the -TS term dominate. Therefore, phases with more freedom of atomic movement, such as liquids and gases are most stable. Hence, in the solid-state transformations, a close packed structure is more stable at low temperatures, while a less close packed structure is most stable at higher temperatures. A metastable state is one in internal equilibrium, that is, within the range of configurations to which there is access by continuous change, the system has the lowest possible free energy. However, if there were large fluctuations (the nucleation of a more stable phase), transformation to the new phase would occur if the change in free energy, ΔG, is negative. A phase is non-equilibrium or metastable if it's Gibbs free energy is higher than in the equilibrium state for the given composition. If the Gibbs free energy of this phase is lower than that of other competing phases (or mixtures thereof), then it can exist in a metastable equilibrium. Consequently, non-equilibrium phases can be synthesized and retained at room temperature and pressure when the free energy of the stable phases is raised to a higher level than under equilibrium conditions, but is maintained at a value below those of other competing phases. Also, if the kinetics during synthesis is not fast enough to allow the formation of equilibrium phase(s), then metastable phases could form.

3. Transformation mechanism

During the mechanical alloying process, continuous fracturing, cold welding and rewelding of the powder particles lead to the reduction of grain size down to the nanometer scale, and to the increase of the atomic level strain. In addition, the material is usually under far-from-equilibrium conditions containing metastable crystalline, quasi-crystalline or amorphous phases. All of these effects, either alone or in combination, make the material highly metastable. Therefore, the transformation behaviour of these powders to the equilibrium state by thermal treatments is of both scientific and technological importance. Scientifically, it is instructive to know whether transformations in ball milled materials take place *via* the same transformation paths and mechanisms that occur in stable equilibrium phases or not. Technologically, it will be useful to know the maximal use temperature of the ball milled material without any transformation occurring and thus, losing the special attributes of this powder product. One of the most useful techniques for studying transformation behaviour of metastable phases is differential scanning calorimetry (DSC) or differential thermal analysis (DTA). Hence, a small quantity of the powder milled for a given time is heated at a constant rate to high temperatures under vacuum or in an inert atmosphere to avoid oxidation. Depending on the phase transformations, DSC/DTA scans exhibit endothermic and/or exothermic peaks related to absorption or evolution of heat, respectively, as shown in Fig. 1.

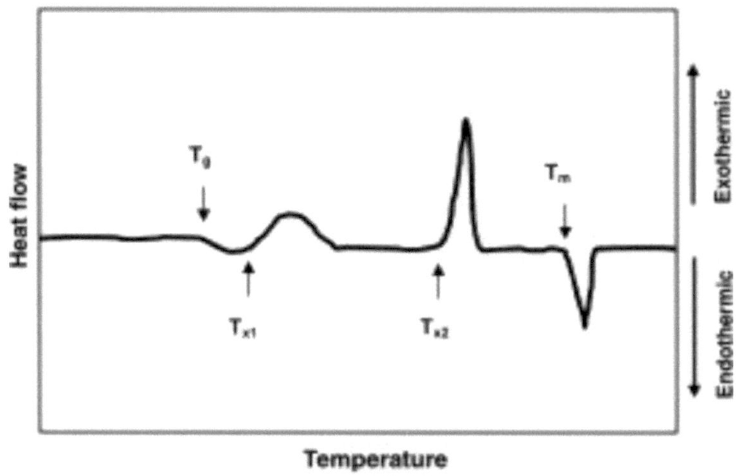

Figure 1. A schematic DSC curve depicting the different stages during crystallization of an amorphous phase where Tg is the glass transition temperature; T_m the melting temperature, T_{x1} and T_{x2} are the onset crystallization temperatures [15].

The values of the peak onset temperature and peak areas depend on the position of the baseline. Therefore, the accurate baseline can be obtained by heating the sample to the desired temperature, then cooled it back to the ambient temperature and then reheated it to higher temperatures. The second DSC scan could be used either as the baseline or subtracted from the first scan to obtain the accurate peak positions and areas. There are two types of transformations: reversible and irreversible. For the former, the product phase will revert back to the parent phase. For example, transformation from one equilibrium phase to another on heating gives rise to an endothermic peak during melting and exothermic peak during cooling. However, during irreversible transformation of metastable phases such as amorphous phases, a peak of the opposite sign is not observed. In fact, there will be no peak at all. Furthermore, because metastable phases are always more energetic than the corresponding equilibrium phases, they often exhibit exothermic peaks in the DSC/DTA curves. If an amorphous alloy powder is heated to higher temperatures, one expects to observe a broad exothermic reaction at relatively low temperatures related to structural relaxation of the amorphous phase, a glass transition temperature as well as one or more exothermic peaks corresponding to crystallization event at higher temperatures. Structural changes that occur during crystallization can be investigated by X-rays diffraction or Mössbauer spectrometry by quenching the sample from a temperature just above the DSC/DTA peak temperature. Transmission electron microscopy investigations can also be conducted to uncover the microstructural and crystal structure changes on a finer scale. In addition, compositional changes can be detected. It may be pointed out, however, that there have not been many detailed crystallization studies of amorphous alloys synthesized by the mechanical alloying process [16].

3.1. Non-isothermal transformation

The crystallization temperature corresponds to the maximum of the exothermic peak,T_p and it increases with increasing heating rate. A relation between heating rate β and position of the transformation peak T_p first described by Kissinger [17], has been extensively used to determine the apparent activation energy for crystallization E_a:

$$\ln\frac{\beta}{T_p^2} = \left(-\frac{E_a}{RT_p}\right) + A \tag{2}$$

Where A is a constant and R is the universal gas constant. The activation energy E_a can be calculated from the slope $\left(\frac{AE_a}{R}\right)$ of the plot $\left(\frac{\beta}{T_p^2}\right)$ against $\left(\frac{1}{T_p}\right)$. Further informations about the transformation temperatures, the number of stages in which the transformation is occurring, details about the product(s) of each individual transformation (crystal structure, microstructure and chemical composition), and the activation energy (and also the atomic mechanism) can be obtained with the combination of DSC/DTA and X-rays diffraction/transmission electron microscopy techniques. The Kissinger method may not be useful in all studies of decomposition. For example, it may not be applicable for metallic glasses which may decompose by nucleation/growth, or a combination of both processes, where the decomposition is seldom described by first-order reaction kinetics [18, 19]. Solid state reactions sometimes exhibit first-order kinetics, this is one form of the Avrami-Erofeev equation (n=1). Such kinetic behaviour may be observed in decompositions of fine powders if particle nucleation occurs on a random basis and growth does not advance beyond the individual crystallite nucleated. The physical interpretation of E_a depends on the details of nucleation and growth mechanisms, and in some cases equation (2) is not valid. For each crystallization peak, the calorimetric results can be explained using the Johnson-Mehl-Avrami-Erofe've kinetics equation [20] for the transformed fraction:

$$\frac{dx}{dt} = K(T)f_n(x) \tag{3}$$

With:

$$f_n(x) = n(1-x)\{-\ln(1-x)\}^{(n-1)/n} \tag{4}$$

$f_n(x)$ gives the transformation rate at time t and temperature T in terms of the rate constant:

$$K(T) = k_0 exp(E/RT) \tag{5}$$

k_0 is the pre-exponential factor; E is the effective activation energy and n is the kinetic exponent. According to the Avrami exponent value, the reaction may be three-dimensional, interface-controlled growth with constant nucleation rate (n=4); three-dimensional, interface-controlled growth with zero nucleation rate (n=3) or diffusion-controlled with growth and segregation at dislocations (n=2/3)[21].

3.2. Isothermal transformation

Isothermal transformation kinetics study at different temperatures can be conducted by the Kolmogorov-Johnson-Mehl-Avrami formalism [22-25] in which the fraction transformed, x, exhibits a time dependence of the form:

$$x(t) = 1 - \exp(-kt)^{\mathrm{n}} \tag{6}$$

Where n is the Avrami exponent that reflects the nucleation rate and/or the growth mechanism; $x(t)$ is the volume of transformed fraction; t is the time, and k is a thermally-activated rate constant. The double logarithmic plot $\ln(-\ln(1-x))$ against $\ln t$ should give a straight line, the slope of which represents the order of reaction or Avrami parameter n. The rate constant k is a temperature-sensitive factor $k = k_0 exp(AE_a/RT)$, where E_a is the apparent activation energy and k_0 a constant. $x(t)$ corresponds to the ratio between the area under the peak of the isothermal DSC trace, at different times, and the total area. Such analysis was conducted on the phase transformation mechanisms in many mechanically alloyed powders since the milling process occurs at ambient temperature for different milling durations [26-31]. If the Kolmogorov-Johnson-Mehl-Avrami analysis is valid, the value of n should not change with either the volume fraction transformed, $\mathrm{V_f}$ or the temperature of transformation. Calka and Radlinski [32] have shown that the usual method of applying the Kolmogorov-Johnson-Mehl-Avrami equation and calculating the mean value of Avrami exponent over a range of volume fraction transformed, may be inappropriate, even misleading, if competing reactions or changes in growth dimensionality occur during the transformation progress. Also, a close examination of the Avrami plots reveals that there are deviations from linearity over the full range of volume fraction transformed [33]. The first derivative of the Avrami plot $\delta\,[\ln(-\ln(1-n))]/\delta \ln t$ against the volume fraction transformed [34], which effectively gives the local value of n with V_f, seems to be more sensitive. Such a plot allows a more detailed evaluation of the data and can emphasize changes in reaction kinetics during the transformation process.

4. Mechanical alloying process

Mechanical alloying has received a great interest in developing different material systems. It is a solid state process that provides a means to overcome the drawback of formation of new alloys starting from mixture of low and/or high melting temperature elements. Mechanical alloying is a ball milling process where a powder mixture placed in the vials is subjected to high-energy collisions from the balls. The two important processes involved in ball milling are fracturing and cold welding of powder particles in a dry high energy ball-mill. The alloying process can be carried out using different apparatus such as planetary or horizontal mills, attrition or spex shaker mill. The elemental or prealloyed powder mixture is charged in the jar (or vial) together with some balls. As a result of the induced heavy plastic deformation into the powder particles during the milling process, nanostructured materials are produced by the structural decomposition of coarser-grained structure. This leads to a continuous refinement of the internal structure of the powder particles down to nanometer scales.

Depending on the microstructure, the mechanical alloying process can be divided into many stages: initial, intermediate, final and complete [35]. Since the powder particles are soft in the early stage of milling, so they are flattened by the compressive forces due to the collisions of the balls. Therefore, both flattened and un-flattened layers of particles come into intimate contact with each other leading to the building up of ingredients. A wide range of particle sizes can be observed due to the difference in ductility of the brittle and ductile powder particles. The relatively hard particles tend to resist the attrition and compressive forces. However, if the powder mixture contains both ductile and brittle particles (Fig. 2a), the hard particles may remain less deformed while the ductile ones tend to bind the hard particles together [10, 36]. Cold welding is expected to be predominant in fcc metals (Fig. 2b) as compared to fracture in bcc and hcp metals (Fig. 2c).

During the intermediate stage of milling, significant changes occur in the morphology of the powder particles. Greater plastic deformation leads to the formation of layered structures (Fig. 2d). Fracturing and cold welding are the dominant milling processes. Depending on the dominant forces, a particle may either become smaller in size through fracturing or may agglomerate by welding as the milling process progresses. Significant refinement in particle size is evident at the final stage of milling. Equilibrium between fracturing and cold welding leads to the homogeneity of the particles at the macroscopic scale as shown in Fig. 2d for the $Fe_{50}Co_{50}$ powder mixture [37, 38]. True alloy with composition similar to the starting constituents is formed at the completion of the mechanical alloying process (Fig. 2e) as evidenced by the energy dispersive X analysis, EDX, (Fig. 2f). The large plastic deformation that takes place during the milling process induces local melting leading to the formation of new alloys through a melting mechanism and/or diffusion at relatively high temperature.

Mechanical alloying is a non-equilibrium process resulting in solid state alloying beyond the equilibrium solubility limit. Several studies of the alloy formation process during mechanical alloying have led to conflicting conclusions such as the interdiffusion of elements, the interactions on interface boundaries and/or the diffusion of solute atoms in the host matrix. Indeed, Moumeni et al. have reported that the FeCo solid solution was formed by the interdiffusion of Fe and Co atoms with a predominance of Co diffusion into the Fe matrix according to the spectrometry results [37]. However, Brüning et al. have shown that the FeCo solid solution was formed by the dissolution of Co atoms in the Fe lattice [39]. Sorescu et al. [40] have attributed the increase of the hyperfine magnetic field to a progressive dissolution of Co atoms in the bcc–Fe phase. Such discrepancies have been attributed to the milling conditions and/or to the fitting procedure of the Mössbauer spectra. The role of grain boundaries, the proportions and the thickness of which are dependent on the milling energy affect thus, the hyperfine structure originating some misinterpretations.

Diffusion in mechanical alloying differs from the steady state diffusion since the balance of atom concentration at the interface between two different components may be destroyed by subsequent fracturing of the powder particles. Consequently, new surfaces with different compositions meet each other to form new diffusion couples when different powder particles are cold welded together. Large difference in composition at the interface therefore promotes interdiffusion. In addition, the change in temperature during the milling process

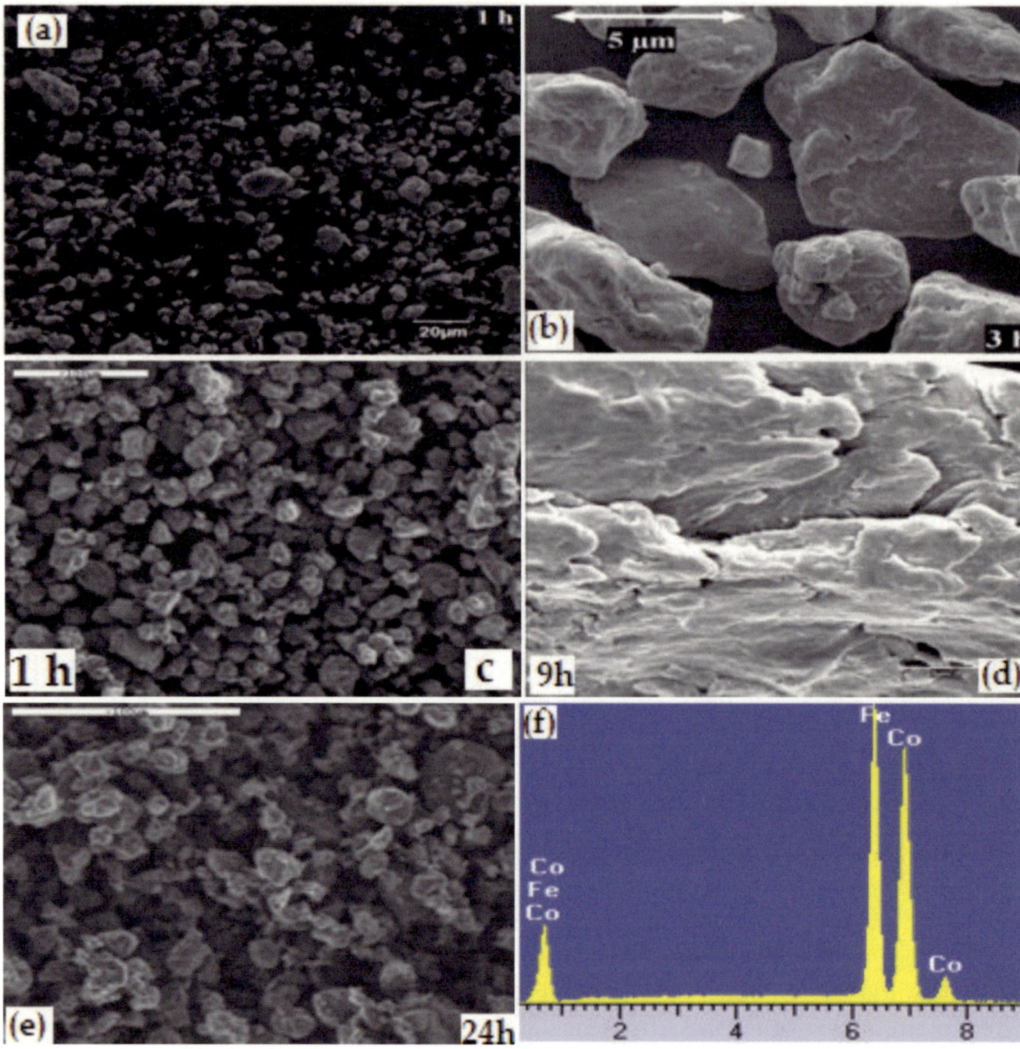

Figure 2. Morphologies of powder particles of the ball-milled $Fe_{75}Si_{15}B_{10}$ (a), $Ni_{20}Co_{80}$ (b), $Fe_{57}Cr_{31}Co_{12}$ (c and d), and $Fe_{50}Co_{50}$ powders (e) with the corresponding EDX analysis (f).

is very significant due to the exothermic reaction causing local combustion. Two major phenomena can contribute to the increase in milling temperature: friction during collisions and localized plastic deformation. At low temperatures, surface diffusion dominates over grain boundary and lattice diffusion. As the temperature is increased, however, grain boundary diffusion predominates, and at higher temperature lattice diffusion becomes the principal mode of diffusion. The first key factor controlling the formation of new alloys is the activation energy which is related to the formation of defects during balls-powder-balls and/or balls-powder-vials collisions. The second key is the vial temperature which is associated with plastic deformation as well as sliding between powder particles and high energetic balls and powder particles. The third key is the crystallite size that is related to the formation of nanometer crystalline structure during the milling process.

5. Experimental section

Mechanical alloying process was used to prepare nanocrystalline and/or amorphous alloys such as Fe, Fe-Co, Fe-Co-Nb-B, Fe-P and Ni-P from pure elemental powders in high-energy planetary ball-mills Fritsch Pulverisette P7 and Retsch PM 400/2, and vibratory ball-mill spex 8000. The milling process was performed at room temperature, under argon atmosphere, with different milling conditions such as rotation speed, ball-to-powder weight ratio, milling time and composition. In order to avoid the temperature increase inside the vials, the milling process was interrupted for 15–30 min after each 30–60 min depending on the raw mixture.

Particles powder morphology evolution during the milling process was followed by scanning electron microscopy. Structural changes were investigated by X-ray diffraction in a (θ–2θ) Bragg Brentano geometry with Cu-K$_\alpha$ radiation (λ_{Cu}=0.15406 nm). The microstructural parameters were obtained from the refinement of the X-rays diffraction patterns by using the MAUD program [41, 42] which is based on the Rietveld method. Differential scanning calorimetry was performed under argon atmosphere. Magnetic and hyperfine characterizations were studied by vibrating sample magnetometer and Mössbauer spectrometry, respectively.

6. Fe and FeCo-based alloys

6.1. Fe and Fe-Co powders

Fe and $Fe_{50}Co_{50}$ were prepared by mechanical alloying from pure elemental iron and cobalt powders in a planetary ball mill Fritsch P7, under argon atmosphere, using hardened steel vials and balls. The milling intensity was 400 rpm and the ball-to-powder weight ratio was 20:1. A disordered bcc FeCo solid solution is obtained after 24 h of milling (Fig 3), having a lattice parameter, a = 0.2861(5) nm, larger than that of the coarse-grained FeCo phase (a = 0.2825(5) nm). Such a difference in the lattice parameter value may be due to heavily cold worked and plastically deformed state of the powders during the milling process, and to the introduction of several structural defects (vacancies, interstitials, triple defect disorder, etc.).

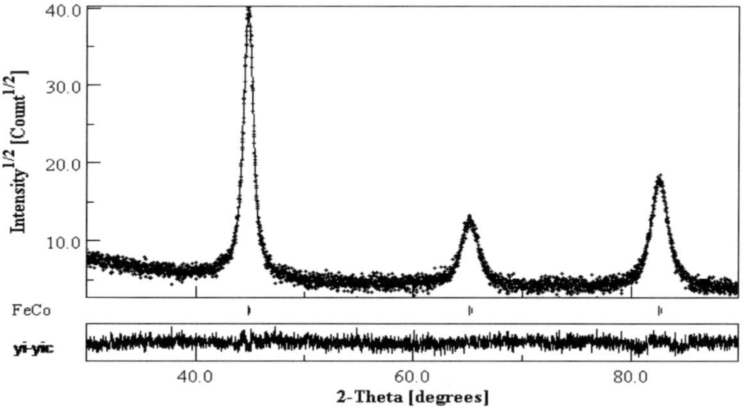

Figure 3. Rietveld refinement of the XRD pattern of the $Fe_{50}Co_{50}$ powders milled for 40 h [7].

With increasing milling time, the crystallite size decreases down to the nanometer scale and the internal strain increases. The double logarithmic plot of the crystallite size versus milling time exhibits two-stage behaviour for both Fe and $Fe_{50}Co_{50}$ powders (Fig. 4). A linear fit gives slopes of –0.65 and –0.20 for short and extended milling times, respectively, in the case of Fe; and slopes of –0.85 and –0.03, respectively, for short and extended milling times in the case of $Fe_{50}Co_{50}$ mixture. The critical crystallite size achievable by ball milling is defined by the crossing point between the two regimes with different slopes [43]. Consequently, the obtained critical crystallite sizes are of about 13.8 and 15 nm for Fe and $Fe_{50}Co_{50}$ powders, respectively. By using different milling conditions (mills type, milling intensity and temperature) to prepare nanostructured Fe powders, Börner et al. have obtained the two-regime behaviour, for the grain refinement by using the Spex mill, with slopes of –0.41 and –0.08 for short and extended milling times, respectively. However, the crystallite sizes show only a simple linear relation with slopes of –0.265 and –0.615 by using the Retsch MM2 shaker and the Misuni vibration mill, respectively. The obtained critical crystallite size value was 19 nm [44].

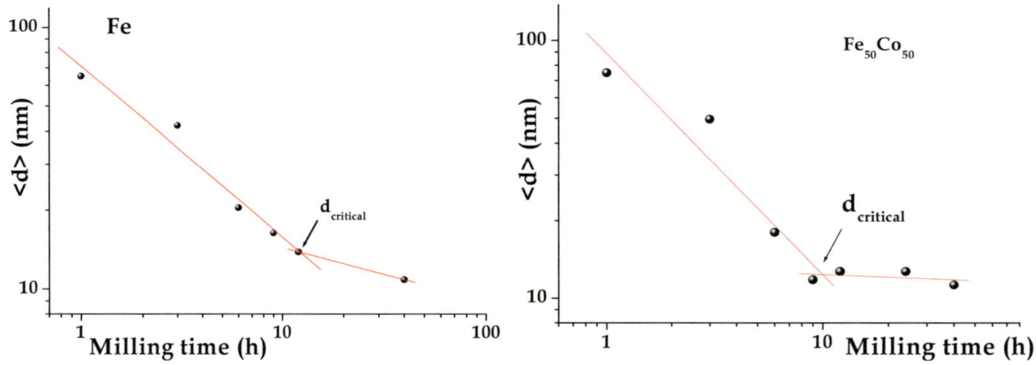

Figure 4. Double logarithmic plot of the crystallite size against milling time for nanostructured Fe and $Fe_{50}Co_{50}$ powders [7].

DSC scans of nanostructured Fe and $Fe_{50}Co_{50}$ powders milled for 40 h are shown in Fig. 5. The non-equilibrium state is revealed by the broad exothermic reaction for both samples, in the temperature range 100–700°C, which is consistent with the energy release during heating due to recovery, grain growth and relaxation processes. As a result of the cold work during the milling process, the main energy contribution is stored in the form of grain boundaries and related strains within the nanostructured grains which are induced through grain boundary stresses [45]. It has been reported that the stored energies during the alloying process largely exceed those resulting from conventional cold working of metals and alloys. Indeed, they can achieve values typical for crystallization enthalpies of metallic glasses corresponding to about 40% of the heat of fusion, ΔH_f [45]. The major sources of mechanical energy storage are both atomic disorder and nanocrystallite boundaries because the transition heats evolving in the atomic reordering and in the grain growth are comparable in value [46].

For the nanostructured Fe powders, the first endothermic peak is linked to the bcc ferro-paramagnetic transition temperature, T_C, and the second peak to the bcc→fcc transition

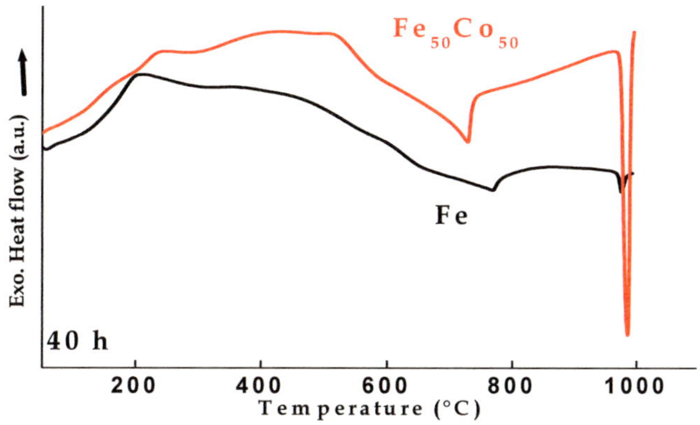

Figure 5. DSC scans of nanostructured Fe and $Fe_{50}Co_{50}$ powders milled for 40 h [7].

temperature, $T_{\alpha\rightarrow\gamma}$. The depression of Curie temperature with increasing milling duration (Fig. 6), which is ascribed to changes in local order, indicates that the nearest-neighbour coordination is essentially changed in the magnetic nanocrystallites. This reflects to some extent that there are more open disordered spaces or the nearest–neighbour coordination distance in the nanometer sized crystallites is increased, caused by lattice distortion. In fact, if the crystallite sizes are small enough, the structural distortions associated with surfaces or interfaces can lower the Curie temperature. This can be correlated to the increase of the lattice parameter and its deviation from that of the perfect crystal. It has been reported on far-from-equilibrium nanostructured metals, that interfaces present a reduced atomic coordination and a wide distribution of interatomic spacing compared to the crystals and consequently, the atomic arrangement at the grain boundary may be considered close to the amorphous configuration and should therefore alter the Curie temperature. The most reported values of Tc do not deviate strongly from that of the bulk materials. For example, the Tc of 360°C for Ni[C] nanocrystals is in good agreement with that of bulk Ni [47]. Host et al. have reported a Tc value of 1093°C for carbon arc produced Co[C] nanoparticles, in good agreement with the 1115°C value for bulk Co [48]. The Curie temperature of 10 nm Gd is

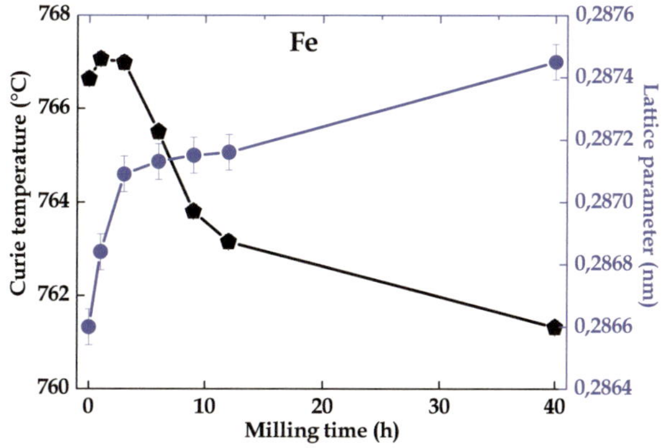

Figure 6. Evolution of the Curie temperature and the lattice parameter of the Fe powders as a function of milling time [7].

decreased by about 10 K from that of coarse-grained Gd while the magnetic transition is broader [49]. According to both Tc and $T_{\alpha\rightarrow\gamma}$ temperature values, the paramagnetic nanostructured bcc α–Fe domain is extended by about 50°C at the expense of both magnetic bcc α–Fe and nonmagnetic fcc γ–Fe as compared to coarse-grained bcc α–Fe.

The disorder-order phase transformation temperature of the nanostructured FeCo powders which is nearly constant (~724°C) along of the milling process (Fig. 7), is comparable to that of bulk Fe-Co alloys. It is commonly accepted that Fe-Co undergoes an ordering transition at around 730°C, where the bcc structure takes the ordered α'–CsCl(B2)-type structure [50]. The ordering effect in the FeCo nanocrystals has been revealed by the changes in the magnetization upon heating and the temperature variation of the coercivity on heating and cooling [51]. Also, the phase transformation temperature from bcc–α to fcc–γ structure in the $Fe_{50}Co_{50}$ powders is rather milling time independent (~982°C). The lower resistivity of $Fe_{50}Co_{50}$ compared to that of pure Fe at 300 K [52] and the higher Curie temperature of $Fe_{50}Co_{50}$ suggest that there is less scattering of the conduction electrons by the magnetic excitations. Thus, the Curie temperature cannot be clearly observed because there is a phase transformation from the bcc to fcc form at 985°C.

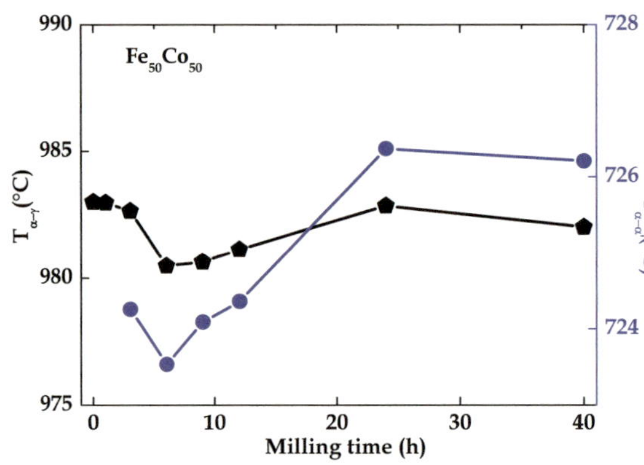

Figure 7. Evolution of the order-disorder, $T_{\alpha\rightarrow\alpha'}$ and the bcc→fcc, $T_{\alpha\rightarrow\gamma}$, temperatures of the $Fe_{50}Co_{50}$ powders as a function of milling time [7].

6.2. Fe-Co-Nb-B powders

Nanostructured and disordered structures obtained by mechanical alloying are usually metastable. Depending on the Nb and B contents, the mechanically alloyed Fe-Co-Nb-B powders structure may be partially amorphous either magnetic and/or paramagnetic. Pure elemental powders of iron (6-8 μm, 99.7%), cobalt (45 μm, 99.8%), niobium (74 μm, 99.85%) and amorphous boron (> 99%) were mixed to give nominal compositions of $Fe_{57}Co_{21}Nb_7B_{15}$ and $Fe_{61}Co_{21}Nb_3B_{15}$ (wt. %), labelled as 7Nb and 3Nb, respectively. The milling process was performed in a planetary ball-mill Fritsch Pulverisette 7, under argon atmosphere, using hardened steel balls and vials. The ball-to-powder weight ratio was about 19/2 and the rotation speed was 700 rpm. For the $(Fe_{50}Co_{50})_{62}Nb_8B_{30}$ mixture, the milling process was performed in a planetary ball-mill Retsch PM400/2, with a ball-to-powder weight ratio of about 8:1 and a rotation speed of 350 rpm. In order to avoid the increase of the temperature inside the vials, the milling process was interrupted after 30 min for 15 min.

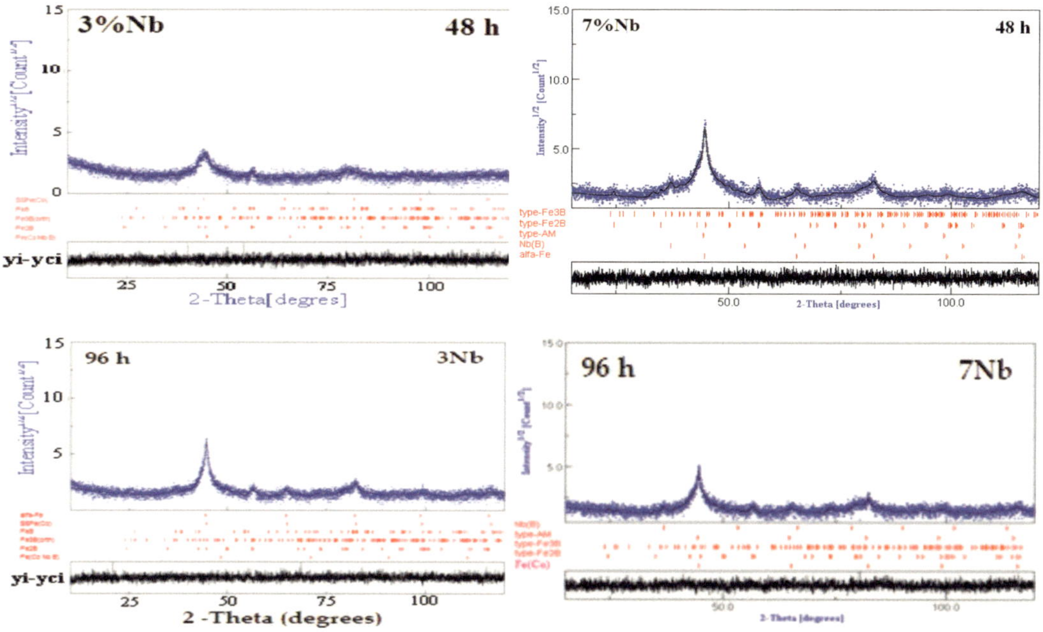

Figure 8. Rietveld refinement of the XRD patterns of 7Nb and 3Nb powders milled for 48 and 96 h [53, 54].

The XRD patterns of 7Nb and 3Nb mixtures milled for 48 h (Fig. 8) are consistent of a large number of overlapping diffraction peaks related to different phases. The Rietveld refinement reveals the formation of a partially amorphous structure of about ~78%, where nanocrystalline tetragonal–Fe_2B, tetragonal–Fe_3B and bcc–FeCo type phases were embedded for 3Nb powders [53]. Whereas, for 7Nb powders, the milling product is a mixture of amorphous (~73.6%), bcc–Nb(B), tetragonal–Fe_2B, orthorhombic–Fe_3B and bcc FeCo type phases [54]. Further milling (up to 96 h) leads to the increase of the amorphous phase proportion for 7Nb and the mechanical recrystallization in the case of 3Nb mixture (Fig. 8) as evidenced by the decrease and the increase of the diffraction peaks intensity, respectively. The formation of the amorphous phase is confirmed by the Mössbauer spectrometry results as shown in Fig. 9. After 48 h of milling, the Mössbauer spectra exhibit more or less sharp absorption lines superimposed upon a broadened spectral component assigned to the structural disorder of the amorphous state [55]. For 3Nb powders, the mechanical recrystallization is confirmed by the emergence of sharp sextet related to the primary crystallization of α–Fe and FeCo after 96 h of milling. However, a stationary state is achieved for 7Nb powders. The increase of the average hyperfine magnetic field, $< B_{hyp} >$, from 19.18 to 23.14 T after 96 h of milling of 3Nb powders is correlated to the decrease/increase of the amorphous/nanocrystalline relative area. The nanocrystalline (NC) component consists of Fe sites with B_{hyp}>31 T and the interfacial (IF) one is related to the nanostructured Fe–borides with B_{hyp} ranged from 24 to 30 T [28, 56, 57].

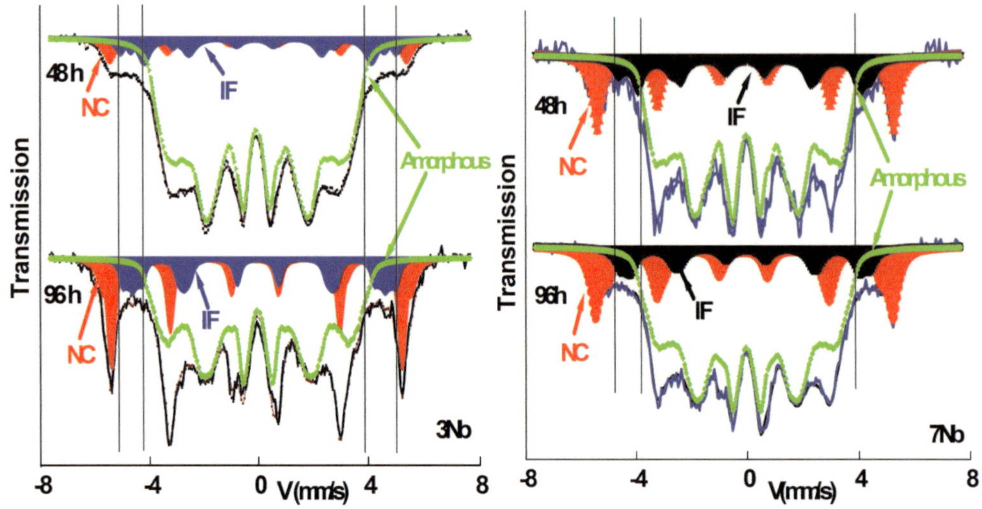

Figure 9. Room temperature Mössbauer spectra of 3Nb and 7Nb powders milled for 48 and 96 h [55].

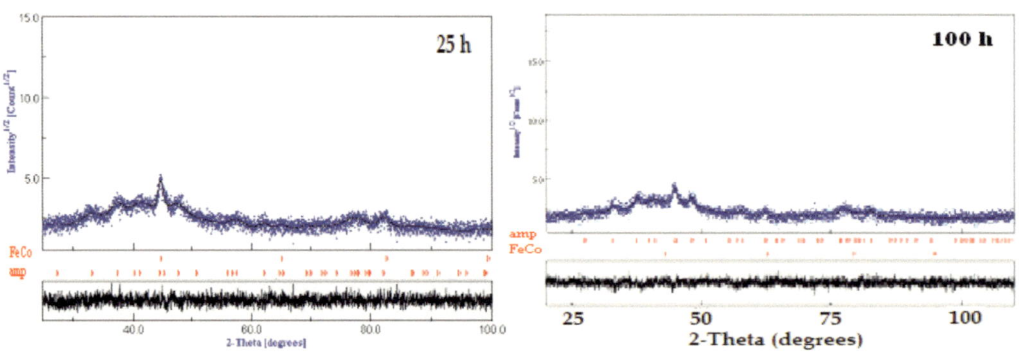

Figure 10. XRD patterns of the $(Fe_{50}Co_{50})_{62}Nb_8B_{30}$ powders milled for 25 and 100 h.

For the $(Fe_{50}Co_{50})_{62}Nb_8B_{30}$ powders mixture milled for 25 and 100 h, the best Rietveld refinements of the XRD patterns were obtained with two components: bcc–FeCo and amorphous phase (Fig. 10). The complete transformation of the heavily deformed FeB and bcc FeCo type phases into an amorphous state is achieved, after 125 h of milling, through the mechanically enhanced solid-state amorphization which requires the existence of chemical disordering, point defects (vacancies, interstitials) and lattice defects (dislocations). Indeed, the severe plastic deformation strongly distorts the unit cell structures making them less crystalline. The powder particles are subjected to continuous defects that lead to a gradual change in the free energy of the crystalline phases above those of amorphous ones, and hence to a disorder in atomic arrangement. The Mössbauer spectra confirm the formation of a paramagnetic amorphous structure, where about 3.8% of FeCo and Fe_2B nanograins are embedded, after 125 h of milling (Fig. 11).

Nanocrystalline $Fe_{72.5}Co_{7.5}Nb_{5+x}B_{15-x}$ with x=0, 5 and 10 at.% labelled as A, B and C, respectively, were prepared by mechanical alloying from pure elemental powders in a planetary ball-mill Retsch PM400, under argon atmosphere, using stainless steel balls and vials. The ball-to-powder weight ratio was about 8:1 and the rotation speed was 200 rpm [58]. The crystallite size decreases with increasing milling duration to about (7.1 ± 0.3) nm for the B-richest alloy (A). The XRD patterns (Fig. 12) reveal the formation of a bcc Fe-rich solid solution after 80 h of milling having an average lattice parameter of about 0.2871 nm for the three alloys.

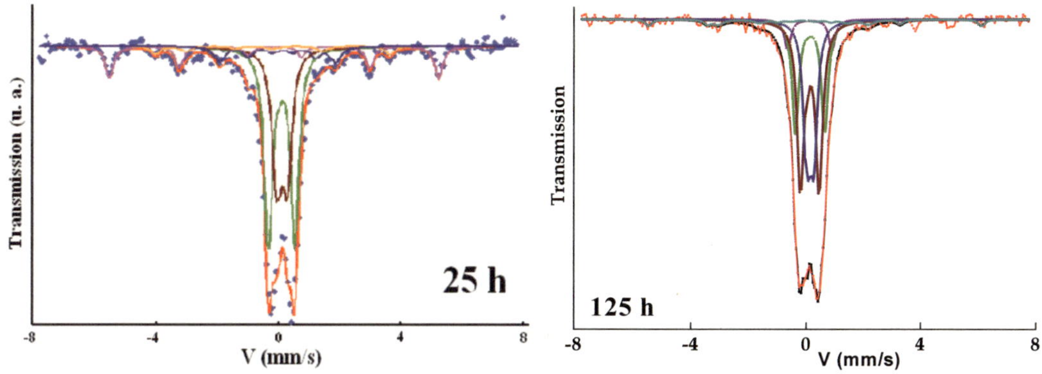

Figure 11. Room temperature Mössbauer spectra of the $(Fe_{50}Co_{50})_{62}Nb_8B_{30}$ powders milled for 25 and 125 h.

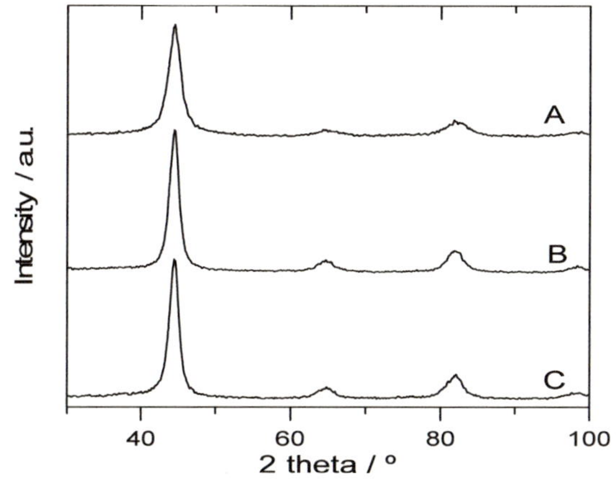

Figure 12. XRD patterns of alloys A, B and C milled for 80 h [58].

Depending on the structural state after each milling time, several exothermic and endothermic peaks appear on heating of the mechanically alloyed Fe-Co-Nb-B powders. Representative DSC scans of 7Nb and 3Nb (Fig. 13) as well as $(Fe_{50}Co_{50})_{62}Nb_8B_{30}$ powder mixtures (Fig. 14) exhibit different thermal effects (Table 1). For all ball milled powders, the first exothermic peak that spreads over the temperature range 100–300°C can be attributed to recovery, strains and structural relaxation. The important heat release (20.56 J/g) for 3Nb powders might be related to the amount of structural defects. The second exothermic peak (2), at 415°C, can be attributed to the α-Fe and/or α-FeCo primary nanocrystallization. This temperature is smaller than that obtained for the ball-milled 7Nb and $(Fe_{50}Co_{50})_{62}Nb_8B_{30}$ powders. Such a difference might be attributed to the Nb content since Co usually increases the onset of crystallization by about 20°C because this atom inhibits atomic movement raising the kinetic barrier for crystallization. The small exothermic peaks centred at ~623.5°C (3) and ~675.7°C (4) in the 3Nb powders can be related to the crystallization of Fe-borides.

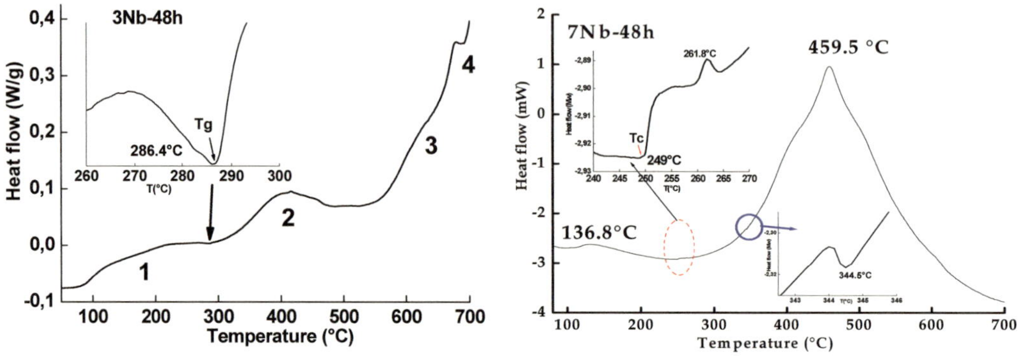

Figure 13. DSC scans of 3Nb and 7Nb powder mixtures milled for 48 h [55].

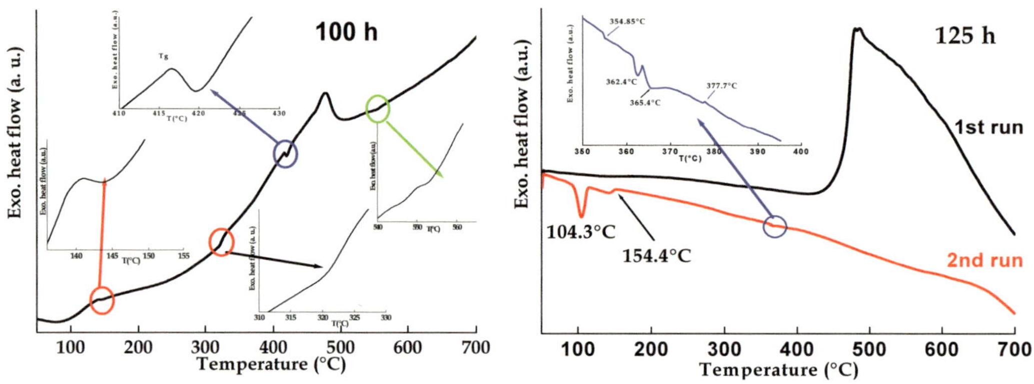

Figure 14. DSC scans of the $(Fe_{50}Co_{50})_{62}Nb_8B_{30}$ powders milled for 100 and 125 h.

Thermal stability of the nanocrystalline phases was investigated by DSC for alloys A, B and C milled for 160 h at a heating rate of 10 K/min (Fig. 15). The broad exothermic process starting at ~400–420 K is due to early surface crystallization (particle surface) and/or internal stress relaxation [58]. In all alloys, an additional exothermic process was detected with a peak temperature between 713 and 743 K. One observes that the peak temperature increases with increasing Nb content from 5 to 15%. This result agrees with those of the ball-milled 3Nb, 7Nb and $(Fe_{50}Co_{50})_{62}Nb_8B_{30}$ mixtures.

The endothermic peak at about 286.4, 344.5 and 420°C for 3Nb, 7Nb and $(Fe_{50}Co_{50})_{62}Nb_8B_{30}$ powders, respectively, that can be attributed to the glass transition temperature, Tg, gives evidence of the amorphous state formation. The glass transition temperature of the $(Fe_{50}Co_{50})_{62}Nb_8B_{30}$ powders increases rapidly up to 25 h of milling, and then remains nearly constant on further milling time (Fig. 16). The increase of Tg might be correlated to the amorphous phase proportion and/or to the change of its composition. The obtained low values compared to those of the amorphous ribbons with the same composition, can be linked to the heterogeneity of the ball-milled samples. The glass transition is not a first order phase transition but a kinetic event dependent on the rearrangement of the system and experimental time scales. Therefore, the transition would be a purely dynamic phenomenon.

Sample Milling time (h)	**Peak**	**T(°C)**	**ΔH (J/g)**	**Tg (°C)**
$(Fe_{50}Co_{50})_{62}Nb_8B_{30}$ (100 h)	1	138.83	7.58	420
	2	**475.76**	**27.98**	
7Nb (48 h)	1	136.8	2.1	344.5
	2	**459.5**	**169.5**	
3Nb (48 h)	1	198.5	20.56	286.4
	2	**415.0**	**35.9**	
	3	623.5	1.4	
	4	675.7	3.9	

Table 1. Peak temperature, T_p, enthalpy release, ΔH, and glass transition temperature, Tg, of 7Nb and 3Nb powders milled for 48 h, and $(Fe_{50}Co_{50})_{62}Nb_8B_{30}$ mixture milled for 100 h [55].

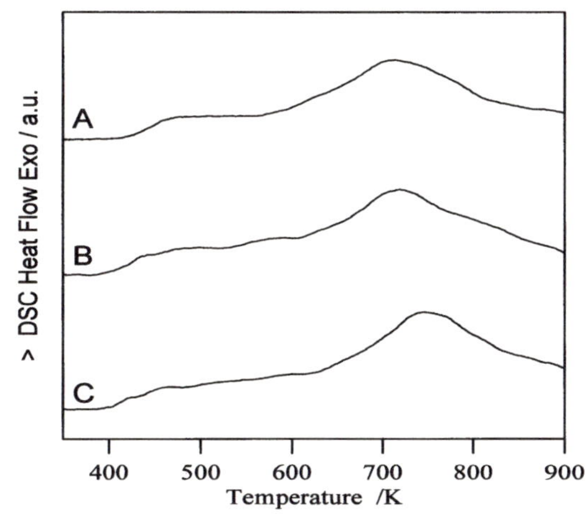

Figure 15. DSC scans at a heating rate of 10 K.min^{-1} of the ball-milled A, B and C powders for 160 h [58].

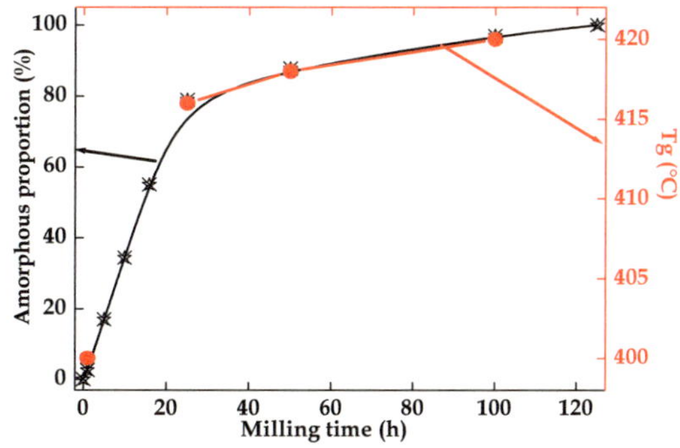

Figure 16. Variation of the glass transition temperature and the amorphous phase proportion of the $(Fe_{50}Co_{50})_{62}Nb_8B_{30}$ powders as a function of milling time.

DSC, which measures heat flow to and from a specimen relative to an inert reference, is the most common thermal analysis method used to measure the glass transition. The heat capacity step change at the glass transition yields three temperature values: onset, midpoint and endset. The midpoint is usually calculated as the peak maximum in the first derivative of heat flow (Fig. 1), although it can be calculated as the midpoint of the extrapolated heat capacities before and after the glass transition. This later is the temperature region where an amorphous material changes from a glassy phase to a rubbery phase upon heating, or *vice versa* if cooling. For example, the glass transition is very important in polymer characterization as the properties of a material are highly dependent on the relationship of the polymer end-use temperature to its Tg. In fact, an elastomer will be brittle if its Tg is too high, and the upper use temperature of a rigid plastic is usually limited by softening at Tg. Therefore, an accurate and precise measure of Tg is a prime concern to many plastics manufacturers and end use designers.

DSC detects the Curie temperature as a change in heat flow and due to the small amount of energy associated with this transition. An endothermic reaction occurs just below the Curie temperature as energy is being absorbed by the sample to induce randomization of the magnetic dipoles. An exothermic event occurs directly after the Curie temperature since no further energy is needed for randomization. Consequently, the line break at about 237°C and 249°C for 3Nb and 7Nb powders (Fig. 13), respectively, can be assigned to the ferro-paramagnetic transition at Curie temperature of the amorphous phase. Those values are comparable to that reported for the amorphous $(Fe_{100-x}Co_x)_{62}Nb_8B_{30}$ bulk metallic glasses [59], where Tc was found to be 245°C for x=0. Accordingly, one can suppose that the amorphous phase composition is Co-free FeBNb-type. Different Tc values of about (157–167)°C and (87–97)°C have been reported for the as-quenched $Fe_{52}Co_{10}Nb_8B_{30}$ and $Fe_{22}Co_{40}Nb_8B_{30}$ alloys [60], respectively. Tc of the residual amorphous phase exhibits antagonist behaviour for both alloys. It decreases with increasing crystalline fraction for the Co–rich $Fe_{22}Co_{40}Nb_8B_{30}$ alloy, and shifts to higher temperature for the Fe–rich $Fe_{52}Co_{10}Nb_8B_{30}$ alloy. Also, lower Tc values in the temperature range (214–230)°C were obtained for the as-cast state and in nanocrystalline $Fe_{77}B_{18}Nb_4Cu$ ribbons annealed at different temperatures [61].

Fig. 17 shows the evolution of Curie temperature of the amorphous phase in the $(Fe_{50}Co_{50})_{62}Nb_8B_{30}$ powders against milling time. Since the amorphous phase Curie temperature is very sensitive to the chemical composition, therefore the progressive decrease of Tc with increasing milling time can be attributed to the increase of B and/or Nb content in the amorphous matrix. It has been reported that the Curie temperature of the FeCoNbB amorphous alloys increases with the B content in the amorphous matrix [62]. Both the first and the second DSC scans of the powders milled for 100 and 125 h, respectively, display many endothermic peaks (see the inset in Fig. 14) that can be attributed to Curie temperatures of different Fe-boride phases and the residual matrix (t=125 h). For example, the endothermic peak at T=579.8°C can be related to the Curie temperature of Fe_3B [63].

The apparent activation energy of the crystallization process in the alloys A, B and C was evaluated by the Kissinger method. The obtained values 2.47±0.07, 2.63±0.05 and 2.71±0.08 eV for alloys A, B and C, respectively, can be associated with grain growth process. The activation energy and the peak temperature variation as a function of Nb content (Fig. 18) reveal that the highest peak temperature and activation energy correspond to the 15%Nb alloy. According to the structural and thermal analysis, it can be concluded that the partial substitution of B by Nb favours the stability of nanocrystalline phase with regard to crystal growth.

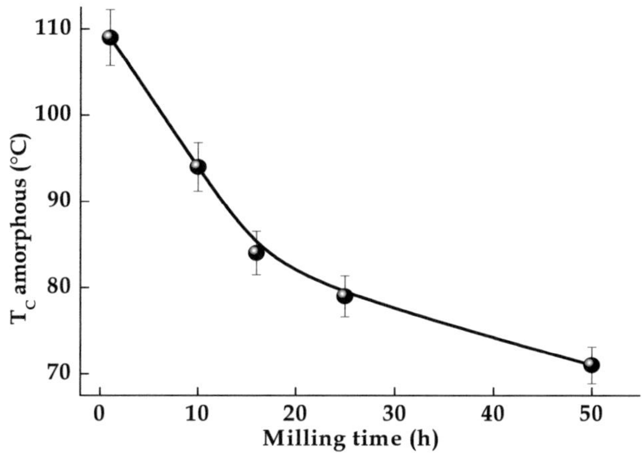

Figure 17. Variation of the amorphous phase Tc in the $(Fe_{50}Co_{50})_{62}Nb_8B_{30}$ powders as a function of milling time.

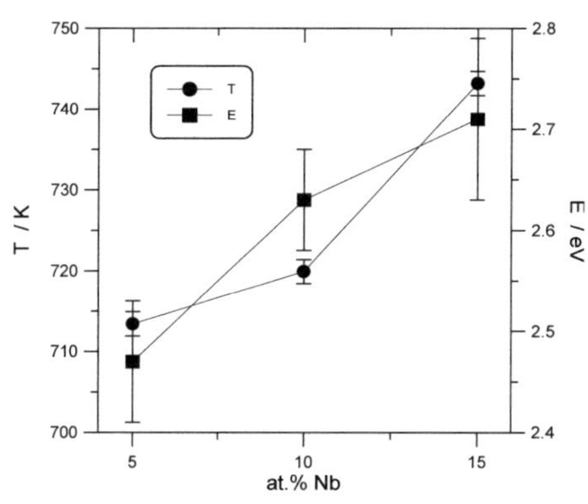

Figure 18. Apparent activation energy and peak temperature of the crystallization process against Nb content for alloys A, B and C milled for 160 h [58].

Stability of the nanostructured Fe-Co-Nb-B powders can be followed by the variation of the magnetic properties such as saturation magnetization, Ms, and coercivity, Hc. The hysteresis loops of ball milled 3Nb powders for 48 h and 7Nb powders for 96 h and heat treated up to 700°C (Fig. 19) display a sigmoidal shape which is usually observed in nanostructured samples with small magnetic domains. This can be correlated to the presence of structural distortions inside grains. One notes that both Ms and Hc values of 3Nb powders are higher than those of 7Nb powders. The increase of Hc from 71 to 115.5 Oe, after heat treatment of the ball milled 3Nb powders for 96 h, points out that the FeCo-rich ferromagnetic grains might be separated by Nb and/or B-rich phase with weaker ferromagnetic properties. Another possible origin for this behaviour is the increase of Fe_2B boride proportion. Nonetheless, for 7Nb mixture Ms increases slightly while Hc remains nearly constant after heat treatment of the powders milled for 48 h. One can conclude that the nanostructured state is maintained after heat treatment.

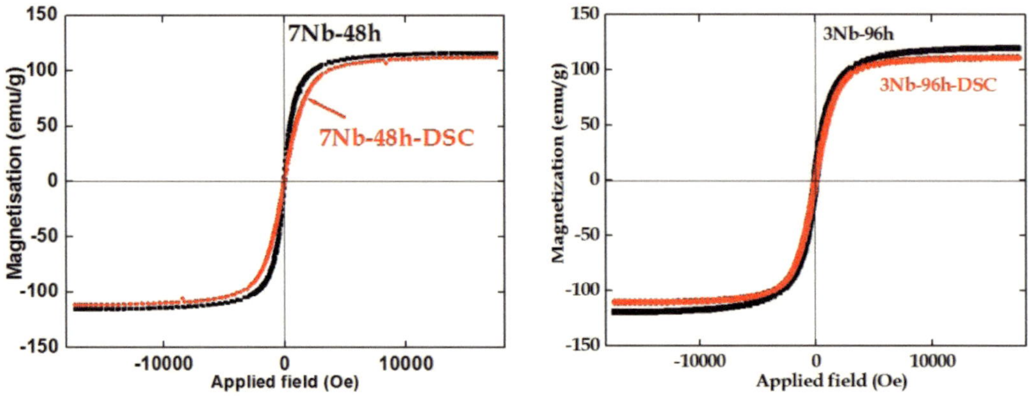

Figure 19. Hysteresis loops of 3Nb and 7Nb powders milled for 96 h and 48 h, respectively, and after heat treatment up to 700°C [55].

7. Ni-P powders

Thermal annealing leads, in general, to the relaxation of the introduced stresses during the milling process. The DSC curves of the ball-milled $Ni_{70}P_{30}$ powders for 3 and 12 h (Fig. 20) display different behaviour on heating at a rate of 10°C.min^{-1}. After the first run up to 700°C (scan a), samples are cooled down to ambient temperature, then reheated in the same conditions. One notes that the DSC signal of the second run (scan b) shows a line without any thermal effect indicating that the phase transformation is achieved during the first run [64]. However, for the first run curve, the enthalpy release spreads over the temperature range (100–650)°C. The large exothermic reactions at temperatures below 300°C can be attributed to recovery and strain relaxation. The DSC curve of the powder milled for 3 h shows a single exothermic peak at 496.4°C. While, after 12 h of milling, the DSC curve reveals several endothermic peaks, and one exothermic peak at 567.6°C. According to the Curie temperature of pure Ni (T_c = 350°C), the endothermic peaks (Fig. 21) can be related to the magnetic transition temperature of dilute Ni(P) solid solutions. However, the exothermic peak might be assigned to a growth process of Ni_2P nanophase. The depression of Tc compared to that of pure Ni indicates that the nearest-neighbour coordinates are essentially changed in the magnetic nanocrystallites by the P additions. The reason for the existence of several magnetic phase states and therefore, several Curie temperatures can be attributed to inhomogeneities since the Curie temperature is sensitive to the chemical short range order and subsequently, to the local Ni environment.

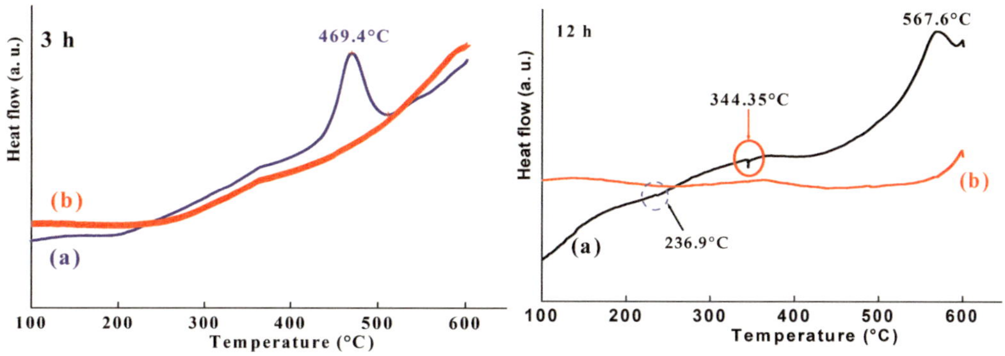

Figure 20. DSC plots of the $Ni_{70}P_{30}$ powders milled for 3 and 12 h at a heating rate of 10°C/min; first (a) and second heating runs (b) [64].

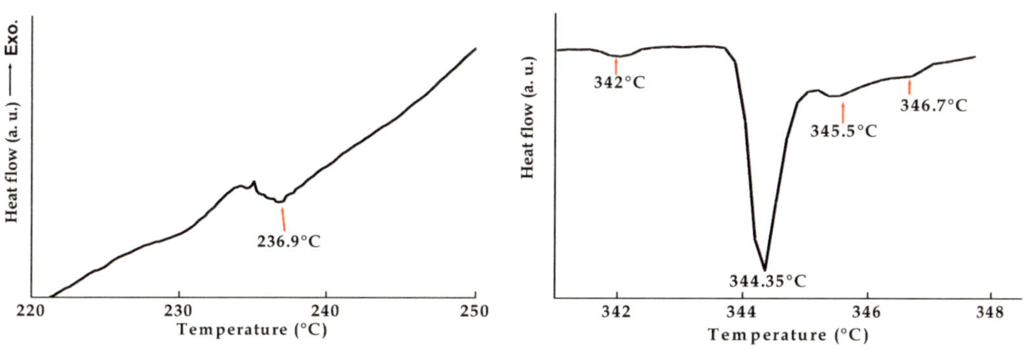

Figure 21. Enlargement of the low temperature regions of the DSC scan of the $Ni_{70}P_{30}$ powders milled for 12 h.

8. Kinetics of powder mixing

8.1. Fe-Mo mixture

The kinetics of Mo dissolution into the α-Fe matrix of the Fe-6Mo mixture has been deduced from the XRD analysis by following the evolution of the (110) diffraction peak intensity of the unmixed Mo as a function of milling time [26]. Since the milling process occurs at room temperature, one can suppose that the temperature is constant. In addition, the milling time can be considered as the necessary time for phase transformation. Consequently, the mixed fraction of Mo which is considered as the fraction transformed, x, can be described by the Johnson-Mehl-Avrami formalism. The double logarithmic plot ln(-ln(1-x)) versus lnt leads to the Avrami parameter n, and the rate constant k. Two stages have been distinguished according to the kinetics parameter values: (i) a first stage with n_1= 0.83 and k_1= 0.34; and (ii) a second stage with n_2= 0.33 and k_2 = 0.73. The former proves that the Mo dissolution is very slow even non-existent in the early stage of milling (up to 6 h), while the later can be linked to the increased diffusivity by decreasing crystallite size and increasing the grain boundaries area on further milling time.

8.2. $Fe_{27.9}Nb_{2.2}B_{69.9}$ mixture

Amorphization kinetics of the $Fe_{27.9}Nb_{2.2}B_{69.9}$ (at. %) powders has been deduced from the Mössbauer spectrometry results by following the variation of the α-Fe transformed fraction as a function of milling time [27]. The amorphization process can be described by one stage with an Avrami parameter of about n~1 (Fig. 22). This value is comparable to those obtained for transformations controlled by the diffusion at the interface and dislocations segregation with 0.45< n < 1.1. This might be correlated to the existence of a high density of dislocations and various types of defects as well as to the crystallite size refinement. Comparable values of the Avrami parameter were obtained for the primary crystallization of the amorphous FeCoNbB alloy prepared by melt spinning [65].

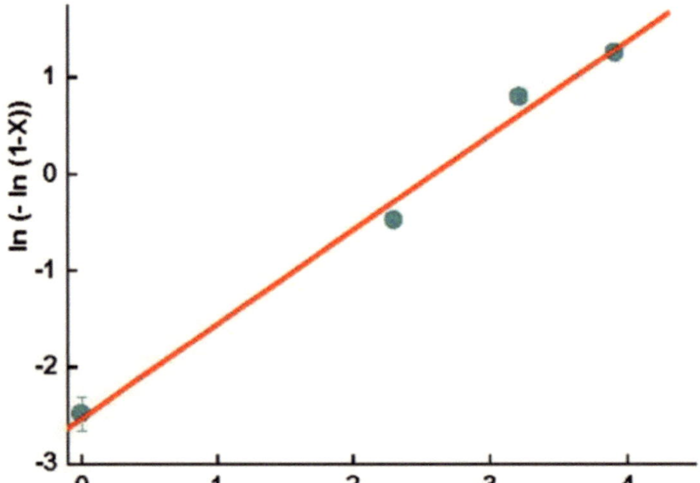

Figure 22. Johnson-Mehl-Avrami plot of the ball-milled $Fe_{27.9}Nb_{2.2}B_{69.9}$ versus milling time [28].

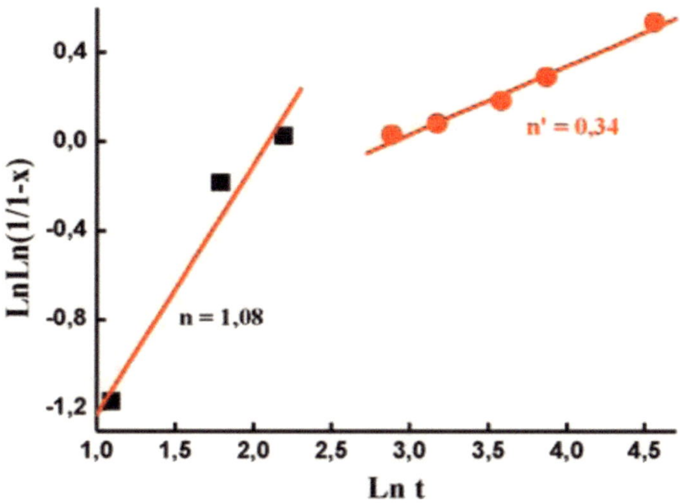

Figure 23. Johnson-Mehl-Avrami plot of the ball-milled $Fe_{57}Co_{21}Nb_7B_{15}$ versus milling time [27, 66].

8.3. FeCoNbB powders

The mixing kinetics of the $Fe_{57}Co_{21}Nb_7B_{15}$ powders can be described by two stages [27, 66] with different Avrami parameters n = 1.08 and n' = 0.34 (Fig. 23). The lower values of the Avrami parameter can be ascribed to the presence of both Nb and B which favour the grain size refinement and the formation of a highly disordered state. For the $(Fe_{50}Co_{50})_{62}Nb_8B_{30}$ mixture, two stages have been obtained with different Avrami parameter values n_1 = 1.41 and n_2 = 0.34 [2]. The former value is comparable to those obtained for the Finemet and Nanoperm [67]. However, it is higher than that obtained during the crystallization of the amorphous FeCoNbB alloy where α-(Fe,Co) nanocrystals with grain size of 15 nm are distributed in the amorphous matrix [65]. Bigot et al. have obtained a value of n = 1.5 for the nanocrystallization of the Finemet [68]. Comparable kinetics parameters have been obtained in the Ni-15Fe-5Mo (n = 1.049 and k = 0.57) [69]. The important fraction of structural defects which is introduced during the milling process favours the phase formation through the diffusion at the surface which is dominant, at lower temperatures, in comparison to the diffusion by the grain boundaries and the lattice parameter (vacancy's diffusion).

9. Conclusion

Thermal analysis is widely used in the reaction study of the mechanically alloyed powder particles because of the obtained metastable disordered structures. Hence, thermal annealing leads to the relaxation of the introduced stresses during the milling process. The heat effects are dependent on the structural and microstructural properties of the ball-milled powders.

Author details

Safia Alleg* and Saida Souilah
Badji Mokhtar Annaba University, Department of Physics, Laboratoire de Magnétisme et Spectroscopie des Solides (LM2S) B.P. 12, 23000 Annaba, Algeria

Joan Joseph Suñol
Dep. De Fisica, Universitat de Girona, Campus Montilivi, 17071 Girona, Spain

Acknowledgement

Prof. Safia Alleg is grateful to the University of Girona-spain for the financial support as invited professor. Financial support from AECID A/016051/08 and AECID A/025066/09 projects is acknowledged. Financial support from WLI Algeria is acknowledged.

10. References

[1] Suryanarayana C (2004) Mechanical alloying and milling. Marcel Dekker. 457-p

[2] Alleg S, Azzaza S, Bensalem R, Suñol JJ, Khene S, Fillion G (2009) Magnetic and structural studies of mechanically alloyed $(Fe_{50}Co_{50})_{62}Nb_{8}B_{30}$ powder mixtures. J. Alloys Compd. 482: 86-89.

[3] Moumeni H, Alleg S, Djebbari C, Bentayeb FZ, Greneche JM (2004) Synthesis and characterization of nanostructured FeCo alloys. J. Mat. Sci. 39: 5441-5443.

[4] Bensebaa N, Alleg S, Bentayeb FZ, Bessais L, Greneche JM (2005) Microstructural characterization of Fe-Cr-P-C powder mixture prepared by ball milling. J. Alloys Compd. 388:41-48.

[5] Bentayeb FZ, Alleg S, Bouzabata B, Greneche JM (2005) Study of alloying mechanisms of ball milled Fe-Cr and Fe-Cr-Co powders. J. Magn. Magn. Mat. 288: 282-296.

[6] Tebib W, Alleg S, Bensalem R, Greneche JM (2010) Structural study of the mechanically alloyed Fe-P powders. Int. J. Nanoparticles 3:237-244.

[7] Azzaza S, Alleg S, Moumeni H, Nemamcha AR, Rehspringer J L, Greneche J M (2006) Magnetic properties of nanocrystalline ball milled Fe and $Fe_{50}Co_{50}$ alloy. J. Phys.: Condens. Matter 18 : 7257-7272.

[8] Tebib W, Alleg S, Bensebaa N, Bentayeb FZ, Suñol JJ, Greneche JM (2008) Structural characterization of nanostructured Fe-8P powder mixture. J. Nanosci. Nanotechnol. 8:2029-2036.

---

* Corresponding Author

[9] Alleg Safia, Bentayeb Fatima Zohra, Djebbari Chafia, Bessais Lotfi, Greneche Jean Marc (2008) Effect of the milling conditions on the formation of nanostructured Fe-Co powders. Phys. Stat. Sol. (a) 205: 1641-1646.

[10] Alleg S, Ibrir M, Fenineche NE, Azzaza S, Suñol JJ(2010) Magnetic and structural characterization of the mechanically alloyed $Fe_{75}Si_{15}B_{10}$ powders. J. Alloys Compd. 494: 109-115.

[11] Bansal C, Gao ZQ, Hong L B, Fultz B (1994) Phases and phase stabilities of Fe_3X alloys (X=Al, As, Ge, In, Sb, Si, Sn, Zn) prepared by mechanical alloying. J. Appl. Phys. 76:5961–5966.

[12] Macrí PP, Enzo S, Cowlam N, Frattini R, Principi G, Hu WX (1995) Mechanical alloying of immiscible $Cu_{70}TM_{30}$ alloys (TM = Fe,Co). Philosophical Magazine Part B 71:249-259.

[13] Bentayeb FZ, Alleg S, Greneche J M (2007) Structural and microstructural study of Fe-31Cr-12Co mixture prepared by ball milling. J. Alloys Compd. 434: 435-477.

[14] Dekhil L, Alleg S, Suñol JJ, Greneche JM (2009) X-rays diffraction and Mössbauer spectrometry studies of the mechanically alloyed Fe-6P-1.7C powders. Adv. Pow. Technol. 20:593-597.

[15] Azzaza S (2006) Magister. Badji Mokhtar Annaba University, Algeria.

[16] Sherif El-Eskandarany M, Saida J, Inoue A (2002) Amorphization and crystallization behaviours of glassy $Zr_{70}Pd_{30}$ alloys prepared by different techniques. Acta Mater. 50:2725–2736.

[17] Kissinger HE (1957) Reaction kinetics in differential thermal analysis. Anal. Chem. 29:1702–1706.

[18] Greer AL (1982) Crystallization kinetics of $Fe_{80}B_{20}$ glass. Acta Metall. 30:171–192.

[19] Henderson D W (1979) Thermal analysis of non-isothermal crystallization kinetics in glass forming liquids. J. Non-Cryst. Solids 30:301–315.

[20] Galwey AK, Brown ME (1998) Kinetic background to thermal analysis and calorimetry. Handbook of thermal analysis and calorimetry: principles and practice. Brown ME, editor. Elsevier Science B.V. pp. 147-224.

[21] Christian JW (1975) The Theory of Transformations in Metals and Alloys. Pergamon, Oxford p. 542.

[22] Avrami M (1939) Kinetics of phase change I. J. Chem. Phys. 7: 1103-1112.

[23] Avrami M (1940) Kinetics of phase change II. J. Chem. Phys. 8: 212-224.

[24] Avrami M(1941) Kinetics of phase change III. J. Chem. Phys. 9: 177-184.

[25] Kolmogorov AN (1937) Statistical theory of crystallization of metals. Bull. Acad. Sci. USSR, Phys. Sci.1:355-359.

[26] Moumeni H, Alleg S, Greneche JM (2006), Formation of ball-milled Fe-Mo nanostructured powders. J. Alloys Compd. 419: 140-144.

[27] Souilah S, Alleg S, Djebbari C, Suñol JJ (2012) Magnetic and microstructural properties of the mechanically alloyed $Fe_{57}Co_{21}Nb_7B_{15}$ powder mixture. Mat. Chem. Phys. 132: 766-772.

[28] Alleg S, Hamouda A, Azzaza S, Suñol JJ, Greneche JM (2010) Solid state amorphization transformation in the mechanically alloyed $Fe_{27.9}Nb_{2.2}B_{69.9}$ powders. Mat. Chem. Phys. 122: 35-40.

[29] Alleg S, Bensalem R (2011) Nanostructured Fe-based Mixtures Prepared by Mechanical Alloying. In: Jason M. Barker, editor, Powder Engineering, Technology and applications, Nova Science Publishers: pp. 81-124.

[30] Louidi S, Bentayeb FZ, Suñol JJ, Escoda L (2010) Formation study of the ball-milled $Cr_{20}Co_{80}$ alloy. J. Alloys Compd. 493: 110-115.

[31] Loudjani Nadia, Bensebaa Nadia, Alleg Safia, Djebbari Chaffia, Greneche Jean Marc (2011) Microstructure characterization of ball-milled $Ni_{50}Co_{50}$ alloy by Rietveld method. Phys. Status Solidi A 208:2124-2129.

[32] Calka A, Radlinski AP (1986) The effect of surface on the kinetics of crystallization of Pd-Si glassy metals. MRS Proceedings 80:195-201.

[33] Gibson MA, Delamore GW (1987) Crystallization kinetics of some iron-based metallic glasses. J. Mater. Sci. 22:4550-4557.

[34] Cao M G, Fritsch HU, Bergmann HW (1985). Thermochim. Acta 83:23.

[35] Lü L, Lai M (1998) Mechanical alloying. Kluwer Academic Publishers. 273 p.

[36] Ibrir M (2011) PhD Thesis. Badji Mokhtar Annaba University, Algeria.

[37] Moumeni H, Alleg S, Greneche JM (2005) Structural properties of $Fe_{50}Co_{50}$ nanostructured powder prepared by mechanical alloying. J. Alloys Compd. 386: 12-19.

[38] Moumeni Hayet, Nemamcha Abderrafik, Alleg Safia, Greneche Jean-Marc (2010) Stacking faults and structure analysis of ball-milled Fe-50%Co powders. Mat. Chem. Phys. 122:439-443.

[39] Brüning R, Samwer K, Kuhrt C, Schultz L (1992) The mixing of iron and cobalt during mechanical alloying. J. Appl. Phys. 72:2978-2983.

[40] Sorescu M, Grabias A (2002) Structural and magnetic properties of $Fe_{50}Co_{50}$ system. Intermetallics 10:317-321.

[41] Lutterotti L (2000) MAUD CPD Newsletter (IUCr) 24.

[42] Lutterotti L, Matthies S, Wenk HR (1999) MAUD: a friendly Java program for material analysis using diffraction. IUCr: Newsletter of the CPD, 21:14-15.

[43] Li S, Wang K, Sun L and Wang Z (1992) Simple model for the refinement of nanocrystalline grain size during ball milling. Scr. Metall. Mater. 27: 437-442

[44] Börner I, Eckert J (1997) Nanostructure formation and steady-state grain size of ball-milled iron powders. Mat. Sci. Eng. A226-228: 541-545.

[45] Fecht HJ (1994) Nanophase Materials. In:Hadjipanayis G C, Siegel R W, editors. 260:125-132

[46] Zhou GF, Bakker H (1994) Atomically disordered nanocrystalline Co_2Si by high-energy ball milling. *J. Phys.: Condens. Matter. 6:4043*–4052.

[47] McHenry ME, Gallagher K, Johnson F, Scott JH, Majetich SA (1996) Recent advances in the chemistry and physics of fullerenes and related materials. In: Kadish KM, Ruoff RS, editors. PV96-10, ECS Symposium Proceedings, Pennington, NJ, p. 703.

[48] Host J J, Teng M H, Elliot B R, Hwang J H, Mason T O, Johnson D L (1997) Graphite encapsulated nanocrystals produced using a low carbon:metal ratio. J. Mat. Res. 12: 1268-1273.

[49] Krill C E, Merzoug F, Krauss W and Birringer R (1997) Magnetic properties of nanocrystalline Gd and *W*/Gd. NanoStruct. Mater 9:455-460.

[50] Massalski T (1990) Binary alloy phase diagrams, Materials Park OH: ASM International.

[51] Turgut Z, Huang MQ, Gallagher K, McHenry ME (1997) Magnetic evidence for structural phase-transformations in Fe-Co alloy nanocrystals produced by a carbon arc. J. Appl. Phys. 81: 4039-4041.

[52] Persiano AIC, Rawlings RD (1991) Effect of niobium additions on the structure and magnetic properties of equiatomic iron cobalt alloys. J. Mat. Sci. 26: 4026-4632.

[53] Bensalem R, Younes A, Alleg S, Souilah S, Azzaza S, Suñol JJ, Greneche JM (2011) Solid state amorphisation of mechanically alloyed Fe-Co-Nb-B alloys. Int. J. Nanoparticles 4: 45-52.

[54] Alleg S, Souilah S, Bensalem R, Younes A, Azzaza S, Suñol JJ(2010) Structural characterization of the mechanically alloyed $Fe_{57}Co_{21}Nb_7B_{15}$ powders. Int. J. Nanoparticles 3: 246-256.

[55] Alleg S, Souilah S, Achour Y, Suñol JJ, Greneche JM (2012) Effect of the Nb content on the amorphization process of the mechanically alloyed Fe-Co-Nb-B powders. J. Alloys Compd. 536S:S394-S397.

[56] Blazquez JS, Conde A, Greneche JM (2002) Mössbauer study of FeCoNbBCu hitperm-alloys. Appl. Phys. Letters 81:1612-1614.

[57] Miglierini M, Greneche JM (1997) Mössbauer spectrometry of Fe(Cu)MB-type nanocrystalline alloys II: the topography of hyperfine interactions in Fe(Cu)ZrB alloys. J. Phys.: Condens. Matter 9:2321-2347.

[58] Suñol JJ, Güell JM, Bonastre J, Alleg S (2009) Structural study of nanocrystalline Fe-Co-Nb-B alloys prepared by mechanical alloying. J. Alloys Compd. 483: 604-607.

[59] Gercsi Zs, Mazaleyrat F, Kane SN, Varga LK (2004) Magnetic and structural study of $(Fe_{1-x}Co_x)_{62}Nb_8B_{30}$ bulk amorphous alloys. Mater. Sci. Eng. A 375–377: 1048-1052.

[60] Gloriant T, Suriňach S, Baró MD (2004) Stability and crystallization of Fe-Co-Nb-B amorphous alloys. J. Non-Crystal. Sol. 333: 320-326.

[61] Hernando A, Navarro I, Gorría P (1995) Iron exchange-field penetration into the amorphous interphases of nanocrystalline materials. Phys. Rev. B 51:3281-3284.

[62] Suzuki K, Cadogan JM, Sahajwalla V, Inoue A, Masumoto T (1996) $Fe_{91}Zr_7B_2$ soft magnetic alloy. J Appl. Phys. 79: 5149-5151.

[63] Liebermann HH, Marti J, Martis RJ, Wong CP (1989) The effect of microstructure on properties and behaviours of annealed $Fe_{78}B_{13}Si_9$ amorphous alloy ribbon. Metall. Trans. A 20:63-70.

[64] Alleg S, Rihia G, Bensalem R, Suñol JJ (2009) Structural evolution of the ball-milled $Ni_{70}P_{30}$ powders. Ann. Chim. Sci. Mat. 34:267-273.

[65] Blazquez JS, Conde CF, Conde A (2001) Crystallization process in $(FeCo)_{78}Nb_6(BCu)_{16}$ alloys. J. Non-Cryst. Solids 287:187-192.

[66] Souilah S (2012) PhD thesis. Badji Mokhtar Annaba University, Algeria.

[67] McHenry ME, Willard MA, Laughlin DE (1999), Amorphous and nanocrystalline materials for applications as soft magnets. Prog. Mat. Sci. 44:291-433.

[68] Bigot J, Lecaude N, Perron JC, Milan C, Ramiarinjaona C, Rialland JF (1994) Influence of annealing conditions on nanocrystallization and magnetic properties in $Fe_{73.5}Cu_1Nb_3Si_{13.5}B_9$ alloy. J. Magn. Magn. Mater. 133: 299-302.

[69] Shen SY, Hng HH, Oh JT (2004). Mater. Letter 58:2824.

Chapter 3

Studies on Growth, Crystal Structure and Characterization of Novel Organic Nicotinium Trifluoroacetate Single Crystals

P.V. Dhanaraj and N.P. Rajesh

Additional information is available at the end of the chapter

http://dx.doi.org/10.5772/53795

1. Introduction

Single crystal growth has a prominent role in the present era of rapid scientific and technical advancement, whereas the application of crystals has unbounded limits. New materials are the lifeblood of solid state research and device technology. Nonlinear optical (NLO) crystals have come upon the materials science scene and are being studied by many research groups around the world. These materials operate on light in a way very analogous to the way of semiconductors which operate on electrons to produce very fast electronic switching and computing circuits.

Organic crystals have compounds with carbon atoms as their essential structural elements. The design and synthesis of organic molecules exhibiting NLO properties have been motivated by the tremendous potential for their applications in the fast developing domains of optoelectronics and photonic technologies. The relevance of the organic materials in the present context is, the delocalized electronic structure of π–conjugated organic compound offers a number of tantalizing opportunities in the applications as NLO materials. Extensive research in the last decades has shown that organic crystals often possess a higher degree of optical nonlinearity than their inorganic counterparts [1, 2]. Some of the advantages of organic materials include inherently high nonlinearity, high electronic susceptibility through high molecular polarizability, fast response time, the ease of varied synthesis, scope for altering the properties by functional substitutions, high damage resistance, relative ease of device processing, etc. Organic materials have another advantage over inorganic materials, in that the properties of organic materials can be optimized by modifying the molecular structure using molecular engineering and chemical synthesis [3]. A very large operating bandwidth modulation in organic electro-optic devices can be obtained through its low dielectric constant at low frequencies. Hence they are projected as forefront candidates for fundamental and applied investigations.

In organic materials, there is a strong correlation between structure and nonlinear properties. Thus, in the case of second order nonlinear effects, it has been established that the macroscopic susceptibility of the materials $\chi^{(2)}$, is related to both the magnitude of the molecular nonlinearities, i.e. the first hyperpolarizability β, and the relative orientation of the molecules in the medium. Therefore, a fundamental limitation for second-order effects to be observed is the non-centrosymmetry requirement, both at the microscopic molecular level and at the macroscopic bulk level. On the other hand, the third order effects described by $\chi^{(3)}$ can be present in any medium. The $\chi^{(3)}$ coefficients are thus essential in centrosymmetric compounds where the second order coefficients equal zero. They are also important in the non-centrosymmetric molecules. Moreover, these $\chi^{(3)}$ coefficients play a part in some experimental determination of $\chi^{(2)}$ coefficients. It is for example the case in electric field induced second harmonic generation [4-6] (EFISHG) experiment or in hyper-Rayleigh technique [7-10] where the two-photon absorption (TPA), which is third order effect, can induce fluorescence thus making imprecise the determination of the β coefficient. An ultimate goal for designing the molecules with large third order nonlinearities is to incorporate them into devices used in all optical signals processing [11, 12]. Nonlinear optical absorption (NOA) has shown its potential application in optical information storage, all optical logic gates, laser radiation protection, and locked laser mode. Interest in searching for NOA materials has been gradually increased. Organic molecules have been the subjects of great attention due to their potential applications in nonlinear optics, optical switching, and light emitting diodes. Indeed, the potential use of organic device materials in optoelectronics is now a very serious matter.

In order to achieve good macroscopic nonlinear response in organic crystals, one requires an increase in the number of π electrons and π delocalization length, so as to lead to high molecular hyperpolarizability and also proper orientation of the molecule in the solid-state structure to facilitate high-frequency conversion efficiency. Effective materials generally contain donor and acceptor groups positioned at either end of suitable conjugation path. The increased effective conjugation and the large π delocalization length have been recognized as the factors leading to the large third order nonlinearities. While the engineering for enhancing second order NLO efficiency is relatively well understood, the need for efficient third order molecules and materials still exists. The design of organic polar crystals for the quadratic NLO applications is supported by the observation that organic molecules containing π electron systems asymmetrized by electron donor and acceptor groups are highly polarizable entities [13]. Donor/acceptor benzene derivatives are sure to produce high molecular nonlinearity. So far, many organic Donor–π–Acceptor (D–π–A) type compounds have been studied theoretically and also experimentally [14]. The studies indicate that the organic D–π–A compounds are highly promising candidates for NLO applications.

Nicotinic acid, a B vitamin also known as niacin, and its derivatives have been studied extensively over the last decade due to their biological and chemical importance. Niacin forms coordination complexes with tin (Sn), which have been found to have better antitumor activity than the well-known cis-platin or doxorubicin [15]. Many of its pharmacological properties are detailed in literature [16-18]. The reported structures of complexes reveal that nicotinic acid and its derivatives acting as bridging ligands through the carboxylate group and pyridyl N atom [19]. We have synthesized the crystalline salt of nicotinium trifluoroacetate and their crystals in monoclinic system were grown by using solution growth technique for the first time. The crystal structure of nicotinium trifluoroacetate in triclinic system has reported by S. Athimoolam and S. Natarajan [20]. Here we report monoclinic polymorph of nicotinium trifluoroacetate, its asymmetric unit contains a protonated nicotinium cation and a trifluroacetate anion. This chapter discusses synthesis, solubility, crystal growth, structural, dielectric and mechanical properties of nicotinium trifluoroacetate (NTF). Thermal properties of NTF were analyzed and compared with that of two nicotinium derivative crystals nicotinium oxalate and nicotinium nitrate monohydrate.

2. Experimental studies

2.1. Synthesis of NTF

NTF was synthesized by the reaction between nicotinic acid (SRL, India) and trifluoroacetic acid (Merck) taken in equimolar ratio. The growth solution was prepared by adding calculated amount of trifluoroacetic acid slowly in saturated aqueous solution of equimolar nicotinic acid at 50 °C. The continuous stirring of the solution for 6 h at constant temperature using a temperature controlled magnetic stirrer yielded the precipitate of crystalline substance of NTF. Repeated crystallization and filtration processes were applied for the purification of the synthesized compound.

2.2. Determination of solubility and metastable zone width

The nucleation studies were carried out in a constant temperature bath (CTB) with cooling facility of accuracy of ±0.01 °C. The solubility at 30 °C was determined by dissolving the recrystallized salt of NTF in 100 ml Millipore water of resistivity 18.2 MΩcm taken in an air tight container. The solution was stirred continuously for 6 h to achieve stabilization using an immersible magnetic stirrer. After attaining the saturation, the concentration of the solute was estimated gravimetrically. The same procedure is repeated for different temperatures (35, 40, 45 and 50 °C).

Metastable zone width is an essential parameter for the growth of large size crystals from solution, since it is the direct measure of the stability of the solution in its supersaturated region. The metastable zone width was measured by adopting the conventional polythermal method [21]. The saturated solution (100 ml) at 30 °C was prepared according to the presently determined solubility data. After attaining the saturation, the solution was filtered by the filtration pump and Whatman filter paper of pore size 11 μm. The solution was preheated to 5 °C above the saturated temperature for homogenization and left at the superheated temperature for about 1 h before cooling. Then it was slowly cooled at a desired cooling rate of 4 °C/h, until the first crystal appeared. The temperature was instantly recorded. The difference between the saturation temperature and nucleation temperature gives the metastable zone width of the system. Then experiment was repeated for different saturation temperatures 35, 40, 45 and 50 °C and the corresponding metastable zone widths were measured. Several runs (3–5 times) were carried out under controlled conditions for the confirmation of the saturation and nucleation points. The measured values of solubility and metastable zone width of NTF are shown in Figure 1. It shows that NTF has good solubility in water and it increases almost linearly with temperature. Hence solution growth could be a better method to grow good quality single crystals of NTF. The value of the metastable zone width depends not only on the temperature but also on the type of the crystal and its physicochemical properties [22]. One can observe that the metastable zone width decreases with increasing temperature.

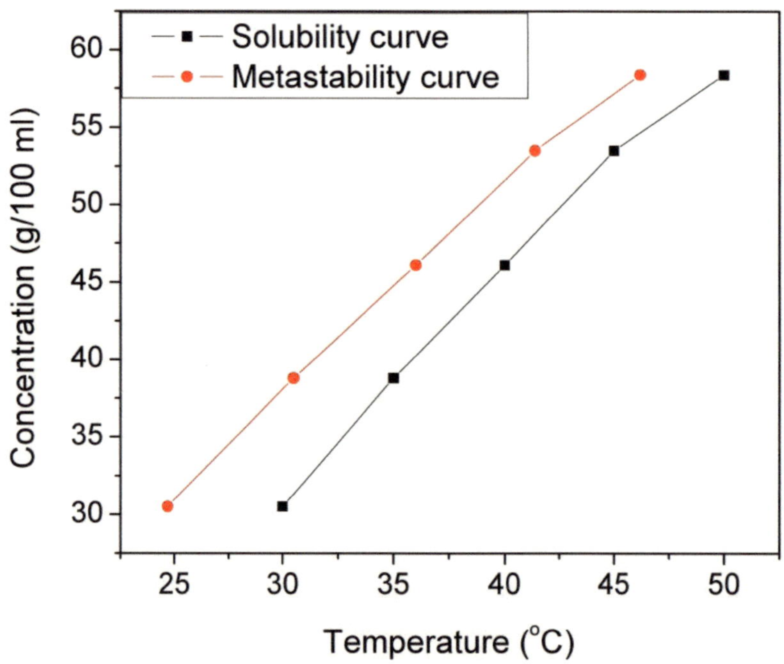

Figure 1. Solubility and metastablity curves of NTF

2.3. Crystal growth

Slow evaporation method was employed for growing the single crystals of NTF. The recrystallized salt was used for the preparation of saturated solution at room temperature (35 °C). The solution was filtered by filtration pump and Whatman filter paper of pore size 11 μm. Then the filtered solution was transferred to a petridish with a perforated lid in order to control the evaporation rate and kept undisturbed in a dust free environment. The single crystal of NTF of size 27 x 12 x 7 mm^3 was harvested from mother solution after a growth period of 22 days. The grown single crystal of NTF is shown in the Figure 2.

Morphology of the grown crystals was identified by the single crystal X-ray diffraction studies (Bruker Kappa APEXII). It establishes that the crystal has eight developed faces, out of which $(\bar{1}12)$ and $(1\bar{1}\bar{2})$ planes are more prominent. For each face, its parallel Friedal plane is also present in the grown crystal and shown diagrammatically in Figure 3.

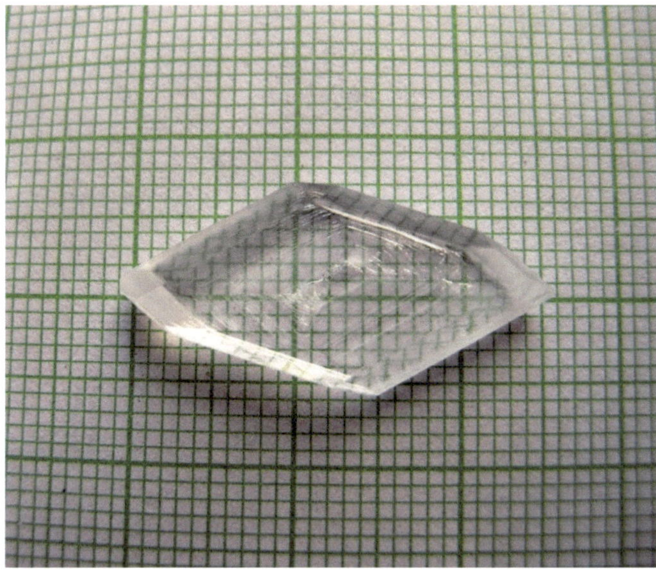

Figure 2. Photograph of as-grown NTF crystal

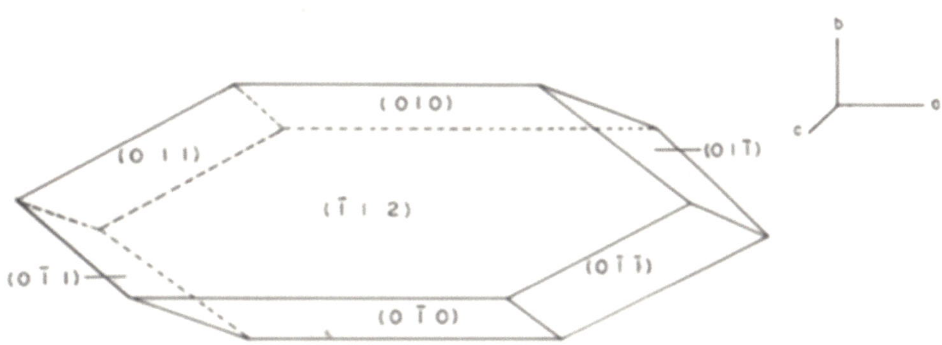

Figure 3. Morphology representation of NTF crystal

3. Analysis of physicochemical studies

3.1. X-ray diffraction analysis

The unit cell parameters and crystal structure of NTF were determined from single crystal X-ray diffraction data obtained with a four-circle Nonius CAD4 MACH3 diffractometer (graphite monochromated, MoKα = 0.71073 Å) at room temperature (293 K). The data reduction was done by using XCAD4 [23] and absorption correction was done by the method of ψ–scan [24]. The structure solution and refinement were performed using SHELXTL 6.10 [25]. The crystal structure of NTF was solved by direct methods, and full-matrix least-squares refinements were performed on F^2 using all the unique reflections. All the non-hydrogen atoms were refined with anisotropic atomic displacement parameters, and hydrogen atoms were refined with isotropic displacement factors. The crystallographic data and structure refinement parameters of NTF crystal are presented in Table 1. The crystal structure and packing diagram of NTF are shown in Figures 4 and 5 respectively. The H atom participating in the N–H bond was located from the difference Fourier map and all other H atoms were positioned geometrically and refined using a riding model with C–H = 0.93 Å and O–H = 0.82 Å with Uiso(H) = 1.2 – 1.5 Ueq (parent atom). The absolute configuration is assigned from the starting materials taken for reaction.

Empirical formula	$C_8H_6F_3NO_4$
Formula weight	237.14
Temperature	293(2) K
Wavelength	0.71073 Å
Crystal system, space group	Monoclinic, *I*2/a
Unit cell dimensions	a = 15.616(5) Å, α = 90° b = 7.455(5) Å, β = 95.74° c = 16.503(5) Å, γ = 90°
Volume	1911.6 $Å^3$
Z, Calculated density	8, 1.648 g/cm^3
Absorption coefficient	0.167 mm^{-1}
F(000)	960
Crystal size	0.18 mm x 0.15 mm x 0.13 mm
Theta range for data collection	2.48 - 24.96°
Limiting indices	$-18 \leq h \leq 18$, $-8 \leq k \leq 1$, $-19 \leq l \leq 19$
Reflections collected / unique	3929 / 1680 [R(int) = 0.0204]
Completeness to theta = 24.96	99.6%
Absorption correction	ψ– scan
Refinement method	Full-matrix least-squares on F^2
Data / restraints / parameters	1680 / 0 / 194
Goodness-of-fit on F^2	1.095
Final R indices [I>2σ(I)]	R_1 = 0.0353, wR_2 = 0.0905
R indices (all data)	R_1 = 0.0495, wR_2 = 0.0976
Extinction coefficient	0.0040(7)
Largest diff. peak and hole	0.210 and -0.201 e.$Å^{-3}$

Table 1. Crystallographic data and structure refinement parameters for NTF

This monoclinic polymorph of NTF crystallized in *I*2/a space group and the asymmetric unit contains a protonated nicotinium cation and a trifluroacetate anion. The angle between the mean plane of the pyridine ring and the mean plane of the acetate group is 48.93° where as in the triclinic form [20], it is 14° and the distance between the anion to cation is 3.134 Å which is 0.463 Å longer than the triclinic form. The structure is stabilized by N–H···O, O–H···O and C–H···O hydrogen bonds and hydrogen bond geometry is given in Table 2.

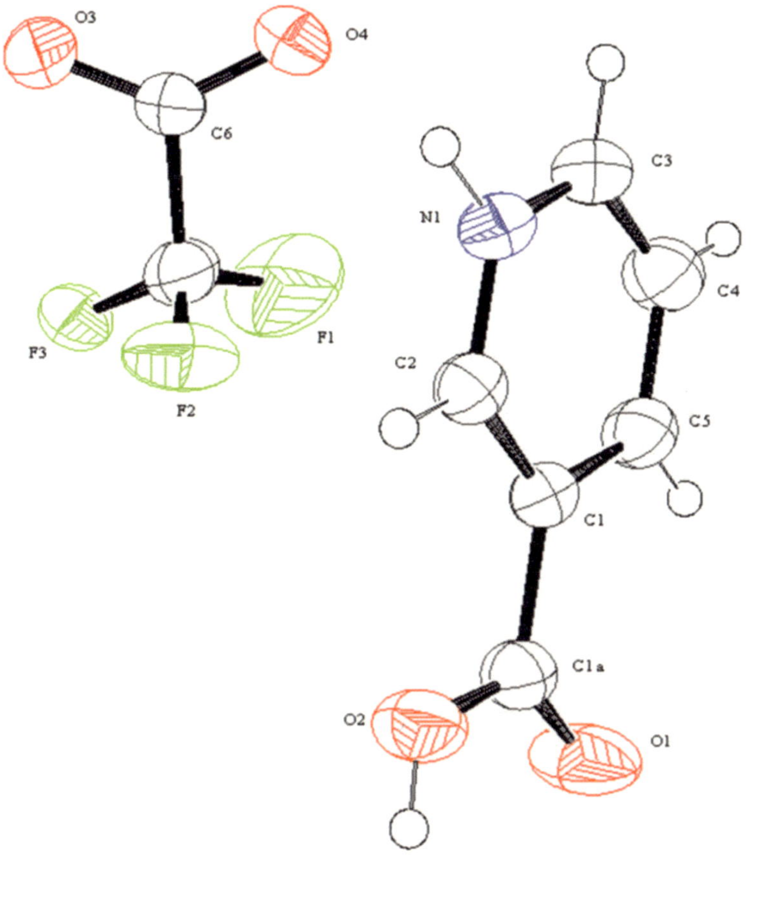

Figure 4. Structure of NTF showing 50% probability displacement ellipsoids with atom numbering scheme (for clarity, only major components of the disordered fluorine atoms are shown)

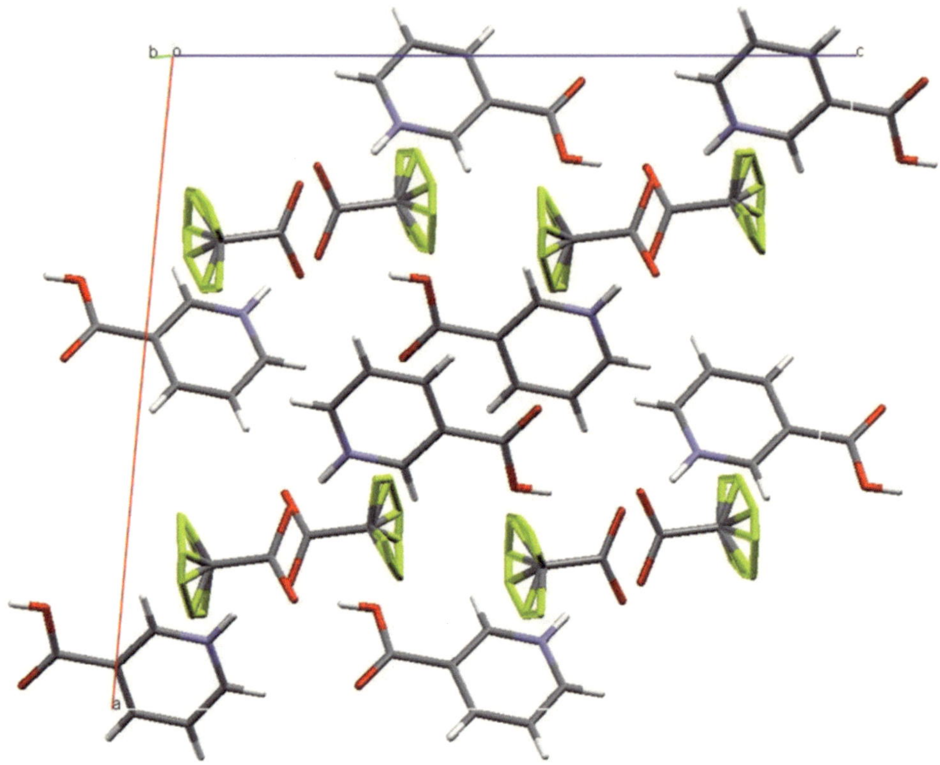

Figure 5. Molecular packing diagram of NTF viewed down the b-axis

D–H···A	d(D–H)	d(H···A)	d(D···A)	<(DHA)
N(1)–H(1N)···O(3)	0.97	1.69	2.6468	167
O(2)–H(2O)···O(4)^{i}	0.82	1.74	2.5460	167
C(3)–H(3)···O(1)^{ii}	0.93	2.44	3.3397	163
C(4)–H(4)···O(3)^{iii}	0.93	2.37	3.2929	174

Symmetry codes: (i) 1/2-x,y,-z, (ii) x,1/2-y,1/2+z and (iii) 1-x,-1/2+y,1/2-z.

Table 2. Hydrogen bonds geometry of NTF

CCDC No. 779179 contains the supplementary crystallographic data for this paper. These data can be obtained free of charge via www.ccdc.cam.ac.uk/data-request/cif, by e-mailing data-request@ccdc.com.ac.uk or by contacting the Cambridge crystallographic data centre, 12 Union Road, Cambridge CB21 EZ, U.K.; Fax: +44 1223 336033.

Powder X-ray diffraction study was carried out by employing SEIFERT, 2002 (DLX model) diffractometer with CuK_{α} (λ = 1.5405 Å) radiation using a tube voltage and current of 40 kV and 30 mA respectively. The grown crystals were finely powdered and have been subjected to powder XRD analysis. The sample was scanned over the range 10–60° at the rate of 1°/min. The indexed powder X-ray diffraction pattern of NTF is given in Figure 6. The well defined Bragg's peaks at specific 2θ angles confirmed the crystallinity of NTF.

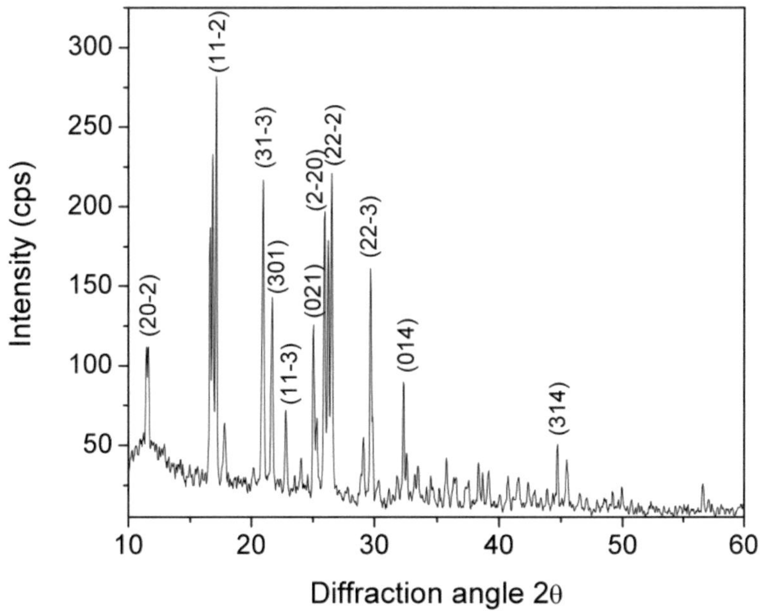

Figure 6. Powder XRD pattern of NTF

3.2. High-resolution x-ray diffraction (HRXRD) analysis

The crystalline perfection of the grown NTF single crystals was characterized by HRXRD analysis by employing a multicrystal X-ray diffractometer designed and developed at National Physical Laboratory [26]. The schematic diagram of this multicrystal X-ray diffractometer is shown in figure 7. The divergence of the X-ray beam emerging from a fine focus X-ray tube (Philips X-ray Generator; 0.4 mm × 8 mm; 2kWMo) is first reduced by a long collimator fitted with a pair of fine slit assemblies. This collimated beam is diffracted twice by two Bonse-Hart [27] type of monochromator crystals and the thus diffracted beam contains well resolved $MoK\alpha_1$ and $MoK\alpha_2$ components. The $MoK\alpha_1$ beam is isolated with the help of fine slit arrangement and allowed to further diffract from a third (111) Si monochromator crystal set in dispersive geometry (+, –, –). Due to dispersive configuration, though the lattice constant of the monochromator crystal and the specimen are different, the dispersion broadening in the diffraction curve of the specimen does not arise. Such an arrangement disperses the divergent part of the $MoK\alpha_1$ beam away from the Bragg diffraction peak and thereby gives a good collimated and monochromatic $MoK\alpha_1$ beam at the Bragg diffraction angle, which is used as incident or exploring beam for the specimen crystal. The dispersion phenomenon is well described by comparing the diffraction curves recorded in dispersive (+, –, –) and non-dispersive (+, –, +) configurations [28]. This arrangement improves the spectral purity ($\Delta\lambda/\lambda << 10^{-5}$) of the $MoK\alpha_1$ beam. The divergence of the exploring beam in the horizontal plane (plane of diffraction) was estimated to be << 3 arc s. The specimen occupies the fourth crystal stage in symmetrical Bragg geometry for diffraction in (+, –, –, +) configuration. The specimen can be rotated about a vertical axis, which is perpendicular to the plane of diffraction, with minimum angular interval of 0.4 arc

s. The diffracted intensity is measured by using an in-house developed scintillation counter. To provide two-theta ($2\theta_B$) angular rotation to the detector (scintillation counter) corresponding to the Bragg diffraction angle (θ_B), it is coupled to the radial arm of the goniometer of the specimen stage. The rocking or diffraction curves were recorded by changing the glancing angle (angle between the incident X-ray beam and the surface of the specimen) around the Bragg diffraction peak position θ_B (taken as zero for the sake of convenience) starting from a suitable arbitrary glancing angle. The detector was kept at the same angular position $2\theta_B$ with wide opening for its slit, the so-called ω scan. This arrangement is very appropriate to record the short range order scattering caused by the defects or by the scattering from local Bragg diffractions from agglomerated point defects or due to low angle and very low angle structural grain boundaries [29].

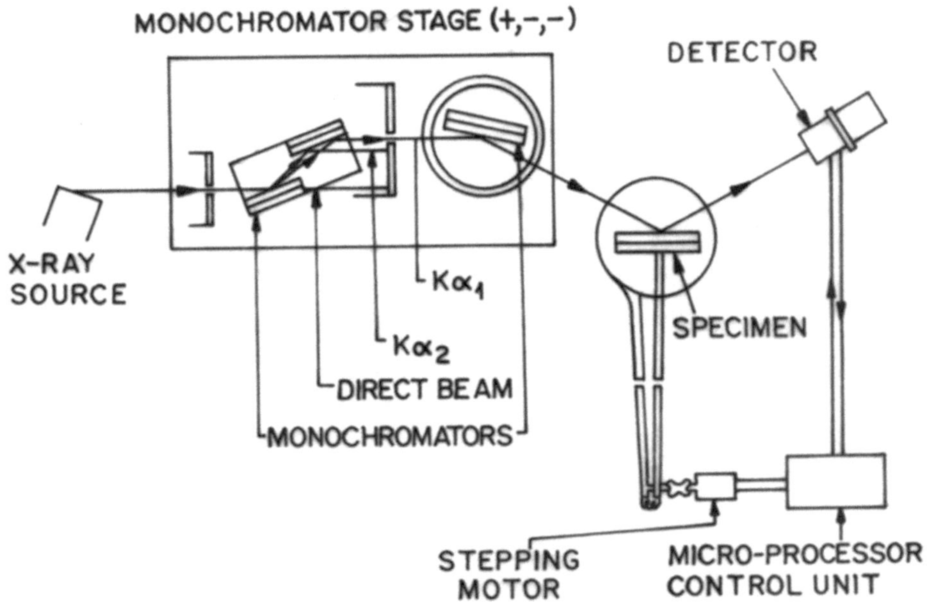

Figure 7. Schematic line diagram of multicrystal X-ray diffractometer designed, developed and fabricated at National Physical Laboratory

Before recording the diffraction curve, to remove the non-crystallized solute atoms remained on the surface of the crystal and also to ensure the surface planarity, the specimen was first lapped and chemically etched in a non preferential etchant of water and acetone mixture in 1:2 ratio.

Figure 8 shows the high-resolution X-ray diffraction curve recorded for NTF specimen crystal using $\left(11\bar{2}\right)$ diffraction planes using MoKα_1 radiation. The solid line (convoluted curve) is well fitted with the experimental points represented by the filled circles. On deconvolution of the diffraction curve, it is clear that the curve contains two additional peaks, which are 18 and 28 arc s away from the main peak (at zero glancing angle). These two additional peaks correspond to two internal structural very low angle (tilt angle ≤ 1 arc min) boundaries [30] whose tilt angles (Tilt angle may be defined as the misorientation angle between the two crystalline regions on both sides of the structural grain boundary) are 18 and 28 arc s from their adjoining regions. The FWHM (full width at half maximum) of the main peak and the very low angle boundaries are respectively 22, 27 and 40 arc s. The low FWHM values and relatively low angular spread of around 200 arc s of the diffraction curve show that the crystalline perfection of the specimen is reasonably good. Thermal fluctuations or mechanical disturbances or segregation of solvent molecules during the growth process could be responsible for the observed very low angle boundaries. It may be mentioned here that such very low angle boundaries could be detected with well resolved peaks in the diffraction curve only because of the high-resolution of the multicrystal X-ray diffractometer used in the present studies.

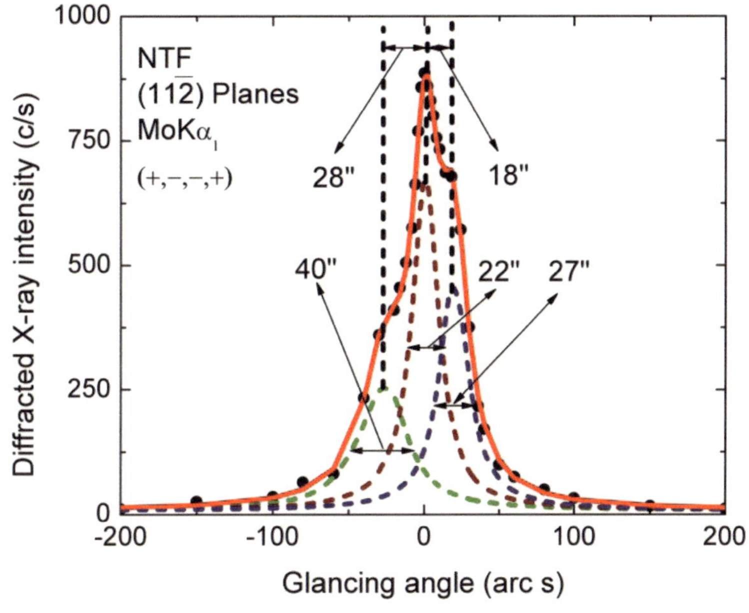

Figure 8. High-resolution X-ray diffraction curve recorded for a typical NTF single crystal specimen using $(11\bar{2})$ diffracting planes

3.3. Thermal analysis

Thermal properties of NTF were analyzed and compared with that of two nicotinium derivative crystals nicotinium oxalate (NOX) and nicotinium nitrate monohydrate (NNM). All these crystals had grown in our laboratory and belong to monoclinic system. Differential thermal analysis (DTA) and thermogravimetric analysis (TGA) of crystals were carried out simultaneously by employing TA instrument Model Q600 SDT thermal analyzer. The sample was heated at a rate of 10 °C/min in inert nitrogen atmosphere. Thermal stability of crystals was further tested using differential scanning calorimetry (DSC). DSC study was performed by using TA instrument Model Q20 in the temperature range 50–300 °C at a heating rate of 10 °C/min in inert nitrogen atmosphere and sample was placed in the Alumina crucible.

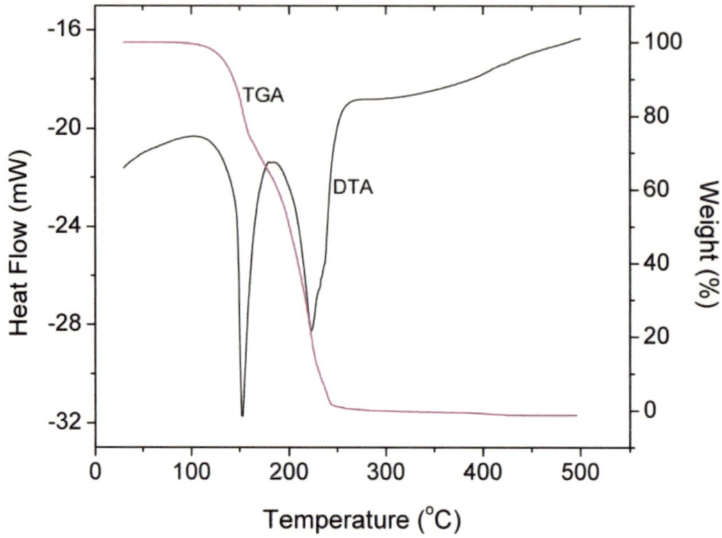

Figure 9. TGA-DTA curves of NTF

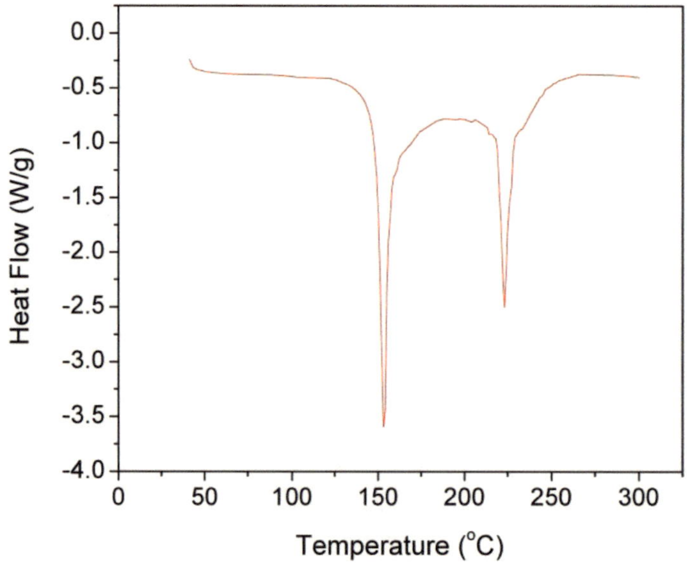

Figure 10. DSC trace of NTF

TG-DTA and DSC curves of NTF were depicted in Figures 9 and 10 respectively. In DTA curve, first endothermic peak at 152 °C is attributed to meting point of the sample, which is also evident in the DSC curve. Another important observation is that, there is no phase transition till the material melts and this enhances the temperature range for the utility of the crystal for NLO applications. The absence of water of crystallization in the molecular structure is indicated by the absence of the weight loss near 100 °C. Further there is no decomposition near the melting point [31]. This ensures the suitability of the material for possible application in lasers, where the crystals are required to withstand high temperatures. The weight loss starts around 110 °C and weight loss corresponding to decomposition of NTF was observed at 223 °C, which takes place over large temperature range (110–252 °C) where almost all the gaseous fragments like carbon dioxide and ammonia might be liberated. The TGA reveals exactly the same changes shown by DTA. The second endothermic peak in the DTA curve shows that the material is fully decomposed at 223 °C as confirmed by DSC. The small difference in the shape of the second peak may be due to the presence of impurities. The sharpness of the endothermic peaks shows good degree of crystallinity of the grown sample.

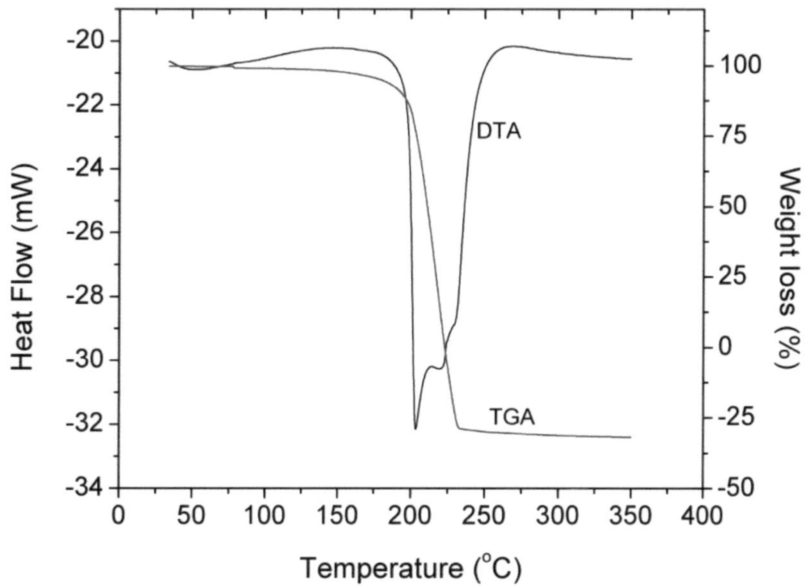

Figure 11. TGA-DTA curves of NOX

In TG-DTA trace of NOX (Figure 11), endothermic peak in DTA trace at 203 °C represents the melting point of the sample. The absence of water of crystallization in the molecular structure is indicated by the absence of the weight loss around 100 °C. TGA trace reveals that weight loss of the sample starts from this region and at 223 °C it shows complete weight loss. The shoulder peaks in DTA after the main peak corresponds to the decomposition of the material. There is no phase transition till the material melts and this enhances the temperature range for the utility of the crystal for applications. Under these conditions, phase transition means common phase transition (e.g., solid-to-liquid, liquid-to-gas etc.).

Figure 12 shows the thermogram for DTA and TGA of NNM. The compound starts to lose single molecule of water of crystallization at about 90 °C and the loss continues up to 102 °C. The weight loss in this temperature range is consistent with the weight of single molecules of water present in the crystal. The DTA curve shows a major endothermic peak, which corresponds to the melting point of NNM at 189 °C. The second weight loss take place over the temperature range 135–235 °C and almost all the compounds decomposed as its gaseous products. The second endothermic peak in the DTA curve at 230 °C attributed to the decomposition temperature of NNM. Summarized results of thermal analysis of NNM are given in Table 3.

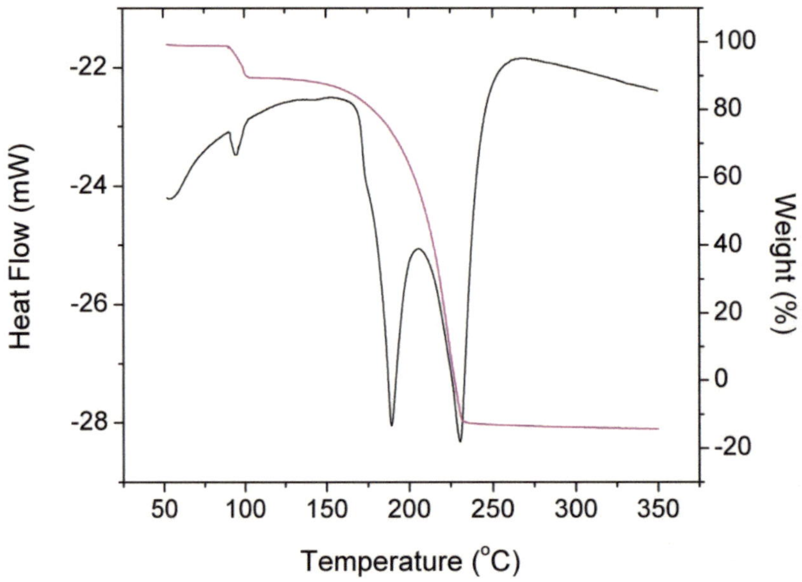

Figure 12. TGA-DTA curves of NNM

Stage	Decomposition temperature range (°C)	Decomposition steps	Weight loss (%) Observed	Calculated
1	90–102	$C_6H_6NO_2.NO_3.H_2O$ ↓	8.58	8.82
2	135–235	$C_6H_6NO_2.NO_3$	91.42	91.18

Table 3. Summarized TGA and DTA results of NNM

The figure 13 shows the comparison of DSC curves of NTF with other nicotinium derivative crystals. The calculated values of enthalpy for vapourisation, melting reaction and decomposition reaction for these three materials show that enthalpy value for NTF is less than that of NNM and NOX. As the temperature increases, initially NNM loses its single molecule of water of crystallization in the range 90-102 °C. NTF melts at 152 °C, NNM at 189 °C and NOX at 203 °C respectively. Thus thermal stability of NTF is higher than that of NNM but low when compared to NOX. The low thermal stability of NNM is due to vapourisation of its water molecule.

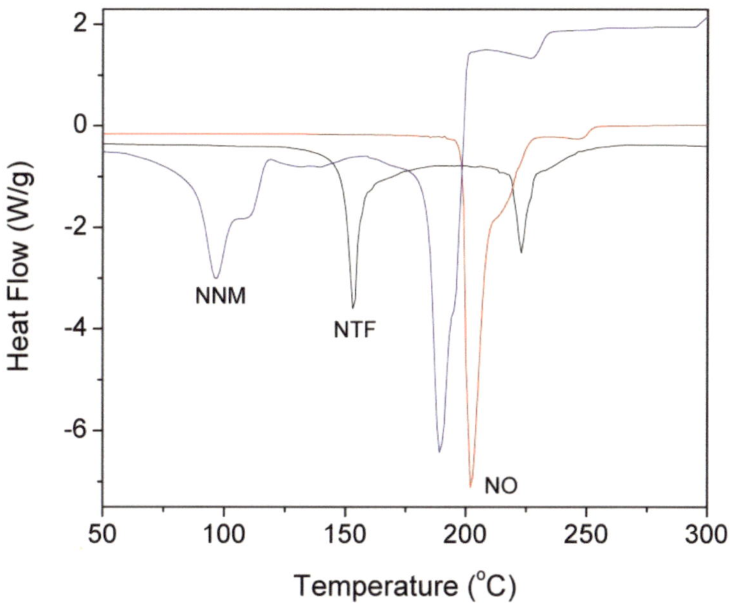

Figure 13. DSC curves of NNM, NTF and NO

3.4. FTIR studies

The FTIR spectrum was recorded using Perkin–Elmer FTIR spectrum RXI spectrometer by KBr pellet technique in the range 400–4000 cm^{-1} at room temperature (35 °C). In the FTIR spectrum of NTF (Figure 14), the strong band at 3439 cm^{-1} is attributed to stretching vibrations of O–H groups. The peak at 3086 cm^{-1} corresponds to the aromatic C–H stretching vibrations in the ring. O–H and C–C stretching vibrations are observed at 2809 cm^{-1} and 1907 cm^{-1} respectively. The existence of COO^- or COOH groups in the studied crystal was deduced on the basis of vibrational spectra. It is clearly seen that the existence of COOH is illustrated by the very strong infrared band located at 1708 cm^{-1}. Asymmetric stretching vibration of COO^- is observed at 1597 cm^{-1}. The peak at 1420 cm^{-1} is due to symmetric stretching vibration of COO^-. The C–H vibrations are occurred at 1369 cm^{-1}. C–H in plane bending vibrational modes in nicotinic acid is assigned to the frequency at 1322 cm^{-1}. It should be noted that the next band at 1194 cm^{-1} in IR spectrum is assigned to C–F stretching, which is the characteristic vibration peak of CF_3 group [32, 33]. The absorption at 1140 cm^{-1} is also due to the stretching type of vibrations of C–F bonds. The band at 517 cm^{-1} is assigned to C–C=O wagging. The characteristics bands, one at 836 cm^{-1} (COO^- rocking), one at 746 cm^{-1} (COO^- scissoring) and the third at 622 cm^{-1} (COO^- wagging) are observed in the IR spectrum.

3.5. Dielectric studies

The dielectric constant is one of the basic electrical properties of solids. Dielectric properties are correlated with the electro-optic property of the crystals [34]. The capacitance (C_{crys}) and dielectric loss (tan δ) of NTF crystal were measured using the conventional parallel plate capacitor method with the frequency range 100 Hz to 1 MHz using the Agilent 4284A LCR meter at various temperatures ranging from 40 to 80 °C. A good quality crystal of size 5 × 5 × 2 mm³ was electroded on either side with graphite coating to make it behave like a parallel plate capacitor. The observations were made during cooling of the sample. The air capacitance (C_{air}) was also measured.

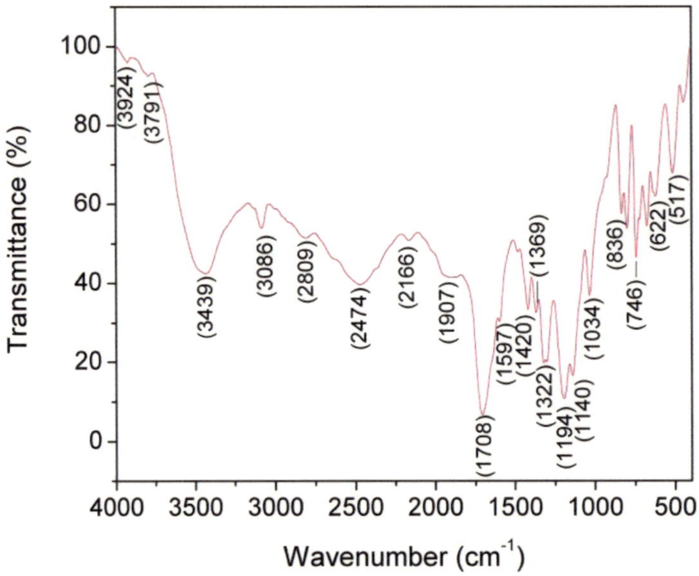

Figure 14. FTIR spectrum of NTF

The dielectric constant (ε_r) of the crystal was calculated using the following relation

$$\varepsilon_r = \frac{C_{crys}}{C_{air}} \tag{1}$$

As the crystal area was smaller than the plate area of the cell, parallel capacitance of the portion of the cell not filled with the crystal was taken into account and, consequently the above equation becomes

$$\varepsilon_r = \left(\frac{C_{crys} - C_{air}\left(1 - \frac{A_{crys}}{A_{air}}\right)}{C_{air}} \right) \left(\frac{A_{air}}{A_{crys}} \right) \tag{2}$$

where A_{crys} is the area of the crystal touching the electrode and A_{air} is the area of the electrode.

The variation of dielectric constant with frequency at different temperatures (Figure 15) shows that dielectric constant decreases with increasing frequency and finally it becomes almost a constant at higher frequencies for all temperatures. It is also indicates that dielectric constant increases with increase in temperature. The measurements of dielectric loss at different frequencies and temperatures show the same trend. This dielectric behavior [35] can be understood on the basis that the mechanism of polarization is similar to that of conduction process. The electronic exchange of the number of ions in the crystals gives local displacement of electrons in the direction of the applied field, which in turn gives rise to polarization namely, electronic, ionic, dipolar and space charge polarization. Space charge polarization is generally active at lower frequencies and high temperatures and indicates the perfection of the crystal. As the frequency increases, a point will be reached where the space charge cannot sustain and comply with the external field and hence the polarization decreases, giving rise to decrease in values of ε_r. At 80 °C, the dielectric constant of NTF crystal at 100 Hz is 10.851, and this value decreases to 1.955 at 1 MHz. Lowering the value of dielectric constant of the interlayer dielectric (ILD) decreases the RC delay, lowers the power consumption and reduces the crosstalk between nearby interconnects [36]. Also the materials with quite low dielectric constant lead to a small RC constant, thus permitting a higher bandwidth in the range of 10^{12} Hz for light modulation. Thus materials with low dielectric constant have considerable advantages in this context.

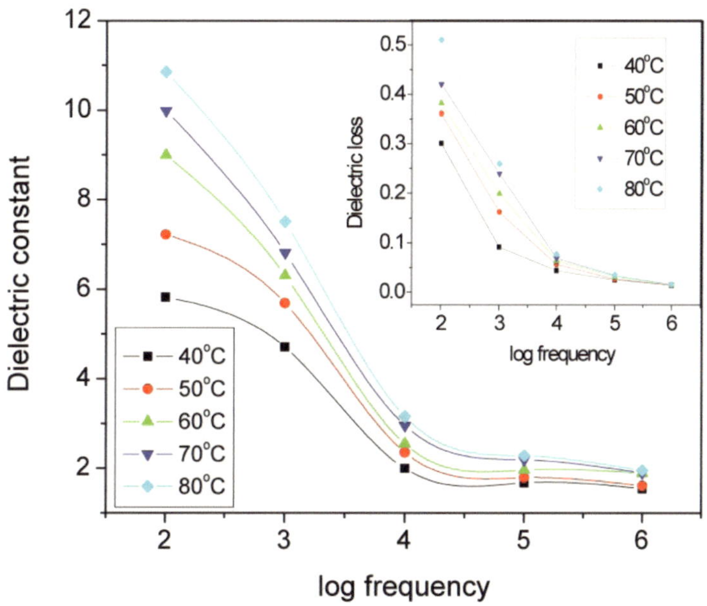

Figure 15. Plot of dielectric constant versus applied frequency. Plot of dielectric loss versus applied frequency is in inset

3.6. Mechanical hardness studies

Mechanical strength of the materials plays a key role in the device fabrication. Vickers hardness is one of the important deciding factors in selecting the processing (cutting, grinding and polishing) steps of bulk crystal in fabrication of devices based on crystals.

Microhardness measurements were done on $(\bar{1}12)$ face of NTF crystal using Leitz-Wetzlar hardness tester fitted with a Vickers diamond indenter at room temperature. The Vickers microhardness number, Hv was calculated using the relation [37]:

$$H_V = 1.8544 \left(p / d^2\right) kg / mm^2 \tag{3}$$

where p is the applied load (g) and d is the diagonal length (μm) of the indentation. The indentation time was kept at 10 s and microhardness value was taken as the average of the several impressions made. Figure 16 shows the variation of Hv as function of applied load in the range 10–100 g on $(\bar{1}12)$ face of NTF crystal.

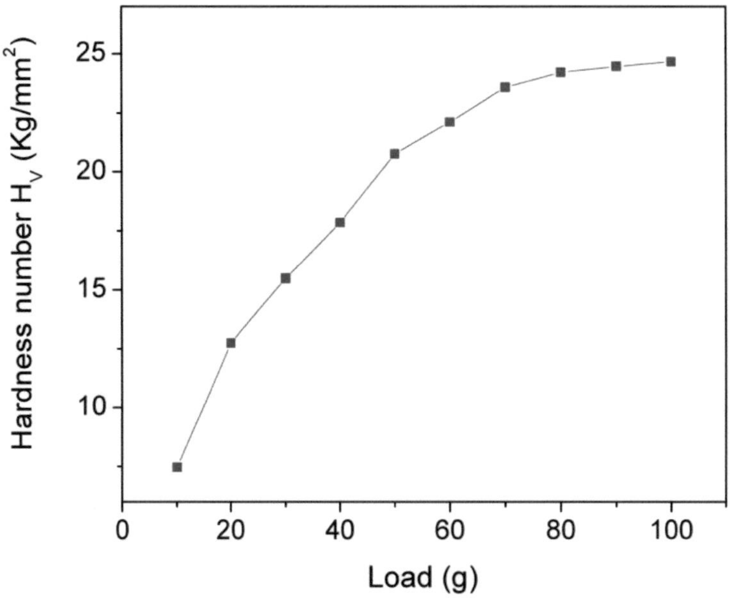

Figure 16. Variation of Vickers microhardness values versus applied load

It reveals that hardness number increases with increasing applied load. This phenomenon is known as reverse indentation size effect (RISE). When the material is deformed by the indenter, dislocations are generated near the indentation site. The major contribution to the increase in hardness is attributed to the high stress required for homogenous nucleation of dislocations in the small dislocation-free region indented [38]. The RISE can be caused by the relative predominance of nucleation and multiplication of dislocations. The other reason for RISE is that the relative predominance of the activity of either two sets of slip planes of a particular slip system or two slip systems below and above a particular load [39]. The RISE phenomenon essentially takes place in crystals which readily undergo plastic deformation [40]. The relation between load and the size of indentation can be interpreted using Meyer's law, $P = k_1 d^n$, where k_1 is a constant and n is the Meyer's number (or index). The slope of log P versus log d gives the n value and it is estimated to be 2.73. According to Onitsch [41] and

Hanneman [42], n should be between 1 and 1.6 for hard materials and above 1.6 for softer ones. Thus NTF crystal belongs to soft material category. Meyer number is a measure of the indentation size effect (ISE). For the normal ISE behaviour, the exponent $n < 2$. When $n > 2$, there is the RISE behaviour [39].

4. Conclusions

Single crystals of NTF in monoclinic system were grown by solution growth technique for the first time and its solubility and metastable zone width were determined. X-ray diffraction analysis reveals that NTF crystallizes in monoclinic system with space group *I*2/a and unit cell parameters are a = 15.616(5) Å, b = 7.455(5) Å, c = 16.503(5) Å, and $\beta = 95.74^{\circ}$. HRXRD analysis substantiates the good quality of the crystals. TG-DTA and DSC studies show that NTF melts at 152 °C. It is observed that thermal stability of NTF is in between that of other nicotinium derivative crystals. The FTIR analysis confirms the presence of functional groups constituting NTF. Dielectric measurements indicate that NTF crystal has low values of dielectric constant and dielectric loss.

Author details

P.V. Dhanaraj
Department of Physics, Malabar Christian College, Kozhikode, India

N.P. Rajesh
Centre for Crystal Growth, SSN College of Engineering, Kalavakkam, India

5. References

[1] Zyss J, Dhenaut C, Van T C, Ledoux I (1993) Quadratic Nonlinear Susceptibility of Octupolar Chiral Ions. Chem. Phys. Lett. 206: 409-414.

[2] Russell V A, Evans C C, Li W, Ward M D (1997) Nanoporous Molecular Sandwiches: Pillared Two-Dimensional Hydrogen-bonded Networks with Adjustable Porosity. Science 276: 575-579.

[3] Matos Gomes E D, Venkataramanan V, Nogueira E, Belsley M, Proenca F, Criado A, Dianez M J, Estrada M D,Perez-Garrido S (2000) Synthesis, Crystal Growth and Characterization of a Nonlinear Optical Material–Urea L- Malic Acid. Synth. Met. 115: 225-228.

[4] Levine B F, Bethea C G (1975) Conjugated Electron Contributions to the Second Order Hyperpolarizability of Substituted Benzene Molecules. J. Chem. Phys. 63: 6-10.

[5] Bosshard Ch, Knopfle G, Pretre P, Gunter J P (1992) Second-order Polarizabilities of Nitropyridine DerivativesDetermined with Electric-Field-Induced Second-Harmonic Generation and a Solvatochromic Method: A Comparative Study. J. Appl. Phys. 71: 1594-1599.

[6] Clays K, Persoons A (1991) Hyper-Rayleigh Scattering in Solution. Phys. Rev. Lett. 66: 2980-2983.

[7] Clays K, Olbrechts G, Munters T, Persoons A, Kim O K, Choi L S (1998) Enhancement of the Molecular Hyperpolarizability by a Supramolecular Amylose–Dye Inclusion Complex, Studied by Hyper-Rayleigh Scatteringwith Fluorescence Suppression. Chem. Phys. Lett. 293: 337-342.

[8] Dhenaut C, Ledoux I, Samuel I D W, Zyss J, Bourgault M, Bozec H L (1995) Chiral Metal Complexes with Large Octupolar Optical Nonlinearities. Nature 374: 339-342.

[9] Sylla M, Giffard M, Boucher V, Illien B, Mercier N, Nguyen Phu X (1999) Nonlinear Optical Properties of Chiral Thiolates. Synth. Met. 102: 1548-1549.

[10] Ferrier J L, Gazengel J, Nguyen Phu X, Rivoire G (1984) Backscattering in the Picosecond Range: an Optical Triggered Switching Effect. Opt. Commun. 51 (4): 285-288.

[11] Somac M, Somac A, Davies B L, Humphery M G, Wong M S (2002) Third-Order Optical Nonlinearities of Oligomers, Dendrimers and Polymers Derived from Solution Z-Scan Studies. Opt. Mater. 21: 485–488.

[12] Natarajan L V, Sutherland R L, Tondiaglia V P, Bunning T J, Adams W W (1996) Electro-Optical Switching Characteristics of Volume Holograms in Polymer Dispersed Liquid Crystals. J. Nonlinear Opt. Phys. Mater. 5: 89- 98.

[13] Pecaut J, Bagieu-Beucher M (1993) 2–Amino–5–nitropyridiniummonohydrogenphos phite. Acta Cryst. C49: 834-837.

[14] Ravindra H J, John Kiran A, Dharmaprakash S M, Satheesh Rai N, Chandrasekharan K, Kalluraya B, Rotermund F (2008) Growth and characterization of an efficient nonlinear optical D–π–A–π–D type chalcone single crystal. J. Cryst. Growth 310: 4169-4176.

[15] Gielen M, Kholufi A E, Biesemans M, Willem R (1992) (2-Methylthio-3-Pyridinecarboxylato)-diethyltin and -di- n-butyltin Compounds: Synthesis, Spectroscopic Characterization and in vitro Antitumour Activity. X-ray Crystal Structure of bis[diethyl(2-methylthio-3-Pyridinecarboxylato)tin] oxide and of diethyltin bis(2-methylthio-3- pyridinecarboxylate), Polyhedron. 11: 1861-1868.

[16] Athimoolam S, Anitha K, Rajaram R K (2005) Nicotinium dihydrogenphosphate. Acta Cryst.E61: o2553–o2555.

[17] Athimoolam S, Rajaram R K (2005) Bis(nicotinic acid) hydrogen perchlorate. Acta Cryst. E61:o2674–o2676.

[18] Athimoolam S, Rajaram R K (2005) Dinicotinium sulfate. Acta Cryst. E61: o2764–o2767.

[19] Gao S, Liu J W, Huo L H, Sun Z Z, Gao J S, Ng S W (2004) Catena-Poly[[diaquabis(2-chloronicotinato-κ ^{2}O,O')cadmium(II)]-μ-2-chloronicotinato-κ^{3}O,O':N]. Acta Cryst. C 60: m363-m365.

[20] Athimoolam S, Natarajan, S (2007) Nicotinium trifluoroacetate. Acta Cryst. E 63: 2656-2657.

[21] Nyvlt J, Rychly R, Gottfried J, Wurzelova J (1970) Metastable Zone-Width of Some Aqueous Solutions. J. Cryst. Growth 6: 151-162.

[22] Sangwal K (1989) On The Estimation of Surface Entropy Factor, Interfacial Tension, Dissolution Enthalpy and Metastable Zone-Width for Substances Crystallizing from Solution. J. Cryst. Growth 97: 393-405.

[23] Harms K, Wocadlo S, XCAD4, University of Marburg, Germany, 1995.

[24] North A C T, Phillips D C, Mathews F S (1968) A semi-empirical method of absorption correction. Acta Cryst. A24: 351-359.

[25] SHELXTL/PC Version 6.10 Madison, WI: Bruker AXS Inc., 2000.

[26] Lal K, Bhagavannarayana G (1989) A High-Resolution Diffuse X-Ray Scattering Study of Defects in Dislocation-Free Silicon Crystals Grown by the Float-Zone Method and Comparison with Czochralski-Grown Crystals. J. Appl. Cryst. 22: 209-215.

[27] Bonse U, Hart M (1965) Tailless X–ray Single Crystal Reflection Curves Obtained by Multiple Reflection. Appl. Phys. Lett. 7: 238-240.

[28] Bhagavannarayana G (1994) High Resolution X-Ray Diffraction Study of As-Grown and BF_2^+ Implanted Silicon Single Crystals, Ph. D. Thesis, University of Delhi, Delhi, India.

[29] Bhagavannarayana G, Kushwaha S K (2010) Enhancement of SHG Efficiency by Urea Doping in ZTS Single Crystals and its Correlation with Crystalline Perfection as Revealed by Kurtz Powder and High-Resolution X-Ray Diffraction Methods. J. Appl. Cryst. 43: 154-162.

[30] Bhagavannarayana G, Ananthamurthy R V, Budakoti G C, Kumar B, Bartwal K S (2005) A Study of the Effect of Annealing on Fe-Doped $LiNbO_3$ by HRXRD, XRT and FT-IR. J. Appl. Cryst. 38: 768-771.

[31] Willard, Merritt, Dean, Settle (1986) Instrumental Methods of Analyses, First Indian Edition: CBS, Delhi.

[32] Fuson N, Josien M L, Jones E A, Lawson J R (1952) Infrared and Raman Spectroscopy Studies of Light and Heavy Trifluoroacetic Acids. J. Chem. Phys. 20: 1627-1635.

[33] Takeda Y, Suzuki H, Notsu K, Sugimoto W, Sugahara Y (2006) Preparation of a Novel Organic Derivative of the Layered Perovskite Bearing $HLaNb_2O_7{\cdot}nH_2O$ Interlayer Surface Trifluoroacetate Groups. Mat. Res. Bull. 41: 834-841.

[34] Aithal P S, Nagaraja H S, Mohan Rao P, Avasti D K, Sarma A (1997) Effect of high energy ion irradiation on electrical and optical properties of organic nonlinear optical crystals. Vacuum 48: 991-994.

[35] Anderson J C (1964) Dielectrics, Chapman and Hall.

[36] Hatton B D, Landskron K, Hunks W J, Bennett M R, Shukaris D, Pervoic D D, Ozin G A (2006) Materials Chemistry for Low k-Materials. Mater. Today. 9: 22-31.

[37] Mott B W (1956) Microindentation Hardness Testing, Butterworths, London.

[38] Kunjomana A G, Chandrasekharan K A (2005) Microhardness Studies of GaTe Whiskers. Cryst. Res. Technol. 40: 782-785.

[39] Sangwal K (2000) On the Reverse Indentation Size Effect and Microhardness Measurement of Solids. Mater. Chem. Phys. 63: 145-152.

[40] Li H, Han Y H, Bradt R C (1994) Knoop Microhardness of Single Crystal Sulphur. J. Mater. Sci. 29: 5641-5645.

[41] Onitsch E M (1947) Mikroscopia 2: 131-134.
[42] Hanneman M (1941) Metall. Manch. 23: 135-139.

Section 2

Application of Isothermal Titration Calorimetry for Analysis of Proteins and DNA

Isothermal Titration Calorimetry: Thermodynamic Analysis of the Binding Thermograms of Molecular Recognition Events by Using Equilibrium Models

Jose C. Martinez, Javier Murciano-Calles, Eva S. Cobos,
Manuel Iglesias-Bexiga, Irene Luque and Javier Ruiz-Sanz

Additional information is available at the end of the chapter

http://dx.doi.org/10.5772/53311

1. Introduction

The revolution achieved during the last decade in the fields of genomics and proteomics has shown the need of going in-depth into the structural, dynamic, energetic and functional knowledge of biological macromolecules, mainly proteins and nucleic acids. Of special interest is the study of the molecular recognition between such kind of molecules or between them and other biological molecules, for example, natural metabolites or designed drugs to alter their functionalities.

Isothermal titration calorimetry (ITC) is a technique that directly measures the heat exchange accompanying a chemical or biochemical reaction. It is the ideal technique for the investigation of the energetics of ligand binding to biological macromolecules because it provides a complete thermodynamic characterization of the macromolecule-ligand interactions, allowing for the measurement of the binding affinity as well as of the changes in enthalpy and entropy of the process. The nature (enthalpic or entropic) and magnitude of the forces directing the interaction are very important factors to be considered in the design of ligands with specific characteristics. Additionally, the heat capacity of binding can be determined by carrying out titration experiments at different temperatures.

From the diverse methodologies that can be applied in the research of binding processes, ITC presents a series of advantages and possibilities, and as such it is considered a very powerful tool. Precluding the structural interpretation, the direct determination of binding thermodynamic parameters becomes necessary to describe the energetic aspects of the binding and, thus, to define and to rationalize macromolecular recognition. Nevertheless, although calorimetry has been widely used as an experimental resource, it has not always been interpreted correctly, mainly due to the difficulty found in extracting thermodynamic information from experimental data. Thus, the rigorous analysis of ITC thermograms should be done under the assumption of theoretical models, able to describe the most significant stages present during the binding process and which application would give rise to valuable thermodynamic information for each of such stages.

2. General aspects of binding equilibrium

Through this Chapter we are going to scrutinise the use of ITC in the study of binding equilibrium processes, as well as how to design and perform the experiments and the correct way to handle the data and achieve the corresponding fit to the proper equilibrium models. Nevertheless, prior to focusing on the different ITC aspects, we will describe briefly some basic features of binding equilibrium, for which it is crucial to introduce some basic concepts and equilibrium formulas.

2.1. Basic concepts

Apart from the capacity of self-copying, biomolecules are characterized by their ability to specifically interact with other molecules within the cell, which defines their biological functionality. Many of the biochemical processes occurring in living systems are based on, or regulated by, binding interactions between biological macromolecules or with other small molecules. Examples of interactions between macromolecules can be found out in interactions between polypeptide chains to form the quaternary structure of multi-subunit proteins, in the close association of protein and RNA molecules in the ribosome, in the binding of transcription regulators to DNA, protein-protein interactions in many signalling cascades, etc. Besides, many biological macromolecules bind small molecules, for example, enzymes that bind substrates and effector molecules, or proteins that bind metabolites in order to transport or store them. Signalling transmission is also based on interactions, as those of hormones with membrane receptors. Additionally, some of the regulation pathways of the transcription and replication of nucleic acids involve the change of their conformations induced by binding of metallic ions.

The interactions that can take place under different backgrounds and contexts from a physico-chemical point of view, can be summarized into three different types: i) at *equilibrium*, ii) at *steady-state conditions*, and iii) at the *transition between different steady-state conditions*. In this Chapter, we will direct attention to the first case, the binding equilibrium process between a biological macromolecule (such as a protein or a nucleic acid) and a small molecule, called a *ligand*, occurring by *specific* interactions, that is, the ligand (L) binds at specific sites of the macromolecule (M). The establishment of such specific interactions is crucial for the correct functioning of the cell, as happens in the most of biological processes, where one or more macromolecule-ligand (ML) interactions are involved, determining and regulating the biological function.

All these ML interactions present some common features:

- Binding of the ligand involves a non-covalent *reversible* interaction to a specific region of the macromolecule, called the *binding site*, usually situated at its surface or close to it.
- The ligand binding process may induce conformational changes that modify the activity of the macromolecule; this phenomenon is known as *alosterism*.
- When the macromolecule has more than one binding site, the binding of one ligand to one of the sites may change the affinity of the ligand for the rest of binding sites; this feature is known as *co-operativity* and is closely related to the alosterism phenomenon.
- In some cases, the binding process can result in a change in the aggregation state of the molecules (*polisterism*) or, even, give rise to a new phase in the system (*poliphasic* processes). These two aspects are not within the scope of this Chapter.

The correct characterization of the binding process requires some experimental work in order to determine a variety of parameters such as:

- **Number of binding sites** per macromolecule for a defined ligand, $\boldsymbol{n}$. The numeric value can be one or higher, sites can be identical or different in terms of affinity into the same macromolecule.
- **Binding parameter,** $\bar{\nu}$**.** Represents the moles of bound ligand by each mole of macromolecule. It ranges between zero and the number of binding sites, n.
- **Saturation fraction, θ.** The fraction of the total number of sites of the macromolecule occupied by ligand molecules, which ranges from zero (no occupancy) to one (all sites occupied). It can be easily deduced that $\bar{\nu} = n \cdot \theta$.
- **Binding affinity** of the ligand to the macromolecule, expressed by means of the equilibrium binding constant, K_b, or the corresponding Gibbs energy change, $\Delta G_b = -RT \cdot lnK_b$. As mentioned above, ITC experiments provide a complete thermodynamic characterization of the macromolecule-ligand interactions, allowing the determination of the binding affinity as well as the changes in enthalpy, ΔH_b, and entropy, ΔS_b, of the binding process, where $\Delta G_b = \Delta H_b - T \cdot \Delta S_b$.
- $\bar{\nu}$ **/(n-**$\bar{\nu}$**)** is the **relationship between occupied (**$\bar{\nu}$**) and empty (n-**$\bar{\nu}$**) sites** in the macromolecule.

Thus, binding studies can provide the answer to some fundamental questions related to the functional aspects of biological macromolecules, such as, for example: How many binding sites in the macromolecule for a defined ligand exist? What is the affinity of the ligand for each binding site? Is there any dependency or inter-connection among the sites? Can affinity be modulated by the proper ligand molecule or by any other metabolites?

The experimental data is ideally expressed in terms of changes in the binding parameter, $\bar{\nu}$, as a function of the free ligand concentration in solution, $[L]$. In practice, it is necessary to move along the whole equilibrium process, starting usually from a solution containing the free macromolecule where the ligand solution is added progressively until the saturation of all sites is achieved. During this titration process, we should measure the binding parameter, generally by using spectroscopic or calorimetric probes. This kind of approach allows us to know the total concentration of both macromolecule, $[M]_T$, and ligand, $[L]_T$, in the solution.

2.2. The Adair's equation

This equation defines the type of equilibrium that can be established between a macromolecule and its ligand upon binding.

2.2.1. *Binding to one site*

In order to establish how the equilibrium constants can be determined from experimental data we are going to develop the simplest binding process, described by the binding of a ligand to a macromolecule which has only one binding site. It can be expressed as following:

$$M + L \xleftrightarrow{K_b} ML \qquad K_b = \frac{[ML]}{[M][L]} \tag{1}$$

Although the thermodynamic equilibrium constant that characterizes the binding process, K_b, must be expressed as a function of the activities of the different species present at equilibrium, it is usual to use concentrations instead of this, as experimental data contains larger errors than the ones derived from this approximation.

As we have stated previously, a very useful parameter obtained by experimental data is the *binding parameter,* $\bar{\nu}$, defined as the average of ligand molecules that are bound per macromolecule, its range from 0 to n (number of binding sites per macromolecule). Mathematically it can be defined as:

$$\bar{\nu} = \frac{[L]_b}{[M]_T} = \frac{[ML]}{[M]+[ML]} = \frac{K_b[L]}{1+K_b[L]} \tag{2}$$

The representation of $\bar{\nu}$ *versus [L]* (free ligand concentration) gives the so called *binding curve*. As shown in Figure 1, our simple example corresponds to a hyperbolic curve trending asymptotically to the number of sites n ($n=1$ in this case), because saturation conditions are reached as the free ligand concentration increases.

The value of the equilibrium constant, K_b, can be determined from the non-linear fitting of the experimental data, represented in the binding curve, to equation 2. In the case of a single binding site, it is also possible to convert equation 2 into a variety of linear equations to obtain the K_b value from the corresponding linear regression.

Such linear representations (Figure 1) can be easily deduced from the previous equations and are named as follows:

Double reciprocal representation:

$$\frac{1}{\bar{\nu}} = 1 + \frac{1}{K_b[L]} \tag{3}$$

Hill representation:

$$\log\left(\frac{\overline{\nu}}{1-\overline{\nu}}\right)=\log\left(K_b\left[L\right]\right) \tag{4}$$

- Scatchard representation:

$$\frac{\overline{\nu}}{\left[L\right]}=K_b-K_b\overline{\nu} \tag{5}$$

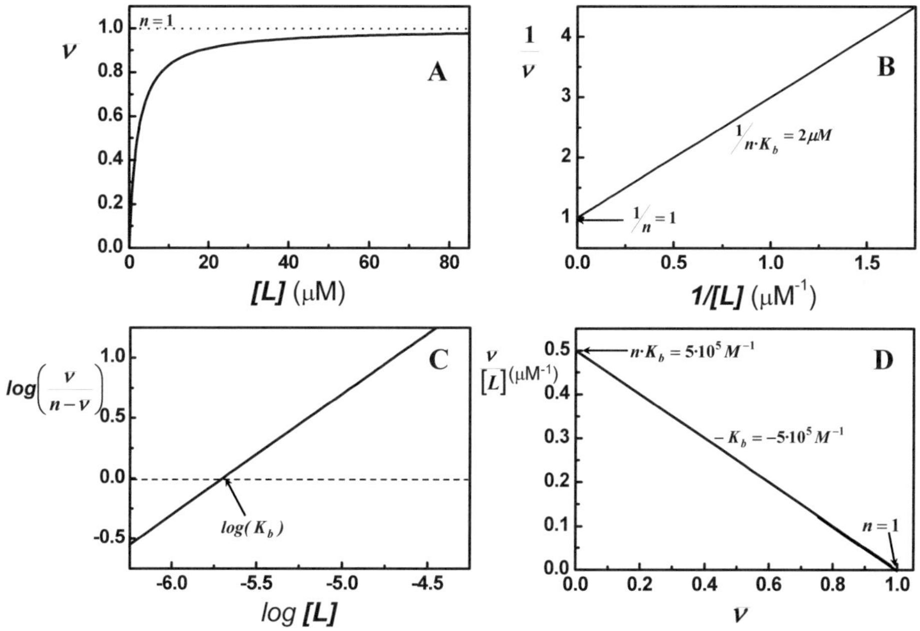

Figure 1. Binding to one site
Different representations of simulated data for the binding equilibrium of a ligand to a macromolecule with a single binding site with $K_b = 5{\cdot}10^5\ M^{-1}$. (A) Binding curve. (B) Double reciprocal representation. (C) Hill representation. (D) Scatchard representation.

Another accessible parameter from binding experiments is the *saturation fraction*, θ, defined as the fraction of binding sites occupied by ligand. It is related to the binding parameter by the expression: $\theta=\overline{\nu}/n$

Depending on the techniques used to obtain the experimental data of the binding equilibrium processes, it is possible to determine either the binding parameter, $\overline{\nu}$, or the saturation fraction, θ, and then, to develop the subsequent analysis to determine the rest of parameters of interest.

The most appropriate technique to determine $\overline{\nu}$ is ITC [1, 2]. It can also be achieved by equilibrium dialysis, which allows the calculation of the concentrations of the free ligand in equilibrium with the different species of the macromolecule (free and bound), in order to construct directly the binding curve. Although this technique provides the complete set of experimental data required, it requires extensive work and also needs large amounts of sample.

The techniques that allow the determination of θ values are based on detecting physical change occurring in the macromolecule or in the ligand during the binding process. Such physical change has to be a linear function, as the ligand is bound to the macromolecule. Furthermore, if the macromolecule presents several binding sites, the change must be the same for all of them or, at least, the change relationship between the binding sites must be known. Depending on the nature of the physical change, different techniques may be used: UV-visible spectroscopy, fluorescence, circular dicroism, nuclear magnetic resonance spectroscopy, etc [1, 2].

2.2.2. *Binding to two equivalent and independent sites*

2.2.2.1. *Macroscopic formulae*

Here, we are going to describe the formulae of the equilibrium processes corresponding to the binding of a ligand to a macromolecule with two binding sites. In this stage, we will focus on the simplest situation, where both sites are equal in affinity and independent, *i.e.*, not influencing each other upon ligand binding. Binding schemes where these basic assumptions do not occur can be useful to describe cooperative interactions and will be described later on in this Chapter. To obtain the binding parameters we can use elementary thermodynamics for the simplest non-cooperative cases, but as the cases turn more complex, this formulae becomes very laborious and a more general formulae is needed.

In order to strengthen the binding concepts, we will start by applying the classical formulae to this simple case, before the description of the general formulae introduced in the biochemical field by Wyman [3], which is useful for the formulae of more complicated binding schemes. Thus, in the case of a macromolecule with two equivalent and independent binding sites, the description of the formulae from a **macroscopic** point of view can be developed in two different but equivalent ways:

- Stage formulae: A first equilibrium stage is considered, where M binds to one ligand molecule, L, followed by a second stage where a second L molecule binds to M, achieving saturation. These two stages are characterized by their corresponding equilibrium constants as follows:

$$M + L \xleftrightarrow{K_{b1}} ML \qquad K_{b1} = \frac{[ML]}{[M][L]} \qquad K_{b2} = \frac{[ML_2]}{[ML][L]} \tag{6}$$
$$ML + L \xleftrightarrow{K_{b2}} ML_2$$

- Global formulae: The equilibriums take place between both the free and the bound M species to either one or two ligand molecules. The equilibrium constants for this formulae are related with the ones of the previous formulae as follows:

$$M + L \xleftrightarrow{\beta_1} ML \qquad \beta_1 = \frac{[ML]}{[M][L]} = K_{b1} \qquad \beta_2 = \frac{[ML_2]}{[M][L]^2} = K_{b1} \cdot K_{b2} \tag{7}$$
$$M + 2L \xleftrightarrow{\beta_2} ML_2$$

The binding equilibrium constants K_b and β are named as *macroscopic constants*.

In this case, where the macromolecule has two equivalent and independent binding sites, the binding parameter, $\bar{\nu}$, can be calculated as:

$$\bar{\nu} = \frac{[ML]+2[ML_2]}{[M]+[ML]+[ML_2]} = \frac{K_{b1}[L]+2K_{b1}K_{b2}[L]^2}{1+K_{b1}[L]+K_{b1}K_{b2}[L]^2} = \frac{\beta_1[L]+2\beta_2[L]^2}{1+\beta_1[L]+\beta_2[L]^2} \quad (8)$$

2.2.2.2. Microscopic formulae

When more than one binding site exists , the binding process can be described using a **microscopic formulae**, that is, distinguishing each binding site. Thus, as it is shown in Figure 2, the first ligand molecule can bind to binding site 1 of M, being the equilibrium process characterized by the microscopic binding constant k_1; or to the binding site 2 and then, characterized by k_2. The second ligand molecule will bind to the free binding site, and the equilibrium will be characterized by k_3 if the free site is the number 2 or k_4 if it is the number 1.

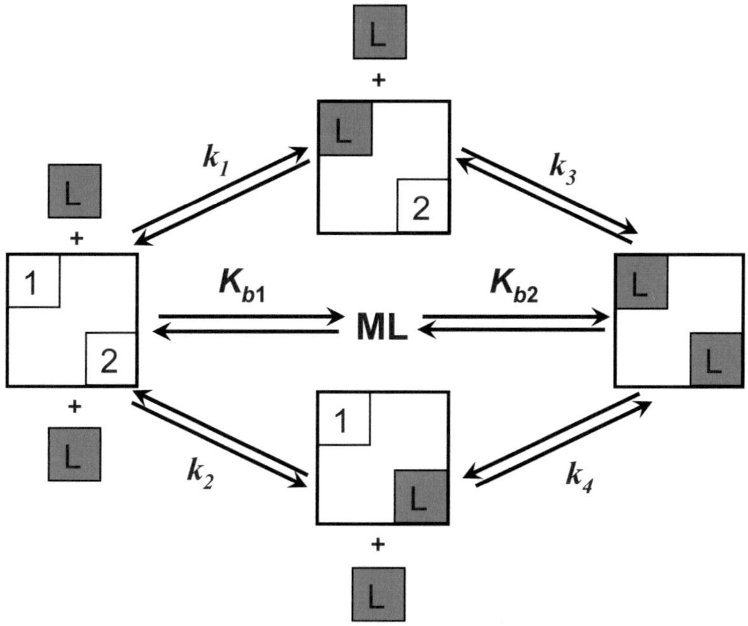

Schematic representation to distinguish between the macroscopic and microscopic formulae for the binding equilibriums of a general ligand L to a macromolecule with two binding sites.

Figure 2. Binding to two sites

For the case we are explaining, the meaning of having equivalent binding sites is that $k_1 = k_2$ and $k_3 = k_4$. Also, independent binding sites imply that $k_1 = k_4$ and $k_2 = k_3$. So, in binding processes where all binding sites are equivalent and independent, all microscopic equilibrium constants will be identical. The relations between macroscopic, K_b, and microscopic, k, equilibrium constants are: $K_{b1} = 2k$ and $K_{b2} = k/2$. Thus, although microscopic constants are identical, the macroscopic ones are different due to statistical factors. By using the microscopic constants instead of the macroscopic ones, a simpler expression for the binding parameter than that given in equation 8 is obtained:

$$\bar{\nu} = \frac{2k[L]}{1+k[L]} \tag{9}$$

Therefore, it is interesting to know the relationships between the equilibrium constants obtained using the different formulae; such relationships between microscopic and macroscopic constants may allow it to be deduced whether the binding sites are independent or not. Meanwhile, the use of microscopic constants will simplify the equation of the binding parameter which, as mentioned above, is the experimentally accessible parameter, besides the fraction saturation.

2.2.3. *Binding to n equivalent and independent sites*

We can obtain the relationship between the different types of binding constants for the general case of a macromolecule having n equivalent and independent binding sites. Firstly, we proceed to apply the macroscopic formulae in its two ways, which are summarized in the next scheme:

$$\begin{array}{llll}
\textit{Stage formulation} & & \textit{Global formulation} & \\
M+L \xrightleftharpoons{K_{b1}} ML & & M+L \xrightleftharpoons{\beta_1} ML & \\
ML+L \xrightleftharpoons{K_{b2}} ML & & M+2L \xrightleftharpoons{\beta_2} ML & \\
\vdots & & \vdots & \\
ML_{i-1} \xrightleftharpoons{K_{bi}} ML_i & & M+iL \xrightleftharpoons{\beta_i} ML_i & \\
\vdots & & \vdots & \\
ML_{n-1} \xrightleftharpoons{K_{bn}} ML_n & & M+nL \xrightleftharpoons{\beta n} ML_n &
\end{array} \tag{10}$$

Looking at the equations defining the macroscopic binding parameter (equations 2 and 8) we can easily deduce that, in the case of having n equivalent and independent binding sites, this variable can be written in general as

$$\bar{\nu} = \frac{[L]_b}{[M]_T} = \frac{\sum_{i=1}^{n} i[ML_i]}{\sum_{i=0}^{n} [ML_i]} \tag{11}$$

where ML_i refers to the macromolecule with i bound ligand molecules. For a general stage i we can obtain the relation between both macroscopic constants, considering that the concentration of ML_i may be expressed as:

$$\left[ML_i\right] = K_i \cdot \left[ML_{i-1}\right] \cdot \left[L\right] = K_i \cdot K_{i-1} \cdot \left[ML_{i-2}\right] \cdot \left[L\right]^2 = K_i \cdot K_{i-1} \cdot \ldots \cdot K_1 \cdot \left[M\right] \cdot \left[L\right]^i \tag{12}$$

So:

$$\beta_i = \frac{\left[ML_i\right]}{\left[M\right]\left[L\right]^i} = K_i K_{i-1} \cdots K_1 = \prod_{j=1}^{i} K_j \tag{13}$$

Taking into account the equation 11, the binding parameter, expressed in terms of the macroscopic constants, is:

$$\bar{\nu} = \frac{\sum_{i=1}^{n} i \cdot \beta_i \cdot \left[L\right]^i}{\sum_{i=0}^{n} \beta_i \cdot \left[L\right]^i} \tag{14}$$

This equation is known as *Adair's general equation*, being the denominator the so called *binding polynomial*.

As equation 14 presents a very high number of macroscopic constants, it is interesting to deduce the relation between microscopic and macroscopic equilibrium constants, to obtain a more simple expression for the binding parameter. Firstly, it is necessary to know the number of possible microscopic states of each macroscopic species. Thus, for ML_i species it will be the number of different ways to arrange i ligands into n binding sites, which corresponds to the combinatorial of n elements taken in groups of i. Since all binding sites are equivalent and independent, all the possible microscopic forms for any macroscopic species are equally probable and, therefore, they will be at the same concentration; then, the concentration of the macroscopic ML_i species expressed as the concentration of its microscopic forms is:

$$\left[ML_i\right] = \frac{n!}{\left(n-i\right)!\,i!}\left[ML_i\right]_{micro} \tag{15}$$

So the microscopic equilibrium constant, k, will be:

$$k = \frac{\left[ML_i\right]_{micro}}{\left[ML_{i-1}\right]_{micro}\left[L\right]} = \frac{(n-i)!\,i!\left[ML_i\right]}{(n-i+1)!\,(i-1)!\left[ML_{i-1}\right]\left[L\right]} = \frac{i}{n-i+1} K_{bi} \tag{16}$$

The macroscopic constant β_i is obtained from equations 13 and 16 as

$$\beta_i = \frac{n!}{\left(n-i\right)!\,i!} k^i \tag{17}$$

The binding parameter expressed in terms of microscopic constants can be obtained from equations 14 and 17 as

$$\overline{\nu} = \frac{\sum_{i=1}^{n} i \frac{n!}{(n-i)!i!} k^i \left[L\right]^i}{\sum_{i=0}^{n} \frac{n!}{(n-i)!i!} k^i \left[L\right]^i} = \frac{nk\left[L\right]\left(1+k\left[L\right]\right)^{n-1}}{\left(1+k\left[L\right]\right)^n} = \frac{nk\left[L\right]}{1+k\left[L\right]} \tag{18}$$

where the resulting expression has been obtained by taking into account the binomial theorem. Once again, the binding parameter, expressed in terms of microscopic constants, gives a quite simple expression with a reduced number of fitting parameters, since a single *k* value is always expected for all microscopic binding constants in the case of a binding process where binding sites are equivalent and independent.

2.3. A general formulae for non-cooperative binding. The binding polynomial

At this point, we have explained the simplest cases of binding equilibrium, where binding sites are equivalent and independent. When more complex schemes are considered, the use of classical thermodynamic formulae to obtain the binding parameter turns complicated and laborious, as was mentioned previously. Thus, it is more convenient to use a general formulae which will allow obtaining $\overline{\nu}$ *([L])* expressions, systematically and independently of the complexity of the case in study. This general formulae is based on the construction of a function, which may be described as the macroscopic analogue of the grand canonical partition function from statistical thermodynamics. This function was introduced in the biochemical field by Wyman [3] and then, applied to ligand binding [4-7].

In order to apply this general formulae to ligand binding systems, firstly, it is necessary to construct the *partition function*, and then, apply it to the system under study. Let us explain briefly the steps to obtain the partition function:

- Identify the different energetic accessible states of the system: in a binding process it would be the different species of the macromolecule (free and bound) in equilibrium with the ligand.
- Determine the energy of each accessible state: it would be equivalent to specify the equilibrium constants given by the mass action law.
- Choose a reference state, that is, a reference specie: it can be chosen any of the identified ones, though is preferable to choose the state (species) with the lowest energy; for a binding process is the free macromolecule (M).
- Calculate the statistical weight of each state (species) with respect to the reference one. The statistical weight for a state *i*, W_i, is

$$W_i = D * \exp\left(-\Delta E_i / RT\right) \tag{19}$$

where *D* is the degeneration of each state, and $\Delta E_i = E_i - E_{ref}$

For a binding process the statistical weight of the specie ML_i is

$$W_i = [ML_i]/[M] = \beta_i [L]^i \tag{20}$$

Construct the partition function, Z, as the sum of the statistical weights of all accessible states (or species): $Z = \sum_i W_i$. For binding processes the partition function is expressed as

$$Z = \sum_{i=0}^{n} \frac{[ML_i]}{[M]} = \sum_{i=0}^{n} \beta_i [L]^i \tag{21}$$

Based on this formalism it is possible to easily obtain interesting expressions such as

Probability of each accessible state, P_i, which is the fraction of each species at equilibrium in the binding process

$$P_i = \frac{W_i}{Z} = \frac{\beta_i [L]^i}{\sum_{i=0}^{n} \beta_i [L]^i} \tag{22}$$

- Average quantities of interest for the system, that is, measured values of any magnitude. For a magnitude a

$$\langle a \rangle = \sum_{i=0}^{n} a_i \cdot P_i \tag{23}$$

where a_i corresponds to the value of the magnitude a for the state i (specie i).

Thus, the binding parameter, $\bar{\nu}$, corresponds to the average of ligand bound to the macromolecule and can be calculated as follows

$$\bar{\nu} = \langle i \rangle = \sum_{i=0}^{n} i \cdot \frac{W_i}{Z} = \sum_{i=0}^{n} i \cdot \frac{\beta_i [L]^i}{\sum_{i=0}^{n} \beta_i [L]^i} = \frac{\sum_{i=0}^{n} i \cdot \beta_i [L]^i}{\sum_{i=0}^{n} \beta_i [L]^i} \tag{24}$$

Of course, the expression obtained is the Adair´s equation (equation 14) and the denominator, which corresponds to the binding polynomial, can be identified with the partition function.

There is also a direct way to calculate the binding parameter, based on the calculation of the partial derivative of lnZ in respect to ln[L] at constant P and T, as is shown below:

$$\bar{\nu} = \left(\frac{\partial (\ln Z)}{\partial (\ln [L])} \right)_{P,T} = \frac{[L]}{Z} \left(\frac{\partial Z}{\partial [L]} \right)_{P,T} = \frac{[L]}{Z} \left(\frac{\partial \left(\sum_{i=0}^{n} \beta_i [L]^i \right)}{\partial [L]} \right)_{P,T} = \frac{\sum_{i=0}^{n} i \cdot \beta_i [L]^i}{\sum_{i=0}^{n} \beta_i [L]^i} \tag{25}$$

If we apply this general formulae to the case of the binding of a ligand to a macromolecule with n equivalent and independent binding sites, we can obtain the same expression than that given in equation 18. It is interesting to note that, since the binding sites are independent, the partition function corresponds to the product of the partition sub-function for each binding site. Therefore, the binding parameter can also be expressed as the sum of the binding parameter for each binding site. Additionally, if the sites are equivalent, such parameters will be equal to n times the value of the binding parameter obtained for one of the sites.

In the case of a macromolecule with n different and independent binding sites, the partition function can also be expressed as the product of the partition sub-functions for each binding site, though as the sites are different these partition sub-functions will not be equivalent. Consequently, the binding parameter will be the sum of the binding parameter corresponding to each site.

2.4. Experimental analysis of binding equilibriums to independent sites

Prior to describing cooperative phenomena, in this Section we will describe how the analysis of the different graphical representations mentioned in Section 2.2.1 may help to rationalize the experimental data to get information about, for example, the existence of different kinds of binding sites for the ligand and how the equilibrium constants describing ligand binding can be estimated.

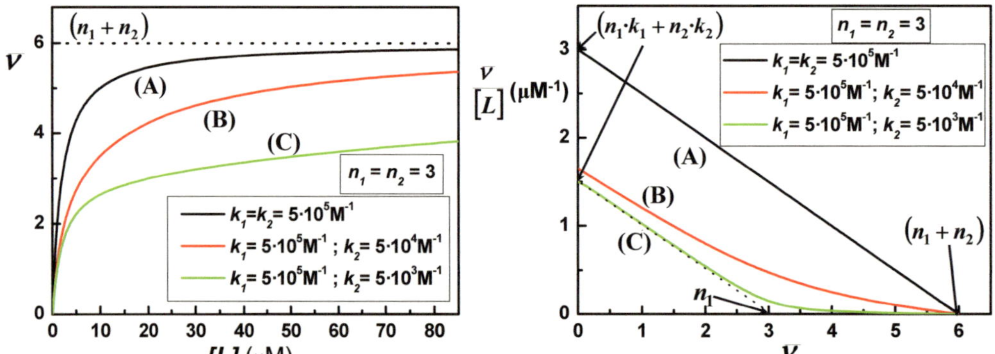

Simulation examples for the binding equilibrium of a ligand to a macromolecule with one or two different classes of sites, to show the differences in the binding curves (left panel) and in the Scatchard representation (right panel). Curve A corresponds to the ligand binding to six equivalent and independent sites, with a microscopic constant of $5 \cdot 10^5$ M^{-1}. Curves B and C correspond to the binding to two different kinds of sites, each class with 3 sites: for curve B the ratio between microscopic constants of the two binding site classes is k_1=10k_2; for curve C the ratio is k_1=100k_2.

Figure 3. Binding to independent sites

The easiest representation of experimental data is the binding curve (Figures 1 and 3). Simulations carried out with the equations described above, by using the different $\bar{\nu}$ ([L]) expressions obtained in the earlier sections indicate that a hyperbolic shape of this curve will represent a binding process corresponding to a macromolecule with equivalent and independent sites, being n estimated from the asymptotic value of the graph (curve A). When the macromolecule displays different kinds of sites, characterized by different values of microscopic constants, the shape of the curve changes, becoming more difficult to distinguish from the former simpler case when such values become similar (curves B and C). The shape will be the equivalent to the sum of two or more hyperbolic binding curves with two or more different k values.

The Scatchard representation would be more helpful to distinguish between equal or different kinds of independent sites (Figure 3). From the intersection with the X-axis the number of binding sites can be easily determined when sites are equal (curve A). In the case of different kind of sites n estimation is difficult, although we can obtain the number of sites of the highest affinity and the respective k value from the extrapolation of the initial linear tendency (see figure).

The Hill representation is fundamentally informative to distinguish between independent (non-cooperative) and dependent (cooperative) sites in the macromolecule, and will be analyzed in detail into the next Section.

2.5. Experimental analysis of binding cooperativity (non-independent sites)

Up to now we have referred only to binding processes where all binding sites are independent. It is time to introduce the effect of binding cooperativity, which means that the interaction of the ligand with one of the sites of the macromolecule produces an alteration of the affinity that the other sites have for such ligand. We can distinguish between *positive cooperativity*, when the binding of a ligand increases the affinity of the rest of binding sites, and *negative cooperativity*, when such affinity is decreased. These changes in affinity are usually related to conformational changes in the macromolecule, that is, what was referred to at the beginning of the chapter as *alosterism*. From a practical point of view, cooperativity can be viewed as a way to regulate the biological activity as a function of ligand concentration.

Although the Adair equation is still a valid approach, as we described in Section 2.4, it could be difficult to distinguish among the different equilibriums and, in addition, it may contain an excessive number of equilibrium constants to be estimated from fitting. Thus, different strategies have been described to analyze cooperativity, as will be explained into the next Section. Prior to this description, let us first describe how to determine cooperativity from analysis of experimental data.

The binding curve reveals an S-shape when cooperativity is positive. From the Scatchard representation positive cooperativity can also be easily discernible from any scheme of independent binding sites, since a concave shape of the experimental data is revealed (Figure 4). However, negative cooperativity can be confused with the scheme of different and independent sites (compared to Figures 3 and 4). The Hill representation is the most useful to distinguish between dependent and independent types of binding sites. In Figure 4 both situations have been simulated. When a non-cooperative behaviour is revealed, a single straight line with a slope equal to one is obtained, while when a positive cooperative behaviour occurs, the Hill analysis shows an increase of the slope at the central region (S-shape). The decrease of this region will indicate negative cooperativity. The slope of this region is known as the *Hill coefficient*, n_H. The explanation is as follows:

At very low ligand concentrations, the binding occurs statistically at different sites allocated in different macromolecule units, all of them free of ligand. Thus, at this stage cooperativity phenomena are not revealed experimentally. This "non-cooperative" initiation of the binding process results in a straight line with slope equal to one. Thus, equation 4 converts into

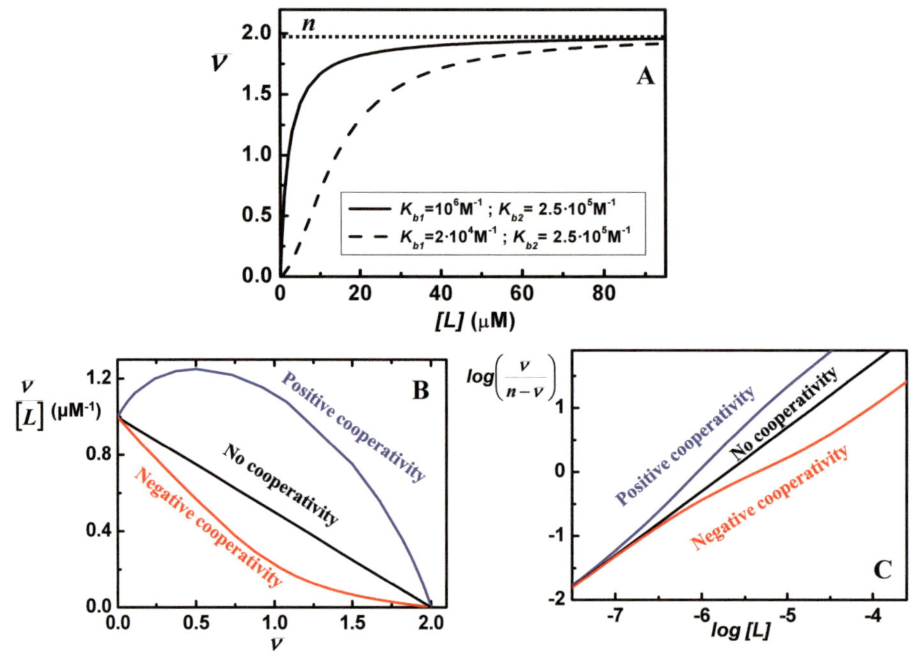

Graph A shows some simulations of the binding curve for the ligand binding to a macromolecule with two binding sites, with a microscopic constant of k=5·10^5 M^{-1}: solid line curve corresponds to equal and independent binding sites, and dashed line curve to equal sites showing positive cooperativity (the binding of a ligand increases fifty times the affinity for the second ligand molecule). Graphs B and C show the Scatchard and Hill representations respectively. In both panels simulated curves correspond to k=5·10^5 M^{-1} and the affinity for the second site is increased (for positive cooperativity) or decreased (negative cooperativity) five times.

Figure 4. Binding to cooperative sites

$$\log\left(\frac{\bar{\nu}}{n-\bar{\nu}}\right)_{[L]\to 0} = \log\frac{K_1}{n} + \log[L] \tag{26}$$

At moderate saturation of the macromolecule, the ligand binds to sites where affinity has changed (second and subsequent sites of the macromolecule). As a result, the slope of this region of the Hill curve will change to values higher than one in the case of positive cooperativity, or lower than one for negative cooperativity. Thus, the slope will achieve its maximum experimental value around half of saturation. The maximum theoretical value in this zone will be hypothetically reached in the case of infinite cooperativity, where only the free and fully saturated M species are significantly populated. In this case, it can be demonstrated that this slope can be equal to n. In the real cases where cooperativity is finite, the value will range between $1 < n_H < n$

$$M + nL \xleftrightarrow{\beta_n} ML_n \qquad \log\left(\frac{\bar{\nu}}{n-\bar{\nu}}\right) = \log K_n + n \cdot \log\left[L\right] \tag{27}$$

- At very high ligand concentrations, reaching saturation of the macromolecule, for almost the totality of macromolecules all sites are occupied by ligand molecules except one of them and the binding to this last site does not influence binding affinity. Therefore, the slope returns again to be equal to 1 (Figure 4) and it can be deduced from the mathematical expression of the Hill equation in the limit of strong binding that

$$\log\left(\frac{\bar{\nu}}{n-\bar{\nu}}\right)_{\left[L\right]\to\infty} = \log\left(n \cdot K_n\right) + \log\left[L\right] \tag{28}$$

2.6. Physico-chemical description of binding cooperativity

2.6.1. Phenomenological description

This approach does not imply the assumption of any structural model for the ML_i species. For a situation where sites are independent it is assumed that only one microscopic binding constant, k, exists, but there are different macroscopic constants, K_i, for the different equilibriums, according to the scheme shown in Figure 5.

Therefore, the binding polynomial can be generally described as

$$Z = 1 + K_1\left[L\right] + K_1K_2\left[L\right]^2 + K_1K_2K_3\left[L\right]^3 + K_1K_2K_3K_4\left[L\right]^4 + \ldots \tag{29}$$

In general, we can also assume that the relationship between macroscopic and microscopic binding constants can be

$$K_i = \frac{Isoforms \;\; of \;\; ML_i}{Isoforms \;\; of \;\; ML_{i-1}} k \tag{30}$$

By replacing this equation in the binding polynomial, we obtain the phenomenological expression for this function, and then, by using equation 25 the binding parameter can be estimated. In the case of cooperative sites, this description assumes that the microscopic binding constant changes upon binding, increasing when positive cooperativity happens, and decreasing for negative cooperativity. This approach does not explain the molecular reasons of such a change, as do the following schemes, but it represents an easy way to estimate the equilibrium constants in these cases.

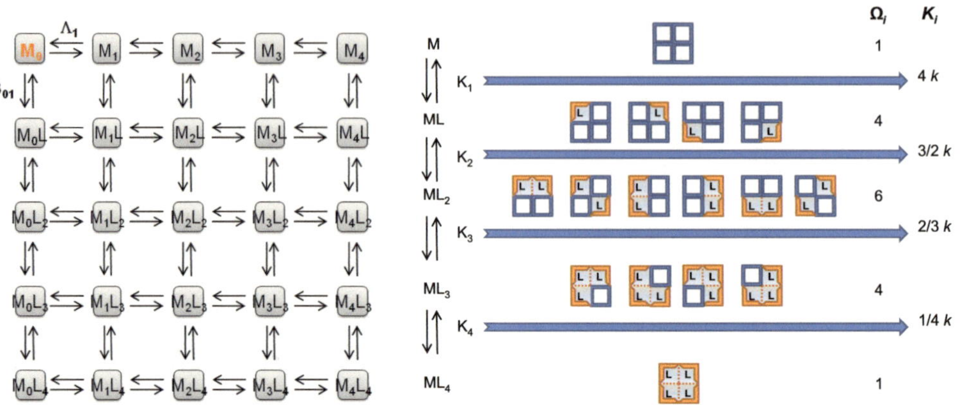

Left side shows the matrix of the model, where columns represent ligand binding equilibriums and rows the conformational changes associated to a hypothetical macromolecule displaying four cooperative binding sites for the ligand. Right side shows the relationship between macroscopic and microscopic constants, where Ω_i represents the number of isoforms of ML_i.

Figure 5. Phenomenological description of binding cooperativity

2.6.2. The Koshland-Nemety-Filmer model

The basics of this model were initially proposed by Pauling to study the cooperative binding of oxygen to haemoglobin [8]. It can explain both, positive and negative cooperative behaviours. More interesting, this model assumes that the different binding sites of the macromolecule are influencing each other through their mutual interconnection by means of a molecular ligature, σ. In Figure 6 we have represented this situation for a macromolecule with four binding sites for the ligand L. It must be considered that when every site is occupied by L, it breaks the ligature with the others, resulting in a modification of their binding affinities. Then, the mathematical expression for the binding polynomial is:

$$Z = \sum_{i=0}^{n} I_i \frac{k^i}{\sigma^{Bi}} \left[L\right]^i \quad (31)$$

where I_i are the number of possibilities of allocating i ligands into the macromolecule, and B_i is the number of broken ligatures for each configuration. The results for the case of a macromolecule with four binding sites are collected into Figure 6 as an example.

This description can be modified as a function of the experimental behaviour of every macromolecule-ligand example. It might be easily developed for the case of different microscopic constants or, even, of different contribution of the ligatures. The main advantage with respect to the phenomenological description is that it can reveal molecular aspects of cooperative phenomena when applied.

2.6.3. The Monod-Wyman-Changeux model

Although this model has been widely used in the literature [4, 9], it can only be used to describe positive cooperativity, which is expressed by assuming that the macromolecule can exists in at least two different conformations, which are under mutual equilibriums, and differ in their affinities for the ligand. Within every conformation, the binding sites behave as if they were equivalent and independent for the binding of L.

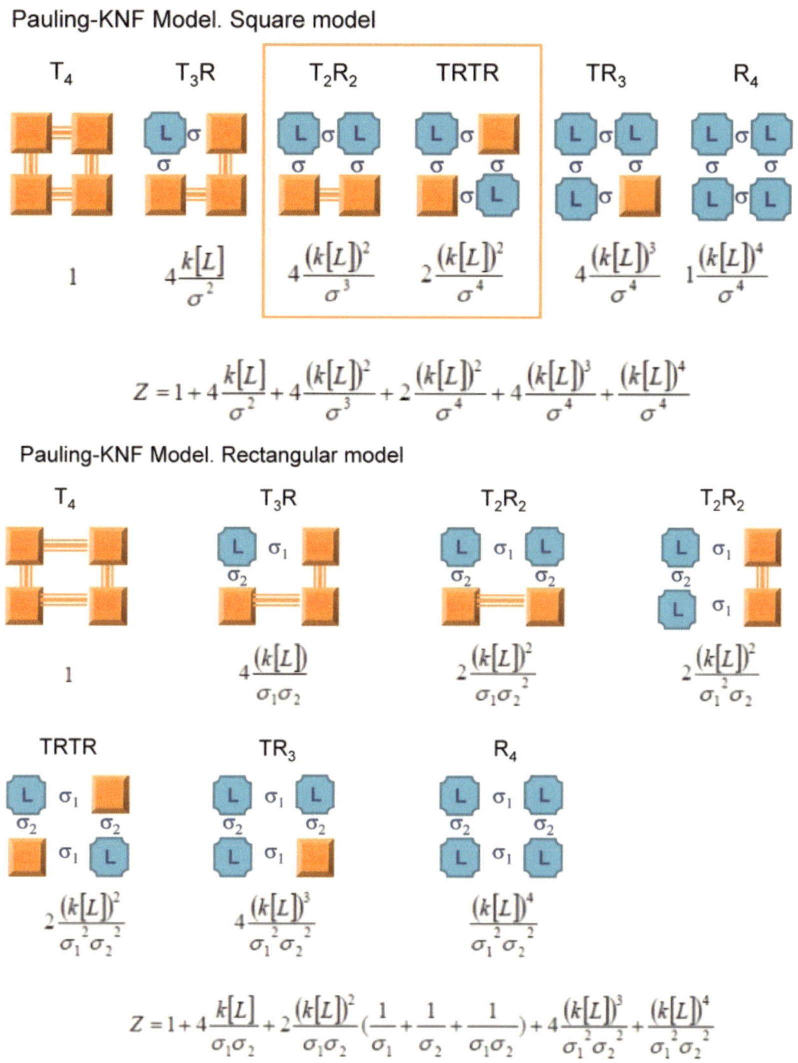

The upper scheme represents the so-called square version of this model for a hypothetical macromolecule displaying fou cooperative binding sites for the ligand and only one kind of ligature. The lower panel shows the rectangular version, with two kinds of ligatures. In both cases we also show the corresponding formulae of the partition function, Z, below each one.

Figure 6. The Koshland-Nemety-Filmer model of binding cooperativity

In Figure 7 a schematic diagram is shown for the case of a macromolecule with four binding sites and two (left side) distinct conformations under equilibrium. In this situation, it is usually assumed that the allosteric equilibrium constant, Λ, is initially big enough to move the equilibrium towards the T-state, considered as the low affinity state. Upon addition of the ligand, the equilibrium moves towards the R-state, of higher affinity than the former. Therefore, this displacement will allow the rest of sites to bind the ligand with higher affinity than the first one. This progressive displacement to the R-state may, thus, explain an increase in affinity (positive cooperative), but not the opposite.

The binding polynomial can be mathematically expressed as

$$Z = \frac{1}{1+\lambda}\left(1+k_R\left[L\right]\right)^n + \frac{\lambda}{1+\lambda}\left(1+k_T\left[L\right]\right)^n \quad (32)$$

This model can be easily generalized to more than two conformations of the macromolecule by adding additional terms to this general equation. For example, for the case of three conformations (right side of Figure 7):

$$Z = \frac{1}{1+\lambda+\varpi}\left(1+k_R\left[L\right]\right)^n + \frac{\lambda}{1+\lambda+\varpi}\left(1+k_T\left[L\right]\right)^n + \frac{\varpi}{1+\lambda+\varpi}\left(1+k_S\left[L\right]\right)^n \quad (33)$$

and so on.

3. Notes on ITC performance and general experimental procedures

As was mentioned in the Introduction, ITC is a thermodynamic technique that directly measures the heat released or absorbed in an intermolecular interaction, such as ligand-protein interactions, protein-protein interactions, etc [10]. An ITC experiment consists of a calorimetric titration of a specific volume of one of the reagents, usually the macromolecule, with controlled quantities of the other reagent, usually the ligand, at constant temperature and pressure. Thus, the measured heat during the titration corresponds to the enthalpy of such interactions [11]. This relatively easy experiment allows a complete and precise thermodynamic characterization of the binding event. Subsequently, if the thermal effect is high enough, and the value of the binding constant is moderately good, a single ITC experiment can establish the equilibrium binding constant, K_b, the apparent enthalpy change, ΔH_{app}, and the stoichiometry of the reaction, n. Additionally, if the experiments are made at different temperatures, the change of heat capacity of the process, ΔC_{pb}, can also be measured.

The most common titration calorimeters are adiabatic and are based on the compensation of the thermal effect generated by the addition of the ligand into the sample cell, which is placed in an adiabatic environment [11]. In the left side of Figure 8 we show a schematic representation of one of these instruments. A thermoelectric device measures the temperature difference between the sample and the reference cells (ΔT_{-1}) and also between each cell and the adiabatic jacket (ΔT_{-2}). As long as the reaction is being developed, ΔT_{-1} value decreases to zero with the heating of the sample cell (if the reaction is endothermic) or the reference cell (if exothermic). This heating generates a spike over the baseline of the stationary power, and the integration of this potential required to get ΔT_{-1} to zero in the time to recover the equilibrium is the heat of each injection (right panel in Figure 8).

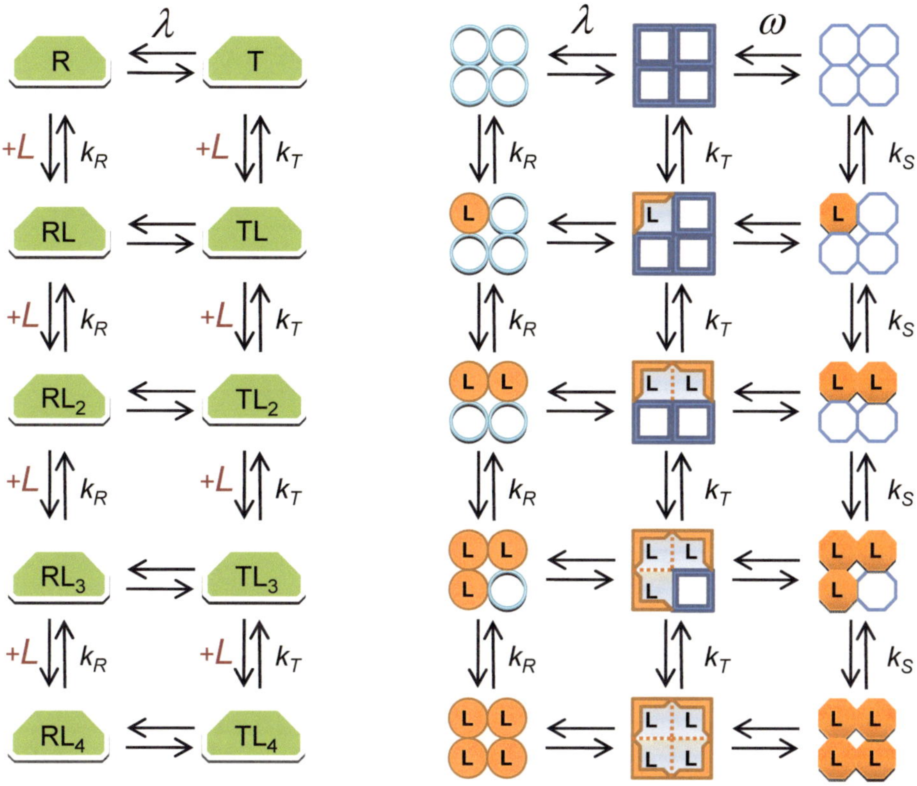

The two schemes show the equilibriums for a hypothetical macromolecule displaying four cooperative binding sites for the ligand and one (left) or two (right) different conformational changes.

Figure 7. The Monod-Wyman-Changeux model of binding cooperativity

As we stated above, a typical ITC experiment consists of a series of injections of determined ligand solution volumes, which can be equal or variable each time, into a macromolecule solution. Such injections have to be separated in between by a time interval, large enough to be sure that the system has reached the equilibrium and all heat absorbed or emitted has been transferred. The titration process is continued until saturation of the macromolecule by the ligand into the cell is reached. In this way, the last additions will not give a significant heat exchange, as shown in Figure 8. The final thermogram is obtained by the individual integration of each peak, setting the integration limits in the baseline that precedes and continues such peak with the equation:

$$Q = \int_{ti}^{tf} W(t) \cdot dt \qquad (34)$$

3.1. Procedures for ITC experiments

As we have mentioned previously, this technique allows us to directly evaluate the heat exchange generated upon binding of two molecules. The correct performance of a titration experiment has to consider two main aspects, first, the samples preparation and, second, the proper measurement of reaction heats in the calorimeter. In this Section we are going to describe both experimental aspects.

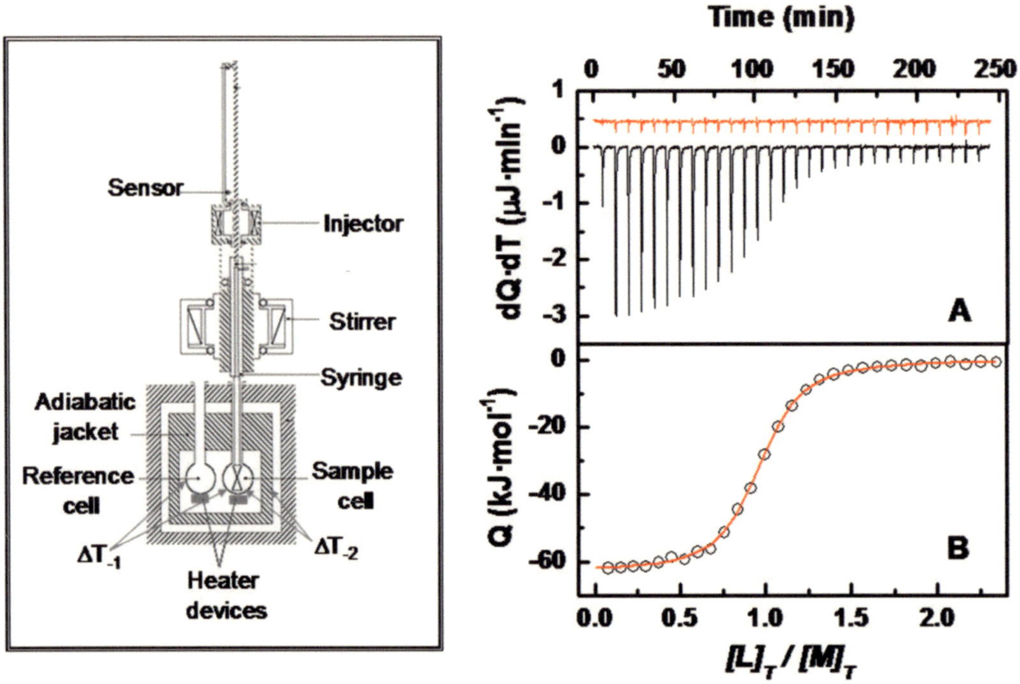

Left side: a schematic diagram of the main components of a titration calorimeter. Right side: a titration calorimetry experiment of a protein with a ligand. In panel A) the titration thermogram is represented as heat per unit of time released after each injection of the ligand into the protein (black), as well as the dilution of ligand into buffer (red). In panel B) the dependence of released heat in each injection *versus* the ratio between total ligand concentration and total protein concentration is represented. Circles represent experimental data and the line corresponds to the best fitting to a model considering *n* identical and independent sites.

Figure 8. Isothermal titration calorimetry instrumentation

3.1.1. Sample preparation

To carry out ITC binding studies the protein must be of higher purity than 95%, and dialyzed against a selected buffer. Moreover, it is recommended to use the buffer solution of the last dialysis change as the reference solution. The process and precautions to prepare the ligand solution are the same as those described for the protein solution. Nevertheless, when ligands are small as is not possible to dialyze them, the lyophilized (solid) ligand can be directly dissolved with the last dialysis change buffer used for protein sample preparation. As an alternative way, and in order to minimize the differences in the composition of solutions of protein and ligand, lyophilized and solid samples (usually the low molecular weight ligands) can be dissolved in Milli-Q water at a double concentration than that required in the experiment, as well as the protein solution. Accordingly, the dialysis buffer and the protein solution may also contain twice the concentration of buffering salts desired in the experiment (2x buffer). The experimental solutions of protein, ligand and reference buffer (1x buffer) are prepared by adding to the protein and ligand solutions in the necessary amounts (1:1 dilution) of 2x buffer from the last change of dialysis and Milli-Q water respectively, following the subsequent pH correction of both solutions.

The three solutions (buffer, protein and ligand) have to be centrifuged and/or filtered prior to filling the calorimetric cells in order to avoid insoluble particles, it is also recommended to degas them in order to avoid bubbles. The exact protein concentration of the solution has to be determined just before filling the calorimetric sample cell. One of the most accurate and used methods is the spectrophotometric one using protein extinction coefficients (described by Gill & von Hippel [12]).

3.1.2. Modeling and performance of an ITC experiment

Once the protein solution is in the calorimetric cell and the ligand solution in the ITC syringe (where no air bubbles are present), it is very important to wait enough time to be sure that everything is properly thermostated; a way to control such thermal equilibration is controlling the signal of the ITC instrument. Once the signal of the ITC is stable, the experiment is ready to start with the series of ligand injections.

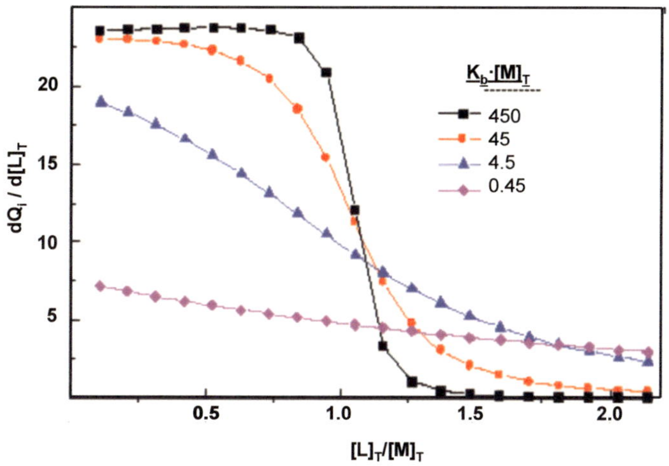

Simulation of heat per added mole of ligand associated to each injection for an ITC experiment with the following experimental parameters: Vc = 1.347 mL; $[M]_T = 1.8x10^{-4}M^{-1}$ in the cell; 20 injections of 5μL; $[L]_T$ = 5mM in the syringe; one binding site and four different values of the association constant, $2.5x10^6$, $2.5x10^5$, $2.5x10^4$ and $2.5x10^3$ M^{-1}.

Figure 9. Simulations in isothermal titration calorimetry

The limits of a correct determination of the binding thermodynamic parameters using this technique are given by the product of the binding constant, K_b, and the total concentration of the macromolecule, $[M]_T$, being $1 < K_b \cdot [M]_T < 1000$ [11]. Different simulations of a conventional ITC experiment with additions of equal volumes in which four different binding constants have been considered, $2.5 \cdot 10^6$, $2.5 \cdot 10^5$, $2.5 \cdot 10^4$ and $2.5 \cdot 10^3$ M^{-1}, are shown in Figure 9. These values are in the range that includes both high and low affinity (the product

$K_b \cdot [M]_T$ ranges from 450 to 0.45). As it can be observed, when the product $K_b \cdot [M]_T$ is within the appropriate range, the sigmoid curve is obtained, which is needed to perform an analysis with acceptable standard errors. When the product $K_b \cdot [M]_T$ is close to the limits, the isotherm can be optimized with some variations of the experimental design, such as the concentration of total macromolecule in the cell or ligand in the syringe, or also by designing profiles of different injected volumes of ligand. Such profiles usually start with lower ligand volumes for the first injections, which increase progressively in a nonlinear way. Another advantage of using an optimal injection volume profile is the increase of the signal/noise ratio at the end of the thermogram, where the heats of binding are quite small. When the product $K_b \cdot [M]_T$ is over the range, titration experiments by displacement can be performed, in which the target ligand competes for the same binding site with another ligand whose interaction has been previously characterized [13-15].

3.2. Previous treatment of ITC experimental data for thermodynamic analysis

Once the ITC experiment has been performed (black titration in panel A of Figure 8), the thermogram can be integrated to obtain the corresponding heats of each injection. Nevertheless, in order to correct the dilution heat effect of the ligand it is necessary to make a baseline ITC experiment (red titration in panel A of Figure 8), which consist of performing an identical ITC experiment of the ligand binding but with buffer instead of protein into the calorimetric cell. Then, the heats for each injection obtained in this baseline experiment are subtracted to the corresponding ones to the ligand binding experiment. Afterwards, we have to normalize the obtained net heats by the total concentration of ligand in the cell after each injection. The binding isotherm can be obtained by the representation of transferred heat per added mole of ligand ($dQ_i/dL_{T,i}$) versus the molar fraction ($[L]_T/[M]_T$) (panel B in Figure 8).

3.3. Corrections to possible additional heat contributions to the binding experiment

The fitting of the experimental data to the equations explained in the next Section, will allow us to obtain, besides the binding constant or the Gibbs energy, the binding enthalpy. Sometimes this enthalpy change obtained from the fittings of experimental data is the result of additional events occurring during the ligand binding process. So it is important to distinguish between the apparent binding enthalpy, ΔH_{app}, and the real or intrinsic binding enthalpy, ΔH_{int}.

One of the possible events associated to the ligand binding process can be a conformational change of the protein and/or the ligand not associated uniquely to the interaction. A typical example is when the free protein is partially denatured at the experimental temperature, so it is important to check the folding of the protein using other techniques, as circular dichroism or differential scanning calorimetry.

Another possibility is that the ligand binding to the protein can be associated to a change in the pK_a of ionisable groups of the protein or/and the ligand [16-18], in such a way that:

$$\Delta H_{app} = \Delta H_{\text{int}} + n_p \cdot \Delta H_{ion} \tag{35}$$

where n_p is the number of protons accepted or liberated due to the ligand binding to the protein and ΔH_{ion} is the ionization enthalpy of the buffer used in the experiment. As ΔH_{app} depends on the buffer used and the working pH, the easiest way to determine ΔH_{int} is to perform several ITC experiments under the same conditions (mainly at the same pH and ionic strength), but using buffers of different ΔH_{ion}, which permits the determination of the net binding enthalpy from the ordinate of the corresponding linear correlation of ΔH_{app} *versus* ΔH_{ion}, i.e., the enthalpy value without buffer ionization contributions ΔH_{int}.

4. Thermodynamic analysis of ITC experiments by using different equilibrium models

The analysis of the isotherms is done by the non-linear fitting of the experimental data using different equations, depending on the way the ligand binds to the macromolecule. In this Section we are going to describe, as an example, the four most common models in the literature. The fittings can be done with the appropriate software, as Origin 7.0 (Microcal Software Inc.) or SigmaPlot 2000 (Jandel Co.).

4.1. Ligand binding to one macromolecule with *n* identical and independent sites

Although the mathematical formulae corresponding to this binding model has been described in Section 2.2.3, we will start defining several functions and parameters. Thus, the binding parameter, $\bar{\nu}$, being the relationship between the concentration of bound ligand, $[L]_b$, and the total macromolecule concentration, $[M]_T$, can be the one given in equation 18:

$$\bar{\nu} = \frac{nk[L]}{1 + k[L]} \tag{36}$$

where k is the microscopic equilibrium constant, which is unique since all binding sites are independent, $[L]$ is the concentration of non-bounded ligand and n is the number of binding sites in the macromolecule.

The heat released or absorbed in any ITC injection, q_i, is related to the binding process as

$$q_i = \Delta H_{app} (\frac{kJ}{molL_b}) \Delta(molesL_b) \tag{37}$$

where ΔH_{app} is the apparent enthalpy change per mole of bound ligand, and $\Delta(molesL_b)$ is the molar amount of ligand bounded in the injection i. If we express the moles of ligand bound in terms of concentrations, the above equation can be written as:

$$q_i = \Delta H_{app} \cdot V_C \cdot ([L]_{b,i} - [L]_{b,i-1}) = \Delta H_{app} \cdot V_C \cdot (\bar{\nu}_i \cdot [M]_{T,i} - \bar{\nu}_{i-1} \cdot [M]_{T,i-1}) \tag{38}$$

where V_c represents the effective volume of the ITC cell and $[M]_T$ is the total concentration of the protein in the cell at injection *i*.

Furthermore, as known parameters the effective volume of the ITC cell, V_c, the injection volume, V_{inj}, and the ligand concentration in the syringe, $[L]_0$, we can express the concentrations of macromolecule, $[M]_{T,i}$, and ligand, $[L]_{T,i}$, at each injection using the equations

$$[M]_{T,i} = [M]_{T,i-1}\frac{V_C - V_{inj}}{V_C} \qquad [L]_{T,i} = \frac{(V_C - V_{inj})\cdot[L]_{T,i-1} + V_{inj}\cdot[L]_0}{V_C} \tag{39}$$

Thus, the total heat accumulated after N injections could be described as

$$Q = \sum_{i=1}^{N} q_i = V_C\cdot[M]_T\cdot\Delta H_{app}\cdot\bar{v} = V_C\cdot[M]_T\cdot\Delta H_{app}\cdot\frac{n\cdot K\cdot[L]}{1 + K\cdot[L]} \tag{40}$$

During the ITC experiment the value of the non-bounded ligand concentration, *[L]*, is an unknown variable and for this reason, it is operationally required to estimate it from the experimental variables $[L]_T$ and *Q* as

$$[L] = [L]_T - [L]_b = [L]_T - \frac{Q}{V_C\cdot\Delta H_{app}} \tag{41}$$

Substituting the above equation in equation 40 we obtain a quadratic equation with Q as unknown variable, whose solution is

$$Q = \frac{Vc\cdot\Delta H_{app}}{2\cdot k}\left[1 + k\cdot[L]_T + n\cdot k\cdot[M]_T - \sqrt{(1 + k\cdot[L]_T + n\cdot k\cdot[M]_T)^2 - 4\cdot n\cdot k^2[M]_T\cdot[L]_T}\right] \tag{42}$$

Finally, deriving this expression with respect to $[L]_T$ we obtain an expression for the heat per mole of ligand added in each injection

$$\frac{1}{V_c}\cdot\frac{dQ}{d[L]_T} \approx \frac{1}{V_c}\cdot\frac{\Delta Q}{\Delta[L]_T} = \frac{\Delta H_{app}}{2}\left[1 - \frac{1 + [L]_T - n\cdot k\cdot[M]_T}{\sqrt{(1 + k\cdot[L]_T + n\cdot k\cdot[M]_T)^2 - 4\cdot n\cdot k^2\cdot[M]_T\cdot[L]_T}}\right] \tag{43}$$

According to these equations, there are two possible ways to analyze the experimental heats from an ITC experiment: one by using equation 43 which considers the heat per mole of added ligand associated with each injection; the second by using equation 42 and considering the total heat accumulated from the beginning to each injection of the ITC experiment. The first approach has the advantage of avoiding experimental errors, since in such analysis is possible to eliminate individual experimental points from the curve (Figure 8), while the second approach imply the sum of all the heats of each injection which accumulates errors.

4.2. Ligand binding to one macromolecule with *m* different and independent classes of sites

In this model, each binding site is defined as an independent site, with different affinity to the other binding sites. The expression "different sites" implies a microscopic equilibrium constant for each binding site, and the term "independent" site means that the binding affinity does not change with the binding of any other ligand to the other sites of the macromolecule. The mathematical formulae that we describe here correspond to a macromolecule with only two different classes of sites (m = 2) with n_1 and n_2 sites for each type, as represented in the following scheme:

$$M + \left(n_1 + n_2\right) L \xleftrightarrow{K1 \cdot K2} ML_{n1+n2} \tag{44}$$

The binding parameter, defined as the ratio of the concentration of ligand bound at any of the two classes of sites, $[L]_{b,i}$, and the total concentration of macromolecule, $[M]_T$, can be expressed now as

$$\bar{\nu} = \sum_{i=1}^{m=2} \bar{\nu}_i = \frac{[L]_{b,i}}{[M]_T} = \sum_{i=1}^{m=2} \frac{n_i \cdot K_i \cdot [L]}{1 + K_i \cdot [L]} \tag{45}$$

Thus, the heat released or absorbed in any injection, q_j, would be

$$q_j = \sum_{i=1}^{m=2} \Delta H_{app,i} \left(\frac{kJ}{molL_{b,i}}\right) \cdot \Delta\left(molesL_{b,i}\right) \tag{46}$$

where $\Delta H_{app,i}$ is the apparent enthalpy change per mole of ligand bound to any of the two classes of sites. If we express the moles of ligand bound in terms of concentrations, then the above equation can be re-formulated as:

$$q_j = \sum_{i=1}^{m=2} \Delta H_{app,i} \cdot V_c \cdot \left([L]_{b,i,j} - [L]_{b,i,j-1}\right) = \sum_{i=1}^{m=2} \Delta H_{app,i} \cdot V_c \cdot \left(\bar{\nu}_j \cdot [M]_{T,j} - \bar{\nu}_{i-1} \cdot [M]_{T,j-1}\right) \tag{47}$$

where V_c represents the effective volume of the ITC cell and $[M]_{T,j}$ is the concentration of protein in the cell after injection j.

Thus, if we substitute equations 39 in the above expression, we obtain the following:

$$Q = \sum_{j=1}^{N} q_j = V_c \cdot [M]_T \cdot \sum_{i=1}^{m=2} \Delta H_{app,i} \cdot \bar{\nu}_{i,N} = V_c \cdot [M]_T \cdot \sum_{i=1}^{m=2} \Delta H_{app,i} \cdot \frac{n_i \cdot k_i \cdot [L]}{1 + k_i \cdot [L]} \tag{48}$$

Solving the summation for two classes of sites, m=2, the expression of the total heat accumulated in N injections can be re-written as

$$Q = V_c \cdot [M]_T \left[\Delta H_{app,1} \cdot \frac{n_1 \cdot k_1 \cdot [L]}{1 + k_1 \cdot [L]} + \Delta H_{app,2} \cdot \frac{n_2 \cdot k_2 \cdot [L]}{1 + k_2 \cdot [L]} \right] \tag{49}$$

Since the value of [L] is unknown, we should express it in terms of total bound ligand ($[L]_{b,T} = [\overline{v_1} + \overline{v_2}]\cdot[M]_T$), as we show in the following equation

$$[L] = [L]_T - [L]_{b,T} = [L]_T - [M]_T \cdot \left[\frac{n_1 \cdot k_1 \cdot [L]}{1 + k_1 \cdot [L]} + \frac{n_2 \cdot k_2 \cdot [L]}{1 + k_2 \cdot [L]} \right] \quad (50)$$

Substituting the previous expression in the equation 49 and re-organizing it, we obtain the following cubic equation:

$$[L]^3 + a_2 [L]^2 + a_1 [L] + a_0 = 0 \quad (51)$$

where the coefficients a_0, a_1 and a_2 are defined as

$$\begin{aligned} a_0 &= -\frac{[L_T]}{k_1 k_2} \\ a_1 &= \left(\frac{n_1}{k_2} + \frac{n_2}{k_1} \right) [M_T] - \left(\frac{1}{k_1} + \frac{1}{k_2} \right) [L_T] + \frac{1}{k_1 k_2} \\ a_2 &= \frac{1}{k_1} + \frac{1}{k_2} + (n_1 + n_2)[M_T] - [L_T] \end{aligned} \quad (52)$$

The only valid solution to the cubic equation 51 can be simply written by just grouping the coefficients a_0, a_1 and a_2 in three new coefficients A, B and C as

$$[L] = \sqrt[3]{A + \sqrt{A^2 + B^3}} + \sqrt[3]{A - \sqrt{A^2 + B^3}} + C \quad (53)$$

where A, B and C are

$$A = \frac{-a_2^3}{27} + \frac{a_1 a_2}{6} - \frac{a_0}{2} \qquad B = \left(\frac{a_1}{3} - \frac{a_2^2}{9} \right) \qquad C = -\frac{a_2}{3} \quad (54)$$

The solution of the cubic equation (using equations 52 to 54) allows us to calculate the non-bounded ligand concentration for a given number of injections. Substituting in equation 49, we can determine the heat associated to each injection using the following expression:

$$dQ \approx \Delta Q(j) = Q_T(j) - Q_T(j-1) + \frac{V_{in}}{V_C} \left(\frac{Q_T(j) + Q_T(j-1)}{2} \right) \quad (55)$$

4.3. Ligand binding by the displacement of another ligand in the single binding site of a macromolecule

To formulate this binding model of displacement, we assume two ligands, A and B, which can bind to the same binding site of a protein, M, with different affinity constants. Then, we can describe the equilibrium binding for each ligand as:

$$M + A \xleftrightarrow{K_A} MA \qquad K_A = \frac{[MA]}{[M]\cdot[A]} \qquad K_B = \frac{[MB]}{[M]\cdot[B]} \tag{56}$$

$$M + B \xleftrightarrow{K_B} MB$$

Because the binding of two ligands takes place in the same binding site of the macromolecule, and having the ligand A tighter affinity than ligand B, $K_A >> K_B$, we should consider the following scheme

$$\begin{aligned} M + B &\xleftrightarrow{K_B} MB \\ MB + A &\xleftrightarrow{K_A} MA + B \end{aligned} \tag{57}$$

According to the schemes 56 and 57 we can express the initial concentration of the ligands A and B as

$$[A]_0 = [MA] + [A] \qquad [B]_0 = [MB] + [B] \tag{58}$$

Substituting these expressions in equations 56, the association constants can be re-written as

$$[MA] = \frac{[M]\cdot[A]_0}{1/K_A + [M]} \qquad [MB] = \frac{[M]\cdot[B]_0}{1/K_B + [M]} \tag{59}$$

It is important to consider that in the case we propose for this binding model, it is usual that the binding of the high-affinity ligand A has been previously analyzed in a simple titration experiment. The displacement titration experiment will allow us to analyze the interaction of the low affinity ligand B, which cannot be determined by direct titration experiments. Thus, initially, the macromolecule M is bounded to ligand B forming the MB complex and during the titration of the ligand A we will shift partially the MB complex formation to the formation of the MA complex.

For the mathematical formulae of this model, we first define the molar fractions of all species containing the macromolecule. Such fractions are

$$x_M = [M]/[M]_T \qquad x_{MA} = [MA]/[M]_T \qquad x_{MB} = [MB]/[M]_T \tag{60}$$

where $[M]_T$ is the total concentration of macromolecule.

If we also write the molar ratios between the initial amounts of A and B relative to the total macromolecule concentration as

$$r_A = [A]_0/[M]_T \qquad r_B = [B]_0/[M]_T \tag{61}$$

we can write the concentrations of A and B bounded ligands during the interaction as products of the association constants

$$c_A = K_A\cdot[M]_T \qquad c_B = K_B\cdot[M]_T \tag{62}$$

Then, substituting these expressions in equations 59, we obtain the following equations for all the species that form the macromolecule, expressed in terms of the molar ratios of macromolecule:

$$x_M + x_{MA} + x_{MB} = 1 \qquad x_{MA} = \frac{r_A \cdot x_M}{1/c_A + x_M} \qquad x_{MB} = \frac{r_B \cdot x_M}{1/c_B + x_M} \tag{63}$$

Substituting the above equations of x_{MA} and x_{MB} into the first one and rearranging, we obtain the following cubic equation:

$$x_M^3 + a \cdot x_M^2 + b \cdot x_M + c = 0 \tag{64}$$

in which we have defined the *a*, *b* and *c* coefficients as:

$$a = \frac{1}{c_A} + \frac{1}{c_B} + r_A + r_B - 1 \qquad b = \frac{r_A - 1}{c_A} + \frac{r_B - 1}{c_B} + \frac{1}{c_A \cdot c_B} \qquad c = -\frac{1}{c_A \cdot c_B} \tag{65}$$

Solving such cubic equation, we obtain the following real solution for the molar fraction of macromolecule:

$$x_M = \frac{2 \cdot (\sqrt{a^2 - 3b}) \cdot \cos(\phi / 3) - a}{3} \tag{66}$$

in which the coefficient ϕ can be written as:

$$\phi = \arccos \frac{-2 \cdot a^3 - 9 \cdot a \cdot b - 27 \cdot c}{2 \cdot (\sqrt{a^2 - 3 \cdot b})^3} \tag{67}$$

Once the molar fraction of free macromolecule, x_M, is determined, we can also know the molar fractions of the other species in which the macromolecule is present (x_{MA} and x_{MB}) by solving the equations 63.

The heat released or absorbed after each injection is proportional to the changes in concentration of MA and MB, [MA] and [MB], and their molar enthalpies of binding. Therefore, the heat after each injection can be written according to the following expression

$$\Delta Q = Vc \cdot (\Delta H_A \cdot \Delta[MA] + \Delta H_B \cdot \Delta[MB]) = Vc \cdot [M]_T \cdot (\Delta H_A \cdot \Delta x_{MA} + \Delta H_B \cdot \Delta x_{MB}) \tag{68}$$

where Vc is the effective volume of the ITC cell. To correct the concentrations of macromolecule and ligand, we define the following infinitesimal change for their concentrations

$$d[X] = -\frac{dV_i}{V_0} \cdot [X] \tag{69}$$

where *[X]* represents the concentration of any species. Integrating the above expression between the limits from $[X]_i$ to $[X]_{i-1}$ and from zero to V_i. The resulting equation is

$$[X]_i = [X]_{i-1} \exp\left(-\frac{V_i}{V_0}\right) = [X]_{i-1} f_i \tag{70}$$

where f_i is the dilution factor that allow us to define the molar ratios after each injection as

$$[A]_i = [A]_{i-1}(1-f_i) \qquad [M]_i = f_i[M]_{i-1} \quad [B]_i = f_i[B]_{i-1} \tag{71}$$

Consequently, the heat absorbed or released after each injection can be expressed as

$$dQ \approx \Delta Q = V_C [M]_T \cdot \{\Delta H_A \cdot (\Delta x_{MA,i} - f_i \cdot x_{MA,i-1}) + \Delta H_B \cdot (\Delta x_{MB,i} - f_i \cdot x_{MB,i-1})\} \tag{72}$$

4.4. Ligand binding to a macromolecule with two dependent (cooperative) binding sites

The scheme described in Figure 2 for a macromolecule with two binding sites was used in Section 2.2.2 to describe the correlation between macroscopic and microscopic equilibrium constants and the binding parameter was obtained for the case of independent sites, characterized by the same microscopic constant. Nevertheless, when cooperativity exists among sites, this simple assumption cannot be considered, and more than one value for such constants must be taken into account. Thus, positive cooperativity would be revealed when $k_3 > k_1$ and $k_4 > k_2$, and the opposite is true for negative cooperativity. In any case, it can be easily deduced that $k_3 \cdot k_1 = k_4 \cdot k_2$.

The most simplified version of this model could be attained for equivalent binding sites, that is, when microscopic binding to the first site is identical, independently if the ligand binds to either site 1 or 2. The occupancy of the first site will drive to the modification of the affinity of the second one, in a similar way in both branches of the scheme. Therefore, $k_1 = k_2$ and $k_3 = k_4$. In parallel, the enthalpy changes associated can be also grouped in only two different values, the changes for the formation of ML and ML_2 species respectively. From here, any other version should be much more complicated from a mathematical point of view. Overcoming these calculation matters, the solution can be achieved in a similar way to the described here for the simplest version, which is also inspired in the models described previously.

The molar fractions of all species containing the macromolecule are in this case

$$\begin{aligned} x_{ML} &= \frac{[ML]}{[M]+2[ML]+[ML_2]} = \frac{k_A[L]}{1+2k_A[L]+k_A k_B[L]^2} \\ x_{ML2} &= \frac{[ML_2]}{[M]+2[ML]+[ML_2]} = \frac{k_A k_B[L]^2}{1+2k_A[L]+k_A k_B[L]^2} \end{aligned} \tag{73}$$

where we have assumed that $k_1 = k_2 = k_A$ and $k_3 = k_4 = k_B$.

Taking into account the equation 40, the total heat accumulated after N injections could be described as

$$Q = \sum_{i\_1}^{N} q_i = V_C \cdot [M]_T \cdot \Delta H_{app} \cdot \bar{v} = V_C \cdot [M]_T \cdot \{x_{ML}(\Delta H_A + \Delta H_B) + x_{ML2}(\Delta H_A + \Delta H_B)\} \tag{74}$$

where ΔH_A and ΔH_B are the enthalpy changes associated to equilibriums characterized by k_A and k_B respectively. Since the value of [L] is unknown, we should express it in terms of total ligand ($[L]_T = [L] + 2[ML] + 2[ML_2]$), as we show in the following equation

$$[L] = [L]_T - 2 \cdot x_{ML}[M]_T - 2 \cdot x_{ML2}[M]_T \tag{75}$$

This can be expressed as the following third-order equation:

$$[L]^3 + a_2[L]^2 + a_1[L] + a_0 = 0 \tag{76}$$

where the coefficients a_0, a_1 and a_2 are defined as

$$a_0 = -\frac{[L]_T}{k_A k_B} \qquad a_1 = \frac{2}{k_B}[M]_T - \frac{2}{k_B}[L]_T + \frac{1}{k_A k_B} \qquad a_2 = \frac{2}{k_B} + 2[M]_T - [L]_T \tag{77}$$

The only valid solution to the cubic equation above can be simply written by grouping the coefficients a_0, a_1 and a_2 in three new coefficients *A*, *B* and *C* as follows:

$$[L] = \sqrt[3]{A + \sqrt{A^2 + B^3}} + \sqrt[3]{A - \sqrt{A^2 + B^3}} + C \tag{78}$$

where *A*, *B* and *C* are

$$A = \frac{-a_2^3}{27} + \frac{a_1 a_2}{6} - \frac{a_0}{2} \qquad B = \left(\frac{a_1}{3} - \frac{a_2^2}{9}\right) \qquad C = -\frac{a_2}{3} \tag{79}$$

The solution of the cubic equation (using equations 77 to 79) allows us to calculate the non bounded ligand concentration for a given number of injections. Substituting in equation 74, we can determine the heat associated to each injection using the following expression:

$$dQ \approx \Delta Q(j) = Q_T(j) - Q_T(j-1) + \frac{V_{in}}{V_C}\left(\frac{Q_T(j) + Q_T(j-1)}{2}\right) \tag{80}$$

4.5. Guidelines for the development of ITC equilibrium models

Following the reasoning given in this Chapter, it is easy to discern the basic rules to build any ITC model. The main point would be to collect any experimental and structural evidence (number of sites in the macromolecule M for ligand L, their dependent or independent character, etc) to develop the basic interaction scheme. This scheme will drive to the construction of the binding polynomial, *Z*, which derives into the binding parameter, $\bar{\nu}$, as it has been described in detail into Section 2.

The examples given into Section 4 for the most common ITC models used reveal that, once the model is described in terms of the binding parameter or by the molar fractions of all species containing the macromolecule, two basic points have to be solved from a mathematical point of view. That is, the total heat accumulated after N injections as a function of the binding parameter (equation 40) and the concentration of free ligand as a function of the well-known total concentrations of macromolecule and ligand. Both solutions can be replaced into an equation such as 80 to determine the heat associated to each injection.

These heats divided by the number of moles of ligand added represent the dependent variable of the curve fitting analysis (dQ_i/dL_i), where the ratio between the total concentrations of ligand and macromolecule $[L]_T/[M]_T$ represents the independent variable of the fitting function. The results of this non-linear regression analysis provide the values of the equilibrium constants and the respective enthalpy changes involved in such equilibriums, according to the proposed model. An example is illustrated in Figure 8.

Author details

Jose C. Martinez, Javier Murciano-Calles, Eva S. Cobos, Manuel Iglesias-Bexiga,
Irene Luque and Javier Ruiz-Sanz
Department of Physical Chemistry and Institute of Biotechnology, Faculty of Sciences,
University of Granada, Granada, Spain

Acknowledgement

This work was financed by grant CVI-5915 from the Andalucian Regional Goverment (Spain), grant BIO2009-13261-C02-01 from the Spanish Ministry of Science and Technology, FEDER and Plan E.

5. References

[1] Langerman N & Biltonen RL (1979) Microcalorimeters for biological chemistry: applications, instrumentation and experimental design. Methods Enzymol 61: 261-286.
[2] Biltonen RL & Langerman N (1979) Microcalorimetry for biological chemistry: experimental design, data analysis, and interpretation. Methods Enzymol 61: 287-318.
[3] Wyman J (1965) Binding Potential a Neglected Linkage Concept. Journal of Molecular Biology 11: 631-&.
[4] Hess VL & Szabo A (1979) Ligand-Binding to Macromolecules - Allosteric and Sequential Models of Cooperativity. Journal of Chemical Education 56: 289-293.
[5] Szabo A & Karplus M (1972) Mathematical-Model for Structure-Function Relations in Hemoglobin. Journal of Molecular Biology 72: 163-&.
[6] Szabo A & Karplus M (1975) Analysis of cooperativity in hemoglobin. Valency hybrids, oxidation, and methemoglobin replacement reactions. Biochemistry 14: 931-940.
[7] Szabo A & Karplus M (1976) Analysis of Interaction of Organic-Phosphates with Hemoglobin. Biochemistry 15: 2869-2877.
[8] Pauling L (1935) The Oxygen Equilibrium of Hemoglobin and Its Structural Interpretation. Proc Natl Acad Sci U S A 21: 186-191.

[9] Cuadri-Tome C, Baron C, Jara-Perez V, Parody-Morreale A, Martinez JC & Camara-Artigas A (2006) Kinetic analysis and modelling of the allosteric behaviour of liver and muscle glycogen phosphorylases. J Mol Recognit 19: 451-457, doi: 10.1002/jmr.772.

[10] Ladbury JE & Chowdhry BZ (1996) Sensing the heat: the application of isothermal titration calorimetry to thermodynamic studies of biomolecular interactions. Chem Biol 3: 791-801.

[11] Wiseman T, Williston S, Brandts JF & Lin LN (1989) Rapid measurement of binding constants and heats of binding using a new titration calorimeter. Anal Biochem 179: 131-137.

[12] Gill SC & von Hippel PH (1989) Calculation of protein extinction coefficients from amino acid sequence data. Anal Biochem 182: 319-326.

[13] Sigurskjold BW (2000) Exact analysis of competition ligand binding by displacement isothermal titration calorimetry. Anal Biochem 277: 260-266, doi: 10.1006/abio.1999.4402S0003-2697(99)94402-0 [pii].

[14] Velazquez-Campoy A & Freire E (2006) Isothermal titration calorimetry to determine association constants for high-affinity ligands. Nat Protoc 1: 186-191, doi: nprot.2006.28 [pii]10.1038/nprot.2006.28.

[15] Zhang YL & Zhang ZY (1998) Low-affinity binding determined by titration calorimetry using a high-affinity coupling ligand: a thermodynamic study of ligand binding to protein tyrosine phosphatase 1B. Anal Biochem 261: 139-148, doi: S0003-2697(98)92738-5 [pii]10.1006/abio.1998.2738.

[16] Baker BM & Murphy KP (1996) Evaluation of linked protonation effects in protein binding reactions using isothermal titration calorimetry. Biophysical Journal 71: 2049-2055.

[17] Mason AC & Jensen JH (2008) Protein-protein binding is often associated with changes in protonation state. Proteins-Structure Function and Bioinformatics 71: 81-91, doi: Doi 10.1002/Prot.21657.

[18] Velazquez-Campoy A, Luque I, Todd MJ, Milutinovich M, Kiso Y & Freire E (2000) Thermodynamic dissection of the binding energetics of KNI-272, a potent HIV-1 protease inhibitor. Protein Sci 9: 1801-1809, doi: 10.1110/ps.9.9.1801.

Chapter 5

Applications of Calorimetric Techniques in the Formation of Protein-Polyelectrolytes Complexes

Diana Romanini, Mauricio Javier Braia and María Cecilia Porfiri

Additional information is available at the end of the chapter

http://dx.doi.org/10.5772/54260

1. Introduction

1.1. Formation of the protein-polyelectrolyte complex

Polyelectrolytes are flexible-chain polymers containing subunits with negative or positive charges. These compounds form soluble or insoluble complexes with proteins with opposite electrical charge.

The different equilibriums present in a solution of protein and polyelectrolyte are shown in figure 1.

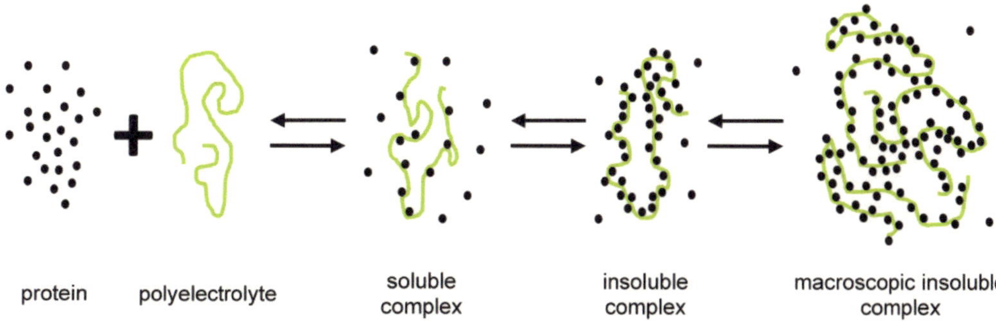

Figure 1. Formation of the protein- polyelectrolyte complex (binary complex).

As can be seen, when a protein interacts with a polyelectrolyte, a soluble complex is formed containing few molecules of the protein. As more molecules of protein interact with the polyelectrolyte, the complex becomes insoluble. Finally, the particles of insoluble complex start to interact with each other to form the macroscopic insoluble complex.

Besides this model of formation of the insoluble complex, there is another model in which the formation of the insoluble complex requires the presence of an inorganic polyion (figure 2).

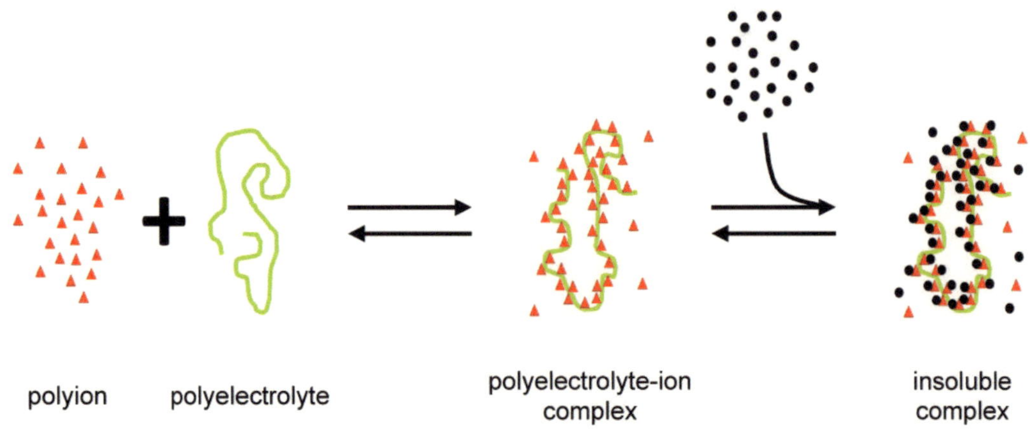

Figure 2. Formation of polyelectrolyte-ion-protein complex (ternary complex).

First, the polyelectrolyte interacts with the polyion and forms a soluble complex with charges oppose to those in the protein. Then, the protein interacts with the complex to form the insoluble complex.

In both cases, the formation and solubility of the complex depends on the pH and the ionic strength of the medium [1], the density of charges in the protein and the polyelectrolyte, the molecular weight and the concentration of the polyelectrolyte [2,3]. Various studies have been directed to understand the formation of these complexes in aqueous medium [4-6]. Equation 1 shows how the density of charges *(σ)* on the surface of the protein and the polyelectrolyte is affected by the pH and the ionic strength of the medium.

$$\sigma = \frac{\partial \sigma}{\partial \mathrm{pH}}\left(\mathrm{pH} - \mathrm{pI}\right) \tag{1}$$

Mattison *et. al.* postulated an equation that correlates the density of charges in the protein *(σ)* and the polyelectrolyte *(ξ)* with the Debye-Hückel constant (κ) that is highly dependent on ionic strength (*a* is a constant that depends on the protein-polyelectrolyte system) [7].

$$\xi\sigma \cong a \cdot \kappa \tag{2}$$

The conditions of the medium determine whether the soluble or the insoluble complex is formed, or if the complex is dissociated.

1.2. Stoichiometry of the protein-polyelectrolyte formation

When studying the interaction of a protein and a polyelectrolyte, it is interesting to know the minimum quantity of protein requires forming the maximum quantity of complex per polyelectrolyte unit. This value is called stoichiometry of the complex *(e)* and it is usually represented as the moles of polyelectrolyte per mol of protein, mass of polyelectrolyte per mass unit of protein, etc.

The stoichiometry of a protein-polyelectrolyte complex might be over or below 1 as shown in figure 3.

A complex with a stoichiometry below 1 contains more protein molecules than polyelectrolyte molecules. Usually, this kind of complex is insoluble, while a complex with a stoichiometry over 1 might be soluble or insoluble. As can be seen in figure 3, a complex with a stoichiometry below 1 can be turned into one with stoichiometry over 1 by adding polyelectrolyte to the medium. Of course, the effect can be reverted by adding protein.

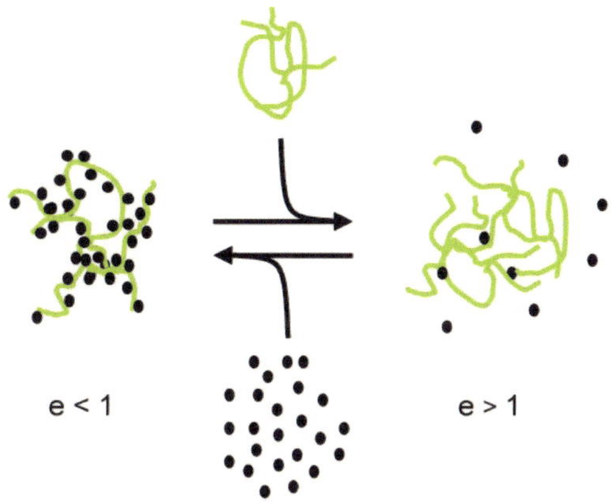

Figure 3. Stoichiometry (*e*) of a protein-polyelectrolyte complex.

1.3. Biotechnological applications of the protein-polyelectrolyte complex

In biotechnology, it is interesting to use insoluble protein-polyelectrolyte complexes with e < 1 since they can be used to purificate and concentrate industrial-interest enzymes [8], to immobilized enzymes in bioreactors or scarefolds.

1.3.1. Bioseparation of proteins from a complex mixture

The development of Biotechnology has allowed the used of enzymes in the production of food, drugs and in many others industries. At the same time, Genetic Engineering and Molecular Biology has allowed the expression of proteins in bacteria and superior microorganisms; however, some proteins must still being isolated from its natural sources due to complex post-traductional modifications that occur during the synthesis of the proteins. In both cases, the protein of interest is in a media containing many other biomolecules and inorganic compounds that need to be separated from the protein before applied it to any industrial process.

Most purification protocols consist on many steps: the first ones have the aim to concentrate the protein of interest and to obtain a high recovery; while the last steps of the protocol are expected to yield a high purification factor.

Precipitation of enzyme-polyelectrolyte insoluble complexes is a very useful technique to apply at the beginning of purification protocols since it requires simple equipment and so is very easy to scale up; needs low concentrations of the polyelectrolyte since it interacts with high affinity with the proteins; and can be based on a wide variety of polyelectrolytes, both natural and synthetic, positively or negatively charged. An important aspect of this technique is that the enzyme usually retains its tertiary structure and its catalytic activity. Even more, usually it is more stable in the presence of the polyelectrolyte [9,10].

1.3.2. Enzyme immobilization

The immobilization of enzymes is a process by which the protein is attached to a solid matrix, synthesized using a polyelectrolyte, in order to enhance the stability of the structure of the protein and to reuse enzyme [11]. Enzymes immobilized, in comparison with enzymes in solution, are more robust and resistant to environmental changes.

Immobilization can be performed by physical or chemical methods. The first results in weak interactions between the enzyme and the solid support and includes adsorption on a water-insoluble matrix and gel entrapment or micelles [12,13]. Chemical methods generate covalent bonds or electrostatic interactions between the enzyme and a water-insoluble support forming reticulated or single-chain particles. Insoluble complexes allow to immobilized enzymes by entrapment in polyelectrolyte solid particles, micelles or by covalent linkage to the support using carboimide as coupling.

The protein-matrix systems are widely used in bio-reactors for industrial process mainly because of the possibility of recycle the enzyme. Bio-reactors are very useful for the synthesis of organic compounds; since immobilized enzymes reduced the steps of the process, enhance the purity of the final products and allow stereo-selective synthesis. These systems also can be applied on the production of micro/nanocapsules for the delivery of proteins (or drugs). In this case, the use of polyelectrolytes sensitive to environmental conditions allows the releasing of the enzyme (or drug) molecules in different parts of the body [14].

1.4. Characterization of the protein-polyelectrolyte complex

The formation of the protein-polyelectrolyte complex can be studied by spectroscopic and calorimetric techniques.

Spectroscopy assays based on turbidimetric measurements allow knowing the effect of pH, ionic strength, time and temperature on the amount of insoluble complex formed. Phase diagrams, turbidimetric titrations and kinetic assays must be performed in order to evaluate the best conditions to obtain the maximum quantity of insoluble protein-polyelectrolyte complex [15;16].

Study the stability of the enzyme when it is part of the complex is also important to understand how the polyelectrolyte affects its catalytic activity since it is expected to use the enzyme in an industrial process. Fluorescence, UV-visible and circular dichroism spectroscopy are very useful techniques to analyze the structure of the protein but also must be performed experiments to study the enzymatic activity.

Although these techniques are helpful, they do not give an idea of the nature of the interaction between the protein and the polyelectrolyte. Calorimetric techniques such as differential scanning calorimetry and isothermal titration calorimetry allow studying the thermal stability of the enzyme in the presence of the polyelectrolyte and the nature of the interaction between the enzyme and polyelectrolyte, respectively.

2. Research course and methodology

In order to study the formation of insoluble complexes between proteins and polyelectrolytes it is necessary to carry out different methodologies in a sequential way.

2.1. Titration curves at different pH in binary systems

The formation of the insoluble polyelectrolyte-protein complexes can be followed by means of turbidimetric titration of the protein with the polyelectrolyte, or vice versa. Taking into account the isoelectric point of the protein and the pK value of the polyelectrolyte, the pH of the medium should be selected so that they have opposite net charge.

Figure 4. shows the turbidimetric titration curve obtained for two systems with different behavior: hyperbolic and sigmoid. Turbidity is usually measured as the absorbance (Abs) at 420 nm.

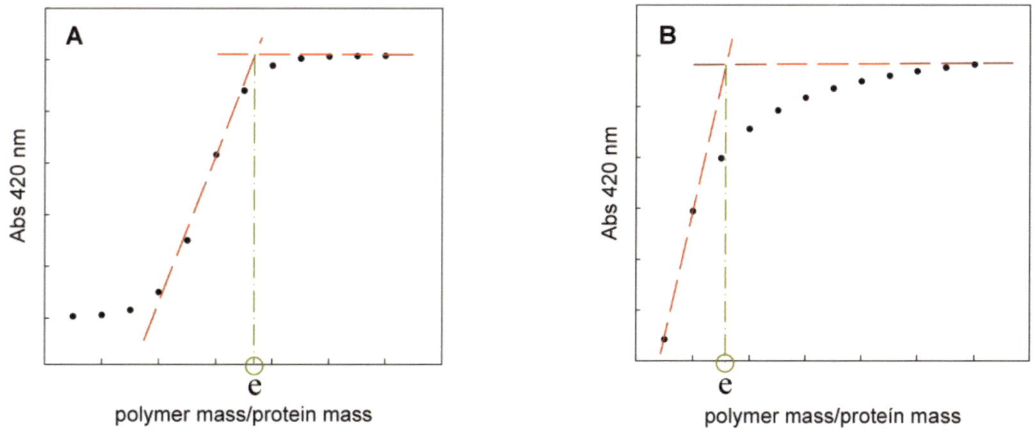

Figure 4. Determination of the polyelectrolyter/protein mass ratio when saturation is reached in a **A**-sigmoid and **B**- hyperbolic graph

These graphs demonstrate a saturation behavior with different mechanism of complex formation, and allow us to determine the *stoichiometric polyelectrolyte/protein ratio "e"*, which is defined as the minimal ratio in which the protein precipitates as an insoluble complex. The value *"e"* is calculated from the plot at the lowest concentration of polyelectrolyte necessary to get the saturation. This value is important because it allows us to determine the minimal amount of polymer needed to fully precipitate the protein, and can be expressed as the number of moles of protein bounded per mol of polyelectrolyte.

2.2. Titration curves at different pH in ternary systems

In ternary systems, the polyelectrolyte forms an insoluble complex with an anion, which associates proteins with opposite net charge. The mixture polyelectrolyte-anion behaves as a pseudo polyampholyte with a characteristic isoelectric point.

The formation of insoluble complexes between the polymer and the anion can be examined by turbidimetric titration of the anion with the polyelectrolyte. When these curves reach the saturation it suggests a complete precipitation of the complex.

The precipitation of polyampholyte -protein complexes is driven by coulombic forces, which are highly dependent on protein and polyampholitic isoelectric pH values [5;17;18]. So, precipitation begins at a critical pH where the attractive forces have just overcome electrostatic repulsion.

2.3. Phase diagrams in binary systems

Phase diagrams, also called solubility curves, show the range of pH in which the complexes are soluble or insoluble. It means that they provide information about the pH of higher interaction between the components and the optimum pH for precipitation and dissolution of the complexes.

To obtain these diagrams, a polyelectrolyte/protein mixture at a ratio close to *"e"* is titrated with acid or alkali and the turbidity of the medium is measured after pH variation.

Figure 5. shows an schematic phase diagram in a binary system.

2.4. Phase diagrams in ternary systems

Classical polyampholytes have both anionic and cationic groups in their molecules. However, the aqueous solution of any polyelectrolyte may behave as a polyampholyte provided there is a small ion with multiple electrical charges (two or more) in the medium which interacts with the opposite charge of the polyelectrolyte to form a pseudo polyampholyte. Under these conditions, it is possible to find a pH interval where the pseudo complex behaves as an ampholyte.

To obtain the phase diagrams, a mixture with fixed polyelectrolyte/anion ratio is titrated with alkali or acid, and the turbidity of the medium is measured as the absorbance at 420 nm after pH variation.

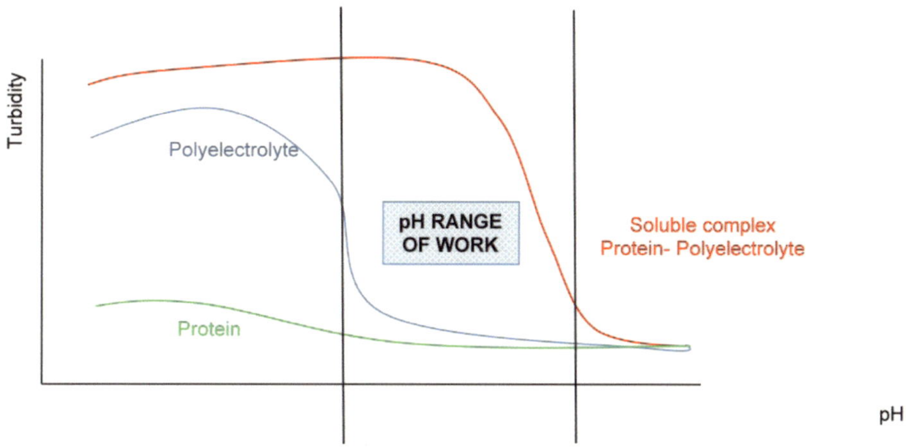

Figure 5. Phase diagram, turbidity vs. pH, for protein (—), polyelectrolyte (—) and binary system (—).

Figure 6 shows a schematic phase diagram between a cationic polyelectrolyte and an anion, in different polymer/anion ratios. Figure 6 A shows a pH range where the turbidity of the medium drastically increases. Each curve has a trapezoidal shape with a plateau, and the height of the trapezium depends on the polyelectrolyte concentration. The pHs that correspond to the edges of the trapezium are the critical pH values, at which the transition from complete dissolution to precipitation occurs. The lower critical pH is usually called acidic, and the higher one is called basic. It is remarkable that the transitions from complete solubility to precipitation occur at the same critical pHs independently of the polyelectrolyte concentration.

Also, phase diagrams can be represented as in figure 6.B. This diagram represents the behavior of the polyelectrolyte-anion complex by filled and open circles: filled circles are drawn at the pH of non-zero absorbance whereas the open circles at the zero absorbance values of the solution.

As mentioned above, insoluble polyelectrolyte-anion complexes behave as an ampholyte. This can be used to precipitate cationic or anionic proteins depending on the pH of work, as indicated in figure 6.B.

2.5. Effect of ionic strength

The coulombic component in the interaction between proteins and polyelectrolytes is closely related with the presence of charges. So, the ionic strength of the medium can alter the forces involved in the interaction and eventually leads to dissociation of the complexes.

In ternary systems, high ionic strength can also inhibit polymer-anion interaction.

Whatever the system, this inhibition of the formation of the precipitates may be directly proportional to salt concentration. So, the effect of the presence of salt in the systems can be evaluated by turbidimetric titration in the presence of different concentrations of NaCl in the medium.

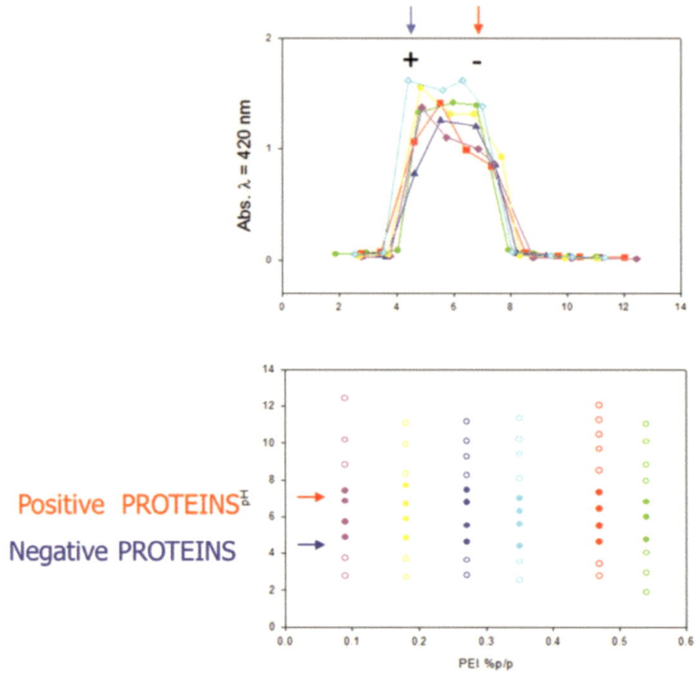

Figure 6. Phase diagram in polyelectrolyte-anion systems. Each color represents a different polyelectrolyte/anion ratio.

2.6. Complex formation kinetics

The interaction between polyelectrolytes and proteins requires time to achieve the maximum quantity of complex (maximum turbidity). Thus, it is necessary a kinetic study by which the turbidity of a mixture polyelectrolyte-protein is measured over time. So, a solution of the polyelectrolyte is added to a solution of the protein, at the pH of precipitation and in an appropriate ratio, and absorbance at 420 nm is measured over time. Finally, a plot of turbidity vs. time is made and the time required to reach the maximum quantity of turbidity is obtained [19].

2.7. Conformational and enzymatic evaluation of the protein in the complex

Several investigations have reported that polymers stabilize the catalytic activity in a variety of enzymes. Besides, it has been suggested that electrostatic interactions between the enzyme and polyelectrolytes play a primary role, also in conformational stabilization [20;21].

2.7.1. Effect of the polyelectrolytes on the far-UV circular dichroism (CD) of proteins

Circular dichroism spectroscopy is a very frequently used technique to evaluate protein conformation in solution. This method is sufficiently simple and allows a rapid determination of protein structure or conformational changes.

In the far-UV region (between 180 and 250 nm) the circular dichroism spectrum provides information one the secondary structure of proteins, due to asymmetrical packing of intrinsically achiral (planar) peptide groups [22].

The effect of polymers on the structure of proteins can be analyzed in terms of its secondary elements. So, far-UV circular dichroism spectra of proteins are recorded in different polymer/protein ratios, with a fixed concentration of protein and varying the amount of polymer. The pH of the medium must be the pH of higher interaction between the protein and the polyelectrolyte.

2.7.2. Effect of the polyelectrolytes on the enzymatic activity of proteins

In order to evaluate the effect of the polyelectrolyte on the enzymatic activity of the protein, enzyme assays are performed at constant protein concentration in the presence of different amounts of polymer. Polyelectrolyte/protein ratios are usually close to the stoichiometry of the complex (“***e***”) or in excess of polymer respect this value.

The stability of the enzyme in the presence of the polyelectrolyte can also be monitored by incubating the mixture polyelectrolyte/enzyme and recording the enzymatic activity over time.

2.8. Precipitation and redissolution of the complexes

After analyzing the conditions of complex formation or dissolution and evaluating the effects of the different variables, we are able to design a methodology of precipitation of the protein with the polyelectrolyte by following the steps shown in figure 7.:

According to this precipitation scheme, an aliquot of the polyelectrolyte is mixed with a solution of the protein, both prepared at the pH of precipitation, to reach a proper polymer/protein ratio. This mixture is incubated the time necessary to form the maximum quantity of insoluble complex and centrifuged to obtain a solid precipitate. Then, the supernatant is separated and the precipitate redissolved in buffer at the pH of dissolution of the complex. Finally, in order to evaluate the effectiveness of the total process, enzymatic activity is measured in both fractions.

This scheme is successfully applied in many systems [15-20] and allows to obtain a protein with a high recovery and catalytically conserved.

2.9. Calorimetric measurements for polymer-protein complex

2.9.1. Differential scanning calorimetry

Thermal desnaturation of proteins was monitored with a high sensitivity differential scanning calorimeter model VP-DSC from MicroCal Inc. Thermograms were obtained between 20-85°C, at scan rate 25-30ºC/h and at constant pressure of 28 psi. All result were averages of, at least, three independent measurements. Buffer versus buffer baseline scans were determined and subtracted from transition scans prior to normalization and analysis of protein denaturation. Finally, the values of the excess heat capacity were obtained after subtraction of the baseline. The calorimetric data were analysed by using the software ORIGIN 7.0, MicroCal Inc., following the methodology recommended by IUPAC. The parameters obtained from this analysis were: temperature at which maximum heat exchange occurs (*Tm*), the area under the peak, which represents the enthalpy of transition for reversible process (ΔH_{cal}) and the van´t Hoff enthalpy (ΔH_{VH}).

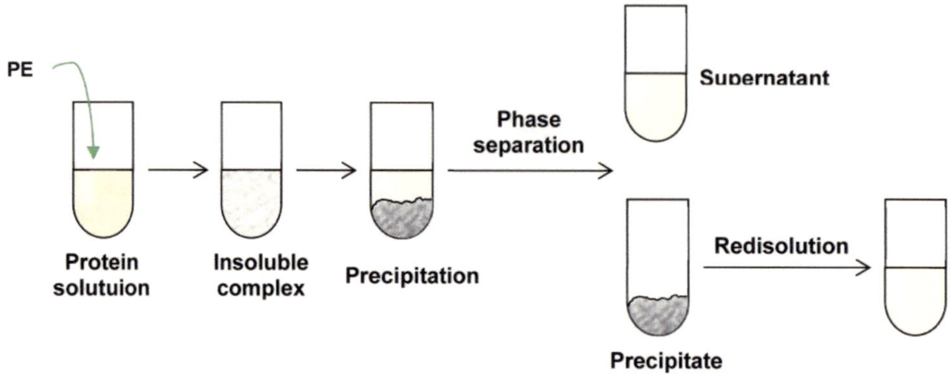

Figure 7. Scheme of precipitation of polyelectrolyte/protein complexes

The evaluation of ΔH_{VH} gives an idea of the mechanism of the unfolding process [23]:

- $\Delta H_{VH} = \Delta H_{cal}$: a two-state process is carried out under equilibrium condition.
- $\Delta H_{VH} > \Delta H_{cal}$: intermolecular cooperation is taking place which may require some degree of molecular association.
- $\Delta H_{VH} < \Delta H_{cal}$: one or more intermediate states of significance in the overall process.

However, in some cases this calorimetric criterion may lead to erroneous conclusion.

2.9.2. Isothermal titration calorimetry

Measurements of the examples presented were performed at 20-25 ºC by using a VP-ITC titration calorimeter (MicroCal Inc. USA). The sample cell was loaded with 1.436 mL of protein solution and the reference cell contained Milli-Q grade water. Titration was carried out using a 300 μL syringe filled with polyelectrolyte solutions. The experiments were performed by adding aliquots of 3-5μL of polyelectrolyte solutions 0.175 % (w/w) to the cell containing the protein solution.

The mathematical model equation selected to fit the ITC data was derived from a model that assumes the polyelectrolyte molecule binding to several protein molecules, all with the same intensity; in other words, the polyelectrolyte was considered as a macromolecule having **n** independent and equivalent sites, all of which have the same affinity constant, K, for the ligand (protein) [24].

The heat associated with the interaction polyelectrolyte-protein (ΔH_b) was calculated by subtraction using the equation 3:

$$\Delta H_b = \Delta H_t - \Delta H_d - \Delta H_{dissol} \quad (3)$$

Where ΔH_t is the total heat associated to each polymer addition, ΔH_d is the heat of dilution of the polyelectrolyte in the buffer in the absence of the protein and ΔH_{dissol} is the heat of polymer dissolution. The heat associated to the dilution of the protein in buffer was negligible. Then ΔH_b was plotted vs polyelectrolyte/protein molar ratio and, by non-linear fitting of these calorimetric curve, the affinity constant (K) for the polyelectrolyte binding to the protein and the number of polymer molecules (n) bound per protein molecule was calculated using the software provided by the instrument.

The resulting data set was fitted using MicroCal ORIGIN 7.0 software supplied with the instrument and the intrinsic molar enthalpy change for the binding, ΔH_b, the binding stoichiometry, n, and the intrinsic binding constant, K, were thus obtained. The equation for determining the heat associated to each injection is:

$$Q = \frac{n\, M_t\, \Delta H_b\, V_o}{2}\left(1 + \frac{1}{nkM_t} + \frac{X_t}{nM_t} - \sqrt{\left(1 + \frac{1}{nkM_t} + \frac{X_t}{nM_t}\right)^2 - \frac{4X_t}{nM_t}}\right) \quad (4)$$

where V_o is the active volume cell, X_t is the bulk concentration of ligand and M_t is the bulk concentration of the macromolecule in V_o [25].

The intrinsic molar free energy change, ΔG^o, and the intrinsic molar entropy change, ΔS^o, for the binding reaction were calculated by the fundamental thermodynamic equations 5 and 6:

$$\Delta G^\circ = -R\, T \ln K \quad (5)$$

$$\Delta S^\circ = \frac{\Delta H^\circ - \Delta G^\circ}{T} \quad (6)$$

3. Results

3.1. Turbidimetric titration curves

Figure 8 shows typical hyperbolic titration curves of a protein with a polyelectrolyte. In this case, trypsin (TRP) with poly vinyl sulfonate (PVS) [26]. As can be seen, two important characteristics were observed: at low polymer/protein ratios, absorbance increases with an increase in the polyelectrolyte total concentration and, at high polyelectrolyte/protein ratio, there is a plateau which depends on the medium pH.

The protein/polyelectrolyte molar ratio which corresponds to the situation where the protein has been precipitated as an insoluble complex was calculated from the intersection of a straight line which corresponds to the prolongation of the linear zone of the curve (at low polymer concentration) with a line which gives a plateau.

Trypsin is a serin-protease found in the digestive system. It is used for numerous biotechnological processes. Its isoelectric point is between 11.0 and 11.4 [27]. The pHs selected in the curves were chosen in the range where TRP and PVS have opposite charges.

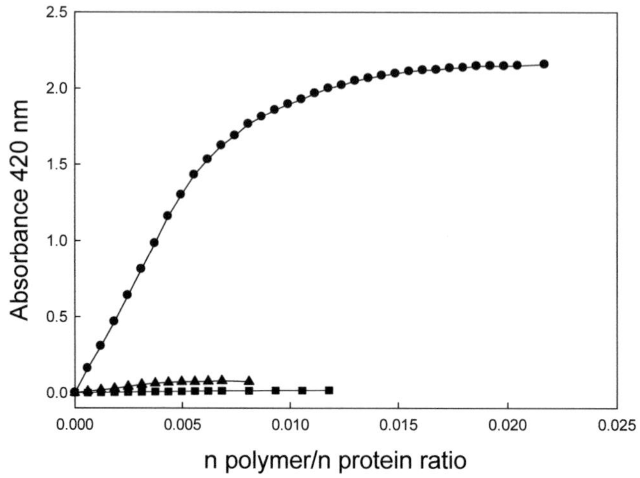

Figure 8. Turbidimetric titration curves of TRP (70μM) solution with PVS (0.25 % w/w) in a medium with phosphate buffer 50mM, pH 3.0 (●), 5.5 (▲) and 7.0 (■). T=20 ºC, [21].

Table 1 shows the molar protein-polyelectrolyte ratio which corresponds to the stoichiometry of the complex formation calculated from the titration curves for the different experiments. These values are important because they allow us to calculate the minimal polymer amount necessary to precipitate the protein in a complete form. The data have been expressed as the number of TRP moles bound per polyelectrolyte mol. Despite the fact that these values were similar, turbidity is much higher at pH 3.00 which suggest a major size of the precipitate particle.

pH	Protein/polyelectrolyte molar ratio
3.00	136 ± 3
5.50	228 ± 21
7.00	147 ± 12

Table 1. TRP/PVS molar ratios at different pHs.

Figure 9 shows titration curves of lysozyme (LYZ) with the polyelectrolyte PVS. LYZ is a basic protein with 19 amino residues, an isoelectrical pH between 11.0 and 11.4 and a molecular mass of 14.3 kDa [28], therefore at the pHs where the turbidimetry titration were assayed the protein has a net positive electrical charge. Formation of LYZ-PVS complex was observed to be influenced by the medium pH, however, at pH 3.1, a minor absorbance maximum value was observed than at pH 5.5, which can be assigned to the loss of the native structure of this protein by influence of the acid medium [26].

Table 2 shows the molar protein-polymer ratios which correspond to the stoichiometry of the complex formation calculate from the titration curves for the different experiments. These values are important because allow to estimate the minimal polyelectrolyte amount needed to precipitate the protein, the data have been expressed as the number of LYZ molecules bound per polyelectrolyte molecule.

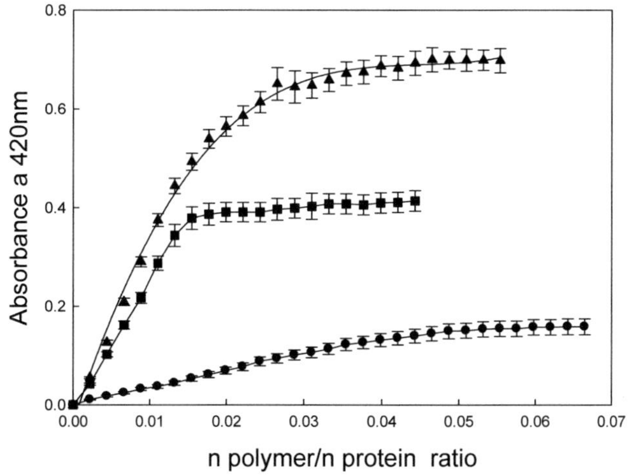

Figure 9. Turbidimetric titration curves of LYZ (0.3mg/mL) solution with PVS in a medium with 50 mM phosphate buffer. pH 5.5 (▲), 7.0 (●), and acetic acid/acetate buffer pH 3.1 (■). T= 20 ºC.

LYZ- PVS	Protein/polyelectrolyte molar ratio
pH 3.1	66
pH 5.5	47
pH 7.0	23

Table 2. Lys/PVS molar ratios at different pHs.

3.2. Turbidimetric titration of ternary complex:

Figure 10 shows turbidimetric titration curves when phosphate (500mM) or citrate (50mM) was titrated by adding a concentrated solution of the polyelectrolyte poly ethylene imine (PEI). Both curves reached a plateau at high polyelectrolyte anion ratios, which suggests a complete precipitation of the complex. It could be seen that the plateau was obtained at a polymer/anion ratio 10 times higher for citrate than for phosphate, suggesting that citrate has a greater precipitation capacity than phosphate. These ternary systems have the capability to precipitate in the absence of protein. Only is required the presence in the medium of the cationic polyelectrolyte (PEI) and a polyanion like phosphate (Pi) or citrate (Cit).

3.3. Phase diagrams of binary systems

Figures 11 show the absorbance dependence (at 420 nm) vs. the pH change by the system LYS with poly acrylate (PAA). These complexes were soluble at basic pH values. At pH lower than 6.50 a significant increase in the turbidity was observed that corresponding to the insoluble complex formation. Similar behavior was reported for the serum albumin titration with anionic polyelectrolyte [7].

The relevance of these phase diagrams are in the possibility to know the insolubility and solubility complex conditions.

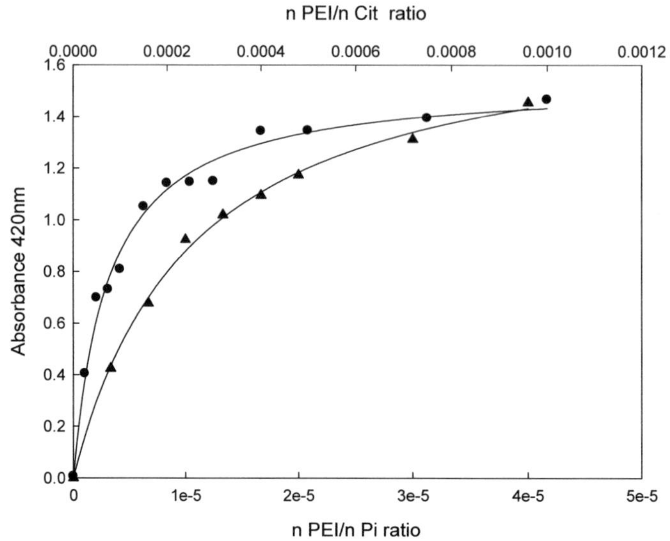

Figure 10. Turbidimetric titration of phosphate (▲) and citrate (•) with PEI. pH 5.5. T= 20º C [6].

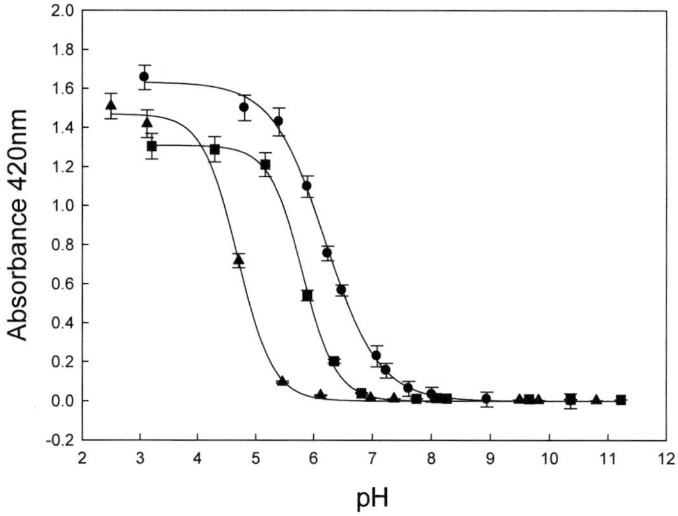

Figure 11. Dependence of the absorbance at 420 nm vs the medium pH at a constant protein-polyelectrolyte molar ratio of LYZ-PAA: (•) 0.0027, (■) 0.0065, (▲) 0.0010. T= 20ºC [25].

3.4. Phase diagrams of ternary systems

Figure 12 shows the pH variation effect on the insoluble complex formation obtained for ternary systems PEI/Pi at different PEI/Pi molar ratios [6]. As can be seen, in all curves there is an increase in the turbidity of the medium, reaching a maximum value and then decreasing in the pH interval 3.5-7. Each curve has a trapezoidal shape and the pH values corresponding to the edges of the trapezium are the critical pHs at which the transition from complete dissolution to precipitation occurs.

In this figure it can be identify both critical pHs: 3.5 and 7. At pH=3.5 the net charge of the complex is positive whereas at pH= 7 is negative. On the other hand, transitions from complete solubility of the insoluble complex are independent of the polyelectrolyte concentration.

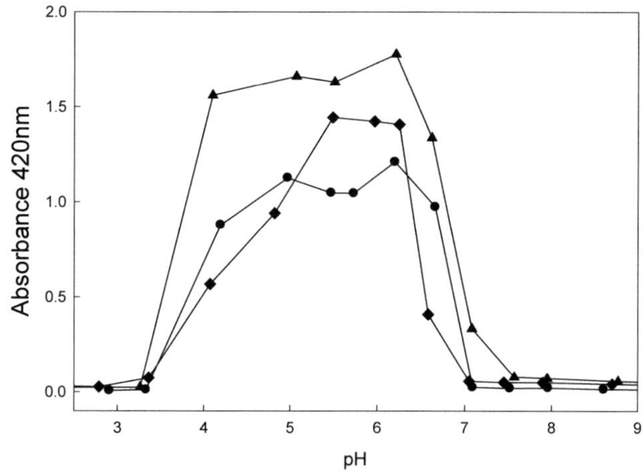

Figure 12. Phase diagram of PEI/Pi systems at different molar ratios.

3.5. Effect of ionic strength

In general, protein/polyelectrolyte insoluble complexes are greatly affected by ionic strength because the molecular mechanism of the interaction is mainly electrostatic in nature. Turbidimetric titrations at pH 7.00 were performed in medium of different ionic strength such as shown Figure. 13. In this case, only 0.1 M of NaCl is enough to avoid formation the insoluble protein-polyelectrolyte complex [26]. This finding may be interesting in the design of an isolation method of protein, allowing in a first step the precipitation by the polyelectrolyte and then the precipitate may be dissolved by NaCl solution addition at low concentration.

Ternary systems like PEI-citrate was dramatically affected by 0.5 for higher ionic strength; in this case, no formation of the insoluble complex was found while the PEI-phosphate system showed to be slightly affected by the NaCl increased concentration. The inhibition of the precipitate formation in both systems was directly proportional to the salt concentration, in agreement with the presence of an important coulombic component in the insoluble complex formation [19].

3.6. Kinetics of the complex formation

In general, the kinetics of complex formation is fast (around 2-10 minutes) [15;17;19]. Figure 14 shows the kinetic studies of different molar ratios of the systems TRP/Eudragit®L100 (EL100). It required less than 2 minutes of incubation to achieve the maximum quantity of complexes. In addition, by increasing the polyelectrolyte concentration increases the turbidity of the system, however the time required achieving the maximum turbidity is independent of the concentration of the molar ratio.

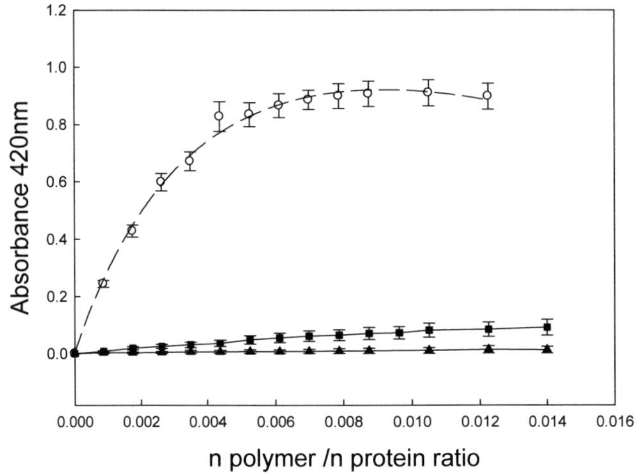

Figure 13. NaCl concentration effect on the turbidity of LYZ-PAA, pH 7.0, NaCl concentration: (O) 0M, (■) 0.1 M and (▲) 0.5M. T= 20ºC.

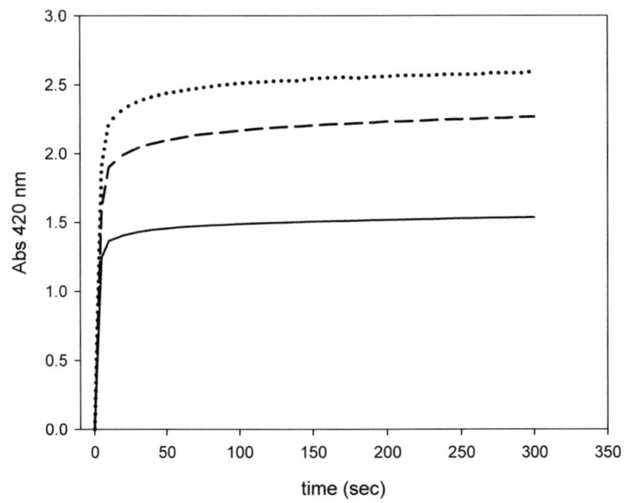

Figure 14. Formation of complex TRP-EL100 through time at three protein/polyelectrolyte molar ratio of TRP/EL100: (——) 32.41, (— —) 16.18, (····) 8.08. [15].

4. Calorimetric techniques of protein-polyelectrolyte complex

4.1. Differential scanning calorimetry by polymer-protein complex

DSC is a useful tool for studying the protein unfolding in which values of excess specific heat capacity (Cp) are obtained as a function of temperature. Two enzymes having different behavior towards charged flexible chain polyelectrolytes are analyzed below.

Lysozyme is a basic protein with 19 amino residues, an isoelectrical pH between 11.0 to 11.4 and a molecular mass of 14.3 kDa. Because LYZ is one of the four proteins whose thermal denaturation is thermodynamically reversible, the equations for systems in thermodynamic equilibrium can be applied to obtain the thermodynamic functions (entropy and enthalpy of unfolding) directly from the thermograms, as described by Privalov [29].

Thermograms of LYS enzyme with PAA and PVS are presented as examples in figure 15 and Table 3 shows the thermodynamics functions and Tm obtained in each case. In these systems DSC measurements demonstrated that the Tm of LYS was not modified by the polyelectrolytes presence only a decrease in the denaturalization heat (ΔH_{cal}) was observed.

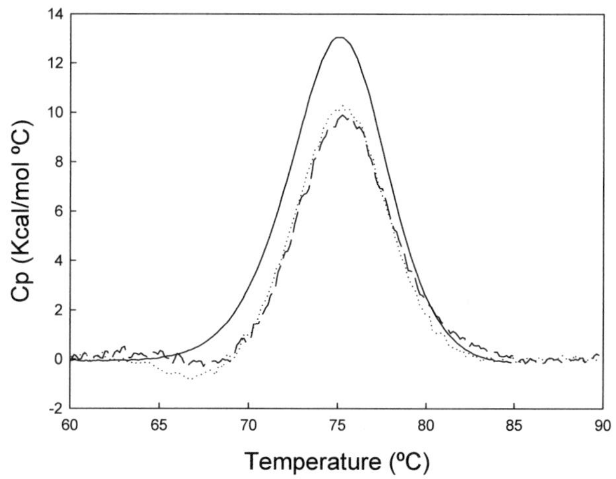

Figure 15. DSC Thermograms of the LYZ in the absence (—) and presence of: PVS (----) and PAA (....) pH 7.00.

	LYZ	LYZ-PVS	LYZ -PAA
ΔHcal (kcal/ mol)	89.4 ± 0.3	72.0 ± 0.3	66.7 ± 0.4
ΔH_{VH} (kcal/ mol)	139.0 ± 0.6	141 ± 0.8	151 ± 1.0
ΔH_{VH} / ΔH_{cal}	1.55 ± 0.05	1.96 ± 0.01	2.26 ± 0.03
Tm (ºC)	75.01 ± 0.01	75.33 ± 0.02	75.2 ± 0.1
ΔS (e.u.)	399 ± 3	405 ± 4	405 ± 5

Table 3. Thermodynamic functions obtained for the thermal LYZ unfolding determined by DSC in the absence and presence of the studied polyelectrolytes.

A Tm constant value is a proof that the protein retains its thermodynamic stability in the presence of both polyelectrolytes. However the polymer presence induced a decrease in the area under the curves, in agreement with a diminution of the heat associated to the denaturation process. The unfolding entropic change showed to be not affected by the polyelectrolyte presence, in accordance with the protein retaining its tertiary structure and no important conformational protein change is occurring.

LYZ is a protein which has only one domain with low molecular mass, its thermal unfolding have been describe as reversible, however the capacity of LYZ to associate in aqueous solution it is well known. $\Delta H_{VH}/\Delta H_{cal}$ ratio greater than 1 is an indication of the intermolecular cooperation presence during the thermal unfolding. The increase of this ratio in the polyelectrolytes presence, suggests more cooperative intermolecular process.

Furthermore, the unfolding entropy was not affected in the protein-polymer complexes (LYZ-PVS and LYS-PAA). It indicates that LYZ in complex follows in the same conformational state that LYZ alone.

Trypsin is a serin-protease found in the digestive system. It is used for numerous biotechnological processes. It is a globular protein which has two domains with similar structures [27]. DSC experimental results for enzyme trypsin are demonstrated a two-state transition model at pH 3.00 [30]. Figure 16.A shows DSC thermograms of TRP. Although the ratio $\Delta H_{VH}/\ \Delta H_{cal}$ is close to 1, however the thermogram clearly shows 3 transitions.

TRP-EL100 complex has a very interesting behavior. As can be seen in Figure 16.B protein thermogram was significantly modified by the polyelectrolyte presence.

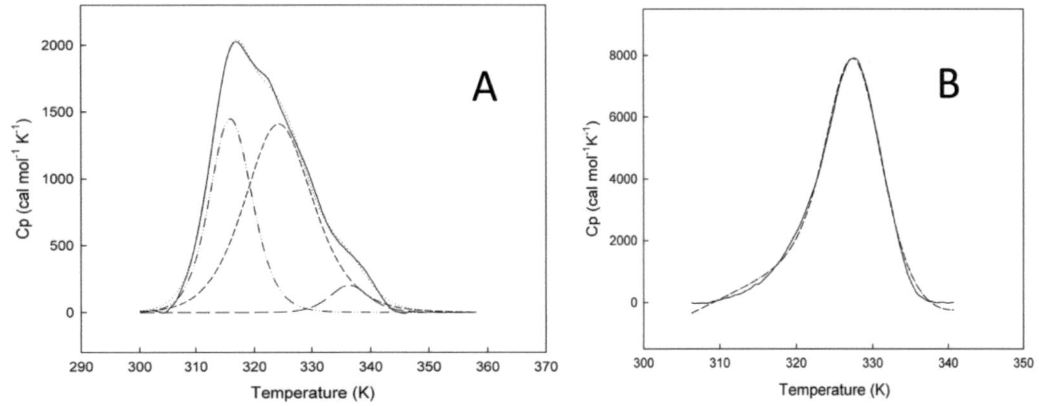

Figure 16. (A) DSC Thermogram of the TRP: (—) experimental data; (---) fit data; (··—) first transition; (- - -) second transition; (— —) third transition. (B) DSC Thermogram of the TRP in the presence of EL100: (—) experimental data; (---) fit data.

	Tm (K)	ΔH_{cal} (kcal/mol)	ΔH_{VH} (kcal/mol)	$\Delta H_{VH}/\ \Delta H_{cal}$
TRP	320.6 ± 0.1	38.7 ± 0.6	42.2 ± 0.8	1.09
TRP (transitions)	316.1 ± 0.1	13.9 ± 0.1	8.3 ± 3	
	324.4 ± 0.3	22.2 ± 0.2	53.1 ± 3	-
	336.4 ± 0.3	1.8 ± 0.5	-	
TRP-EL100	327.8 ± 0.1	82.0 ± 0.1	81.6 ± 0.1	0.99

Table 4. Thermodynamics functions obtained for the thermal TRP unfolding determined by DSC in the absence and presence of the EL100.

The comparison of the two figures and table evidences two main differences

- Tm of the TRP curve is 320.6 K (47.5°C), whereas for the TRP-EL100 complex the same parameter is 327.8 K (54.6°C). The shift of the Tm of TRP in the presence of the polymer means that the EL100 stabilizes the structure of the enzyme against thermal denaturation.
- The different unfolding model in the absence (3 transitions of independent domains) and in the presence of polymer (two-state model)

4.2. Isothermal titration calorimetry

ITC technique gives the direct heat associated to the complex formation (**ΔH**), a number of protein molecules bounded to polyelectrolyte molecule (**n**), the affinity constant (**K**). Before performed ITC experiment is important to known which is the number *"e"* obtained by turbidimetric titration because is a good estimation of **n**.

Figure 17 shows the ITC measurements of the LYS titration with PVS and Table 5 summarizes the parameters obtained by two anionic polyelectrolytes (PVS and PAA).

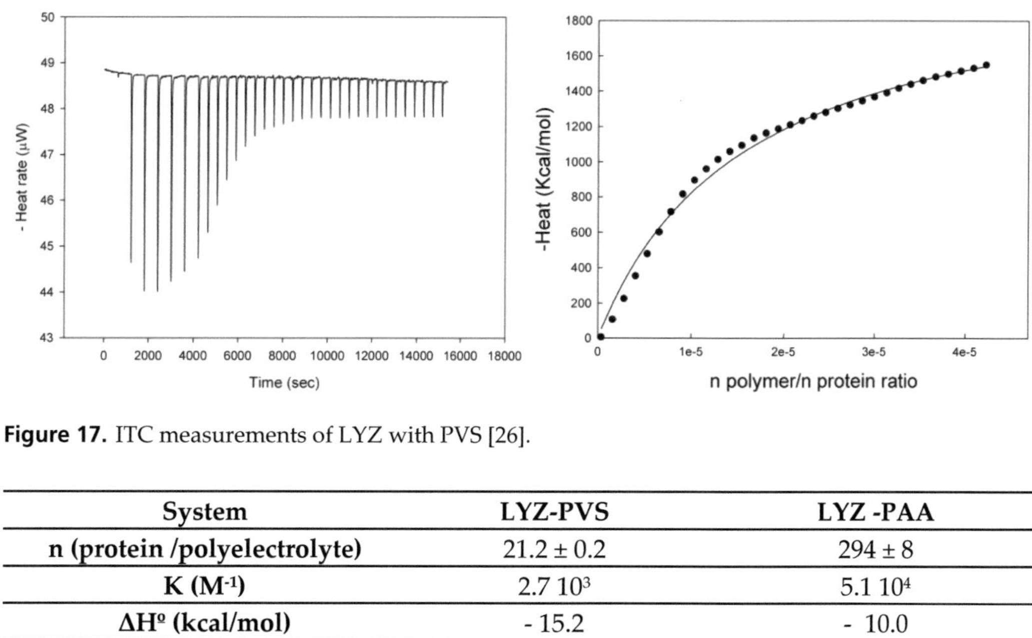

Figure 17. ITC measurements of LYZ with PVS [26].

System	LYZ-PVS	LYZ -PAA
n (protein /polyelectrolyte)	21.2 ± 0.2	294 ± 8
K (M^{-1})	2.7 10^3	5.1 10^4
ΔHº (kcal/mol)	- 15.2	- 10.0
ΔSº (e.u.)	-1103	-1033

*The enthalpic change is expressed per mol of protein bound.

Table 5. Thermodynamic and binding parameters of the interaction LYZ-polyelectrolyte from ITC experiments.

The interaction LYZ-PVS and LYZ-PAA is exothermic. The mechanism of bond is carried out between the electrically charged groups of both. The differences found between complexes were the affinity (K) and the number of molecules of protein bonded to polymer molecule. Because the size of the polyelectrolytes are 10-fold larger than the protein, the number of protein molecules bound per polymer molecule is high.

The heat associated to the complex formation were extremely high, but when they are normalized per protein molecule bound to the polyelectrolyte the heat associate yielded 10-15 kcal/mol which is a normal heat amount for a coulombic interaction between two charge groups in solution. These low interaction heats are in agreement with the low NaCl concentration needed to induces the dissolution of the insoluble complex (around 0.1 M) Other important parameters to know are the thermodynamically stability of the protein in the polyelectrolyte presence. It is desirable that the protein retains its tertiary structure.

TRP- EL100 complex is an interesting example. Although the polymer and protein present opposite electrical charge, however the interaction is endothermic.

Figure 18 shows the binding isotherm obtained when consecutive aliquots of EL100 were added to a solution of trypsin [15]. The parameters calculated are summarized in Table 6.

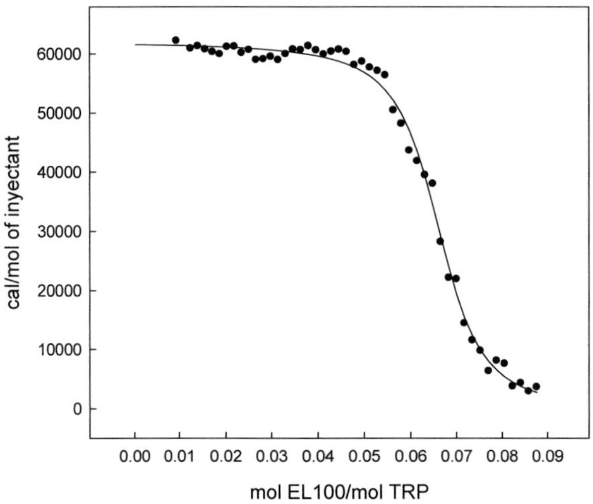

Figure 18. ITC measurements of TRP with EL100.

Binding parameter	**Value**
n (molar ratio) [mol TRP/mol EL-100]	15.22 ± 0.05
K (affinity constant) [M^{-1}]	$9.8\ 10^{6} \pm 7.\ 10^{5}$
ΔH° (kcal/mol)	62.1 ± 0.3
ΔS°(e.u.)	240 ± 10
ΔG° (kcal/mol)	- 9.59 ± 0.05

Table 6. Binding parameters of the TRP- EL100 interaction from ITC experiments.T= 25°C.

A value of 15 mol of protein per mol of polyelectrolyte was found for the complex EL100-TRP formation. The high value of the affinity constant demonstrated that both molecules interact strongly with each other. The ΔH was normalized per mol of protein; therefore, heat value of 62.14 Kcal/mol of protein is yielded. The positive value of ΔH indicates that the interaction between EL100 and TRP requires consuming of heat form de medium. The ΔS° value obtained was positive as a result of the increase of the disorder of the system due to release of structured water molecules.

EL100 is a charged polymer which also contains a hydrophobic framework in its linear chain. For such a complicated system it is not clear to what extent non-electrostatic forces contribute to the observed complexation behavior. Besides, the value of ΔS° was positive indicating that the disorder of the system increased.

ITC experiment performed in presence of NaCl confirmed the results obtained by turbidimetry (data not shown). The values of heat measured during the experiment of titration are similar that obtained when studying the dilution of the polyelectrolyte. This result is indicating that the TRP and the EL100 are not interacting when NaCl 1.00 M is added to the buffer.

Thermodynamic parameters were according to hydrophobic interactions between TRP and EL100. However, ITC and turbidimetric titrations experiments were altered in salt presence. It would demonstrate that the mechanism of interaction between these two molecules involves both hydrophobic and electrostatic interactions.

5. Conclusions

Experimental conditions of charged polyelectrolyte-protein complex formation may be determined by turbidimetric measurements, but are necessary to complement it for calorimetric techniques.

DSC measurements show that the Tm and denaturalization heat of some proteins may increase or not change in the polymer presence and the complex unfolded according to a two-state model.

In general, ΔH° and ΔS° of complex formation obtained by ITC have negative when protein and polyelectrolyte are oppositely charged (electrostatic interaction). Nevertheless, the thermodynamic functions can be positive as a result of the interaction between hydrophobic backbone of polymers and aromatic amino acids. Moreover, if ionic strength modifies this insoluble complex formation, a mechanism of interaction may involve both hydrophobic and electrostatic interactions.

The calorimetric techniques (ITC and DSC), turbidimetry and enzymatic activity studies provide useful quantitative information about the interaction of proteins and charged polyelectrolytes in aqueous solution. The knowledge of the nature of this interaction is essential for the application of the complex formation in protocols as proteins isolation strategy, immobilization or in purification of a target protein.

Author details

Diana Romanini, Mauricio Javier Braia and María Cecilia Porfiri
Laboratory of Physical Chemistry Applied to Bioseparation. College of Biochemical and Pharmaceutical Sciences, National University of Rosario (UNR), Rosario, Argentina

Acknowledgement

We thank Prof Watson Loh, Institute of Chemistry, State University of Campinas (UNICAMP), Campinas, SP, BRAZIL for performing DSC and ITC measurements. We also thank Prof. G. Picó, Prof. B. Nerli and Prof. B. Farruggia for useful discussions.

6. References

[1] Kumara, A Srivastavaa A, Yu Galaevb I, Mattiasson B, (2007) Smart polymers: Physical forms and bioengineering applications. Progress in Polymer Science 32: 1205-1237.

[2] Wang, Y.; Gao, Y.; Dubin, P., (1999) Protein Separation via Polyelectrolyte Coacervation: Selectivity and Efficiency, Biotechnology Progress, 12: 356-362.

[3] Arvind, L.; Aruna, N.; Roshnnie, J.; Devika, T. (2000) Reversible precipitation of proteins on carboxymethyl cellulose, Poocess Biochemistry 35: 777-785.

[4] Weinbreck, F.; De Vries, R.; Schrooyen, P.; De Kruif, C.G. (2003) Complex Coacervation of Whey Proteins and Gum Arabic, Biomacromolecules, 4: 293-303.

[5] Gupta V., Nath S., Chand S., (2002) Estimation of proteins in the presence of polyethylenimine. Biotechnol. Lett., 22: 927.

[6] Manzur A, Spelzini D, Farruggia B, Romanini D, Picó G (2007) Polyethyleneimine phosphate and citrate systems act like pseudo polyampholytes as a starting method to isolate pepsin. Journal of Chromatography B. 860: 63-68.

[7] Mattison, K.W.; Dubin, P.L.; Brittain, I.J., (1998) Complex Formation between Bovine Serum Albumin and Strong Polielectrolytes: Effect of Polymer Charge Density, Journal of Physical Chemistry B, 102: 3830-3836.

[8] Cooper C, Dubin P , Kayitmazer A, Turksen S,Current, (2005) Polyelectrolyte–protein complexes Opinion in Colloid & Interface Science 10: 52–78.

[9] Pessoa Jr., A.; Vahan Kilikian, B.;Purificação de Produtos Biotecnológicos, Ed. Manole Ltda., cap. 2. (2005).

[10] Hilbrig F, Freitag R, (2003) Protein purification by affinity precipitation, J Chrom B 790: 79–90.

[11] Arroyo M., (1998) Inmobilized enzymes: Theory, methods of study and applications. Ars Pharmaceutica, 39: 23-39.

[12] Krajewska B., (2004) Application of chitin- and chitosan-based materials for enzyme immobilizations: a review Enzyme Microb Tech 35: 126–139.

[13] Saskia Lindhoud. (2009) Polyelectrolyte Complex Micelles as Wrapping for enzymes – Tesis, 206 p.

[14] Esposito E., Cervellati F., Menegatti E., Nastruzzi C., Cortesi R., (2002) Spray dried Eudragit microparticles as encapsulation devices for vitamin C, Int J Pharm 242: 329–334.

[15] Braia, M Tubio, G Nerli, B Loh W, Romanini, D., (2012) Analysis of the interactions between eudragit® l100 and porcine pancreatic trypsin by calorimetric techniques. Int J Biol Macromol 50: 180-186.

[16] Porfiri M. C., Picó G., Farruggia B., Romanini D., (2010) Insoluble complex formation between alpha-amylase from Aspergillus oryzae and polyacrylic acid of different molecular weight. Proc. Biochem. 45: 1753-1756.

[17] Tsuboi A., Izumi T., Hirata M., J. Xia, P. Dublin E. Kokufuta, (1999) Complexation of Proteins with a Strong Polyanion in an Aqueous Salt-free System Langmuir 12: 6295-6303.

[18] Fornasiero F., Ulrich J., Prausnitz J.,(1999) Molecular thermodynamics of precipitation. Chem. Eng. Process 38: 463-475.

[19] Patrickios, C, Sharma, L, Armes, S, Billingham, N. (1999) Precipitation of a Water-Soluble ABC Triblock Methacrylic Polyampholyte: Effects of Time, pH, Polymer Concentration, Salt Type and Concentration, and Presence of a Protein. Langmuir 15: 1613-1620.

[20] Foreman T, Khalil M, Meier P, Brainard J, Vanderberg L, Sauer N (2001). Effects of charged water-soluble polymers on the stability and activity of yeast alcohol dehydrogenase and subtilisin carlsberg. Biotechnol Bioeng 76: 241–246.

[21] Braia, M. , Porfiri, M.C., Farruggia, B., Picó G., Romanini, D. (2008) Complex formation between protein and poly vinyl sulfonate as a strategy of proteins isolation. Journal of Chromatography B, 873: 139-143.

[22] Fasman G. D. (1996) Circular dichroism and the conformational analysis of biomolecules. Plenum press 738p.

[23] Sturtevant J, (2001) Biochemical applications of differential scanning calorimetry. Annu. Rev. Phys. Chem. 38 463-488.

[24] Jha N, Kishore N., (2009) Binding of streptomycin with bovine serum albumin: Energetics and conformational aspects.Thermochim. Acta 482: 21-29.

[25] Kim W., Yamasaki Y., Kataoka K., (2006) Development of a Fitting Model Suitable for the Isothermal Titration Calorimetric Curve of DNA with Cationic Ligands. J. Phys. Chem. B 110: 10919-10925.

[26] Romanini D, Braia M, Giatte Angarten R, Loh W, Picó G, (2007) Interaction of lysozyme with negatively charged flexible chain polymers. J Chrom B, 857: 25-31.

[27] Beynon R, Bond J.S., Proteolytic Enzymes, Practical Approach, Oxford University Press, 2001.

[28] Aschaffenburg R., Blake C., Dickie H., Gayen, S., Keegan R., Sen A. (1980). The crystal structure of tortoise egg-white lysozyme at 6 Å resolution .Biochim Biophys Acta 625: 64-71.

[29] Privalov P.L..(1979) Stability of Proteins Small Globular Proteins. Adv. Protein Chem. 33: 167-241.

[30] Santos A. Santana M., Gomidea, F. Miranda A, Oliveira, J., Vilas Boas F., Vasconcelos, A., Bemquerer M. ,Santoro M., (2008) Physical-chemical characterization and stability study of α-trypsin at pH 3.0 by by differential scanning calorimetry Int. J. Biol.Macromol. 42: 278–284.

Chapter 6

Insights into the Relative DNA Binding Affinity and Preferred Binding Mode of Homologous Compounds Using Isothermal Titration Calorimetry (ITC)

Ruel E. McKnight

Additional information is available at the end of the chapter

http://dx.doi.org/10.5772/54061

1. Introduction

1.1. Drug-DNA interactions

Many biologically significant compounds have been known, for several decades now, to bind non-covalently to nucleic acids.[1-7] Ever since the discovery of the structure of DNA in the 1950s, DNA has been a target for many therapeutic compounds. Several of these compounds have been found to bind to DNA while interfering with the activity of many vital enzymes and protein factors involved in DNA metabolism. Others cleave DNA or cause DNA cross-linking (for example, cisplatin) interfering with cell division and leading to apoptosis. As a result, several DNA-binding compounds have been identified as therapeutic agents in especially the anti-cancer and anti-pathogenic classes. Some of the most notable members of these classes include the Streptomyces derived anthracyclins e.g., daunomycin (daunorubicin) and doxorubicin, have been used for decades, initially as antibiotics, then mainly as antitumor agents.[8] Other known DNA binding agent include mitoxantrone, which has been particularly useful in the treatment of breast cancers, the glycopeptide antibiotic bleomycin which has been used in the treatment of Hodgkin's lymphoma and testicular cancer, amsacrine, bisantrene and various porphyrin derivatives. Even though many of these compounds have exhibited therapeutic potency, there still exist the accompanying unwanted side-effects, due mainly to the lack of selectivity and DNA targeting. Now, even after decades of studies of drug-DNA interactions, the existence of deleterious side-effects remains a huge area of concern and presents the main barrier for progress within the field. So, the question of whether a certain molecule will bind to a specific DNA sequence is currently being probed by several research groups. If we are to approach the problem from a fundamental level, such efforts must rely heavily on a fundamental understanding of the predominant contributions to drug-DNA interactions. Although ligand-DNA interactions have been studied, so far there have only been a handful of studies that have probed the factors that govern DNA-binding using homologous series of compounds. This information is especially relevant to the rationale design of novel therapeutics with improved efficacy and specificity. The proposed chapter is designed to yield an understanding of how various features of small molecules govern their binding to DNA and will provide insights into ligand-DNA interactions by studying binding trends within homologous series of compounds. Several studies have suggested that some DNA binding molecules exhibit more than one binding mode while binding in a sequence specific manner. In fact, some researchers have proposed that the therapeutic efficiency of these drugs may be linked to their ability to exhibit mixed binding modes.[9,10] These modes primarily involve intercalation, where planar aromatic molecules slide between adjacent DNA base pairs resulting in significant perturbation of the DNA, and/or minor-groove binding, where molecules with the requisite flexibility and isohelicity with the DNA minor groove are able to fit into the DNA groove, usually with no significant change in the structure of the DNA.

For many years now, microcalorimetry has been utilized to decipher the complete thermodynamic profiles for a number of drug-DNA complexes.[11] Isothermal titration calorimetry (ITC) has been successfully used to parse the thermodynamics of the interactions between drug molecules and DNA.[2,3,11] ITC is regarded as the "gold standard" approach for the determination of binding affinity data in biomolecular interactions. ITC has been used to determine the comprehensive thermodynamic profile of these interactions, by determining enthalpy change (ΔH) directly (usually in the presence of an excess of the macromolecular binding sites), while determining equilibrium binding constant (K), and number of binding sites (n) by model-fitting routines. Free energy change (ΔG) and ultimately entropy change (ΔS) are determined from the known thermodynamic relationships ($\Delta G = -RT\ln K$) and ($\Delta G = \Delta H - T\Delta S$), respectively. Furthermore, heat capacity change (ΔCp) may be determined from ITC measurements of ΔH over a range of different temperatures ($\Delta Cp = d\Delta H/dT$).[11]

In this chapter, we show how isothermal titration calorimetry can be successfully utilized to determine relative DNA binding efficacy, as well as the preferred DNA binding mode for a selection of homologous series of compounds. By comparing the DNA binding characteristics of homologous compounds under identical conditions, we can make robust conclusions as to the most important driving force governing the interaction of ligands to DNA. The chapter will describe two classes of homologous compounds; the naphthalene diimides and chalcogenoxanthyliums. However, the chapter will mainly focus on the naphthalene diimide series. The NDI scaffold has been used by several researchers to design therapeutically significant candidates [12-20] and are used in our studies as model systems to gain additional insight into the binding of "threading" intercalators to DNA. These symmetrical molecules have two substituents on either side of the intercalating moiety, thus necessitating the threading through or involvement of the side chain during binding

(**Figure1**).[13,14,17-19] In this geometry, one side chain occupies the minor groove, while the other lies in the major groove. According to some researchers, a threading intercalator has a number of potential advantages since the contact with both DNA grooves provide additional potential sites for recognition and targeting.[12-20] The NDI scaffold has provided a versatile template for the design of many promising derivatives.[12-20]

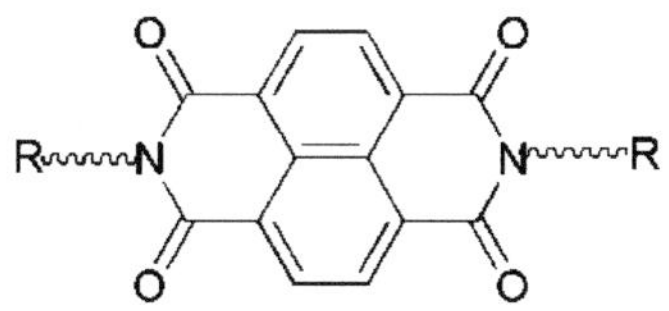

(R = ethyl- or propyl-amino side chain)

Figure 1. General structure of the naphthalene diimides in this study.

Some NDI derivatives have also been found to selectively bind non-standard structural forms of DNA such as triplexes and G-quadruplexes, which are normally transient and unstable.[21-25] Stabilization of DNA triplexes formed when oligonucleotides (normally referred to as triplex formation oligonucleotide or TFO) bind to DNA duplexes, have been explored in anti-gene therapeutics where expression of deleterious DNA sequences are suppressed by the binding and stabilization of complimentary TFO sequences.[15,21] Formation of transient G-quadruplexes in G-rich sequences have been found to be prominent in telomeres, G-rich ends on chromosomes that protects indispensable genes from being depleted, as well as preventing unwanted chromosomal fusions.[23-25] As a result, some compounds (e.g., certain NDI derivatives) can bind to and stabilize these telomeric G-quadruplexes can block access to these sequences by telomerase enzymes, which are responsible for extending and protecting telomeres and have been found to be over-expressed in 80% of cancers cells.[24,25] G-quadruplexes have also been found to be prominent in promoter regions, especially in the promoters of oncogenes such as the *c-myc and Ras genes, were,* found to be directly linked to the formation of certain cancers.[24,25] Stabilization of these G-quadruplexes in oncogene promoter regions can block access by RNA polymerase, and ultimately blocking expression of these deleterious genes. It is therefore important that we continue to probe ligands systems in order to increase our understanding of the driving force behind ligand–DNA interactions, and to use this knowledge to control their preferred binding mode and sequence.

The NDI compounds were synthesized as previously described.[26] As mentioned above, the NDI scaffold has been used by several groups to design biologically significant compounds.[12-20] In the current series, the quaternary amino group in each side chains is close enough (ethyl- and propyl-amino linker) to the naphthalene core group to allow electrostatic contact with the DNA. Therefore, the cationic quaternary amino groups are close to the DNA when the core ring system intercalates between DNA base pairs. As a result, there is a greater probability for electrostatic interaction with the phosphates in the DNA backbone. The NDI molecules of this study have two substituents on either side of the central naphthalene moiety and differ mainly in substituent size and hydrophobicity. That means, each compound should adopt a threading molecular geometry when bound to DNA via intercalation. Threading NDI compounds analogous to the ones in this study have been under investigation for several years as potential therapeutic (especially anticancer) compounds that bind to DNA with improved sequence-selectivity due to their interactions with both DNA grooves.

1.2. Chalcogenoxanthyliums

Although stored blood used during surgery and in blood transfusion is generally safe due to improved screening procedures, there is still a chance (a slight risk) that pathogens within the stored blood may be transmitted from donor to recipient.[27,28] This can occur if the blood was collected from an infected individual before there were detectable levels of the causative pathogen. As a result, there remains a need to develop protocols in which to reduce the risk of pathogen transmission, if only in a precautionary or preventative role.

Photodynamic therapy (PDT) is one approach that has been considered as a viable means in which to purge stored blood samples of deleterious pathogens.[27-32] In PDT, light is used along with endogenous oxygen and an appropriate photosensitizer (a molecule that has the ability to absorb light energy, i.e., photoexcitation, and transfer this energy to another chemical entity inducing a change) to treat or reduce an affliction. Photosensitizers are effective mainly because they are able to absorb appropriate light energy and produce excited triplet states at which time they can transfer energy to ground state oxygen (which is also triplet state) via intersystem crossing producing very toxic singlet oxygen species. PDT has been used for years in the treatment of certain cancers and lesions, as well as age-related macular degeneration. Photofrin, a hematoporhyrin belonging to the porphyrin class of compounds, is probably the most well-known and has been used for many years to treat bladder cancers. Other photosensitizers include those in the clorin class (e.g., photochlor), as well as dyes such as phthalocyanine.

PDT can be applied in pathogen reduction, especially in the removal of microbial material from blood products. In this application, PDT is normally referred to as photodynamic antimicrobial chemotherapy (PACT). Compounds containing the xanthylium core (rhodamines and rosamines), are among some of the most highly touted class of compounds being considered for PACT and have been explored by Wagner, Detty and coworkers.[27,31,32] These compounds have been found to selectively accumulate in cancer cells and mitochondria, and have also been considered as p-glycoprotein inhibitors and mitochondrial stains.[33,34] However, the parent rhodamines and rosamines have been mostly ineffective due to short-lived and low yield of triplet excited state upon photo-excitation. Detty and coworkers have synthesized a group of related chalcogenoxanthyliums (**Figure 2**) that are based on the parent compounds.[33,34]

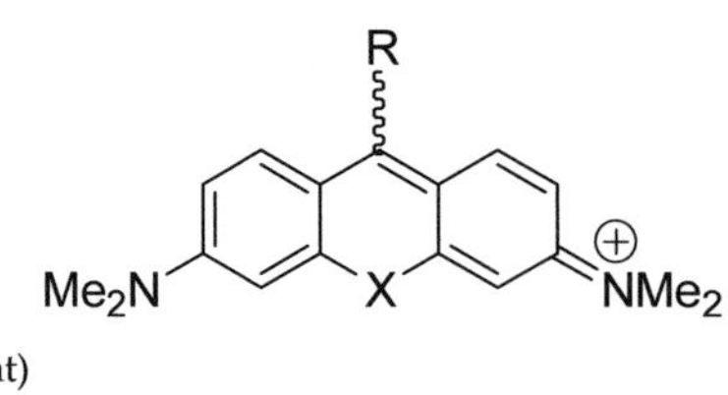

(X = chalcogen, R = 9-aryl substituent)

Figure 2. General structure for the chalcogenoxanthylium derivatives.

These chalcogenoxanthylium derivatives represent an improvement over the parent rhodamine and rosamine since the inclusion of the heavier chalcogen (e.g.,S and Se instead of O) provides the known "heavy atom effect" which increases the production of long-lived excited triplet states.[33] Furthermore, the substituents (for example, a 2-thienyl instead of a phenyl) in the 9-position can be "tuned" such that they absorb light at wavelength that avoids hemoglobin attenuation. [33-35]

To date, PACT has been mostly unsuccessful due largely to 1) low efficacy against pathogens, and 2) unwanted background hemolysis of red blood cells.[32] Both these shortcomings are mostly due to the non-specific actions of the photosensitizers when exposed to the requisite light. To circumvent these problems, photosensitizers that are able to target the pathogenic DNA relative to the red blood cells are currently being explored.[32,35] One approach to target these pathogens in the presence of red blood cells is to use photosensitizers that bind strongly to the pathogenic DNA, since mature red blood cells do not contain organelles or genomic nucleic acids.[32,35] The chalcogenoxanthylium derivatives are advantageous to use since their substituents can be tuned such that 1) they absorb light in a spectral region where light attenuation by hemoglobin absorption is avoided, 2) increased yield of singlet excited state that are responsible for destruction of pathogens, and 3) their planarity and hydrophobicity can be altered to monitor the effects on their interaction with DNA. Thus, offering greater opportunity to potentially reduce the incidence of background hemolysis. The DNA binding efficacy and preferred mode of binding of a series of chalcogenoxanthylium dyes were investigated by isothermal titration calorimetry (ITC).[35]

1.3. Preference for AT-rich vs GC-rich DNA

In an effort to decipher the preferred DNA binding mode for compounds in this study, a preference for an AT- vs GC-rich sequence will be determined. In order to differentiate preferences for intercalation and/or groove binding, the binding of the compounds of this study to $[poly(dAdT)]_2$ and $[poly(dGdC)]_2$ were examined by ITC. Figure 3 shows the structure of $[poly(dAdT)]_2$ and $[poly(dGdC)]_2$ used in this study. It has long been established that known groove binding compounds (e.g., distamycin, berenil, and DAPI) show a strong preference (an order of magnitude or greater) for binding to $[poly(dAdT)]_2$ relative to $[poly(dGdC)]_2$.[6] The lower affinity for GC-rich sequences shown by groove binders is largely due to their restricted access to the minor groove of GC sequences caused by the protruding 2-NH_2 group of guanine. Intercalators are only expected to be affected by this if a substituent is placed into the minor groove during formation of the intercalation complex. It is however expected that compounds that exhibit mixed binding mode (i.e., intercalation and groove binding) will exhibit less (<10) of a preference for the AT sequence.[28,35,36]

5'-ATATATATATATATATATAT-3'   5'-GCGCGCGCGCGCGCGCGCGC-3'
3'-TATATATATATATATATATA-5'   3'-CGCGCGCGCGCGCGCGCGCG-5'

Figure 3. AT-rich and GC-rich DNA sequences used in this study.

In this chapter, calorimetric data of naphthalene diimide derivatives binding to both calf thymus DNA (ctDNA), as well as AT- and GC-rich DNA sequences will be described. The binding characteristics of selected chalcogenoxanthylium derivatives will also be compared. In an effort to gain insight into the involvement of a minor groove vs. intercalative binding mode, the binding of the compounds to [poly(dAdT)]$_2$ and [poly(dCdG)]$_2$ sequences (using ITC) will be discussed. The calorimetric approach will be validated using known/classical DNA intercalating and minor groove binding compounds. Although the main focus of the chapter will be analysis of calorimetric data, the data will also be compared to studies on the same systems using ITC-independents approaches such as a gel electrophoresis based topoisomerase I DNA unwinding assays and fluorescence-based ethidium bromide displacement studies.

2. Methods and materials

2.1. Isothermal titration calorimetry

In general, calorimetric titrations were carried out on a MicroCal VP-ITC (MicroCal Inc., Northampton, MA), an instrument specifically suited for studying biomolecular interactions. The MicroCal VP-ITC is a highly sensitive microcalorimeter that operates on a power compensation method, whereby heat exchange processes occurring in a sample cell is compared to a reference cell as the instruments keeps the two cell temperatures identical. This results in exothermic processes yielding negative (less than zero) peaks as the instrument decreases the power (μcal/s) supplied to the sample cell relative to the reference cell, while endothermic processes yield positive (greater than zero) peaks as the instrument increases the power supplied to the sample cell compared to the reference cell. The intensity of each peak corresponds to the quantity of the heat exchange. The data was analyzed using the Origin 7.0 software provided by the manufacturer. Experiments were typically run at either 25-30 °C in MES00 buffer (1×10^{-2} M MES (2(*N*-morpholino) ethanesulfonic acid) containing 1×10^{-3} M EDTA, with the pH adjusted to 6.25 with NaOH) for runs involving calf thymus DNA (ctDNA, ultrapure, Invitrogen). Due to the relative instability of the shorter DNA sequence (particularly the AT-rich sequence), experiments using the [poly(dAdT)]$_2$ and [poly(dCdG)]$_2$ sequences (Midland Certified Reagents, Midland , TX) were done in MES40 (i.e., MES00 with 40 mM NaCl). Note, the MES00 buffer was selected for the ctDNA studies due to its low concentration of salt; this would presumably promote stronger binding interactions which would yield more intense peaks and thus better signal/noise ratios. Typically, either 5 or 12 μL of the drug solution (typically 5-7 $\times 10^{-5}$ M) was injected into a buffered solution of DNA (typically 10-15 $\times 10^{-6}$ M in bp, 1.4 mL) over 20-24 s at 240 s intervals using a 250 μL syringe rotating at 300 rpm. The initial delay (hold period before injections) was set at 240 s. Before use, samples were degassed at 20 °C using the ThermoVac accessory (provided by MicroCal Inc.). During the isothermal titration experiments, all injections manifested in a peak that corresponded to the decrease in the power (μcal/s) supplied to keep the temperatures of the sample and reference cells (containing either water or MES buffer) the same for each injection and represented the heat given off. Note, in all cases, titration peaks corresponded to negative power compensation resulting from exothermically driven processes. In each case, response signals were corrected for the small heat of dilution associated with the titration of the drug into the MES buffer. The heat of dilution for titrating MES buffer into DNA was found to be negligible. The heat released (i.e., area associated with negative peaks) on binding of the drug to DNA sites was directly proportional to the amount of binding. A binding isotherm of heat released (kcal/mol of injectant) versus the molar ratio ([drug]/[DNA] in bp) was constructed and the data fitted by non-linear least square fitting analysis to an appropriate model.

2.2. Topoismerase I DNA unwinding assay

Typically, 0.24 μg of supercoiled pUC19 plasmid DNA was incubated with human topoisomerase I (Topo I) enzyme (Invitrogen) for 5 min at 37 ºC. An appropriate amount of the compound of interest was then added (all except for the first two tubes, which serves as controls) and the reaction mixture incubated for a further 1 h at 37 ºC. After incubation, the reaction was terminated using 0.5% SDS and 0.5 mg/mL proteinase K. Both the enzyme and compound of interest was then extracted using a mixture of phenol:chloroform:isoamyl alcohol (25:24:1). The remaining DNA sample was then run on an agarose gel (1%) at 75 V for 3 h, stained with ethidium bromide for 45 min and photographed.

2.3. Ethidium bromide displacement assay

A solution of ethidium bromide (EtBr, 5×10^{-6} M, 1.0 mL) was pre-incubated with ultrapure calf thymus DNA (1×10^{-5} M in base pairs, 1.4 mL) obtained from Invitrogen. at room temperature (22-23 °C) for 15 min in MES00 buffer, pH 6.3. Aliquots of exactly 3 μL of the compound (7×10^{-5} M) were then titrated into the EtBr-DNA solution and the change in fluorescence measured (Photon Technology International fluorometer), after 3 min incubation periods (excitation 545 nm and emission 595 nm). The addition of 3 μL aliquots was continued until the DNA was saturated (i.e., no further change in fluorescence due to EtBr displacement). [28,36] Control experiments showed that the compounds (free or DNA-bound) had no significant background fluorescence at the excitation (545 nm) and emission (595 nm) wavelengths of EtBr.

3. Results and discussion

3.1. Using relative binding affinity for AT- vs GC-DNA to evaluate binding mode

In order to validate the approach of using relative preferences for AT vs GC to ascertain the preferred DNA binding mode, several known/classical DNA binding compounds were investigated using ITC. These include two compounds known to bind DNA via the minor groove, distamycin A and berenil, (Figure 4) and two compounds known to bind DNA via intercalation (ethidium bromide, normally regarded as the classical DNA intercalator, and daunomycin) (Figure 5).[2,3,6]

Isothermal titration calorimetric data for distamycin, berenil, daunomycin and ethidium bromide binding to the AT- and GC-rich sequences are shown in Figure 6. As can be seen from the raw data, both minor groove binders distamycin A and berenil show a strong preference for the AT-rich sequence relative to the GC-rich sequence. In fact, ITC signals for each compound binding to the GC-rich sequence was found to be negligible, showing only background signal that was associated with the heats of dilution when the compound was titrated into the cell buffer. Binding constants found for distamycin A and berenil binding to the AT-rich sequence were 2.20±0.4 x 10^7 M^{-1} and 1.76±0.3 x 10^6 M^{-1}, respectively.

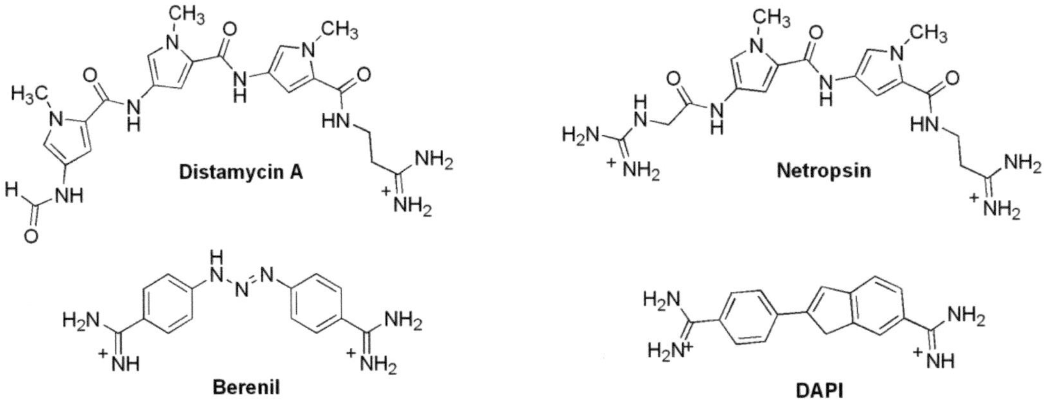

Figure 4. Structures of some common DNA minor groove binding compounds.

A different result was observed with the classical DNA intercalator, ethidium bromide and the known chemotherapeutic DNA intercalator, daunomycin. The isothermal calorimetric data for ethidium bromide and daunomycin showed binding to both the AT- and GC-rich sequences and indicated no significant preference for either sequence. Binding constants obtained for the AT-rich and GC-rich sequence were 1.78±0.5 x 10^5 M^{-1} and 3.38± 0.8 x 10^5 M^{-1}, and 2.93±0.63 x 10^6 M^{-1} and 3.24±0.60 x 10^5 M^{-1}, for ethidium bromide and daunomycin, respectively.

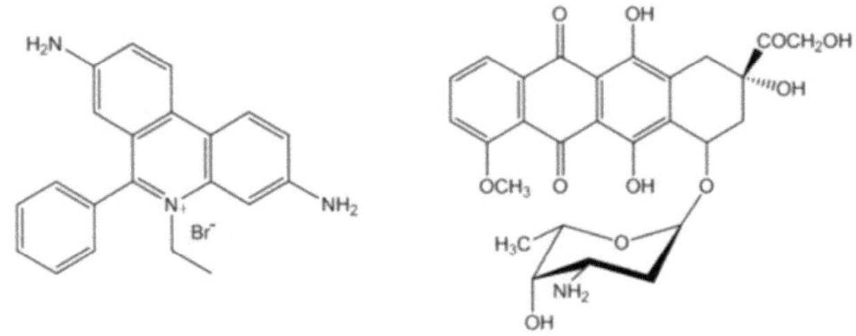

Ethidium bromide **Daunomycin**

Figure 5. Structures of two common DNA intercalators.

The results observed for distamycin A, berenil, ethidium bromide and daunomycin are consistent with both distamycin A and berenil binding via the minor groove, since each compound showed a significant preference for the AT-rich sequence, while as expected, ethidium bromide and daunomycin bind DNA via intercalation, since neither exhibited a significant preference. This is suggested from the fact that the minor groove in the GC-rich sequence is partially blocked by the protruded 2-NH_2 group of guanine, preventing a compound that uses the minor groove for DNA binding to be blocked.[6] This is not the case for the AT-rich sequence. On the other hand, a compound such as ethidium bromide and daunomycin which intercalates into DNA by sliding between adjacent base pairs, will essentially be unimpeded from binding to either the AT or GC-rich sequences. The reported binding modes for distamycin A, berenil, ethidium bromide and daunomycin herein are also consistent with the wealth of literature reports on the binding mode for all four compounds, thus validating our approach.[2,3,6,8,10,37,38]

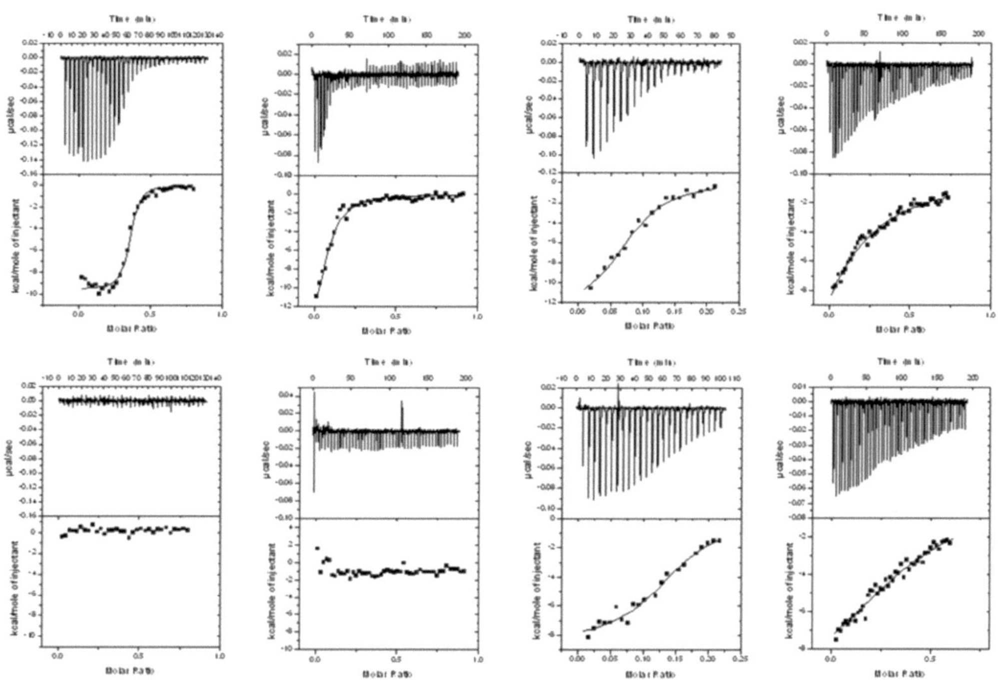

Figure 6. Calorimetric data for the titration of 60 μM of the compounds (from left to right): distamycin A, berenil, daunomycin and ethidium bromide into 15 μM of AT-rich DNA (top), GC-DNA (bottom) at 30 °C. Binding isotherms (heat change vs drug/DNA molar ratio) were obtained from the integration of raw data and fitted to a "one-site" model

4. Binding of the NDI derivatives to DNA using ITC

As was mentioned earlier, the NDI class of compounds is an excellent model system to study DNA binding interactions especially since it offers a useful platform for the syntheses of many homologous series. These molecules are threading intercalators in which side chains on either side of the main intercalating moiety provides the potential for specific recognition sites on the DNA.[12-19] The specific roles of a variety of substituents will be studied with a focus on identifying differential contributions from each moiety. A quaternary amino group will also be incorporated into each NDI side chain to provide electrostatic interaction with the negatively charged DNA backbone. The NDI derivatives in this chapter (Figure 7 and 8) were synthesized by Dixon and coworkers and have three main motifs.[26,36]

Ring Size: Compounds that contains a ring (*N*-methyl pyrrolidine or *N*-methyl piperidine) at the distal end of the side chain, as well as possessing different ring size. To date, the effect of ring size on intercalator-DNA interaction has been mostly unexplored. We have studied two homologous types of NDI that differ by a single carbon with five- vs six-membered heterocyclic rings. These are at identical distances from the main intercalating moiety. The rings are non-aromatic and are not expected to stack with the DNA bases. However, they differ in steric bulk which should have implications during binding. One could predict that **NDI-3** will show relatively lower binding affinity than **NDI-4**, however, the increase in bulkiness might have only kinetic consequences.[26] We are interested in determining whether these substituent variations might have an effect on both the preferred DNA binding mode adopted by these compounds, and consequently their relative DNA binding affinity. We also compare the effect of having a cyclic structure in the side chain vs. acyclic alkyl substituents.

Linker length: Insights into the effect of changing the linker length for two sets of NDI derivatives (acyclic aliphatic and cyclic aliphatic substituents) will be discussed. In both sets of compounds, the side chain linker length differ by one carbon (ethyl vs propyl). This means the quaternary amino group (present in all the NDI compounds) is one carbon further from the main intercalating core for the propyl linker. For the acyclic aliphatic derivatives, we compare the trimethyl-propylamino (**NDI-1p**) and dibutylmethyl-propylamino (**NDI-2p**) derivatives (that are one carbon further from the main intercalating core) to the trimethyl-ethylamino (**NDI-1e**) and dibutylmethyl-ethylamino (**NDI-2e**) derivatives. For the cyclic aliphatic compounds, the ethyl-linker-containing compound, **NDI-3**, is compared to the propyl-linker-containing **NDI-5**. Given the difference in steric bulk of the cyclic aliphatic compared to the acyclic derivatives, there may be steric consequences. We will also be able to gain insights into acyclic vs. cyclic substituent effects on DNA binding.

Substituent length/size: In order to gain additional insights into the role of the side chain size, an analysis of the DNA binding characteristics of NDI compounds that differ in the size and side chain linker-length of their alkyl-amine side chain will also be done. As the length and size of the substituent increases, so does the steric bulk. Of course, hydrophobicity also increases with substituent size. We seek to investigate the effects of steric bulk and hydrophobicity on DNA binding of these derivatives. Hydrophobicity has been reported to be a significant driving force in DNA binding interactions with binding increasing with hydrophobicity.[2,3] We have investigated the relative importance of this factor using a model NDI series in which size/steric contributions should also be a factor. Both hydrophobicity and molecular size increases along the series. If hydrophobicity is the predominant driving force, then one might expect binding to increase with size/hydrophobicity. However, if a size/steric effect dominates, binding should decrease.

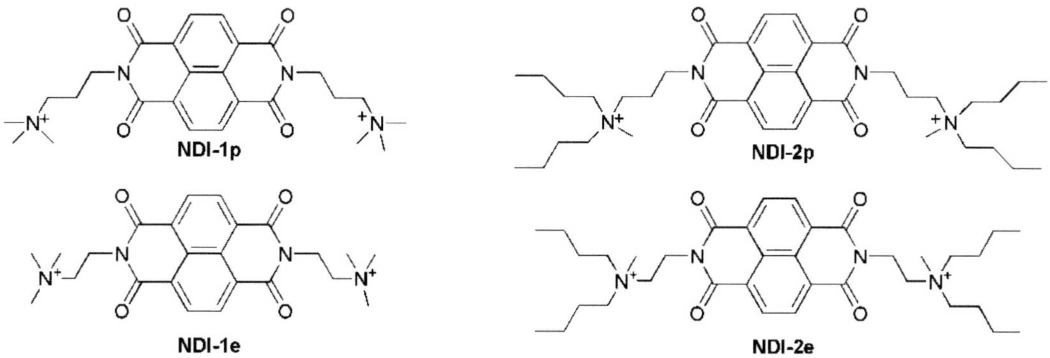

Figure 7. Representative structure for the acyclic NDI derivatives, showing the ethylamino (ethyl linker) side chain derivatives [**NDI-1e** (bottom, left) and **NDI-2e** (bottom, right)] and the propylamino (propyl linker) derivatives [**NDI-1p**,(top, left) and **NDI-2p** (top, right)].

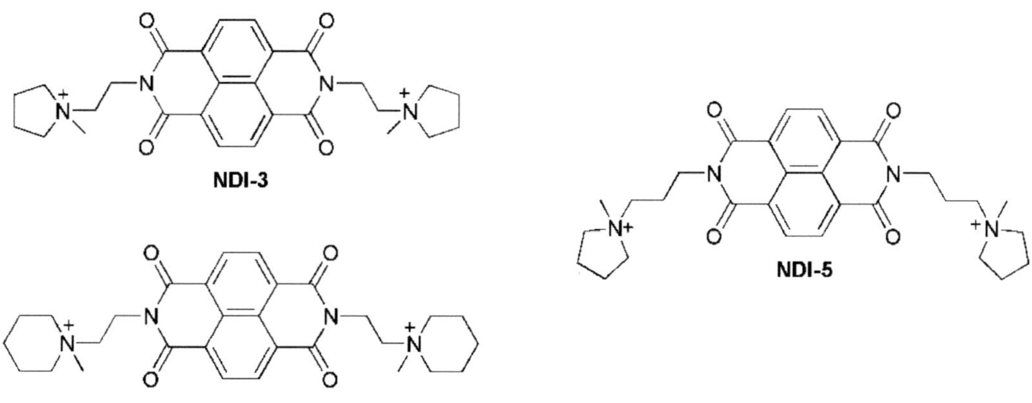

Figure 8. Structures for the cyclic ethylamino NDI derivatives [**NDI-3** (top, left) and **NDI-4** (bottom, left)] and the cyclic propylamino derivatives **NDI-5** (right). Both **NDI-3** and **NDI-5** contain a side chain *N*-methyl pyrrolidine five-membered ring, while **NDI-4** contains a six-membered *N*-methyl piperidine ring.

5. Effect of the side chain ring size and linker length

In general, the inclusion of a cyclic component in the side chain resulted in a biphasic raw calorimetric data for each cyclic NDI compound binding to DNA (**Figure 9**). The raw calorimetric data for the cyclic compounds binding to ctDNA were best defined by a model that assumes two types of binding sites (K_1, K_2) and argues for the involvement of at least two different types of binding modes for the compounds with ring-containing substituents. This biphasic binding mode has been reported by us for larger members of an acyclic substituent NDI series and will be briefly discussed below.[36] In general, the higher binding constant (K_1) for the cyclic NDI derivatives was in the order (~10^7-10^8 M^{-1}), while a lower binding constant (K_2) was in the order of (~10^6 M^{-1}) for compounds possessing the *N*-methyl pyrrolidine ring (**NDI-3** and **NDI-5**) binding to ctDNA. The DNA binding constant for the *N*-methyl piperidine derivative (**NDI-4**) showed strong but significantly lower binding constants compared to the *N*-methyl pyrrolidine derivatives. Calorimetric data for the two compounds that differed only in ring size (*N*-methyl pyrrolidine vs *N*-methyl piperidine) showed that **NDI-3** (*N*-methyl pyrrolidine substituent) exhibited larger binding constants (K_1 = 1.17± 0.3 x 10^8 M^{-1}, K_2 = 5.6±0.65 x 10^6 M^{-1}) as compared to larger **NDI-4** (K_1=1.70 ±0.4 x 10^7 M^{-1} and K_2=3.26 ±0.54 x 10^6 M^{-1}) when binding to ctDNA. Thus both binding constants were lower for the larger *N*-methyl piperidine derivative. Given that **NDI-4** possesses a more bulky *N*-methyl piperidyl substituent suggest that steric hindrance may play a role here. Studies on a series of NDI containing acyclic substituents also found two binding constants; one binding constant was found to be as a result of intercalation, while the other was found to via a non-intercalative mode, presumably via the DNA minor groove.[36] Assuming that the two binding modes found for the cyclic substituents here are similar (given the similarities between the two sets of compounds), the two binding modes found here for the cyclic derivatives are presumed to also be via intercalation (lower binding constant , K_2) and minor groove binding (higher binding constant, K_1). In which case, **NDI-4** with its larger more bulky substituent may find difficulty in sliding itself through adjacent base pairs. This is of course a requirement for intercalation. Furthermore, given that these compounds possess two substituents on either side of the main intercalating moiety (i.e., threading), one substituent must "thread" through DNA base pairs if it is to adopt an intercalating geometry. Since both binding constant decrease for the *N*-methyl piperidine derivative, the second binding mode (i.e., presumed to be via the minor groove) is also affected sterically.

According to the calorimetrically determined binding constants, the linker length did not appear to have significant role for these cyclic side chain containing derivatives since **NDI-5** (ethylamino/ethyl linker) and **NDI-3** (propylamino/propyl linker) both had very similar binding constants for both the higher and lower binding sites (K_1 = 1.08 x 10^8 M^{-1}, K_2 =5.1±0.72 x 10^6 M^{-1} and K_1 = 1.17± 0.3 x 10^8 M^{-1}, K_2=5.6±0.65 x 10^6 M^{-1}, respectively). It therefore appears that the size of the cyclic substituent plays a greater role than the substituent linker in determining the DNA binding affinity.

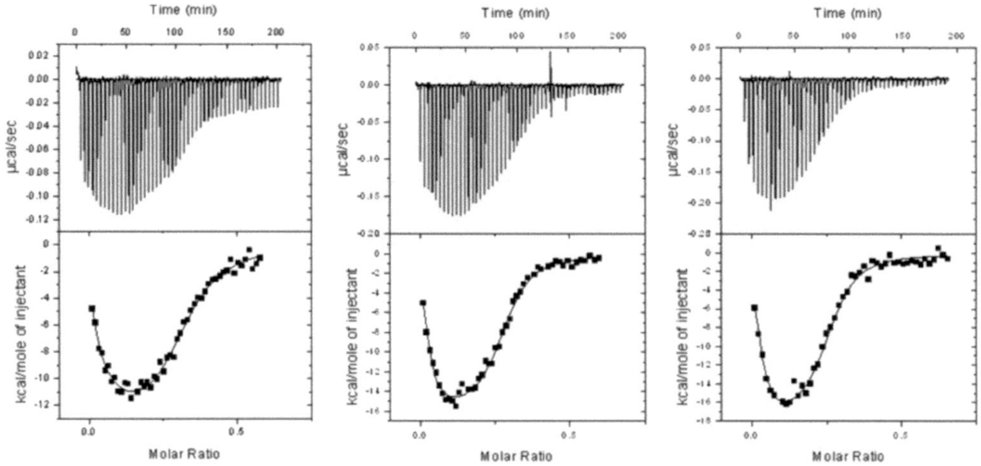

Figure 9. Calorimetric data for the titration of 60 µM **NDI-4** (left), **NDI-3** (middle) and **NDI-5** (right) into 12.5 µM of ctDNA at 30 °C. Binding isotherms (heat change vs drug/DNA molar ratio) were obtained from the integration of raw data and fitted to a "two-site" model.

6. NDI binding mode determination via AT vs GC preference of the cyclic NDI derivatives

Calorimetric studies were carried out to evaluate preferences for AT vs GC-rich sequences, in an effort to detect a possible minor groove binding mode, implied by the above result (**Figure 10**). In general, the cyclic NDI derivatives possessing the ethylamino linker (**NDI-3** and **NDI-4**) exhibited a roughly two-fold preference (2.0x for **NDI-3**, 2.4x for **NDI-4**) for the AT-rich sequence relative to the GC-rich sequence (**Table 1**). The difference in affinity for the AT- vs GC-rich sequence is similar to at least one of the acyclic substituent NDI compounds (a dipropylmethyl ethylamino side chain) reported in an earlier study (see section on the acyclic derivatives below), and which was suggested to have a second minor groove binding mode.[36] We therefore suggest here that the cyclic NDI derivatives **NDI-3** and **NDI-4** does have a minor groove binding mode. It is interesting to note that **NDI-4** showed a slightly greater preference for the AT-rich DNA sequence compared to **NDI-3**, implying a greater involvement of minor groove binding for **NDI-4**.

The cyclic derivative with the propylamino linker (**NDI-5**) exhibited even less of a preference (~1.4x). However, the difference between the **NDI-5** binding constant for AT vs GC-rich sequences could be considered as the same within experimental error. This result may imply that there is a greater contribution from non-intercalative binding from the cyclic ethylamino derivatives relative to the propylamino derivatives. This result is somewhat similar to what was observed in the series of acyclic substituent NDI derivatives. However, given the small differences in AT vs GC-sequences, this would warrant additional studies to confirm.

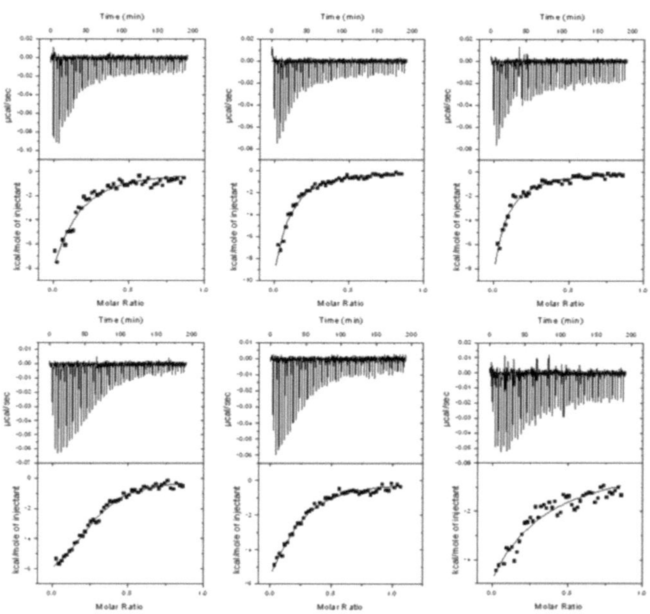

Figure 10. Calorimetric data for the titration of 60 µM **NDI-5** (left), **NDI-3** (middle) and **NDI-4** (right) into 15 µM of AT-rich DNA (top), GC-DNA (bottom) at 30 °C. Binding isotherms (heat change vs drug/DNA molar ratio) were obtained from the integration of raw data and fitted to a "one-site" model.

7. DNA binding mode determination using ITC-independent approaches

Two additional approaches were also utilized to determine the binding mode involved for the compounds in this study. These were a topoisomerase I DNA unwinding assays (topo assay) and ethidium bromide (EtBr) displacement studies. A brief description of the two techniques is in order. Briefly, the topo assay exploits the ability of topoisomerase I enzyme to relax supercoiled DNA, such as the plasmid pUC19 used in all our studies.[39,40] Under the conditions of our topo assay, supercoiled plasmid pUC19 DNA is first relaxed by using excess topoisomerase I enzyme and then is exposed to the compound under study. After extraction of the compound and enzyme, a compound that was bound via intercalation will cause re-supercoiling of the plasmid DNA. Re-supercoiling is due to the change in DNA linking number that accompanies relaxation by the topoisomerase enzyme and occurs to the extent to which the intercalator molecule was initially bound.[39,40] An intercalating molecule will perturb the DNA such that the DNA will unwind, causing the topoisomerase enzyme (which is present in excess) to relax the DNA, thus changing the linking number. The extent to which DNA unwinding occurs will be dependent upon the extent to which DNA binding occurs, thus the minimum concentration needed to cause complete re-supercoiling will be indicative of how much compound was initially bound and thus the relative binding affinity. Conversely, minor groove binders should not induce appreciable re-supercoiling due to negligible DNA unwinding upon binding, and negligible change in DNA linking number.

With the EtBr displacement assays, EtBr, a known intercalator is first bound to DNA, occupying its intercalative sites. The compound of interest is then added to determine whether it is able to displace EtBr from its intercalative sites. Displacement is monitored by a decrease in EtBr–DNA fluorescence.[28,36,41] It is well established that the fluorescence yield of EtBr is enhanced significantly when it binds to DNA. This occurs as EtBr occupies its intercalative sites between bases in the DNA molecule. However, in the presence of another intercalator, there is competition for a limited/defined number of intercalation sites. As the other intercalator molecules are added, they begin to displace EtBr from these intercalative site, increasing the amount of free (unbound) EtBr. This is usually observed as a decrease in EtBr-DNA fluorescence.

Both the topo assay and ETBr displacement assays has been used by our group, as well as other groups, to determine DNA binding mode of DNA binding compounds.[28,35,36,41,42] To validate the topo assay approach, we have run assays on several known DNA binding compounds. These include the classical DNA intercalator, EtBr, and known minor-groove binding compounds such as distamycin A, berenil. **Figure 11** shows representative topo assay for EtBr and berenil.[42] As is expected, the classical DNA intercalator, EtBr, was able to elicit significant re-supercoiling back to the levels of the control (lane 1), whereas, the known minor groove binding compound was unable to do so, even at the high concentrations. In fact, essentially no re-supercoiling was observed for berenil, confirming its known minor groove binding mode. Similarly, we have done validation studies of our EtBr displacement assay, by running studies on DNA binding compounds in which their binding modes have already been established (e.g., distamycin A and berenil). As expected, none of the compounds known to be minor groove binders were able to cause appreciable displacement of EtBr from its intercalative sites, consistent with these compounds binding to non-intercalative sites (**Table 1**). However, the intercalating molecules were able to displace ethidium bromide effectively, as was evident by the significant decrease in EtBr-DNA complex fluorescence.

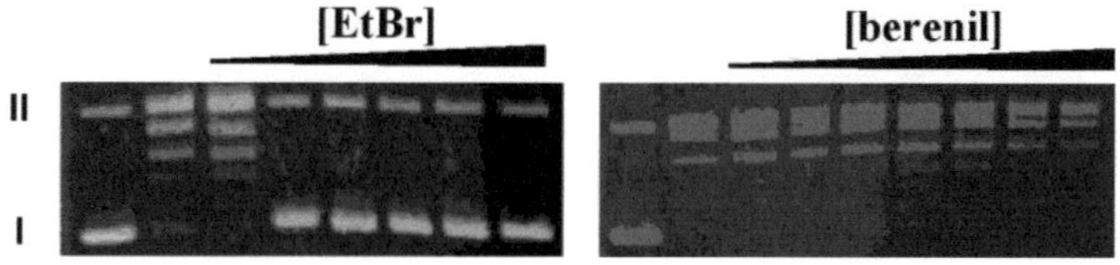

Figure 11. Topo I assay of of the classical DNA intercalator **EtBr** (left) and the known minor groove binder **berenil** (right) using 5 units of the topoisomerase enzyme. From left of each gel, lanes 1 contain only DNA (no compound nor topoisomerase) and serve as controls. Lanes 2 contain DNA and topoisomerase, but no compound. Remaining lanes contain DNA, topoisomerase and increasing concentrations of compound (taken from [42]).

8. Binding mode determination of cyclic NDI derivatives via ITC-independent approaches

When topo assays were done on the NDI derivatives containing the cyclic amino side chains (**NDI-5**, **NDI-3**, and **NDI-4**), each compound was able to cause re-supercoiling, indicating that intercalation is indeed involved in the binding of each compound to DNA. This was not surprising since NDI compounds are known to bind to DNA via intercalation.[17-19] However, **NDI-3** was better able to elicit re-supercoiling than **NDI-5**, which was in turn better than **NDI-4**. That is, while **NDI-3** was able to cause complete re-supercoiling of our plasmid DNA at ~6 μM, **NDI-4** requires >10 μM for complete re-supercoiling (**Table 1**). This suggests that the binding of **NDI-3** involves more of an intercalative mode than either **NDI-5** or **NDI-4** and is consistent with what was observed in the ITC studies for these compounds described above. That is, the strength of the lower binding constants (K_2) was in the order **NDI-3>NDI-5>NDI-4**. The lower binding constant (K_2 in this report), has been found to be that of the intercalative binding mode for a similar series of NDI.[36] It appears that the bulkier *N*-methyl piperidine is either sterically hindering intercalation, or forcing **NDI-4** into a more non-intercalative binding mode, while **NDI-5**, with its propylamino linker, exhibits lower affinity for the DNA as compared to **NDI-4**. The lower binding affinity associated with the propylamino linker will be addressed later.

The behavior of the cyclic substituent NDI compounds in the ITC studies and topo assays were also consistent with our EtBr displacement studies which showed that **NDI-3** was better able to displace EtBr from its intercalative sites; thus **NDI-3** caused a greater decrease in EtBr fluorescence compared to **NDI-4** (**Table 1**). Our EtBr displacement assays also showed that **NDI-5** was able to displace EtBr to the same extent as **NDI-3**, suggesting that both have a similar intercalative strengths. Again, this is consistent with what we observed in the ITC and topo assay studies described above. That is, **NDI-5** and **NDI-3** having very similar K_2 (ITC), and both eliciting re-supercoiling of the plasmid DNA at roughly similar concentrations.

Compound	K_b (ctDNA) (10^6 M^{-1}) (ITC)[a]	K_b (AT) (10^6 M^{-1}) (ITC)[b]	K_b (GC) (10^6 M^1) (ITC)[b]	Topo assay (10^{-6} M)[c]	EtBr displacement Assay (ΔF/μL)[d]
distamycin A	---	2.20 ± 0.4	---	---	6
Berenil	---	1.76 ±0.3	---	---	31
EtBr	---	0.18±0.05	0.34±0.08	2	---
daunomycin	---	2.9 ±0.6	3.24±0.6	---	---
NDI-1e	15±3	1.11±0.27	1.17±0.10	3.5	400
NDI-2e	78±23 3.9±1.1	1.38±0.15	0.38±0.09	>6.7	358
NDI-1p	1.22±0.16	10.1 ±0.7	---	3	---
NDI-2p	0.57 ± 0.2	8.7 ±0.4	---	5	---
NDI-3	117±30 5.66±0.65	0.5±0.09	0.25±0.05	6	949
NDI-4	17.0±4 3.26±0.54	0.39±0.08	0.16±0.04	>10	777
NDI-5	104± 35 5.10±0.72	1.16±0.24	0.85±0.09	>6	1030

[a] MES00 buffer, pH 6.25
[b] MES40 buffer, pH 6.25.
[c] Minimum concentration required for complete re-supercoiling.
[d] Decrease in EtBr fluorescence per μL of compound added.
Data for acyclic **NDI-#**e series are from reference [36].
Data for the acyclic **NDI-#**p series are from reference [42].

Table 1. Representative DNA binding affinity data for the compounds in this study.

9. Effect of the length/size of the substituent and linker length (Ethyl vs propyl)

As was reported by us, data obtained from calorimetric measurements show that the length/size of the substituent plays a significant role in both the preferred binding mode and relative binding affinity of the compounds of these studies.[36] The compounds of this study showed tight binding to DNA with values of K_b between 10^5 to 10^8 M^{-1}, presumably dependent on their preferred mode of binding to DNA. Figure 12 shows the calorimetric data for the four acyclic NDI derivatives (with ethylamino side chain linkers) binding to ctDNA. In that report, we found only a single type of binding constant (binding mode) for the smallest compound in the series (containing a trimethyl-ethylamino side chain).[36] This

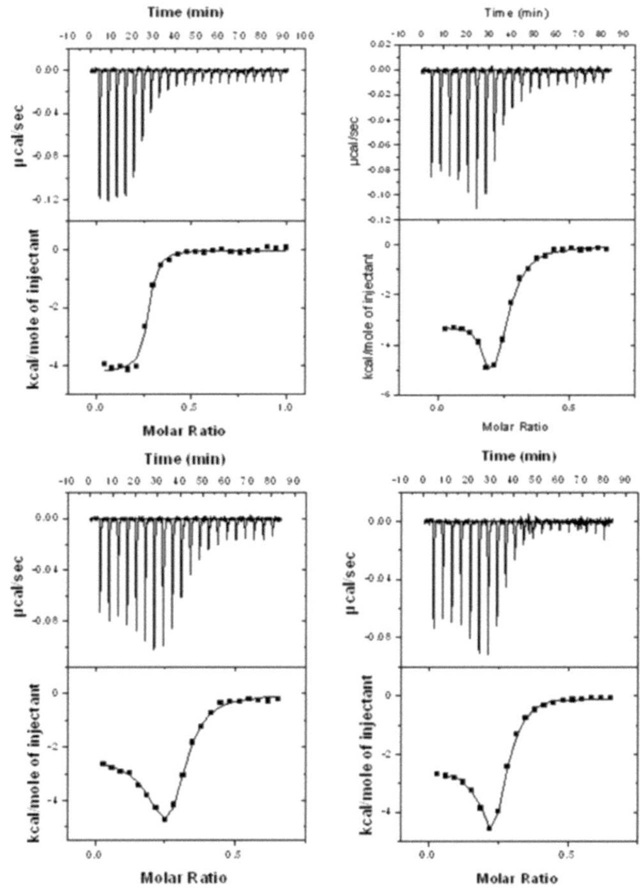

Figure 12. Calorimetric data (raw) for the acyclic ethylamino derivatives binding to ctDNA. In each case, 70 μM of the NDI was titrated into ctDNA (12.5 μM) at 30 °C. Data is shown for the trimethyl ethylamino derivative **NDI-1e** (top, left), diethylmethyl ethylamino derivative (top, right), dipropylmethyl ethylamino derivative (bottom, left), and the dibutylmethyl ethylamino derivative **NDI-2e** (bottom,right). Binding isotherm (heat change vs drug/DNA molar ratio) was obtained from the integration of raw data and fitted to either a "one-site" model (**NDI-1e**) and a "two-site" model (all others). The plot for **NDI-1e** and **NDI-2e** were taken from [36].

was indicated by a single-phased binding isotherm that was well-defined by a one-site binding model. Larger members of the ethylamino series (diethylmethyl-, dipropylmethyl- and dibutylmethyl-ethylamino substituents) adopted two binding modes; a lower affinity binding mode between 3-4 x 10^6 M^{-1} and an additional higher affinity binding mode of between 31 - 78 x 10^6 M^{-1}.[36] This was indicated by a biphasic binding isotherm that was fitted well to a two-site model; one site associated with intercalation and the other associated with minor groove binding. If we compare the results found for the smallest compound in that study, with that of the smallest compound in another study done by us with a similar NDI series with propylamino linker instead,[42] we find that only a single type of binding mode and binding constant (**NDI-1e**, K = 15±3 x 10^6 M^{-1} and **NDI-1p**, K = 1.2

±0.16 x 10^6 M^{-1}) is found for the smallest member of the series whether the side chain is ethylamino or the one-carbon longer, propylamino substituent.[36,42] However, whereas larger members in the acyclic ethylamino series exhibited a dual binding mode, neither compound in the acyclic propylamino series (referred to as **NDI-1p** or **NDI-2p** in this chapter) was found to exhibit more than one binding mode. Additionally, in the ethylamino series, we observe that the relative binding affinity trend for the ethylamino series increased with substituent size. However, this feature was not observed in the propylamino series (one carbon longer on both sides of the main intercalating moiety), since **NDI-2p** with its dibutylmethyl-propylamino substituent exhibited a lower binding constant (0.57±0.17 x 10^6 M^{-1}) compared to smaller homolog (**NDI-1p**) which had a binding constant of (1.2±0.16 x 10^6 M^{-1}), a binding mode attributed to intercalative binding.[42] It is also clear that DNA binding affinity was in general greater for the ethylamino derivative, although some of this difference may be attributed to slightly different experimental conditions used in the two studies. It therefore implies that this small structural difference may (1) enable an additional mode of binding, i.e., a linker length that is one carbon shorter resulted in an additional binding mode, as well as (2), enhance the DNA binding mode by greater than an order of magnitude. An explanation for this could be that steric effects may dominate for the propyl-amino series, resulting in lower DNA binding, especially for the larger members, while hydrophobic and binding mode preferences may be dominant in the ethyl-amino series. The propyl amino derivatives are of course longer especially since the additional carbon linker is on both sides of the molecule, given that these are threading compounds. The longer (more dangling) molecular structure may make it more difficult to thread through adjacent base pairs. However, in the case of the ethylamino series, the solution for the larger substituents appear to be adoption of an additional DNA binding mode. Hydrophobic contributions may also play a role.

10. Comparison of binding mode for NDI derivative with ethyl vs propyl linker using topo assay

Comparing the two series with different linker-length (i.e., ethylamino vs propylamino derivatives), it is also interesting to note that generally higher concentrations of the ethylamino derivatives were required for re-supercoiling, despite having higher binding constants as determined by ITC.[36,42] A striking example of this is seen from the fact that more than 6.5 uM of **NDI-2e** (K_1= 78±23 x 10^6 M^{-1} and K_2=3.9±1.1 x 10^6 M^{-1}) was required for supercoiling, while the corresponding propylamino derivative **NDI-2p** with a significantly lower binding affinity (K = 0.57±0.17 x 10^6 M^{-1}) required only 5 uM. Again, some of this may also be attributed to different experimental conditions. For example, a greater excess of the topoisomerase enzyme was used in the assays for the ethylamino series. However, this factor alone cannot account for the lack of associated re-supercoiling ability given the disproportionately higher DNA binding constants for the ethylamino derivatives. Overall, a side by side comparison of the topo assay results for the two series (ethylamino vs propylamino) suggests that the ethylamino derivatives displays relative re-supercoiling capabilities that are less than expected based on their significantly higher binding affinities. Since the ability to elicit re-supercoiling is primarily based on an intercalative ability, this argues for a greater involvement of non-intercalative binding for ethylamino derivatives relative to their propylamino counterparts.

11. Binding of the chalcogenoxanthylium derivatives to DNA

In an effort to further corroborate our DNA binding characterization approach used for the NDI derivatives discussed above using a different/independent homologous series, we will also briefly describe DNA binding studies of a homologous series of chalcogenoxanthylium derivatives to DNA, reported by our group.[35] The chalgenoxanthylium derivatives in this study were synthesized by Detty and coworkers and have been implicated as potential candidates for therapy against blood-borne pathogens.[27,31-33,35].

Using this independent system as a comparison, we have also found that the results obtained from ITC were consistent with that found using topo assay and EtBr displacement studies. These studies have found that the nature of the substituent attached to the main xanthylium core plays a directing role in the preferred binding mode and accompanying DNA binding affinity.[35] While some of the compounds bind to DNA either through intercalation or via the minor groove, some exhibited mixed-binding modes.[35] Excerpts from the DNA binding studies for selected chalcogenoxanthylium derivatives (**Figure 13**) will now be discussed.

In that report, ITC studies suggested that both the 9-substituent and the identity of the chalcogen play a role in the preferred binding mode and ultimately, the relative DNA binding constant.[35] With a 9-2 thienyl substituent attached to the main xanthylium core (e.g., 2-Se), there appeared to be a preference for intercalation. This was implied from the fact that compounds containing the 9-2 thienyl substituent showed no preference for the AT-rich sequence, a feature that would be typical for a minor-groove binder. The 9-2 thienyl also bound to calf thymus DNA with lower affinity as compared to the 9-phenyl derivatives (e.g.,1-Se).[35] DNA intercalators are known to have lower DNA binding affinity as compared to minor-groove binders,[2] so this result may be due to a greater contribution from minor groove binding (i.e., less contribution from intercalation) with the 9-phenyl series. In addition to exhibiting a 2-3 higher binding constant compared to the corresponding 9-2 thienyl derivative, the 9-phenyl series exhibited a slight preference (2-3 times) for binding to $[poly(dAdT)]_2$ as compared to the $[poly(dGdC)]_2$. Here again, a possible minor groove binding was implied, since it is known that compounds that bind solely to the DNA minor groove generally show a preference for binding to AT-rich sequences relative to GC-rich sequences due to the occlusion from the GC-rich minor groove by the protruded 2-NH_2 group of guanine.[6] As mentioned for the NDI series discussed earlier, it is expected that compounds that bind both via the DNA minor groove and by intercalation (i.e., mixed binding modes) will show a factor of <10 preference for AT-rich sequences, depending on the relative contribution from intercalation (i.e., the difference will be less as contributions from intercalation increases). The chalcogenoxanthylium derivative bearing a 9-(2-thienyl-5-diethylcarboxamide) substituent (compound 10) exhibited the strongest preference for the $[poly(dAdT)]_2$ sequence. In fact, compound 10 showed essentially no binding to the $[poly(dGdC)]_2$ sequence, while binding to $[poly(dAdT)]_2$ with a K of $2.3 \pm 0.4 \times 10^6\ M^{-1}$.[35]

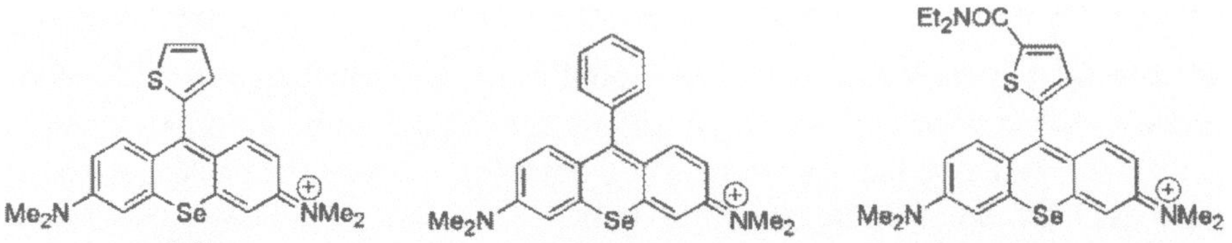

Figure 13. Structures of selected chalcogenoxanthylium derivatives reported in [35]. The 9-2 thienyl derivative (**2-Se,** left) shown bind mostly via intercalation, while **1-Se** derivative (middle) is a mix-binder, and compound **10** binds primarily via the DNA minor groove.

12. Binding mode determination of chalcogenoxanylium derivatives via ITC-independent approaches

As was done for the NDI series discussed earlier, several independent (non-ITC) studies (ethidium bromide displacement and topo assay) were also carried out on the chalcogenoxanthylium derivatives in this study.[35] This was done in an effort to gain additional insights into the preferred DNA binding mode suggested by ITC.

Results from topo assays have been reported by us.[35] These results were in general consistent with the ITC studies on these compounds. We will now report new EtBr data on chalcogenoxanthylium derivatives discussed in this chapter that supports both ITC and topo assay studies.

Further evidence for the preferred DNA binding modes were also observed during ethidium bromide displacement assays on selected members of the chalcogenoxanthylium compounds binding to DNA. These were the seleno derivatives from the 9-2 thienyl series (**2-Se**), the 9-phenyl series (**1-Se**), and compound **10** (suggested to have primarily a non-intercalative binding from the ITC studies). While compound **2-Se** and **1-Se** were both able to cause dislodgement of ethidium bromide from DNA, **2-Se** was markedly better able to do so (decrease in fluorescence per µL of compound added was: **2-Se** = 711, **1-Se** = 581, compound **10** = 350. Considering that part of the change is fluorescence for the compounds was due to accompanying dilution during the titration, we see here that the order of intercalative ability is **2-Se>1-Se>10**. This order mirrors the results from both ITC and topo assay which showed that **2-Se** was a better intercalator than **1-Se**, which was in turn better than compound **10.** This implies that **2-Se** is a stronger intercalator than **1-Se**, consistent with both the ITC and topo assay studies. Compound **10** caused relatively small decreases in ethidium bromide fluorescent (less than any of the NDI derivatives in this study) indicating that it is not a potent displacer of ethidium bromide from its intercalative sites, suggesting that compound **10** is not a strong intercalator, again consistent with the minor groove binding mode implied by both the ITC and topo assay. Given the higher binding constant found for **1-Se** relative to **2-Se** using ITC, if **1-Se** was primarily a DNA intercalator, it would exhibit a greater ability (compared to **2-Se**) to dislodge the classical DNA intercalator EtBr from its binding sites. The fact that it did not, strongly supports the idea that the binding of **1-Se** to DNA involves other binding modes. Also, the fact compound **10** showed little ability to dislodge ethidium bromide from DNA, while having the highest binding constant (as determined by ITC studies in an earlier study [35]), supports the idea of compound **10** involving significant non-intercalative DNA binding (presumably, via the minor-groove).

13. Conclusions

In this chapter, we have shown how ITC can be successfully used to characterize both the preferred DNA binding mode for series of compounds, as well as their relative DNA binding affinity. For this, we have selected two homologous series of compounds; series of symmetrical NDI threading intercalators in which the side chains are mandatorily involved in DNA binding, and a series of chalcogenoxanthilum derivatives. Both classes of compounds have been shown to have biological activity.

While the homologous NDI derivatives in this study all exhibit DNA intercalative abilities, the substituent on either side of the main intercalating core does play a significant role in determining whether or not additional modes are adopted. This occurs because these compounds require a threading geometry when intercalating between DNA base pairs, i.e., there is a necessity for the side chain to "thread" DNA. The side chains are therefore forced to direct DNA binding. We have found that the cyclic (non-aromatic) substituent at the distal end of a side chain play a significant role in both the DNA binding affinity and the preferred mode of binding. Larger ring sizes face steric barriers and have lower DNA binding affinity. The larger rings may however force additional (non-intercalative) binding modes to be involved. Additional studies may be needed to fully understand the full effects of ring size. Future studies may involve attachment of aromatic rings instead of non-aromatic rings in this study. Having flat aromatic rings on the substituent may enhance site recognition and DNA binding due to the ability to stack. We have also found that even a small modification in the linker length in NDI side chain play a significant role during binding of NDI derivatives of acyclic aliphatic side substituents to DNA. In fact, on comparing side chains with an ethyl linker vs those with a propyl linker, it was found that the ethyl linker could enhance DNA binding by more than an order of magnitude. Possession of the ethyl linker also enabled an additional DNA binding mode of higher affinity. The NDI scaffold therefore represent a versatile template for the design of many promising derivatives with enhanced DNA affinity and have implications in the rationale design of DNA binding compounds with improved site recognition capabilities.

Using an independent system for comparison, the approach of using ITC to study binding to both ctDNA and AT vs GC-rich sequences, was shown to be an efficient and consistent approach in the determination of relative DNA binding affinity and preferred DNA binding mode. The ITC studies were well corroborated by ITC-independent studies such as topo assays and EtBr displacement studies, thus exhibiting the efficacy of our approach.

Author details

Ruel E. McKnight
Department of Chemistry, State University of New York at Geneseo, 1 College Circle, Geneseo, NY, USA

Acknowledgement

The author is very grateful to Professors Dabney Dixon and Michael Detty for providing the naphthalene diimide and chalcogenoxanthylium compounds, respectively, for this study. I would also like to acknowledge the very diligent students who have contributed to this work over the years (Douglas Jackson, Luke Marr, Kevin Siegenthaler, Eric Reisenauer, Sadia Sahabi, Shivani Polasani, Bilgehan Onogol, Manuel Pintado, Aaron Gleason, and James Keyes).

14. References

[1] Bailly C, Colson P, Hénichart J-P, Houssier C (1993) The different binding modes of Hoechst 33258 to DNA studied by electric linear dichroism. Nucleic Acids Res. 21:3705-3709.
[2] Chaires, JB (1997) Energetics of Drug-DNA Interactions. Biopolymers 44: 201-215.
[3] Haq I (2002) Thermodynamics of Drug-DNA Interactions. Arch. Biochem. Biophys. 403:1-15.
[4] Barcelo, F.; Capo, D.; Portugal, J. (2002) Thermodynamic characterization of the multivalent binding of chartreusin to DNA. Nucleic Acids Res. 30:4567-4573.
[5] Tse WC, Boger DL (2004) A Fluorescent Intercalator Displacement Assay for Establishing DNA Binding Selectivity and Affinity. Acc. Chem. Res. 37:61-69.
[6] Ren J, Chaires JB (1999) Sequence and Structural Selectivity of Nucleic Acid Binding Ligands. Biochemistry 38:16067-16075.
[7] Denny WA (2002) Acridine Derivatives as Chemotherapeutic Agents. Curr. Med. Chem. 9: 1655-1665.
[8] Chaires, JB (1996) Molecular Recognition of DNA by Daunomycin Advances in DNA Sequence Specific Agents. 2:141-167.
[9] Chen AY, Liu LF (1994) DNA Topoisomerases: Essential Enzymes and Lethal Targets. Annu. Rev. Pharmacol. Toxicol. 34, 191-218.
[10] Pilch DS, Kirolos MA, Liu X, Plum GE, Breslauer KJ (1995) Berenil [1,3-bis(4'-amidinophenyl)triazene] Binding to DNA Duplexes and to a RNA Duplex: Evidence for Both Intercalative and Minor Groove Binding Properties. Biochemistry 34:9962-9976.
[11] Haq I, Jenkins T, Chowdhry B, Ren J, Chaires JB (2000) Parsing Free Energies of Drug-DNA Intercalation. Methods Enzymol. 323: 373-405.

[12] Hampel SM, Sidibe A, Gunaratnam M, Riou JF, Neidle S (2010) Tetrasubstituted Naphthalene Diimide Ligands with Selectivity for Telomeric G-Quadruplexes and Cancer Cells. Bioorg. Med. Chem. Lett. 20:6459-6463.

[13] Sato S, Hirano A, Takenaka S (2010) Selective Immobilization of Double Stranded DNA on a Gold Surface Through Threading Intercalation of a Naphthalene Diimide having Dithiolane Moieties. Analytica Chimica Acta. 665:91-97.

[14] Lee J, Guelev V, Sorey S, Hoffman DW, Iverson BL (2004) NMR Structural Analysis of a Modular Threading Tetraintercalator Bound to DNA. J Am Chem Soc. 126:14036-14042.

[15] Gianolio DA, Segismundo JM, McLaughlin LW (2000) Tethered Napthalene-based Intercalators for Triplex Stabilization. Nucleic Acids Res. 28: 2128-2134.

[16] Liu ZR, Hecker KH, Rill RL (1996) Selective DNA Binding of (N-alkylamine)-Substituted Naphthalene Imides and Diimides to G+C-rich DNA. J Biomol Struct Dyn.14:331-339.

[17] Tanious FA, Yen SF, Wilson WD (1991) Kinetic and Equilibrium Analysis of a Threading Intercalation Mode: DNA Sequence and Ion Effects. Biochemistry 30:1813-1819.

[18] Wilson, W. D. DNA Intercalators. In DNA and Aspects of Molecular Biology; Kool, E. T., Ed.; Elsevier: New York, 1999; pp 427-476.

[19] Yen S, Gabbay E, Wilson WD (1982) Interaction of Aromatic Imides with Deoxyribonucleic Acid. Spectrophotometric and Viscometric Studies Biochemistry, 21:2070-2076.

[20] Sato S, Kondo H, Takenaka, S, (2006) Linker Chain Effect of Ferrocenylnaphthalene Diimide Derivatives on a Tetraplex DNA Binding. Nucleic Acid Symposium Series 50:107-108.

[21] Rusling D, Peng G, Srinivasan N, Fox K, Brown T, (2009) DNA Triplex Formation with 5-Dimethylaminopropargyl Deoxyuridine. Nucleic Acid Res 87:1288-1296.

[22] Cuenca F, Greciano O, Gunaratnam M, Haider S, Munnur D, Nanjunda R, Wilson W, Neidle S (2008) Tri- and tetra-substituted Naphthalene Diimides as Potent G-Quadruplex Ligands. Bioorg. Med. Chem. Lett.18:1668–1673.

[23] Laronze-Cochard M, Kim Y-M, Brassart B, Riou J-F, Laronze J-Y, Sapi J (2009) Synthesis and Biological Evaluation of Novel 4,5-Bis(dialkylaminoalkyl)-Substituted Acridines as Potent Telomeric G-Quadruplex Ligands, Eur. J. of Med. Chem. 44:3880–3888.

[24] Gonzalez, V, Hurley, L (2010) The c-Myc NHE III: Function and Regulation, Annu. Rev. Pharmacol. Toxicol. 50:111-129.

[25] Luedtke, N, (2009) Targeting G-Quadruplexes with Small Molecules. Chimia 63: 134-139.

[26] Steullet V, Dixon, DW (1999) Self-Stacking of Naphthalene bis(dicarboximide) Probed by NMR. Perkin Trans. 2:1547-1558.

[27] Wagner, S, Skripchenko, A, Donnelly, D, Ramaswamy, K, Detty, M. (2005), Bioorg. Med. Chem. 13:5927-5935.

[28] McKnight, RE, Ye M, Ohulchanskyy, TY, Sahabi S, Wetzel, BR, Wagner, SJ, Skripchenko A, Detty MR (2007) Synthesis of Analogues of a Flexible Thiopyrylium Photosensitizer

for Purging Blood-Borne Pathogens and Binding Mode and Affinity Studies of their Complexes with DNA. Bioorg. Med. Chem. 15:4406-4418.

[29] Wainwright M (1998) Photodynamic Antimicrobial Chemotherapy. J. Antimicrob. Chemother 42:13-28.

[30] Dougherty T, Gomer C, Henderson B, Jori G, Kessel D, Korbelik M, Moan J, Pend Q (1998) Photodyanamic Therapy. J. Natl. Caner. Inst. 90:880-905.

[31] Detty MR, Gibson SL, Wagner SJ (2004) Current Clinical and Preclinical Photosensitizers for use in Photodynamic Therapy. J. Med. Chem 47, 3897-3915.

[32] Wagner, SJ, Skripchenko A, Cincotta L, Thompson-Montgomery D, Awatefe H (2005) Use of a Flexible Thiopyrylium Photosensitizer and Competitive Inhibitor for Pathogen Reduction of Viruses and Bacteria with Retention of Red Cell Storage Properties. Transfusion 2005, 45, 752-760.

[33] Calitree B, Donnelly D, Holt J, Gannon M, Nygren C, Sukumaran D, Autschbach J, Detty M. (2007) Tellurium Analogues of Rosamine and Rhodamine Dyes: Synthesis, Structure, ^{125}Te NMR, and Heteratom Contribution to Excitation Energies. Organometallics 26:6248-6257.

[34] Gibson S, Hilf R, Donnelly D, Detty M (2004) Analogues of Tetramethylrosamine as Transport Molecules for and inhibitors of P-Glycoprotein-Mediated Multi-Drug Resistance. Bioorg. Med. Chem. 12:4625-4631.

[35] McKnight, RE, Onogul B, Polasani SR, Gannon MK, Detty MR (2008) Substituent Control of DNA Binding Modes in a Series of Chalcogenoxanthylium Photosensitizers as Determined by Isothermal Titration Calorimetry and Topoisomerase I DNA Unwinding Assay. Bioorg. Med. Chem. 16:10221-10227.

[36] McKnight, RE, Reisenauer, E, Pintado, MV, Polasani, SR and Dixon, DW (2011) Substituent Effect on the Preferred DNA Binding Mode and Affinity of a Homologous Series of Naphthalene Diimides, Bioorg. Med. Chem. Lett. 21:4288-4291.

[37] Barcelo F, Portugal J, (1993) Berenil Recognizes and Changes the Characteristics of Adenine and Thymine Polynucleotide Structures. Biophys. Chem. 47:251-260.

[38] Remata D, Mudd C, Berger R, Breslauer K (1993) Thermodynamic Characterization of Daunomycin-DNA Interactions: Comparison of Complete Binding Profiles for a Series of DNA Host Duplexes. Biochemistry 32:5064-5073.

[39] Pommier, Y, Covey J-M, Kerrigan D, Markovits J, Pham R (2007) DNA Unwinding and Inhibition of Mouse Leukemia L1210 DNA Topoisomerase I DNA Intercalators. Nucleic Acids Res. 15:6713-6731.

[40] Dziegielewski J, Slusarski B, Konitz A, Skladanowski A, Konopa (2002) Intercalation of Imidazoacridinones to DNA and its Relevance to Cytotoxic and Antitumor Activity. J. Biochem. Pharmacol. 63:1653-1662.

[41] Boger DL, Fink BE, Brunette, SR, Tse WC, Hedrick, MP (2001) A Simple, High-Resolution Method for Establishing DNA Binding Affinity and Sequence Selectivity. J. Am. Chem. Soc. 123:5878-5891.

[42] McKnight, R. E.; Gleason, A. B.; Keyes, J. A.; Sahabi, S. (2007) Binding Mode and Affinity Studies of DNA Binding Agents Using Topoisomerase I DNA Unwinding Assay. Bioorg. Med. Chem. Lett. 17:1013-1017.

Chapter 7

Thermodynamic Signatures of Macromolecular Complexes – Insights on the Stability and Interactions of Nucleoplasmin, a Nuclear Chaperone

Stefka G. Taneva, Sonia Bañuelos and María A. Urbaneja

Additional information is available at the end of the chapter

http://dx.doi.org/10.5772/54062

1. Introduction

Nucleoplasmin (NP) is a nuclear chaperone that mediates chromatin remodeling processes, such as sperm decondensation at fertilization [1]. In *Xenopus laevis* eggs, where it was first isolated, this highly acidic protein is thought to be in charge of nucleosomal core histones H2A/H2B storage. Upon fertilization, NP decondenses the densely packed sperm chromatin by means of extracting its specific basic proteins and replacing them with H2A/H2B, therefore enabling the assembly of somatic-type nucleosomes. NP is additionally involved in chromatin remodeling during early development, in particular it is required in the replication licensing mechanism, probably to extract linker-type histones from somatic chromatin, and can facilitate pluripotent cell reprogramming. NP (also designated NPM2) belongs to the nucleophosmin/nucleoplasmin family of histone chaperones [2]. Whereas NP roles have been particularly related to fertilization and embryogenesis, nucleophosmin (or NPM1) is ubiquitously and abundantly expressed in adult cells. It is enriched in the nucleolus, and serves multiple functions that affect cell growth and apoptosis, therefore disregulation of NPM1 is linked to several human cancers. Particular mutations of NPM1, that destabilize its structure, and cause its mislocalization to the cytoplasm, trigger acute myeloid leukaemia (AML). Apart from nucleophosmin and NP, the family includes the less known NPM3 and an invertebrate NPM-like.

NP is a homopentameric protein, composed of 200 residues, each subunit being built of two domains, namely core and tail. The core domain, corresponding to the N-terminal 120 residues, adopts an eight-stranded β-barrel structure, and is responsible for oligomerization, forming a ring with pentameric symmetry of 60 Å (diameter) and 40 Å (height) [3]. This compact core, shared by all NPM family members, confers an extreme stability to NP (see below). Probably, this is mainly due to a conserved network of hydrophobic interactions between the subunits, which, acting as a belt, firmly secures the pentamer. The C-terminal tail domain, instead, is conformationally flexible [4,5], therefore NP is considered "partially disordered" [6]. The tail harbors a segment rich in acidic residues (20 Asp and Glu within residues 120-150) termed "polyGlu", probably involved in histone binding , and a nuclear localization signal (NLS) that directs NP import into the nucleus.

The function of NP is activated through phosphorylation of up to 7 - 10 residues per monomer. NP phosphorylation degree correlates with *Xenopus* egg maturation, so that at the time of fertilization the protein is heavily phosphorylated and displays a maximal chromatin decondensing activity [5,7]. We have identified by mass spectrometry eight phosphorylation sites in natural NP: these phosphoresidues accumulate in flexible regions and loops, along both the core and tail domains, and cluster on a particular pole of the protein, known as distal face [8]. Phosphorylation causes a significant destabilization of the protein and we have made use of calorimetry (differential scanning calorimetry (DSC)) to dissect this effect, and its correlation with NP activation mechanism [9].

To fulfil its chromatin remodeling role, NP has to bind histones, basic proteins needed for packing of DNA. It acts as a reservoir for nucleosomal histones H2A/H2B, and is able to extract sperm specific basic proteins as well as linker-type histones, such as H1 from chromatin. The NP-mediated exchange of these more basic proteins with H2A/H2B results in a looser condensation state of chromatin [1,2]. We have thermodynamically described NP recognition of H2A/H2B and H5, a linker-type histone, by isothermal titration calorimetry (ITC) [10].

NP is the most abundant nuclear protein of *Xenopus* oocytes. Its nuclear import is mediated by the importin α/β heterodimer; in fact, NP is the prototypical substrate of this "classical" pathway, which is in charge of transport of most nuclear proteins [11]. Importin α recognizes a nuclear localization signal (NLS) in its substrates, which consists of a sequence segment with conserved basic residues, and itself associates to importin β [11,12]. The complex formed by importin α/β bound to the NLS cargo, traverses the nuclear envelope through the nuclear pore complexes. The transport relies on a gradient of the small GTPase Ran for directionality. The GTP-bound state of Ran is mostly nuclear and promotes the disassembly of the import complex once it reaches the nucleus, whereas in the cytoplasm, in the presence of Ran-GDP, the import complex formation is favoured [11,12].

Both importin α and β, belong to the karyopherin family of transport receptors, and their structures are constituted by a series of helical repeats, called ARM in the case of importin α and HEAT in importin β, that generate curved, flexible surfaces to bind their ligands. Importin α displays additionally a short N-terminal region for importin β binding (IBB domain) [12]. Most studies on the molecular basis of NLS recognition by nuclear transport receptors are so far limited to isolated domains of the proteins involved (e.g. using peptides corresponding to the NLS of NP and IBB of importin α) [13,14]. We have approached the thermodynamical characterization of the assembly of the complete complexes made of the full length proteins and have additionally built structural models of those import complexes [15]. NP loaded with histones can additionally incorporate importin α, generating large assemblies that could represent putative NP/histones co-transport complexes [16].

2. Calorimetry: Protein folding/unfolding and binding energetics

Calorimetry (DSC and ITC) is the most precise tool in the study of energetics of thermally-induced conformational transitions of proteins and their assembly with other molecules, small ligands or macromolecules. Extensive reviews have been published on the basic thermodynamic formalism, calorimeters' design and application of DSC [17-20] and ITC [18,20-24]. Moreover, surveys on ITC application are published annually since 2002 [25-28]. Calorimetry on proteins in general will be briefly summarized here and examples for NP will be thoroughly reviewed.

The excess heat capacity of a protein in solution, as a function of temperature, and the heat released or absorbed upon binding interactions are the quantities registered in the DSC and ITC experiments. Both the folding and binding events are described by the Gibbs free energy (ΔG), which determines the stability of the protein and the strength of association of molecules, respectively. The partitioning of ΔG into enthalpic (ΔH) and entropic (TΔS) terms is given by the basic thermodynamic equation:

$$\Delta G = \Delta H - T\Delta S \quad (1)$$

From the experimentally observed calorimetric curve, the DSC thermogram (typically an endothermic peak) and the ITC binding isotherm (exotherm or endotherm), a complete set of thermodynamic parameters of the studied folding/unfolding and binding phenomena is provided.

In DSC the values of the thermodynamic parameters of the folded-unfolded state equilibrium: transition midpoint temperature T_m (the temperature at the maximum of the excess heat capacity curve), the enthalpy of unfolding ΔH, calculated by the integral of the excess heat capacity function:

$$\Delta H = \int_{T_o}^{T_u} c_P dT \quad (2)$$

where T_o and T_u are the temperatures of the onset and completion of the transition, respectively, and c_P (the heat capacity change associated with unfolding) can be determined in a model-independent way [29]. In addition, the width at half-height of the transition $T_{m1/2}$ is a measure of the cooperativity of the transition from folded to unfolded state.

In ITC, the binding affinity K_b ($K_b = e^{-\Delta G/RT}$, R is the gas constant and T is the absolute temperature), the enthalpy change ΔH and the stoichiometry N of the binding interactions are determined by fitting the experimentally obtained binding isotherms assuming a model that well describes the binding process. While one binding constant describes a simple 1:1 molecular interaction, complex macromolecular recognition processes are described by model-independent macroscopic and model-dependent microscopic association constants, that account for the overall binding behaviour and for the association at each binding site, respectively [30,31]. Model independent analysis of more complex binding data, with two or more binding sites, based on general binding polynomial formalism, developed by Freire et al. [31], allows the type, independent or cooperative, of the binding interactions to be assessed. Methodology and analysis for heterotropic ligand binding cooperativy, i.e. for two or more different ligands binding to one protein has also been elaborated by Velázquez-Campoy et al. [32,33].

ITC is a suitable technique for characterizing allosteric interactions and conformational changes in proteins [32,34-37].

ITC also allows determination of the heat capacity change of binding interactions, Δc_P, from the temperature dependence of the enthalpy change:

$$\Delta c_P = \partial \Delta H / \partial T \tag{3}$$

with the assumption that the apparent heat capacities of the free molecules and the complex are constant over the temperature range under study. The changes in the heat capacity associated with protein-protein binding originate mostly from changes in the hydration heat capacity due to burial of polar and nonpolar groups upon complex formation and the loss of conformational degree of freedom upon binding [38-40]. Hence, Δc_P can be calculated in terms of the change in the accessible surface areas (apolar (ASA_{ap}) and polar (ASA_{pol})) upon the formation of protein-protein complex using the semi-empirical relationship [39,41-42]:

$$\Delta c_P = 0.45\Delta ASA_{ap} - 0.26\Delta ASA_{pol} \text{ cal } K^{-1}mol^{-1} \tag{4}$$

A good correlation between the experimentally determined and the calculated from structural data Δc_P values has been found in some cases [43-45], however significant difference was reported in other cases [42,46], suggested to be a consequence of changes in the conformational states and significant dynamic restriction of vibrational modes at the surface of the complex, as well as folding transitions coupled to the association event.

The values of Δc_P can also be used to estimate the entropic component due to desolvation of the surfaces of both interacting proteins buried within the binding interface:

$$\Delta S_{solv} = \Delta c_P \ln(T / T^*) \tag{5}$$

where T* = 385.15 K is the temperature of entropy convergence [47,48] and to further decompose the entropic term, which besides the solvation term contains two more contributions (conformational, ΔS_{conf}, associated with changes in conformational degree of freedom and rotational-translational $\Delta S_{rot\text{-}tr}$, ($\Delta S_{rot\text{-}tr}$ = -7.96 cal mol^{-1}K^{-1} [49] which accounts for changes in rotational/translational degrees of freedom):

$$\Delta S = \Delta S_{solv} + \Delta S_{conf} + \Delta S_{rot-tr} \quad (6)$$

Additional information on protonation/deprotonation effects coupled to the binding interactions can be provided by titration experiments in various buffers of different ionization enthalpy, ΔH_{ion}. The number of protons, n_{H+}, exchanged between the macromolecular complex and the bulk solution, and the binding enthalpy, ΔH_{bind}, can be calculated from the dependence of the calorimetrically observed enthalpy change, ΔH_{obs}, and ΔH_{ion} [42,50,51]:

$$\Delta H_{obs} = \Delta H_{bind} + n_{H+}\Delta H_{ion} \quad (7)$$

To decompose the free energy of binding into electrostatic and non-electrostatic contributions one has to study the ionic strength effect on the binding thermodynamics and analyse the data according to the Debye-Hückel approximation [52].

Valuable information on the hydration or solvent exposure of a polypeptide can be obtained by the absolute heat capacity [29]. A DSC method to accurately determine the absolute heat capacity of a protein from a series of calorimetric thermograms obtained at different protein concentrations has been described in [53]. The slope of a plot of the excess heat capacity versus the protein mass in the calorimetric cell is related to the absolute C_p:

$$m = C_p - \upsilon_p \quad (8)$$

where m is the slope of the linear regression of the plot and υ_p is the partial specific volume of the protein. This information can be related to the integrity of the native state or the presence of residual structure in the denatured state.

The calorimetric transitions in many cases are irreversible and scanning rate dependent, suggesting that the denaturation process is kinetically controlled [54-56]. Appropriate kinetic models were applied to analyse the irreversible unfolding process, after obtaining a set of thermograms at various scanning rates. An irreversible protein denaturation event can be described in some cases by a simplified "two-state irreversible" kinetic model [54,57], assuming that only the native/folded and denatured/unfolded states are significantly populated during the denaturation. Mathematical expressions were derived to calculate the activation energy, E_a, of the denaturation transition [54-58], using diverse experimental information from the calorimetric transition:

i. the values of the rate constant of the transition, k, at a given temperature:

$$k = A \exp\left(-E_a / RT\right) \quad (9)$$

where E_a is the activation energy and A is the frequency factor.

The rate constant of the reaction at a given temperature T is given by:

$$k = \nu\, c_p / \left(Q_t - Q\right) \quad (10)$$

where v is the scanning rate (K/min), c_P the excess heat capacity at a given temperature, Q is the heat evolved at that temperature and Q_t the total heat of the calorimetric transition.

ii. the dependence of the heat capacity evolved with temperature expressed as:

$$\ln\left[\ln Q_t / \left(Q_t - Q\right)\right] = E_a / R \left(1/T_m - 1/T\right) \quad (11)$$

iii. the heat capacity c_P^m at the transition temperature T_m, where the activation energy can also be calculated by the following eq.:

$$E_a = eRT_m^2 c_P^m / Q_t \quad (12)$$

This "two-state" kinetic model has described the unfolding of bacteriorhodopsin [58], rhodopsin [59], plastocyanin [60], the major light harvesting complex of photosystem II [61], nucleoplasmin (see below, [9]) and some other proteins [62,63]. This model however cannot describe all cases of irreversible protein denaturation [64]. On the other hand, Davoodi et al. [65] have shown that scanning-rate dependence of DSC thermograms is not limited to irreversible processes only.

DSC can also be used to indirectly study ligand binding to proteins and for analysis of very tight binding that can not be analysed by ITC or other spectroscopic methods [66]. In addition, more comprehensive description of the binding energetics can be derived combining the two techniques, ITC and DSC [18].

Recently DSC was also recognized as a novel tool for disease diagnosis and monitoring [67-70]. Calorimetric studies of blood plasma/serum have revealed a typical DSC thermogram for healthy individuals, whereas pronounced changes in thermograms for diseased subjects, including oncopatients, have been reported. Validation of the technique as an efficient tool for disease diagnostics needs further investigations of a large number of diseases.

Besides the classical application of ITC in studies of binding interactions, it has been proven as a powerful technique in diverse fields like drug discovery and lead optimization, nanotechnology, enzyme kinetics, etc. [71,73].

Kinetics of ligand binding to RNA and the subsequent RNA folding have recently been characterized by the so called kinITC [74]. ITC has also permitted documentation of the energy landscape of tertiary interactions along the RNA folding pathway [75]. Thermodynamic parameters, ΔH and Δc_P, of rigid amyloid fibril formation from monomeric β-microglobulin, associated with degenerative disorders have also been determined by ITC [76]. Recently a protocol for novel application of the technique has been elaborated, in which ITC is used as a tracking tool, combined with chromatography, for identification of target protein in biomolecular mixture [77] and it has been suggested to be valuable when the target protein or ligand is unknown. References for the wide spectrum and examples of novel applications of ITC can be found in the surveys published each year in the Journal of Molecular Recognition [25-28].

3. Nucleoplasmin thermal stability. Differential scanning calorimetry

NP is remarkably stable against chemical and physical challenges, including heat; e.g. the T_m of recombinant, non phosphorylated protein is 110.1°C [4,9]. This extreme stability, which is related, as previously mentioned, to the structural scaffolding role of the core domain, is *per se* an attractive issue to be thermodynamically described. We have characterized by DSC and other techniques the stability properties of NP and how they are related to the functionality of the protein. It should be mentioned that the overpressure used in the calorimeter allows to asses melting points above water boiling temperature, by contrast to other spectroscopic techniques.

We have shown that the stability of NP is solely due to the core domain, the T_m of the isolated core (117.6°C) being still higher than that of the full length protein [9]. The slight destabilizing influence of the tail is explained by its strong acidic character, with negatively charged clusters, such as "polyGlu"; electrostatic repulsion is expected to occur between tails and the also negatively charged core domain. This is reflected by the fact that the full length protein is most stable at pH close to its theoretical isoelectric point (pI=5.1), when its stability equals that of the core domain [9].

Analysis of the chemical denaturation of NP by fluorimetric and biochemical techniques has allowed to describe the unfolding mechanism of NP, in terms of a two-state process, where the pentamer dissociation is coupled with unfolding of the monomers, with no evidence of (partially) folded subunits ($N_5 \leftrightarrow 5U$) [78]. Both chemical and thermal unfolding of NP are reversible processes, while denaturation of the isolated core domain is reversible if chemically induced but irreversible upon heating [9,78]. This different behaviour suggests that the charged tail domains favour the solubility of the full length protein after thermal unfolding.

4. Effect of NP activation. Interplay between function and stability

NP activation, mediated by phosphorylation of multiple residues, implies an energetic cost for the protein. We have observed that NP extracted from *Xenopus* oocytes, corresponding to an intermediate phosphorylation state, is significantly destabilized with respect to recombinant, non phosphorylated NP (T_m ~ 94.4°C, ΔH ~ 80 kcal/mol) [4]. Egg NP, which represents the most active protein in the final stage of egg maturation, exhibits a further destabilization (T_m ~ 75°C, ΔH ~ 50 kcal/mol) [9]. This correlation between phosphorylation degree and loss of stability has been also characterized by chemical unfolding experiments [78].

In order to explore the conformational consequences of NP activation, we assessed the impact of phosphorylation in particular sites on the protein stability. Apart from the experimental evidences pointing to CKII and mitosis promoting factor (MPF) as probable kinases that modify NP [7,79], the amount and identity of kinases phosphorylating NP has not been elucidated. On the other hand, NP can be phosphorylated *in vitro* only with very low yield. As an alternative approach to obtain homogeneous preparations of active NP with a defined modification level, we designed a series of phosphorylation mimicking mutants, in which different Ser and/or Thr residues representing phosphorylatable residues were substituted for Asp [8,9,80]. Most mutation sites correspond to phosphoresidues identified by mass spectrometry analysis of egg NP [9]. However, taking into account that some phosphoresidues might have remained undetected by the proteomic analysis, due to incomplete sequence coverage and/or heterogeneity of the NP natural samples, additional residues were mutated on the basis of prediction software, N-terminal amino acid analysis, sequence comparison within the NP family and structural considerations [8,80].

The mutation sites, which are indicated in Figure 1, can be classified in three groups: 1) mutations in the flexible, N-terminal segment of the protein, not visible in the 3D structure of the core domain (residues 2, 3, 5, 7, 8), 2) mutations of residues located in loop regions of the core domain distal face (15, 66 and 96) and 3) mutations in the tail domain (residues 159, 176, 177, 181 and 183). Apart from the group of three residues within the structured core domain, at least group 1 is expected to face also the distal pole of the protein, which is most probably implicated in histone binding [10,80]. A collection of NP mutants (full length and core domains) were generated, combining the three groups of mutations, as indicated in Figure 1.

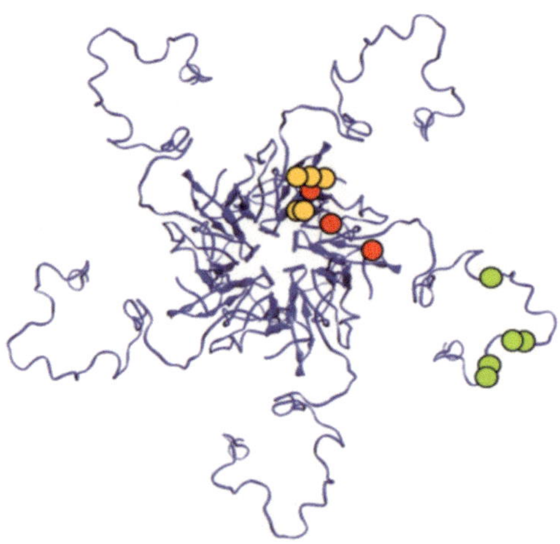

Figure 1. Activation of recombinant NP achieved through phosphomimicking mutations. Their location is highlighted on our model of full length NP based on the crystal structure of the core domain [3] and SAXS data [10]. Orange circles denote substitutions of residues 2, 3, 5, 7, 9 for Asp in the N-terminal segment ("group 1"); red ones correspond to substitutions at 15, 66 and 96 ("group 2"), and green ones to mutations in residues 159, 176, 177, 181, 183 of the tail domain ("group 3"). NP5D carries only mutations in the tail; NP8D harbors groups 1 and 2; NP10D groups 1 and 3; and NP13D comprises all mutations. For the sake of clarity, the positions are shown in only one monomer

By contrast to recombinant, non-phosphorylated NP, which shows negligible ability to decondense chromatin, phosphorylation mimicking mutations render the protein active to varying extents depending on the number and position of mutations. The mutants are capable of decondensing *Xenopus* demembranated sperm nuclei and extracting sperm-specific basic proteins, as well as linker-type histones from chromatin. The core domain isolated from natural, hyperphosphorylated, egg NP is (partly) active in decondensing chromatin, and a recombinant core domain with 8 substitutions (CORE8D, with groups 1 and 2) resembles these functional properties [80]. Nevertheless, full activity can only be attained through accumulation of negative charges (or phosphoresidues) along both the core and tail domains of NP: the mutant NP13D reproduces the functionality of egg NP [8].

Comparison of the thermal unfolding profiles of wild type and mutant core domains reveals that the activating mutations strongly decrease the thermal stability of the protein (Figure 2). Destabilization is probably due to the electrostatic repulsion in the oligomer (already negatively charged at neutral pH), which becomes more intense in the mutants (see inset in Figure 2).

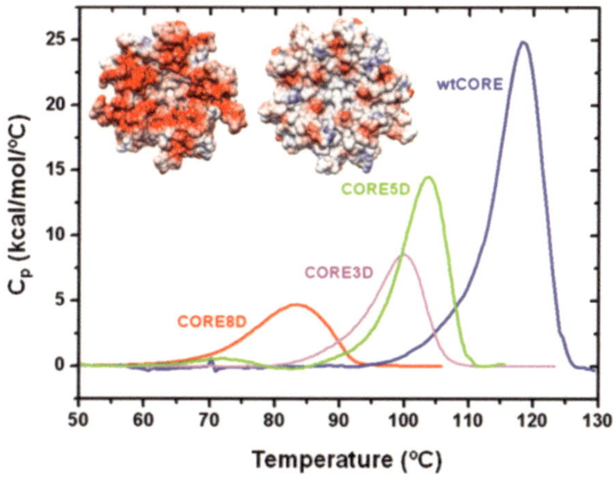

Figure 2. Phosphorylation mimicking mutations decrease the thermal stability of NP core domain. The effect of the mutations on the charge of the protein is also shown, by comparing the surface of the crystal structure of the mutant CORE8D [9] (left) and wild type CORE [3] (right), viewed from the distal face, and colored according to the electrostatic potential

The mutant CORE3D (with group 2 mutations) is less stable, in spite of harbouring fewer substitutions than CORE5D (group 1 mutations), which could be due to the fact that the three residues 15, 66 and 96 locate in structured regions of the protein, whereas the five N-terminal mutations are in a flexible segment that could be re-arranged to alleviate the electrostatic repulsion in NP. The combination of both groups of mutations makes CORE8D the most unstable mutant core, as expected. The strong destabilizing effect (e.g. ΔT_m of 34.5°C in the case of CORE8D) suggests a conformational change in the protein. Phosphorylation does not induce, however, significant changes in the secondary structure of NP [5,9]. Furthermore, we have solved the crystal structure of CORE8D and found that surprisingly enough it is almost identical to that of wild type core domain [9] (see Figure 2).

On the other hand, the activating mutations seem to affect the dynamics of the core domain [9]. The irreversible thermal unfolding of the core can be described as a scanning rate-dependent transition between two states, native and irreversibly denatured. From different mathematical expressions making use of diverse parameters from the calorimetric transition (eqs. 9-12), the activation energy E_a of the denaturation was calculated and compared for wild type CORE and CORE8D. We obtained a higher E_a value for wtCORE (69.8 ± 2.6 kcal/mol) than for CORE8D (52.1 ± 2.4 kcal/mol), indicating that the mutations destabilize also kinetically the core domain, reducing the energy barrier of the transition to the unfolded state [9]. In addition, to further characterize the conformational change associated with protein activation, the excess heat capacity c_P of wtCORE and CORE8D was measured at various protein concentrations (at 37°C), in order to calculate the absolute heat capacity, C_p, of their native states, which is related to solvent exposure of protein hydrophobic groups (eq. 8, see above). The obtained C_p values were 0.23 and 0.42 cal K^{-1} g^{-1} for wtCORE and CORE8D, respectively, suggesting faster dynamics or faster conformational fluctuations in the mutant [9]. Furthermore, the activation process affects the hydrodynamic properties of the protein (see below).

To understand the contribution of both NP domains to its activation mechanism, we also characterized the function and stability of full length NP, with the three groups of mutations and combinations thereof (Figure 3). Substitutions located in the core domain (NP8D) affect the protein stability to a greater extent than those in the tail domain (NP5D), highlighting again that addition of charges in structurally well defined locations is more deleterious for the stability of the protein than in flexible regions. The most active mutant, NP13D, is also the most unstable (T_m ~ 55.2°C, ΔH ~ 17.9 kcal/mol). Taking into account that at neutral pH aspartic acid has one negative charge, while a phosphoryl group would display an average negative charge of -1.5 [81], 13 Asp would be a reasonable approximation of 7-10 phosphates per monomer in egg NP; however, the fact that this mutant is less stable than egg NP reflects that the conformational properties of phosphorylated NP may not be exactly reproduced [9].

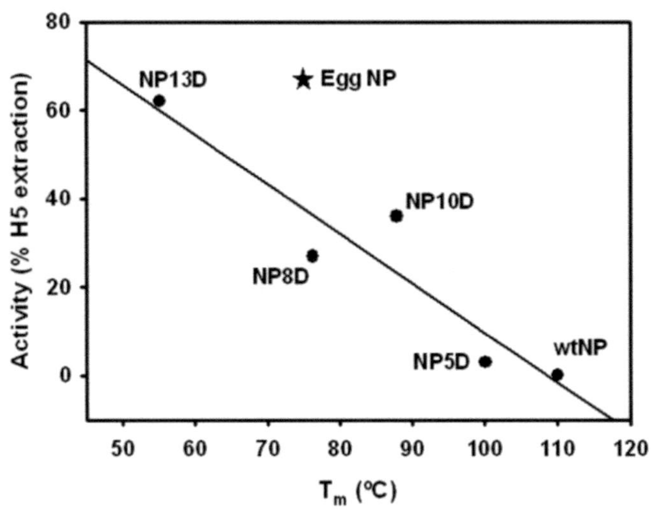

Figure 3. Inverse correlation between NP activity (expressed as percentage of histone H5 extracted from chicken erythrocyte chromatin, in a solubilization assay [8]) and thermal stability (T_m as measured by DSC). Linear regression of the phosphomimicking mutants data is shown

The mutations-induced destabilization of NP is also readily observed by chemical unfolding experiments [9]. Since NP denaturation proceeds through pentamer dissociation intimately coupled to unfolding of the monomers, the activation mechanism, in spite of not affecting conformationally the protein at the level of secondary structure and tertiary structure of the core domain, seems to weaken its quaternary interactions. In support of this notion, we have observed, by size exclusion chromatography and dynamic light scattering, that the activating mutations induce an expansion of the NP pentamer dimensions in solution, both in the core domains (from an average diameter of 64.5 Å to 68.8 Å for CORE8D, as measured by DLS) and the full length mutants (from 93.7 to 99.5 Å in the case of NP13D) [9]. Considering the similarity between the crystal structures of inactive, wild type, and active, mutant core domain, this "swelling" must affect mainly flexible regions of the protein, such as the N-terminal segment, loops of the core domain, and the tail domain.

Therefore, in NP, an inverse correlation exists between activity and stability (Figure 3), the higher the histone chaperone activity performed by NP, the lower its thermal and chemical stability, and larger its dimensions in solution.

In summary, we observed that NP activation mechanism, that depends on the accumulation of negative charge, probably on flexible regions of the distal pole of the protein, implies a destabilizing cost and an expansion of the oligomer in solution. The destabilizing mechanism seems to be the electrostatic repulsion in the pentamer, that weakens the quaternary interactions (tending to "open" apart the structure), which are essential for the stability of this protein. However, the loss of stability does not compromise, under physiological conditions, NP function or folding, which is granted by the extremely stable core domain. Moreover, the activation penalty may explain why this protein, from a mesophilic organism, displays such a remarkable thermal stability: it is necessary to afford the strong destabilization upon activation.

5. Nucleoplasmin chaperoning function studied by isothermal titration calorimetry

The high number of positive charges that histone proteins carry makes them prone to unproductive interactions with nucleic acids and other cellular components. Therefore free histones eventually do not exist within the cellular context and need to be escorted by histone chaperones, which shield their charge, and facilitate their controlled transfer during nucleosome assembly or reorganization. To perform its function, nucleoplasmin has to bind both linker-type and nucleosomal histones. Thermodynamics provided a detailed knowledge of NP-histone complex formation and elicited how NP carries its chaperoning activity [10].

The experimental isotherms of the binding interactions of NP with histones, H5 and H2A/H2B, and the enthalpic and entropic contributions to the Gibbs free energy for the first binding site are summarized in Figure 4.

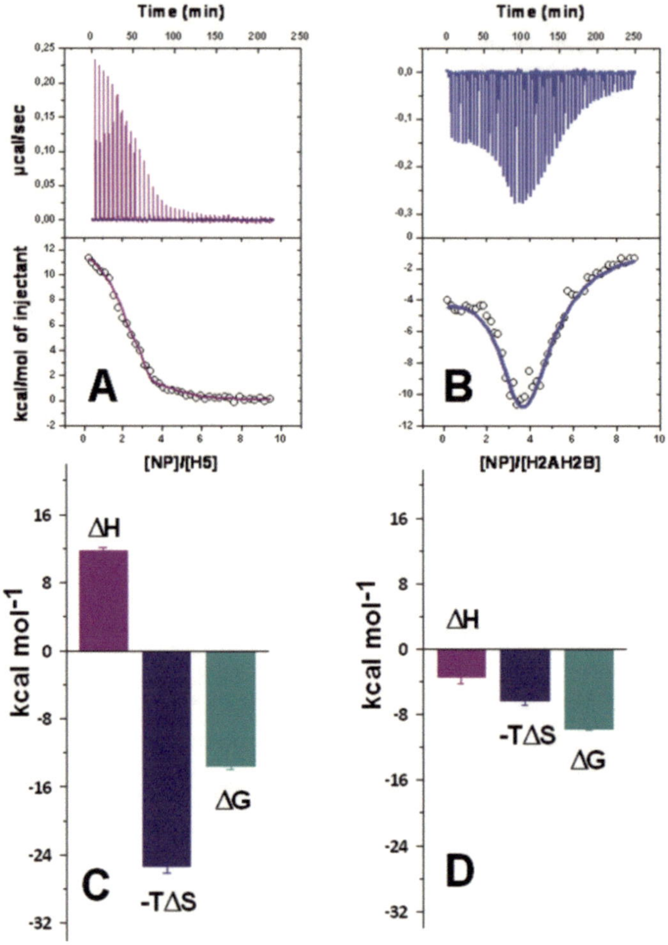

Figure 4. Binding data of NP interactions with the linker, H5, and nucleosomal core, H2A/H2B, histones. (A, B) Baseline-corrected instrumental response of NP titration with successive additions of H5 and H2A/H2B (upper panels); integrated data and the fits of the binding isotherms (solid lines) according to a negative cooperativity model (see text) for H5 and H2A/H2B (lower panels). (C, D) Thermodynamic parameters (ΔG, ΔH, -TΔS) of the assembly of the first histone, H5 and H2A/H2B, molecule with NP

ITC data reveal that NP can accommodate five histone molecules utilizing a negative cooperative binding mechanism with dramatic difference in the binding strength. The binding affinity of histones for the first site is moderate for nucleosomal core ($\Delta G = -9.8 \pm 0.1$ kcal/mol, $K_b = 1.5 \times 10^7$ M^{-1}) and extremely high for linker ($\Delta G = -13.6 \pm 0.4$ kcal/mol, $K_b = 10^{10}$ M^{-1}) histones (Figure 4, C and D), which can provide the basis for its histone exchange capabilities. The binding isotherms of the complex formation of histones with NP were analyzed using a site specific cooperative binding model. The model, developed especially for NP-histone interactions, considers negative cooperative interactions for both adjacently and non-adjacently bound histones and fit the experimental data better than an independent binding sites model and a general model based on the overall association constants. Eight thermodynamic parameters: four association constants (intrinsic association constant (K) and cooperativity binding parameters: k_1 (associated with the binding of an additional ligand), k_2 (binding of a ligand with contact to one nearest-neighbour) and k_3 (binding of a ligand with contact to two nearest-neighbours)) and four enthalpies were defined in the cooperativity model (details on the model can be found in [10] and in Supplementary Material of [10]).

The binding of histone molecules upon occupancy of the first binding site progresses with an energetic penalty, with exception of H2A/H2B molecules that bind to a non-adjacent site ($k_1 = 1$). Therefore, negative cooperativity was observed for all four additional H5 molecules ($k_1 < 1$), while only for H2A/H2B histone dimers bound to NP adjacently. Since the source of the cooperativity interaction may be an allosteric conformational change in NP induced by histone binding or a direct histone-histone interaction upon binding, or/as well as a combination of both, our results indicate different origin of the negative cooperativity for the binding of H5 and H2A/H2B to NP. Hence the main source of the cooperative binding interaction of H2A/H2B dimers is H2A/H2B-H2A/H2B interaction, whereas a conformational change in the NP pentamer upon binding of the first H5 molecule should provide a less favourable binding interface for the next histone molecules through energetic communication.

Somewhat surprising, considering the strong opposite charge of NP and histones, the binding of both histone types to NP is dominated by a favourable entropic term indicating a strong contribution of the hydrophobic effect to the binding affinity (Figure 4, C and D). The enthalpic term also contributes favourably to the binding energy of H2A/H2B, while unfavourable enthalpy changes counterbalance the entropic contribution to the free energy of H5 binding (Figure 4, C and D).

Furthemore, and contrary to the generally accepted major determinant of tail "polyGlu" tract in histone binding, the thermodynamic analysis as well as the low resolution structural models of NP/histone complexes, constructed by small angle X-ray scattering (SAXS) [10], demonstrate clearly that both NP domains are involved in the interaction with histones.

This was evidenced by comparing the binding energetics of the full-length protein with that of isolated core domain (CORE). Interestingly, NP core domain contributes equally to the intrinsic binding energy of H5 and H2A/H2B (ΔG = -8.2 kcal/mol). The tail domain of NP provides an additional thermodynamic driving force (estimated as the difference between the binding free energies of histones to NP and CORE, $\Delta\Delta G_{NP\text{-}CORE}$) (Figure 4 and Figure 5) for the much stronger binding of H5 ($\Delta\Delta G_{NP\text{-}CORE}^{H5}$ = -5.5 kcal/mol) compared to H2A/H2B ($\Delta\Delta G_{NP\text{-}CORE}^{H2AAH2B}$ = -1.6 kcal/mol) suggesting that this domain is particularly essential in the binding to H5 molecules.

To approach an activity/energetics relationship, we analysed the energetics of histone association with NP variants with phosphorylation-mimicking mutations in both the core and tail domains (NP8D, NP13D, CORE, CORE8D).

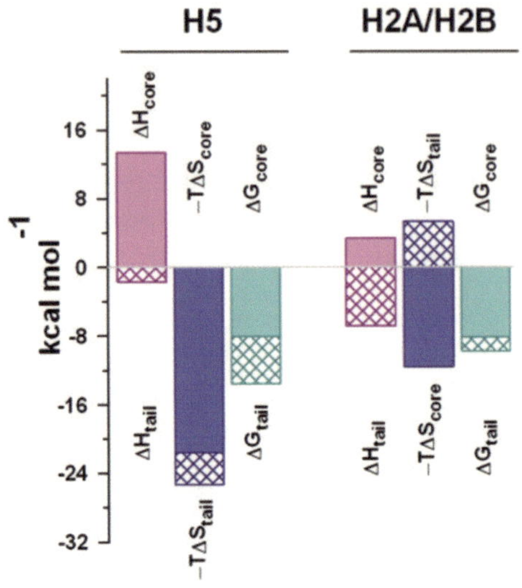

Figure 5. NP core (full bars) and tail (crossed bars) domain contributions to the intrinsic ΔG of their binding to linker and nucleosomal histones

As mentioned above the NP activity is regulated by its phosphorylation state. Insertion of mutations (8 and 13) gradually enhanced the binding affinity and affected to different extent the changes in the Gibbs energy contributors, the entropic and enthalpic terms. This reflects a strong impact of phosphorylation mimicking mutations in both core and tail domains of NP on its recognition by histones (Figure 6). The strongest affinity observed for the NP variant with the highest number of mutations, NP13D, is compatible with the fact that it mimicks the activity of the hyperphosphorylated native protein and can explain the protein activation through post-translational modifications.

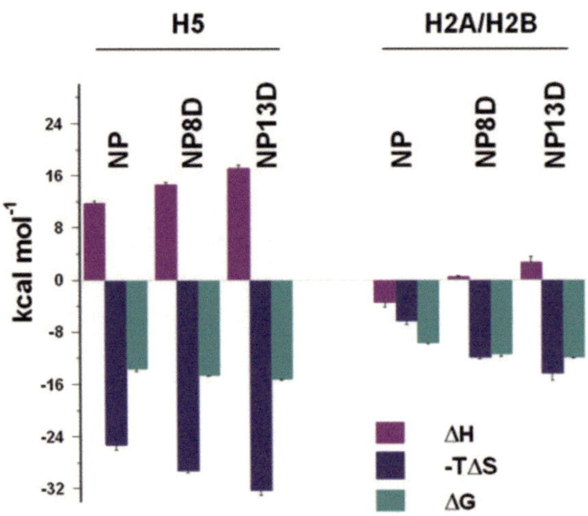

Figure 6. Effect of phosphorylation mimicking mutations on the binding energetics. Bar graphs comparing the intrinsic Gibbs energy, enthalpy and entropy changes, for the intrinsic binding of the two histone types, H5 and H2A/H2B, to NP and the phosphomimicking mutants NP8D and NP13D

Although the hydrophobic interactions are the major source of NP/histone binding free energy (about 80% of the intrinsic free energy for H2A/H2B and about 60% of that for H5), electrostatic and polar interactions between the acidic NP and basic histones also play an important role, either in direct binding or helping in orienting properly the binding partners, given the structural features of NP and histones. In order to get more insight into the nature of binding interactions we studied the ionic strength effect on the binding energetics (Figure 7, left panel). We found that despite the highly charged nature of H5 and NP, the non-electrostatic interactions contribute stronger to the stabilization of NP/histone complexes than the electrostatic ones (Figure 7, right panel). The significantly lower observed free energy of binding ΔG for H2A/H2B compared to H5 originates from lower ΔG_{el} (electrostatic) term (the non-electrostatic term ΔG_{nel} is comparable for H2A/H2B and H5), that should reflect the distinct number of positively charged residues in each histone type. For H2A/H2B and H5 binding to the NP13D mutant ΔG is -9.8 ± 0.1 and -11.9 ± 0.14, and the ΔG_{el} term −2.6 kcal/mol and −5.1 kcal/mol, respectively (Figure 7, right panel).

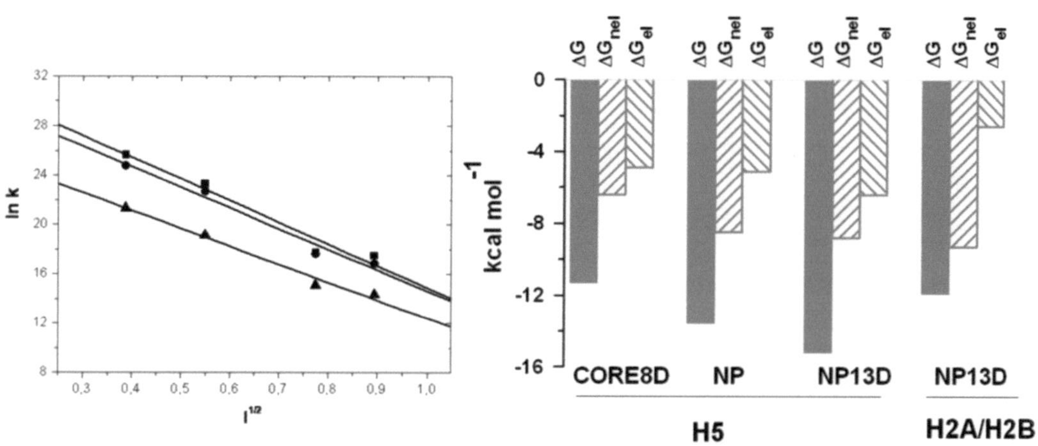

Figure 7. Ionic strength dependence of the association constant of H5 binding to the first (■), non-adjacent (•) and adjacent (▲) binding sites of NP13D variant (left panel). Extrapolation of $\partial \ln(K) / \partial I^{1/2}$ to 1M NaCl yields the non-electrostatic contribution ΔG_{nel} to the binding energy ΔG. Contribution of ΔG_{nel} and the electrostatic contribution, ΔG_{el}, to ΔG for H5 binding to CORE8D, NP and NP13D, and of H2A/H2B to NP13D (right panel)

ITC data also show that the NP flexible tail domain undergoes a histone binding-induced transition to a more structured or ordered state. This follows from the conformational entropy difference between full length proteins and core domains. We estimated from the heat capacity change, that there is a conformational entropy loss of ca. -20 kcal/mol upon H5 binding to the full-length protein as compared to the core domain (and even higher ca. -33 kcal/mol for H5 binding to the mutant proteins, NP8D and CORE8D), that can be attributed to the ordering of the intrinsically disordered nucleoplasmin tails [5] when bound to histones and indicates that NP tails do establish contacts with the histone molecules.

On the other hand, Δc_P (obtained from the temperature dependence of ΔH, presented for the NP13D variant in Figure 8) is smaller for the core domain NP variants compared to the full-length NP that would indicate a smaller molecular surface area involved in the binding of H5 to the core domain fragments than to full length NP.

Since no high resolution structural data are available for the NP/histone complexes the experimental Δc_P heat capacity changes cannot be compared with the ones estimated from structural data. We therefore roughly estimated the area buried within the binding interface from the SAXS data, in terms of "dummy" atoms of the corresponding "phase" that are in contact with the atoms of another "phase" in MONSA models (for details on SAXS experiments and data analysis see ref.[10]).

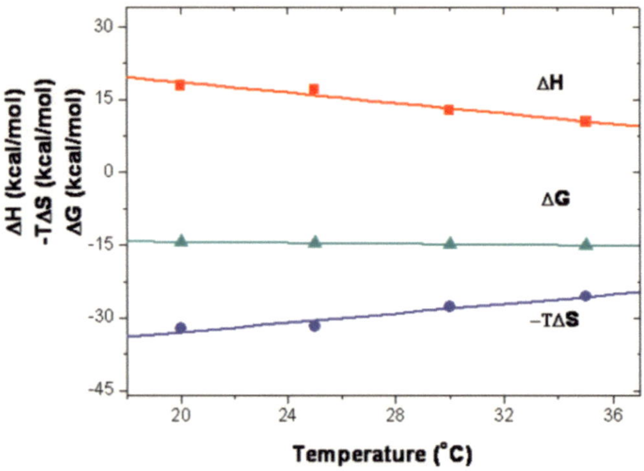

Figure 8. Temperature dependence of the thermodynamic parameters of the binding of the first H5 molecule to NP13D mutant. Intrinsic enthalpy (ΔH, ■), entropy (-TΔS, •) and free energy (ΔG, ▲) of binding. The heat capacity change Δc_P ($\Delta c_P = (\partial \Delta H/\partial T)$) is determined from linear regression analysis of ΔH data (solid line). The intrinsic free energy of binding is almost independent of temperature reflecting compensation of the enthalpic and entropic terms

The interaction interface area corresponding to NP tail/H5 is approximately double that of NP core/H5, whereas NP tail/H2A/H2B is half of that of NP core/H2A/H2B, which reflects strong difference in the binding of NP tails to both histone types. Although the ratio of the interaction interface areas NP core/NP tails is a rough estimate, it well compares with the ratio estimated from the experimentally determined heat capacity changes, $\Delta c_P^{NPcore}/\Delta c_P^{NPtails}$ ($\Delta c_P^{NPtails} = \Delta c_P^{NP} - \Delta\ c_P^{core}$).

The significant differences between the intrinsic association constants and the cooperative character of NP binding to the nucleosomal and the linker histones defines different "affinity windows" for NP binding from picomolar to nanomolar and from nanomolar to micromolar for H5 and H2A/H2B, respectively. This difference in recognition of nucleosomal and linker histones might provide an efficient mechanism for regulation of the dynamic histone exchange and might allow NP to fulfill its histone chaperone role, simultaneously acting as a reservoir for the core histones and a chromatin decondensing factor. Our data are compatible with the traditional model where NP facilitates nucleosome assembly by removing the linker histones and depositing H2A/H2B dimers onto DNA [1,2,82].

These data provided new insight into NP/histones assembly and interactions. It should be emphasized that the binding affinity of NP is enhanced upon insertion of phosphorylation-mimicking mutations that explains the protein activation through post-translational modifications. Importantly the data reveal a negative cooperativity-based regulatory mechanism for the linker histone/nucleosomal histone exchange, that in general renders proteins operative in a wider concentration range [83], with significantly populated intermediate liganded states. The employed site-specific cooperativity model, an extension of a previous one that analyses the interaction of another pentameric protein (cholera toxin, with the oligosaccharide portion of its cell surface receptor) considering only nearest-neighbour cooperative interactions [84], has potential application in studies of other macromolecular complexes between proteins sharing structural complexity with NP and their ligands.

6. Recognition of nucleoplasmin and histones by nuclear transport receptors

Nucleoplasmin, that possesses a classical bipartite NLS targeting sequence in each tail domain, is a prototypic substrate of the best characterized route for protein import into the nucleus, which is mediated by importin α/β heterodimer. However, as mentioned in Introduction, most structural and energetic approaches on cargo-import receptor recognition have been achieved merely using peptides carrying the corresponding NLS or IBB sequence [13, 85-91], and to date only two studies (besides ours [10]) deal with assembly of a macromolecular transport complex of full-length proteins [92,93]. On the other hand there has been paid little attention to nuclear import of oligomeric proteins. Therefore understanding the molecular basis of recognition of an oligomeric cargo as nucleoplasmin by its transport receptors, importin α/β heterodimer, would shed light on the arrangement of a large macromolecular nuclear import complex.

We obtained saturated NP/importin α/β complexes proving that all five available NLS binding sites of NP can be occupied by importins. Whereas *in vivo* binding of one α/β heterodimer to any protein should be enough to deliver it to the nucleus, it has been reported that the presence of multiple NLSs in NP [94] enhances its nuclear accumulation, suggesting that the number of NLSs might govern the traffic rate, which would play an advantage for oligomeric nuclear proteins, provided with multiple recognition sites. The binding isotherms of the NP/α/β complex formation (Figure 9A) were well fitted with an independent binding sites model, reflecting that NP makes use of different energetic scheme for assembling with histones and importins, most likely due to the involvement of dissimilar binding surfaces. The binding reaction is enthalpy-driven and counterbalanced by an unfavorable entropy change (Figure 9B) resulting in a relatively high-affinity interaction, $K_b = 18.5 \times 10^6\ M^{-1}$ ($K_d = 57 \pm 15$ nM). The entropic penalty most probably reflects an ordering effect on the otherwise flexible and mobile NLS motifs [5] upon the interaction event. This loss of conformational flexibility of the NLS segment in the NP tails [5] is estimated to correspond to conformational entropy change ΔS_{conf} = -282 cal mol^{-1} K^{-1} (using eq. 6) that dominates the entropic penalty. This conformational entropy change is unfavorable and greater than the favorable solvation entropy (ΔS_{solv} = 223 cal mol^{-1}K^{-1}, calculated from eq. 5), associated with hydrophobic interactions, thus resulting in an unfavorable entropy contribution to the Gibbs free energy of binding (Figure 9C).

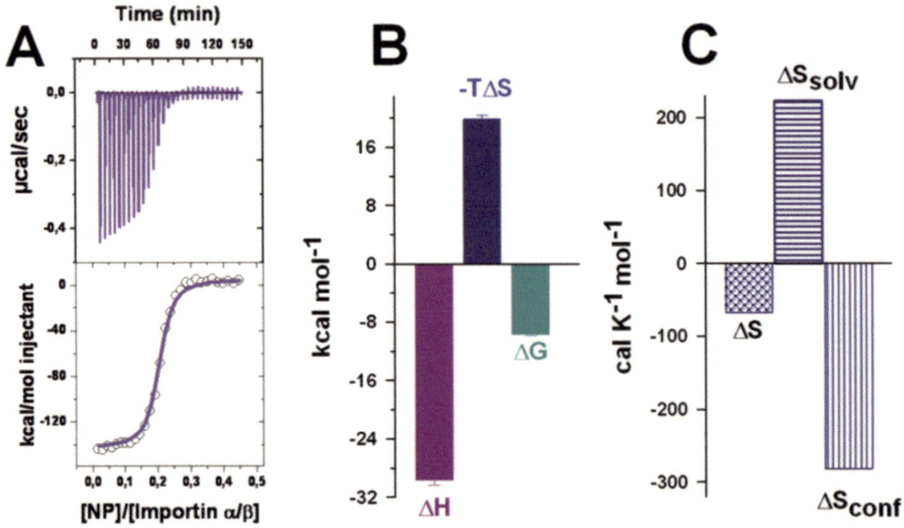

Figure 9. Energetics of NP assembly with the nuclear transport receptor importin α/β. Binding isotherms: the upper panel represents baseline-corrected instrumental response of importin α/β titration with NP; the lower panel shows the integrated data and the fit of the binding isotherm (solid line) by an independent binding site model (A). Enthalpic (ΔH) and entropic (-TΔS) contributions to the free energy (ΔG) of binding (B). Dissection of the binding entropy, ΔS, into solvation, ΔS_{solv}, and conformational, ΔS_{conf}, terms (C)

Similar binding mode and thermodynamic parameters (ΔG = -8.67 ± 0.1 kcal/mol, ΔH = -15 ± 0.3 kcal/mol and stoichiometry 5 per NPM pentamer) characterize the recognition of nucleophosmin (NPM), an abundant nucleolar protein, structurally homologous to nucleoplasmin (as mentioned in the Introduction) and related to oncogenic transformation, by the nuclear transport machinery (unpublished data), which supports an equivalent import mechanism for both chaperones.

Full-length importin α, not assembled with importin β, is also able to bind NP, albeit with a lower apparent affinity (K_d = 513 ± 87 nM) and with a lower enthalpic contribution to the free energy of binding. The loss of apparent affinity comes from the fact that the importin α N-terminal domain, the IBB domain, which contains a similar sequence to the NP-NLS, exerts an autoinhibitory role in the binding process [90,91] because in the absence of importin β, it occupies the NLS binding site and therefore it must be displaced by NP-NLSs. In this regard a truncated importin mutant lacking the IBB domain, ΔIBB-importin α or Δα, shows a similar affinity for NP (K_d = 54 ± 6 nM) as importin α/β [15].

Since protein phosphorylation is one of the mechanisms that up- or down-regulate nuclear transport [95,96] and it had been described that phosphorylated nucleoplasmin presents higher import rate than its unphosphorylated form [79], we studied how phosphorylation of NP affects its interaction with the import receptor. Nevertheless, phosphorylation mimicking mutations in residues close to NLS sequence, as in mutant NP13D, which shows high binding affinity to histones and is active in histone chaperoning, do not modulate the interaction with importin. NP13D mutant displays the same binding strength (ΔG), though different ΔH and –TΔS terms, compared to unphosphorylated protein (Figure 10). No effect has been observed when phosphorylated monopartite NLS from simian virus 40-large T antigen interacts with importin α [97]. Altogether, phosphorylation-mediated regulation of nuclear import must involve interactions other than post-translationally modified NLS with α/β importin.

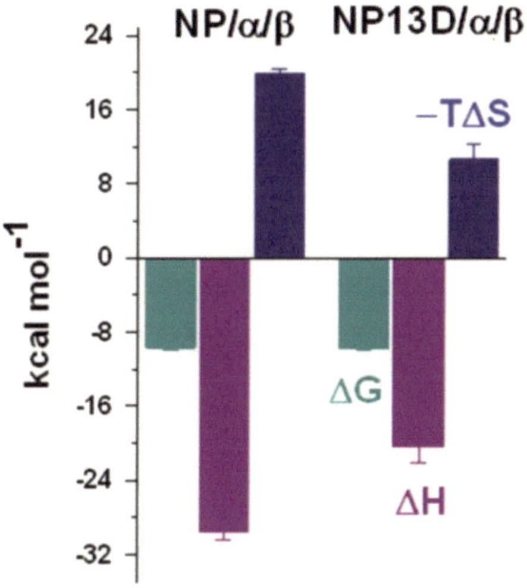

Figure 10. Comparison of the energetics (ΔG, ΔH and -TΔS) of importin α/β binding to NP and NP13D mutant

Importin binds similarly to a peptide corresponding to the NLS sequence when the latter is isolated or in the context of the full-length NP macromolecule, suggesting that no other regions of NP contribute significantly to the binding. The similar large negative Δc_P values, -817 and -796 cal mol^{-1} K^{-1} for NLS and full-length NP respectively, also suggest that the surface area buried within the binding interface is comparable in both cases. It is not surprising that the NLS recognition is not significantly affected by the protein context considering the flexibility displayed by NP tail domains harboring the NLS segments. The same notion applies to the interaction of importin α IBB domain with importin β. The former domain binds likewise to importin β independently of whether it is an isolated peptide or connected to the ARM domain of importin α, as is evidenced by the good correspondence of the heat capacity Δc_P value of -727.4 cal mol^{-1} K^{-1}, predicted from the X-ray structure of importin β bound to the IBB domain of importin α [98], with buried polar and apolar solvent accessible areas (ΔASA) of 1402.5 and 2426.7 Å^2, respectively (eq. 4), and the experimentally determined Δc_P = -840 cal mol^{-1} K^{-1} value for full length α/β interaction. Moreover, both proteins behave as independent units when they form the heterodimer complex since they display the same thermal stability as the one they exhibit when free in solution, thanks to flexibility of the linker between the IBB and the rest of importin α [15]. These data support the idea that α residues which act as link between both importins exhibit such flexibility that allows each of the importin entities to interact with a wide range of ligands during the nuclear translocation process.

Given the fact that the NP/α/β complexes are formed by multiple proteins that present flexible domains, SAXS technique has provided valuable information about the structure of those assemblies. Multiple models of NP fit equally well the experimental SAXS data reflecting the inherent flexibility of the particle, due to the adaptable linkers between the NP core domain and the NLS (residues 121-154 of NP) [15], which allows the accommodation of five bulky α/β heterodimers per NP pentamer. This 3D in solution structural model, the first one for a complete nuclear transport complex with an oligomeric cargo, is consistent with the notion that the canonical binding elements (NLS and IBB) are the ones determining the molecular basis of the recognition. The multidomain NP/α/β complex remains stable by virtue of two attachment points: recognition of the NLS by importin α and recognition of the IBB domain by importin β, which otherwise allow for conformational flexibility. This modular and articulated architecture might facilitate the passage of such a large particle through the nuclear pore complex.

Due to their highly basic nature histones need nuclear import receptors to be transferred to the nucleus, and most of the pathways described are mainly mediated by karyopherins of β family [99,100]. On the other hand, histones present multiple NLS-like motives and are also recognized by importin α family members for nuclear targeting [101]. Accordingly, we observed that both nucleosomal and linker histones bind to importin β (unpublished data), as previously demonstrated, and to importin Δα [16]. The high affinity exothermic binding interactions (Figure 11) suggest specific recognition events of importin Δα by H5 and H2A/H2B. Regardless of the different stoichiometry, two importin Δα per H5 and one per H2A/H2B, the thermodynamic parameters are quite similar, the apparent binding affinity and the enthalpy are in the order of 9 and 28 nM, and −20 and −17 kcal/mol for H2A/H2B and H5, respectively. Similar binding energetics, though higher stoichiometry (five per NP pentamer), characterizes the assembly of importin Δα with NP (K_d = 54 nM and ΔH = -18.5 kcal/mol, Figure 11). This suggests that similar molecular interactions are involved in the complex formation of importin Δα with the binding motifs of the two histone types and of NP.

Importantly, ITC together with fluorescence anisotropy and centrifugation in sucrose gradients show that NP, histones and importin α can associate and form co-complexes, NP/H5/importin Δα and NP/H2A/H2B/importin Δα of discrete size, that would support a co-transport of histones and NP to the nucleus, mediated by the classical import pathway. Depending on the histone type, linker or core, and the amount of bound histones, different number of importin Δα molecules can be loaded on NP/histone complexes, in which Δα can bind both to NP-NLS as to histones-NLS-like binding sites, as was demonstrated using a NP mutant with impaired binding to Δα. The binding is an enthalpy driven process and it is characterized by nanomolar affinity [16].

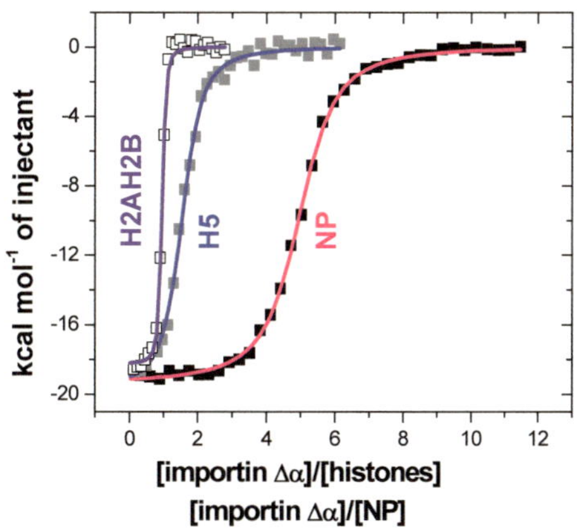

Figure 11. ITC isotherms for the binding interactions of the nucleosomal, H2A/H2B (violet) and linker, H5 (blue) histones, and NP (pink) with importin α ΔIBB (a truncated form lacking the autoinhibitory N-terminal domain)

We have also described the formation of quaternary NP/H5/importin α/β complexes by means of fluorescence and centrifugation, which makes conceivable that α/β heterodimer might "pull" NP/histones complexes into the nucleus, importin α binding either to NP, histones or both. Since importin β binds to both linker and core histones, NP/histones co-transport mediated by importin β, could also be expected. However, this hypothetical route seems unlikely since importin β competes with NP for histones, inducing the release of the latter from NP (unpublished data). Even though no detailed study has been performed, the comparable thermodynamic signature for H5/importin α and H5/importin α/β interaction supports the notion that H5 always binds through importin α in the presence of α/β heterodimer, which would explain the formation of quaternary complexes. Therefore the assembly of NP/histone/importin α/β complexes might have physiological meaning since it supports the existence of a putative and redundant histone import pathway in which positively charged histones would be protected against unspecific interactions by the histone chaperone nucleoplasmin.

7. Conclusion and future prospects

In summary, we have highlighted the importance of calorimetry in the study of nuclear chaperones. Detailed analysis demonstrated that the nuclear chaperone NP can associate with the two histone types and the transport machinery, and that co-complexes of NP, histones, and importins can assemble proving that ITC is suitable to study biological recognition in complex macromolecular assemblies. Notably, a link between NP phosphorylation state, its stability and the strength with which it assembles with histones is demonstrated.

One key feature of NP assembly with histones is the negative cooperative interactions, that render the protein operative in a wider concentration range and is an effective mechanism of regulation of the activity of macromolecular complexes. We have dissected the thermodynamic cooperativity of NP and its variants, core domain fragments and phosphorylation mimicking mutants, and presented strong evidence of the involvement of both NP domains in binding of histones, the NP tail domain being particularly essential in the assembly with H5 molecules. Only the nuclear localization signal NLS, however, is the recognition site in the multi-component NP/importin α/β complex. The significant differences in the enthalpic and entropic terms of the Gibbs free energy of NP association with histones and importins reflect different energetic strategy for NP chaperoning functions and its recognition for nuclear trafficking.

Both the experimental results and the methodological approach, ITC complemented with SAXS, allow a mechanistic understanding of nucleosome assembly/disassembly and its nuclear trafficking. The NP/histone complexes, which were modeled using five-fold symmetry, have a much more compact shape than the NP/importin α/β complex, reconstructed with multiple models, reflecting inherent flexibility.

Future work should focus towards description of the energetics of NPM export mechanism and the molecular recognition between NPM and nuclear export machinery (exportin), as well as with other proteins and peptides/small molecules. NPM is overexpressed in solid cancers (gastric, colon, ovarian and prostate), while genetic modifications of *NPM1* gene by chromosomal translocation, mutation and deletion are involved in lymphomas and leukemias [102-105]. Mutations of *NPM1* gene result in aberrant cytoplasmic localization of NPM in about 35% of acute myeloid leukemia (AML) patients [102]. The involvement of NPM in human cancer has received an increasing research interest during the last years, but the molecular mechanism of NPM implication in leukaemia and tumorigenesis is not understood yet. Studying the energetics of NPM binding with different (de)stabilizing ligands/drugs would help to regulate its interaction with cellular partners and thereby control its localization and function. This will entail, on one hand knowledge about the NPM nucleo-cytoplasmic shuttling and on the other is expected to provide a strategy for molecular therapeutics.

Author details

Stefka G. Taneva
Unidad de Biofísica (CSIC/UPV-EHU), Departamento de Bioquímica y Biología Molecular, Universidad del País Vasco, Bilbao, Spain,
Institute of Biophysics and Biomedical Engineering, Bulgarian Academy of Sciences, Sofia, Bulgaria

Sonia Bañuelos and María A. Urbaneja
Unidad de Biofísica (CSIC/UPV-EHU), Departamento de Bioquímica y Biología Molecular, Universidad del País Vasco, Bilbao, Spain

8. References

[1] Prado A, Ramos I, Frehlick LJ, Muga A, Ausió J (2004) Nucleoplasmin: a nuclear chaperone. Biochem. Cell Biol. 82, 437-445.
[2] Frehlick LJ, Eirín-López JM, Ausió J (2007) New insights into the nucleophosmin/nucleoplasmin family of nuclear chaperones. Bioessays 29, 49-59.
[3] Dutta S, Akey IV, Dingwall C, Hartman KL, Laue T, Nolte RT, Head JF, Akey CW (2001) The crystal structure of nucleoplasmin-core: implications for histone binding and nucleosome assembly. Mol. Cell 8, 841-853.
[4] Hierro A, Arizmendi JM, Bañuelos S, Prado A, Muga A (2002) Electrostatic interactions at the C-terminal domain of nucleoplasmin modulate its chromatin decondensation activity. Biochemistry 41, 6408-6413.
[5] Hierro A, Arizmendi JM, De las Rivas J, Urbaneja MA, Prado A, Muga A (2001) Structural and functional properties of Escherichia coli-derived nucleoplasmin. A comparative study of recombinant and natural proteins. Eur. J. Biochem. 268, 1739-1748.
[6] Sickmeier M, Hamilton JA, Le Gall T, Vacic V, Cortese MS, Tantos A, Szabo B, Tompa P, Chen J, Uversky VN, Obradovic Z, Dunker AK (2007) DisProt: the Database of Disordered Proteins. Nucleic Ac. Res. 35, D787 – D793.
[7] Cotten M, Sealy L, Chalkley R (1986) Massive phosphorylation distinguishes Xenopus laevis nucleoplasmin isolated from oocytes or unfertilized eggs. Biochemistry 25, 5063–5069.
[8] Bañuelos S, Omaetxebarria MJ, Ramos I, Larsen MR, Arregi I, Jensen OM, Arizmendi JM, Prado A, Muga A (2007) Phosphorylation of both nucleoplasmin domains is required for activation of its chromatin decondensation activity. J. Biol. Chem. 282, 21213-21221.
[9] Taneva SG, Muñoz I, Franco G, Falces J, Arregi I, Muga A, Montoya G, Urbaneja MA, Bañuelos S (2008) Activation of nucleoplasmin, an oligomeric histone chaperone, challenges its stability. Biochemistry 47, 13897-13906.
[10] Taneva SG, Bañuelos S, Arregi I, Falces J, Konarev P, Svergun D, Velázquez-Campoy A, Urbaneja MA (2009). A mechanism for histone chaperoning activity of nucleoplasmin: thermodynamic and structural models. J. Mol. Biol. 393, 448-463.
[11] Görlich D, Kutay U (1999) Transport between the cell nucleus and the cytoplasm. Annu. Rev. Cell Dev. Biol. 15, 607–660.
[12] Stewart M (2007) Molecular mechanism of the nuclear protein import cycle. Nat. Rev. Mol. Cell Biol. 8, 195–208.

[13] Fontes M R, Teh T, Kobe B (2000) Structural basis of recognition of monopartite and bipartite nuclear localization sequences by mammalian importin-α. J. Mol. Biol. 297, 1183–1194.

[14] Catimel B, Teh T, Fontes M R, Jennings IG, Jans DA, Howlett GJ, Nice EC, Kobe B (2001) Biophysical characterization of interactions involving importin-α during nuclear import. J. Biol. Chem. 276, 34189–34198.

[15] Falces J, Arregi I, Konarev P, Urbaneja MA, Taneva SG, Svergun D, Bañuelos S (2010) Recognition of nucleoplasmin by the nuclear transport receptor importin α/β: Insights into a complete transport complex. Biochemistry 49, 9756-9769.

[16] Arregi I, Falces J, Bañuelos S, Urbaneja MA, Taneva SG (2011) The nuclear transport machinery recognizes nucleoplasmin–histone complexes. Biochemistry 50, 7104-7110.

[17] Sánchez-Ruiz JM (1995) Differential scanning calorimetry of proteins. Subcell. Biochem. 24, 133–176.

[18] Jelesarov I, Bosshard HR (1999) Isothermal titration calorimetry and differential scanning calorimetry as complementary tools to investigate the energetics of biomolecular recognition. J. Mol. Recognit.12, 3-18.

[19] Cooper A (1999) Thermodynamics of protein folding and stability. In: Allen G, editor. Protein: A Comprehensive Treatise. JAI Press Inc.Volume 2, pp. 217-270.

[20] Schön A, Velázquez-Capmoy A (2005) Calorimetry In: Jiskoot W, Crommelin DJA, editors. Methods for structural analysis of protein pharmaceuticals. Arlington, VA, AAPS Press. pp. 573-589.

[21] Freire E, Mayorga OL, Straume M (1990) Isothermal titration calorimetry. Anal. Biochem. 179, 131-137.

[22] Laldbury JE, Chowdhry BZ (1996) Sensing the heat: the application of isothermal titration calorimetry to thermodynamic studies of biomolecular interactions. Chem. Biol. 3, 791-801.

[23] Pierce MM, Raman CS, Nall BT (1999) Isothermal titration calorimetry of protein-protein interactions. METHODS 19, 213-221.

[24] Leavitt S, Freire E (2001) Direct measurement of protein binding energetics by isothermal titration calorimetry. Curr. Opin. Struct. Biol. 11, 560-566.

[25] Cliff MJ, Ladbury J (2003) A survey of the year 2002 literature on applications of isothermal titration calorimetry. J. Mol. Recognit. 16, 383-391.

[26] Bjelic S, Jelesarov I (2008) A survey of the year 2007 literature on applications of isothermal titration calorimetry. J. Mol. Recognit. 21, 289-312.

[27] Falconer RJ, Collins BM (2009) Survey of the year 2009: applications of isothermal titration calorimetry. J. Mol. Recognit. 24, 1-16.

[28] Ghai R, Falconer RJ, Collins BM (2011) Applications of isothermal titration calorimetry in pure and applied research - survey of the literature from 2010. J. Mol. Recognit. 25, 32-52.

[29] Freire E (1995) Thermal denaturation methods – study of protein folding. Methods in Enzymol. 259, 144-168.
[30] Brown A (2009) Analysis of cooperativity by isothermal titration calorimetry. Int. J. Mol. Sci. 10, 3457-3477.
[31] Freire E, Schön A, Velázquez-Campoy A (2009) Isothermal titration calorimetry: general formalism using binding polynomials. Methods Enzymol. 455, 127-155.
[32] Velázquez-Campoy A, Goñi G, Peregrina JR, Medina M (2006) Exact analysis of heterotropic interactions in proteins: characterization of cooperative ligand binding by isothermal titration calorimetry. Biophys. J. 91, 1887-1904.
[33] Martinez-Julvez M, Abian O, Vega S, Medina M, Velázquez-Campoy A (2012) Studying the allosteric energy cycle by isothermal titration calorimetry. Methods in Molecular Biology 796, 53-70.
[34] Shiou-Ru T, Charalampos GK (2009) Dynamic activation of an allosteric regulatory protein. Nature 462, 368-374.
[35] Moro F, Taneva SG, Velázquez-Campoy A, Muga A (2007) GrpE N-terminal domain contributes to the interaction with DnaK and modulates the dynamics of the chaperone substrate binding domain. J. Mol. Biol. 374, 1054-1064.
[36] Taneva S, Moro F, Velázquez-Campoy A, Muga A (2010) Energetics of nucleotide-induced DnaK conformational states. Biochemistry 49, 1338–1345.
[37] Seldeen KL, Deegan BJ, Bhat V, Mikles DC, McDonald CB, Farooq A (2011) Energetic coupling along an allosteric communication channel drives the binding of Jun-Fos heterodimeric transcription factor to DNA. FEBS J. 278, 2090-2104.
[38] Sturtevant JM (1977) Heat capacity and entropy changes in processes involving proteins. Proc. Natl. Acad. Sci. U.S.A. 74, 2236-2240.
[39] Spolar RS, Record MT (1992) Coupling of local folding to site-specific binding of proteins to DNA. Science 263, 777-784.
[40] Makhatadze GI, Privalov PL (1995) Energetics of protein structure. Advan. Protein Chem. 47, 307-425.
[41] Murphy KP, Freire E (1992) Thermodynamics of structural stability and cooperative folding behavior in proteins. Adv. Protein Chem. 43, 313-336.
[42] Murphy KP, Xie D, Garcia KC, Amzel LM, Freire E (1993) Structural energetics of peptide recognition: angiotensin II/antibody binding. Proteins: Struct. Funct. Genet. 15, 113-120.
[43] Ladbury JE, Wright JG, Sturtevant JM, Sigler PB (1994) A thermodynamic study of the trp repressor-operator interaction. J. Mol. Biol. 238, 669-681.
[44] Pearce KHJr, Ultsch MH, Kelley RF, de Vos AM, Wells JA (1996) Structural and mutational analysis of affinity-inertcontact residues at the growth hormone-receptor interface. Biochemistry 35, 10300-10307.
[45] Holdgate GA, Tunnicliffe A, Ward WH, Weston SA, Rosenbrock G, Barth PT, Taylor IW, Pauptit RA, Timms D (1997) The entropic penalty of ordered water accounts for

weaker binding of the antibiotic novobiocin to a resistant mutant of DNA gyrase: a thermodynamic and crystallographic study. Biochemistry 36, 9663-9673.

[46] Baker BM, Murphy KP (1996) Evaluation of linked protonation effects in protein binding reactions using isothermal titration calorimetry. Biophys. J. 71, 2049-2055.

[47] Baldwin RL (1986) Temperature dependence of the hydrophobic interaction in protein folding. Proc. Natl. Acad. Sci. U.S.A. 83, 8069–8072.

[48] Murphy KP, Xie D, Thompson KS, Amzel LM, Freire E (1990) Entropy in biological binding processes: estimation of translational entropy loss. Science 247, 559–561.

[49] Luque I, Freire E (1998) Structure-based prediction of binding affinities and molecular design of peptide ligands. Methods Enzymol. 295, 100–127.

[50] Gómez J, Freire E (1995) Thermodynamic mapping of the inhibitor site of the aspartic protease endothiapepsin. J. Mol. Biol. 252, 337-350.

[51] Fukada H, Takahashi K (1998) Enthalpy and heat capacity changes for the proton dissociation of various buffer components in 0.1 M potassium chloride. Proteins: Struct. Func. Genet. 33, 159-166.

[52] Petrucci S (1972) Ionic Interactions 1, 117–177.

[53] Kholodenko V, Freire E (1999) A simple method to measure the absolute heat capacity of proteins. Anal. Biochem. 270, 336–338.

[54] Sánchez-Ruiz JM, López-Lacomba JL, Cortijo M, Mateo PL (1988) Differential scanning calorimetry of the irreversible thermal denaturation of thermolysin. Biochemistry 27, 1648-1672.

[55] Sánchez-Ruiz JM (1992) Theoretical analysis of Lumry-Eyring models in differential scanning calorimetry. Biophys. J. 61, 921-935.

[56] Kurganov BI, Lyubarev AE, Sánchez-Ruiz JM, Shnyrov VL (1997) Analysis of differential scanning calorimetry data for proteins. Criteria of validity of one-step mechanism of irreversible protein denaturation. Biophys. Chem. 69, 125–135.

[57] Freire E, Murphy KP, Sánchez-Ruiz JM, Galisteo ML, Privalov PL (1992) The molecular basis of cooperativity in protein folding. Thermodynamic dissection of interdomain interactions in phosphoglycerate kinase. Biochemistry 31, 250-256.

[58] Galisteo ML, Sánchez-Ruiz JM (1993) Kinetic study into the irreversible thermal denaturing of bacteriorhodopsin. Eur. Biophys. J. 22, 25-30.

[59] Landin JS, Katragadda M, Albert A (2001) Thermal destabilization of rhodopsin and opsin by proteolytic cleavage in bovine rod outer segment disk membranes. Biochemistry 40, 11176-11183.

[60] Milardi D, La Rosa C, Grasso D, Guzzi RC, Sportelli L, Carlo F (1998) Thermodynamic and kinetics of the thermal unfolding of plastocyanin. Eur. Biophys. J. 27, 273-282.

[61] Krumova SB, Todinova SJ, Busheva MC, Taneva SG (2005) Kinetic nature of the thermal destabilization of LHCII Macroaggregates. J. Photochem. Photobiol. B 78, 165-170.

[62] Guzzi R, La Rosa C, Grasso D, Milardi D, Sportelli L (1996) Experimental model for the thermal denaturation of azurin: a kinetic study. Biophys. Chem. 60, 29–38.

[63] Meijberg W, Schuurman-Wolters GK, Boer H, Scheck RM, Robillard GT (1998) The thermal stability and domain interactions of the mannitol permease of *Escherichia coli*. A differential scanning calorimetry study. J. Biol. Chem. 273, 20785-20794.

[64] Lubarev AE, Kurganov BI (2001) Study of irreversible thermal denaturation of proteins by differential scanning calorimetry. Recent Res. Devel. Biophys. Chem. 2,141-165.

[65] Davoodi J, Wakarchuk WW, Surewicz WK, Carey PR (1998) Scan-rate dependence in protein calorimetry: The reversible transitions of *Bacillus circulans* xylanase and a disulfide-bridge mutant. Protein Sci 7, 1538-1544.

[66] Brandts JF, Lin L-N (1990) Study of strong to ultratight protein interactions using differential scanning calorimetry. Biochemistry 29, 6927-6940.

[67] Garbett NC, Miller JJ, Jenson AB, Miller DM, Chaires JB (2007) Interrogation of the plasma proteome with differential scanning calorimetry. Clin. Chem. 53, 2012-2014.

[68] Garbett NC, Miller JJ, Jenson AB, Chaires JB (2008) Calorimetry outside the box: A new window into the plasma proteome. Biophys. J. 94, 1377–1383.

[69] Garbett NC, Mekmaysy C, Helm CV, Jenson AB, Chaires JB (2009) Differential scanning calorimetry of blood plasma for clinical diagnosis and monitoring. Exp. Mol. Pathol. 86, 186-191.

[70] Todinova S, Krumova S, Gartcheva L, Robeerst C, Taneva SG (2011) Microcalorimetry of blood serum proteome – a modified interaction network in the multiple myeloma case. Anal. Chem. 83, 7992-7998.

[71] Ruben AJ, Kiso Y, Freire E (2006) Overcoming roadblocks in lead optimization: A thermodynamic perspective. Chem. Biol. Drug Des. 67, 2–4.

[72] Cai L, Cao A, Lai L (2001) An isothermal titration calorimetric method to determine the kinetic parameters of enzyme catalytic reaction by employing the product inhibition as probe. Anal. Biochem. 299, 19–23.

[73] Chaires JB (2006) A thermodynamic signature for drug–DNA binding mode. Arch. Biochem. Biophys. 453, 26–31.

[74] Burnouf D, Ennifar E, Guedich S, Puffer B, Hoffmann G, Bec G, Disdier F, Baltzinger M, Dumas P (2012) kinITC: A new method for obtaining joint thermodynamic and kinetic data by isothermal titration calorimetry. J. Am. Chem. Soc. 134, 559-565.

[75] Reymond C, Bisaillon M, Perreault J-P (2009) Monitoring of an RNA multistep folding pathway by isothermal titration calorimetry. Biophys. J. 96, 132-140.

[76] Kardos J, Yammamoto K, Hasegawa K, Naiki H, Goto Y (2004) Direct measurement of thermodynamic parameters of amyloid formation by isothermal titration calorimetry. J. Biol. Chem. 279, 55308-55314.

[77] Zhou X, Manjunatha K, Sivaraman J (2011) Application of isothermal titration calorimetry and column chromatography for identification of biomolecular targets. Nature protocols 6,158-165.

[78] Franco G, Bañuelos S, Falces J, Muga A, Urbaneja MA (2008) Thermodynamic characterization of nucleoplasmin unfolding: interplay between function and stability. Biochemistry 47, 7954-7962.

[79] Vancurova S, Paine TM, Lu W, Paine PL (1995) Nucleoplasmin associates with and is phosphorylated by casein kinase II. J. Cell Sci. 108,779–787.

[80] Bañuelos S, Hierro A, Arizmendi JM, Montoya G, Prado A, Muga A (2003) Activation mechanism of the nuclear chaperone nucleoplasmin: role of the core domain. J. Mol. Biol. 334, 585-593.

[81] Tholey A, Lindermann A, Kinzel V, Reed J (1999) Direct effects of phosphorylation on the preferred backbone conformation of peptides: a nuclear magnetic resonance study. Biophys. J. 76, 76–87.

[82] Philpott A, Leno GH (1992) Nucleoplasmin remodels sperm chromatin in *Xenopus* egg extracts. Cell 69, 759-767.

[83] Koshland DEJr (1996) The structural basis of negative cooperativity: receptors and enzymes. Curr. Opin. Struct. Biol. 6, 757-761.

[84] Schön A, Freire E (1989) Thermodynamics of intersubunit interactions in cholera toxin upon binding to the oligosaccharide portion of its cell surface receptor, ganglioside GM1. Biochemistry, 28, 5019–5024.

[85] Conti E, Uy M, Leighton L, Blobel G, Kuriyan J (1998) Crystallographic analysis of the recognition of a nuclear localization signal by the nuclear import factor karyopherin alpha. Cell 94, 193–204.

[86] Conti E, Kuriyan J (200) Crystallographic analysis of the specific yet versatile recognition of distinct nuclear localization signals by karyopherin alpha. Structure 8, 329–338.

[87] Fanara P, Hodel MR, Corbett AH, Hodel AE (2000) Quantitative analysis of nuclear localization signal (NLS)-importin alpha interaction through fluorescence depolarization. Evidence for auto-inhibitory regulation of NLS binding. J. Biol. Chem. 275, 21218–21223.

[88] Fontes MR, Teh T, Jans D, Brinkworth RI, Kobe B (2003) Structural basis for the specificity of bipartite nuclear localization sequence binding by importin-α. J. Biol. Chem. 278, 27981–27987.

[89] Yang SN, Takeda AA, Fontes MR, Harris JM, Jans DA, Kobe B (2010) Probing the specificity of binding to the major nuclear localization sequence-binding site of importin-α using oriented peptide library screening. J. Biol. Chem. 285, 19935–19946.

[90] Kobe B (1999) Autoinhibition by an internal nuclear localization signal revealed by the crystal structure of mammalian importin. R. Nat. Struct. Biol. 6, 388–397.

[91] Harreman MT, Cohen PE, Hodel MR, Truscott GJ, Corbett AH, Hodel AE (2003) Characterization of the auto-inhibitory sequence within the N-terminal domain of importin alpha. J. Biol. Chem. 278, 21361–21369.

[92] Tarendeau F, Boudet J, Guilligay D, Mas PJ, Bougault CM, Boulo S, Baudin F, Ruigrok RW, Daigle N, Ellenberg J, Cusack S, Simorre JP, Hart DJ (2007) Structure and nuclear import function of the C-terminal domain of influenza virus polymerase PB2 subunit. Nat. Struct. Mol. Biol. 14, 229–233.

[93] Dias SM, Wilson KF, Rojas KS, Ambrosio AL, Cerione RA (2009) The molecular basis for the regulation of the cap-binding complex by the importins. Nat. Struct. Mol. Biol. 16, 930–937.

[94] Dingwall C, Sharnick SV, Laskey RA (1982) A polypeptide domain that specifies migration of nucleoplasmin into the nucleus. Cell 30, 449-458.

[95] Jans DA (1995) The regulation of protein transport to the nucleus by phosphorylation. Biochem. J. 311, 705-716.

[96] Hood JK, Silver PA (1999). In or out? Regulating nuclear transport. Curr. Opin. Cell Biol. 11, 241-247.

[97] Fontes MR, Teh T, Toth G, John A, Pavo I, Jans DA, Kobe B (2003) Role of flanking sequences and phosphorylation in the recognition of the simian-virus-40 large T-antigen nuclear localization sequences by importin-α. Biochem. J. 375, 339-349.

[98] Cingolani G, Petosa C, Weis K, Muller CW (1999) Structure of importin-β bound to the IBB domain of importin-alpha. Nature 399, 221–229.

[99] Jäkel S, Albig W, Kutay U, Bischoff FR, Schwamborn K, Doenecke D, Görlich D (1999) The importin beta/importin 7 heterodimer is a functional nuclear import receptor for histone H1. EMBO J. 18, 2411-2423.

[100] Jäkel S, Mingot JM, Schwarzmaier P, Hartmann E, Görlich D (2002) Importins fulfil a dual function as nuclear import receptors and cytoplasmic chaperones for exposed basic domains. EMBO J. 21, 377-386.

[101] Johnson-Saliba M, Siddon NA, Clarkson MJ, Tremethick DJ, Jans DA (2000) Distinct importin recognition properties of histones and chromatin assembly factors. FEBS Lett. 467, 169-174.

[102] Falini B, Mecucci C, Tiacci E, Alcalay M, Rosati R, Pasqualucci L, La Starza R, Diverio D, Colombo E, Santucci A, Bigerna B, Pacini R, Pucciarini A, Liso A, Vignetti M, Fazi P, Meani N, Pettirossi V, Saglio G, Mandelli F, Lo-Coco F, Pelicci PG, Martelli MF (2005) Cytoplasmic nucleophosmin in acute myelogenous leukemia with a normal karyotype.Engl. J. Med. 352, 254-266.

[103] Falini B, Bigerna B, Pucciarini A, Tiacci E, Mecucci C, Morris SW, Bolli N, Rosati R, Hanissian S, Ma Z, Sun Y, Colombo E, Arber DA, Pacini R, La Starza R, Verducci Galletti B, Liso A, Martelli MP, Diverio D, Pelicci PG, Lo Coco F, Martelli MF (2006) Aberrant subcellular expression of nucleophosmin and NPM-MLF1 fusion protein in acute myeloid leukaemia carrying t(3;5): a comparison with NPMc+ AML. Leukemia 20, 368-371.

[104] Falini B, Bolli N, Liso A, Martelli MP, Mannucci R, Pileri S, Nicoletti I (2009) Altered nucleophosmin transport in acute myeloid leukaemia with mutated NPM1: molecular basis and clinical implications. Leukemia 23,11731-1743.

[105] Grisendi S, Mecucci C, Falini B, Pandolfi PP (2006) Nucleophosmin and cancer. Nature Reviews 6, 493-505.

Section 3

Application of MicroCalorimetry to Study Protein Stability and Folding Reversibility

Chapter 8

Determination of Folding Reversibility of Lysozyme Crystals Using Microcalorimetry

Amal A. Elkordy, Robert T. Forbes and Brian W. Barry

Additional information is available at the end of the chapter

http://dx.doi.org/10.5772/51131

1. Introduction

An important aspect in the preparation of proteins as pharmaceutical products is stabilisation of the native protein conformation (folded, three-dimensional, tertiary state), which is required for biological activity. Moreover, it is not enough for this conformation to be stable, but the protein must be able to find the state or folding pathway in a short time from a denatured, unfolded conformation [1]. Folding minimises exposure of non-polar groups and maximises exposure of polar groups to the solvent [2].

Lysozyme, a globular protein, molecular weight 14,300 Da, was chosen as a model protein; it consists of a single 129 amino acid chain divided into two domains cross linked by four disulfide bridges. The hydrophilic groups tend to concentrate on the surface and the hydrophobic groups in the core [3]. The goal of this study was to investigate the influences of crystallisation on folding reversibility of lysozyme in solution as assessed calorimetrically.

Many literature reports cited the value of High Sensitivity Differential Scanning Calorimetry (HSDSC) for determining thermodynamic parameters (transition temperature, T_m, and enthalpies, ΔH) that describe the folded and unfolded states [4-6]. Furthermore, HSDSC was used to measure thermal transition reversibility that is no less important than T_m and ΔH [7-8]; a protein transition is considered reversible if the molecule renatures upon cooling after heat treatment. Thermodynamic or conformational stability is defined as the difference in free energy (ΔG) between the folded and unfolded state. This stability is the sum of weak non-covalent interactions including hydrogen bonds, van der Waal interactions, salt bridges and hydrophobic forces [9]. Thermodynamic stability is divided into biophysical, which includes the study of thermodynamics, protein denaturation and renaturation (as is discussed in this Chapter) and biochemical, which involves comparative studies of protein conformation and stability of two or more proteins to establish various structural features that deduce stability within a given biomolecule [10].

Differential scanning calorimetry has the ability to provide detailed information about both the physical and energetic properties of a substance [11]. The HSDSC technique was described in detail by Cooper and Johnson (1994) [12]. In summary, all equilibrium processes involving molecules are governed by free energy changes (ΔG), made up of enthalpy (ΔH) and entropy (ΔS). The relationship between (ΔH) and (ΔS) is given by the Gibbs free energy equation:

$$\Delta G = \Delta H - T.\Delta S \quad (1)$$

where T is the temperature. Differential scanning calorimeters measure enthalpies, which provide the basis for determining the thermodynamic properties of a system. Both enthalpy and entropy are related to the heat capacity of the system. The enthalpy is the total energy (at constant pressure) required to heat the system from absolute zero to the required temperature:

$$H = \int_0^T C_P(T).dT \quad (2)$$

where C_P (T) is the temperature-dependent heat capacity at constant pressure. The total entropy of the system can be expressed as:

$$S = \int_0^T [C_P(T)/T].dT \quad (3)$$

Accordingly, differences in H and S can be expressed as:

$$\Delta H = \int_0^T \Delta C_P(T).dT \quad (4)$$

and

$$\Delta S = \int_0^T \Delta[C_P(T)/T].dT \quad (5)$$

Thermodynamic parameters depend on conditions, such as temperature, pressure, concentration and composition. Thus, it is necessary to correct experimental results to standard conditions (standard states) denoted by the superscript 0 (*e.g.* ΔG^0) for simplicity of comparison of data from different situations. The standard state of solutes in dilute solutions is a concentration of 1 M. The standard temperature and pressure are usually 25°C and 1 atm, respectively. The standard free energy change (ΔG^0), which is the free energy change during the reaction where all components are in their standard states, can be measured from:

$$\Delta G^0 = - RT.\ln K \quad (6)$$

where R is the gas constant and K is the equilibrium constant for the process and is related to (ΔH^0) and (ΔS^0) and to the absolute temperature by:

$$\ln K = - (\Delta H^0/RT) + (\Delta S^0/R) \quad (7)$$

assuming that (ΔH^0) and (ΔS^0) do not vary significantly with temperature over the range of interest.

The HSDSC provides an insight into the thermal stability and instability (e.g. formation of soluble and insoluble aggregates) of solutions of different formulations. HSDSC was used to assess the thermal stability of lysozyme solutions after storage at stressed conditions [13]. Consecutive heating scans indicated the folding reversibility of thermal transitions [8, 14] and the validity of calorimetrically measured protein folding reversibility [15-17]. Creighton (1994) [18] reported mechanisms and thermodynamic factors controlling protein folding-and-unfolding.

In the present study, HSDSC investigated thermal changes; in particular protein refolding performance, of crystallised samples (in low and high protein concentrations) upon heat treatment. The thermal structural transition of lysozyme involves two thermodynamic states, native and denatured [19] as for other globular proteins [20]. However, Hirai et al. (1999) [21] indicated that folding-and-unfolding kinetics of proteins depend on the number of amino acid residues. Proteins with residues above ~100 do not follow simple two state kinetics in a folding-and-unfolding process as a single cooperative unit. Accordingly during HSDSC analysis of lysozyme, there might be formation of intermediates between native and denatured states. The thermodynamic stability of proteins not only requires that the transition temperature (T_m) and other thermodynamic parameters remain constant but also implies reversibility of protein from unfolded (denatured) to folded (native) state after removing the effect of an external condition such as heat. Anfinsen (1973) [22] reported that denaturation of Ribonuclease A, by heat or urea, was reversible when denatured molecules returned to a normal environment of temperature and solvent. Hence, both structure and enzymatic activity were regained.

Consequently, formation of the native state is a global property of the protein as described [1]. This state is necessary for stability and activity; proteins are marginally stable and achieve stability only within narrow ranges of conditions of solvent and temperature. The free energy of stabilization of proteins under ordinary conditions is ~ 5-15 kcal mol^{-1} [1].

Proteins undergo various structural changes if physiological conditions alter. Accordingly, they may denature and the denatured protein tends to adsorb to surfaces and aggregate with other protein molecules. Katakam et al. (1995) [23] proposed that denaturation of recombinant human growth hormone involves unfolding of the molecule; the unfolded part adsorbs to surfaces and aggregates with neighbouring molecules. Shaking and exposure to an air/water interface, heating, lyophilisation or reconstitution of lyophilised protein may aggregate protein with subsequent loss of stability and activity.

The combination of HSDSC and enzymatic activity determined if refolding of denatured crystallised lysozyme after thermal denaturation in HSDSC arises from the nativeness, three-dimensional folded state, of the initial lysozyme structure. This means that enzymatic activity was employed to investigate if folding reversibility of the thermal transition reflects the renaturation of the unfolded protein to folded native structure.

2. Materials and methods

2.1. Materials

Chicken egg white lysozyme (purity 95%, 5% sodium chloride and sodium acetate), sodium chloride (99.5%), sodium phosphate (99.3%) and *Micrococcus lysodeikticus* were purchased from Sigma Chemical Company (St. Louis, Mo). Sodium acetate anhydrous (98%), potassium dihydrogen orthophosphate (> 99%) were obtained from BDH Chemicals Ltd. Poole, UK. Water was deionised, double distilled.

2.2. Preparation of crystallized lysozyme

Lysozyme was crystallised using a published method [24]. Crystals formed were filtered, dried and kept in a freezer (-15°C) until tested.

2.3. High Sensitivity Differential Scanning Calorimetry (HSDSC)

Solution samples of crystallised lysozyme were analysed with a Microcal MCS differential scanning calorimeter (Microcal Inc., MA, USA). Degassed samples (5 and 20 mg product / 1 mL 0.1M sodium acetate buffer, pH 4.6) and reference (0.1M sodium acetate buffer, pH 4.6) were loaded into cells using a gas tight Hamilton 2.5 mL glass syringe. The folding reversibility of lysozyme denaturation was assessed by temperature cycling using two scan calorimetric methods. The upscan-upscan method (UU) employed two consecutive upscans from 20-90°C at 1°C/min. After the first upscan, the sample was immediately cooled in the calorimeter (downscan) to 20°C at 0.75°C /min (the fastest cooling rate allowed by the instrument) and the heating cycle was immediately repeated. Transition reversibility was measured as ratio (%) of enthalpy change of second upscan (ΔH_2) over that of first upscan (ΔH_1). The upscan-downscan method (UD) involved heating of protein solution from 20-90°C at 1°C/min immediately followed by downscan (cooling) from 90-20°C at a cooling rate of 0.75°C/min. Enthalpies were measured and downscan (ΔH_3) / upscan (ΔH_1) enthalpy ratios were calculated as a measure of folding reversibility. The calorimeter was temperature- and heat capacity-calibrated using sealed hydrocarbon standards of known melting points and electrical pulses of known power, respectively.

Experiments were performed under 2 bar nitrogen pressure. A base line was run before each measurement by loading the reference in both the sample and reference cells; this base line was subtracted from the protein thermal data and the excess heat capacity was normalized for lysozyme concentration. Data analysis and deconvolution employed ORIGIN DSC data analysis software. The T_m (mid point of the transition peak) values for all transitions were calculated.

2.4. Enzymatic assay

Biological activities of thermally denatured crystallised lysozyme were determined after cooling (in HSDSC) to determine whether the renaturation is due to the nativeness of the protein structure i.e. to correlate the folding reversibility with biological activity. In this assay, a bacterial suspension was prepared by adding 20 mg of *Micrococcus lysodeikticus* to 90 mL of phosphate buffer 0.067 M, pH 6.6, and 10 mL of 1% NaCl. The biological reaction was initiated by addition of 0.5 mL of each enzyme solution to 5 mL of the bacterial suspension. The activity unit of lysozyme is defined as the amount of enzyme decreasing the absorption rate at 450 nm at 0.001 /min at 25°C and pH 6.6. Rates were monitored using a UV/Vis. spectrophotometer (Pu 8700, Philips, UK) at 25°C.

All data were presented as mean of three determinations ± standard deviation. The Student's *t*-test assessed significance.

3. Results and discussion

3.1. High Sensitivity Differential Scanning Calorimetry (HSDSC)

Differential scanning microcalorimetry experiments can thermodynamically characterise the unfolding transition by determing heat capacities, enthalpies and melting temperatures of native and denatured protein [25]. HSDSC monitored thermal stability and folding reversibility of reconstituted lysozyme preparations. For samples, traces for thermal denaturation and folding reversibilities, using (UU) method, of reconstituted crystallised lysozyme are illustrated in Figure 1(a) and (b) for 5mg/mL and 20mg/mL protein concentrations, respectively. Thermodynamic parameters and enzymatic activities are in Table 1. Figure 2 shows an example for folding reversibility of unprocessed lysozyme using (UD) method. As is evident in Figures 1 and 2, HSDSC profiles of all samples showed a single endothermic peak (first upscan). Lysozyme crystals started to unfold at ~65°C with a mean T_m of 76.1°C (T_{m1}).

It is noticeable that rescan profiles, whether endothermic (second heating cycle, Figure 1a and b dotted lines) or exothermic (downscan upon cooling, Figure 2 lower trace) showed two peaks, a main one and a small peak or shoulder. Deconvolution of the data (using ORIGIN DSC data analysis software) revealed two transition regions characterised by T_m at ~76.1°C (T_{m2}) for the main peak and at ~66°C for the shoulder, indicating that the lysozyme transition is not a two state transition. This may be explained on the basis that lysozyme consists of a single polypeptide chain divided into two structural domains (α-helix and β-sheet). During refolding each domain may be refolded separately with different pathways. Consequently, two peaks appear instead of one; this explanation agrees with Remmele et al. (1998) [14] who attributed the three T_m peaks to the three domains of immunoglobulin-type domains that make up the Interleukin-1 receptor.

The other reasonable explanation is that lysozyme, when its folding process is analysed using circular dichroism, does not obey a single co-operative transition, but the process involves several parallel folding pathways. Each of the two domains stabilises with different kinetics [26]. In particular, the amides in the α-helix are involved in the formation of stable helical structure and assembly of the hydrophobic core. Then a stable hydrogen bonded structure in the β- domain forms. Accordingly, partially structured intermediates develop during the folding of lysozyme. This explanation is supported by Buck et al. (1993) [27] who reported that lysozyme consists of two structural domains that are stabilised by different pathways.

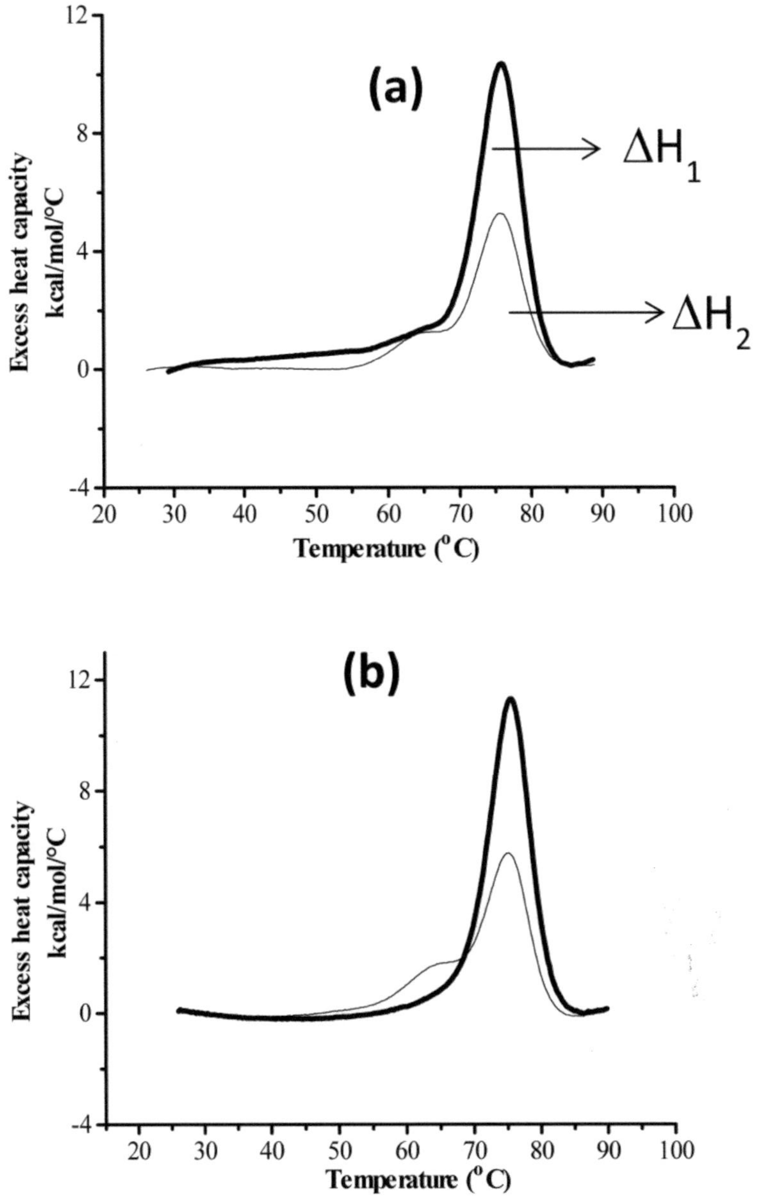

Figure 1. Normalised consecutive calorimetric upscans, first upscan (solid line) and second upscan (dotted line) of crystallised lysozyme. Conditions: (a) 5mg/mL protein, 0.1 M sodium acetate buffer, pH 4.6, heating rate 1°C/min and (b) 20mg/mL protein, 0.1 M sodium acetate buffer, pH 4.6, heating rate 1°C/min.

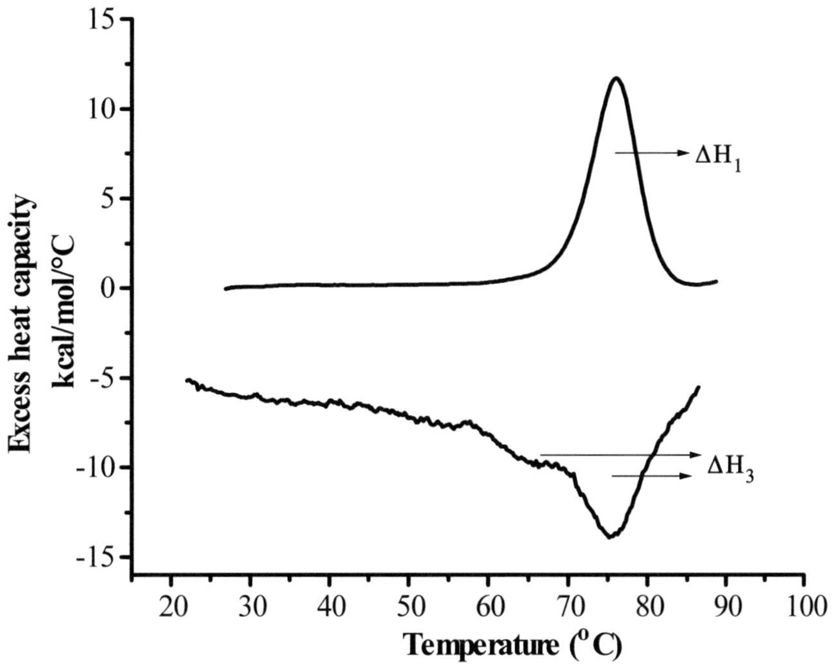

Figure 2. Normalised calorimetric upscan (upper trace) and downscan (lower trace) of lysozyme, as received. Conditions: 5mg/mL protein, 0.1 M sodium acetate buffer, pH 4.6, heating rate 1°C/min, cooling rate 0.75°C/min.

Lysozyme Sample	T_{m1} (°C)	T_{m2} (°C)	% folding reversibility ($\Delta H_2/\Delta H_1$)	% enzymatic activity
Crystallised				
5 mg mL^{-1}	76.1±0.21	75.9±0.15	66.5±1.4	65.7±1.4
20 mg mL^{-1}	75.6±0.07	75.2±0.02	52.4±1.6	52.7±2.2

[a] T_{m1}, T_{m2} are mid-point peak transition temperatures of first and second upscans; ΔH_1, ΔH_2 are calorimetric enthalpies of transitions of first and second upscans; % enzymatic activity is the activity of each sample relative to fresh material of that sample, ±S.D., n= 3.

Table 1. Thermodynamic parameters for the thermal denaturation, folding reversibilities using consecutive upscan method (UU) and enzymatic activities of crystallised lysozyme samples[a].

Lysozyme Sample	% folding reversibility ($\Delta H_3/\Delta H_1$)
Crystallised	
5 mg mL^{-1}	43.6 (1.6)
20 mg mL^{-1}	48.6 (3.8)

[a] Values between brackets are S.D., n= 3.

Table 2. Folding reversibilities using upscan-downscan (UD) method of crystallised lysozyme samples [a].

A comparison of calorimetric enthalpy (ΔH_{cal}) and the theoretical enthalpy (ΔH_{VH}, van't Hoff enthalpy) changes, to judge the validity of a two-state mechanism for the unfolding of lysozyme, reveals the presence of intermediates [25]. It was reported that, in the unfolded state, proteins aggregate and react chemically with amino acid residues exposed to the solvent; this can lead to misfolding or irreversible denaturation [28]. Also, the formation of any irreversible component alters the shape of a HSDSC thermodynamic peak over the temperature range at which it forms [29].

For low protein concentration (5mg/mL, Figure 1a), Table 1 shows no significant difference between T_{m1} (transition temperature of first upscan) and T_{m2} (transition temperature of second upscan) for protein samples.

It was reported that determination of the mechanism and pathway of unfolding and refolding depends on the identification of the intermediates that may not be stable at the equilibrium [18]. Thus, detection and characterisation of kinetic folding intermediates is complex. This intricacy can arise from accumulation of intermediates or from subpopulations of the unfolded state refolding at different rates. Also, events in folding are obscure [1]. With respect to samples with high protein concentration (20mg/mL, Figure 1b), Table 1 demonstrates that T_{m2} decreased compared to its corresponding T_{m1} .

On comparing low and high protein concentrations, thermal stabilities (T_{m1} and T_{m2}) of lysozyme crystals at high concentration significantly decreased ($p < 0.05$). Accordingly, high protein concentration influences thermal stability, this is in agreement with previously published data for dried proteins [17]. Moreover, folding reversibilities and enthalpies of first upscan of all samples (Figure 1 and Table 1) decreased with increasing concentration. Enthalpy values correlated with the content of ordered secondary structure of protein [30]. The decrease in enthalpy of protein may be attributed to denaturation at high protein concentration because a partially unfolded protein needs less heat energy to denature than a native form [17]. In general, crystals are chemically and physically pure substances (atoms or molecules within crystals are arranged in highly ordered patterns in three dimensional structures); the other possible reason for the observed reversibility of lysozyme at high protein concentration is that the water in lysozyme crystal lattices inhibits protein-protein interactions and aggregation, to some extent, which may take place at high lysozyme concentration after denaturation by heat in the HSDSC. Consequently, the crystallisation maintains the three-dimensional folded structure of lysozyme and enhances the renaturation of the protein. Water molecules play an important role in the function of proteins through maintaining their tertiary structure. The structure of biological macromolecules in an aqueous solution is similar to that in a crystalline state [31]. There are two kinds of hydration shell of biomolecules in aqueous solution; the primary and secondary hydration shells. Water molecules in the primary hydration shell are directly bound to molecules. The water molecules in the secondary hydration shell have a character intermediate between those of the primary hydration shell and bulk water. On the other hand, crystal water is classified into groups that correspond to the hydration shells in solutions. These water molecules correspond mainly to those in the primary hydration shell and partly to those in the secondary hydration shell [31]. Consequently, there is strong interaction between water and protein molecules in crystalline states. Takano et al. (1999) [6] used differential scanning calorimetry to examine the contribution of hydrogen bonds to the conformational stability of mutant human lysozyme. The authors commented that hydrogen bonding between human lysozyme atoms and water bound with the protein molecules in crystals contributes to the protein conformational stability. The net contribution of one intramolecular hydrogen bond to protein stability in terms of Gibbs energy was ~8.9 kJ/mol. On the basis of Takano et al. (1999) [6] study, hydrogen bonds are one of the important factors stabilising the folded conformations of proteins. From these results crystals were capable of refolding and hence lysozyme crystals not only maintained thermal stability and conformational integrity as suggested previously [24], but also improved refolding ability, which is necessary in protein formulation and processing. Refer to a study by Elkordy et al. [17] for folding reversibility of lyophilised and spray dried lysozyme. Also, a study by Forbes et al. [16] reported the folding reversibility of spray dried and crystallised trypsin.

For folding reversibility calculated by (UD) method, Table 2 above summarises the results of folding reversibilities of crystallised lysozyme at low and high protein concentrations. From Table 2, it is apparent that the percentage folding reversibility calculated by (UU) method (Table 1) was significantly higher ($p < 0.05$) than that derived from the (UD) method (Table 2). This implies that the latter method underestimates the apparent folding reversibility of samples.

3.2. Enzymatic assay

Lysozyme solutions upon cooling in the HSDSC after thermal denaturation were assayed for biological activity towards *Micrococcus lysodeikticus.* Based on the HSDSC results, all samples renatured to some extent after thermal stress. Thus, enzymatic assay should answer an important question. Is this renaturation or folding reversibility related to regain of the native structure of lysozyme (which is essential for biological activity), or does it result from misfolding, i.e. folding of the protein in a manner different from the original natured structure which subsequently leads to loss of activity?

Table 1 presents percentage enzymatic activities of preheated solutions, in HSDSC, of crystallised lysozyme (relative to an aqueous solution of a fresh sample). It is evident that the biological activity of lysozyme was maintained by crystals (5mg/mL and 20mg/mL). The results were consistent with data of folding reversibilities. This answers the question posed previously in that folding reversibility was related to the native structure of lysozyme that is required for its activity, as the greater the folding reversibility, the higher the enzymatic activity. The results illustrated that lysozyme crystals maintained structural integrity even after heating in the HSDSC. A review by Jen and Merkle (2001) [32] showed that hydrated protein within crystals is present in a folded, native form.

From the HSDSC and enzymatic activity results, the folding reversibility, calculated by consecutive upscans (UU, Tables 1), correlated with enzymatic activity of lysozyme, confirming that the upscan-downscan method (UD, Table 2) underestimates the magnitude of folding reversibility. However, proteins are diverse molecules and the presence of correlation between folding reversibility and biological activity of lysozyme, as demonstrated in this study, may not be applicable to other proteins.

4. Conclusions

The overall results suggested that reconstituted lysozyme crystals were able to refold after heating. The folding reversibility arises from the nativeness of the initial lysozyme structure as demonstrated by biological activity data. The results indicated that the upscan-downscan method underestimated the extent of folding reversibility. Consequently, it is preferable to calculate this reversibility, employing high sensitivity differential scanning calorimetry, by the consecutive heating upscan method.

Author details

Amal A. Elkordy*
Department of Pharmacy, Health and Well-being, University of Sunderland, Sunderland, UK

Robert T. Forbes and Brian W. Barry
School of Pharmacy, University of Bradford, Bradford, UK

5. References

[1] Lesk A M 2001. In vivo, in vitro, in silicio. In: Lesk, A.M. (Ed.), Introduction to protein architecture. Oxford University Press, Oxford, New York, pp.: 15-35, 143.

[2] Rupley J A, Gratton E, Careri G 1983. Water and globular proteins. Trends in Biochem Sci. 8: 18-22.

[3] Rosenberger F 1996. Protein crystallization. J Cryst Growth. 166: 40-54.

[4] Pikal M J, Lukes A L, Lang J E, Gaines K 1978. Quantitative crystallinity determination for b-lactam antibiotics by solution calorimetry: correlations with stability. J Pharm Sci. 67: 767-772.

[5] Freire, E. (1995). Thermal denaturation methods in the study of protein folding. In *Methods in Enzymology*, Vol. 259 (Johnson, M. L. and Ackers, G. K., Eds.) Academic Press, San Diego, New York, Boston, London, Sydney, Tokyo, Toronto, pp. 144-168.

[6] Takano K, Yamagata Y, Kubota M, Funahashi J, Fujii S, Yutani K 1999. Contribution of hydrogen bonds to the conformational stability of human lysozyme: Calorimetry and x-ray analysis of six ser →ala mutants. Biochem. 38: 6623-6629.

[7] Maneri L R, Farid A R, Smialkowski P J, Seaman M B, Baldoni J M, Sokoloski T D 1991. Preformulation of proteins using high sensitivity differential scanning calorimetry (DSC). Pharm Res. 8: S-48.

[8] Boye J I, Alli I, Ismail A A 1997. Use of differential scanning calorimetry and infrared spectroscopy in the study of thermal and structural stability of α-lactalbumin. J Agric Food Chem. 45: 1116-1125.

* Corresponding Author

[9] Daniel RM 1996. The upper limits of enzyme thermal stability. Enzyme Microb. Technol. 19: 74-79.
[10] Quinn ÉÁ 2000. The stability of proteins in hydrofluoroalkane propellants. Ph. D. Thesis, University of Bradford, UK.
[11] Clas S.-D, Dalton CR, Hancock BC 1999. Differential scanning calorimetry: applications in drug development. PSTT 2: 311-320.
[12] Cooper A, Johnson CM (1994). Introduction to microcalorimetry and biomolecular energetics. In Methods in Molecular Biology: Microscopy, Optical Spectroscopy, and Macroscopic Techniques, Vol. 22 (Jones, C., Mulloy, B. and Thomas, A. H., Eds.) Humana Press Inc., Totowa, NJ, pp. 109-124.
[13] Elkordy A A, Forbes R T, Barry B W 2004. Stability of crystallised and spray-dried lysozyme. Int J Pharm. 278: 209-219.
[14] Remmele R L, Nightlinger N S, Srinivasan S, Gombotz W R 1998. Interleukin-1 receptor (IL-1R) liquid formulation development using differential scanning calorimetry. Pharm Res. 15: 200-208.
[15] Branchu S, Forbes R T, Nyqvist H, York P 1998. The relationship between the folding reversibility and enzymatic activity of trypsin. J Pharm Sci. 1 (suppl.): 541.
[16] Forbes R T, Barry B W, Elkordy AA 2007. Preparation and characterisation of spray-dried and crystallised trypsin: FT-Raman study to detect protein denaturation after thermal stress. European J Pharm Sci. 30: 315-323.
[17] Elkordy A A, Forbes R T, Barry B W 2008. Study of protein conformational stability and integrity using calorimetry and FT-Raman spectroscopy correlated with enzymatic activity. European J Pharm Sci. 33: 177-190.
[18] Creighton T E 1994. The protein folding problem. In: Pain, R.H. (Ed.),. Mechanisms of protein folding. Oxford University Press, Oxford, New York, Tokyo, pp. 1-25.
[19] Schmid F X 1992. Kinetics of unfolding and refolding of single-domain proteins. In: Creighton, T.E. (Ed.), Protein folding. Freeman, W.H. and Company, New York, pp. 197-241.
[20] Catanzano F, Giancola C, Graziano G, Barone G 1996. Temperature-induced denaturation of ribonuclease S: A thermodynamic study. Biochem. 35: 13378-13385.
[21] Hirai M, Arai S, Iwase H 1999. Complementary analysis of thermal transition multiplicity of hen egg-white lysozyme at low pH using X-ray scattering and scanning calorimetry. J Phys Chem B. 103: 549-556.
[22] Anfinsen C B 1973. Principles that govern the folding of protein chains. Sci. 181: 223-230.
[23] Katakam M, Bell L N, Banaga A K 1995. Effect of surfactants on the physical stability of recombinant human growth hormone. J Pharm Sci. 84: 713-716.
[24] Elkordy A A, Forbes R T, Barry B W 2002. Integrity of crystalline lysozyme exceeds that of a spray dried form. Int J Pharm. 247: 79-90.
[25] Matouschek A, Serrano L, Fersht A R 1994. Analysis of protein folding by protein engineering. In: Pain, R.H. (Ed.), Mechanisms of protein folding. Oxford University Press, Oxford, New York, Tokyo, pp. 137-159.

[26] Radford S E, Dobson C M, Evans P A 1992. The folding of hen lysozyme involves partially structured intermediates and multiple pathways. Nature. 358: 302-307.

[27] Buck M, Radford S E, Dobson C M 1993. A partially folded state of hen egg white lysozyme in trifluoroethanol: Structural characterization and implications for protein folding. Biochem. 32: 669-678.

[28] Creighton T E 1993. Proteins in solution and in membranes. In: Creighton, T. E. (Ed.),. Proteins: Structures and molecular properties, 2nd Ed., Freeman, W.H. and Company, New York, pp. 287-323.

[29] Lepock J R, Ritchie K P, Kolios M C, Rodahl A M, Heinz K A, Kruuv J 1992. Influence of transition rates and scan rate on kinetic simulations of differential scanning calorimetry profiles of reversible and irreversible protein denaturation. Biochem. 31: 12706-12712.

[30] Koshiyama I, Hamano M, Fukushima D 1981. A heat denaturation study of the 11 S globulin in soybean seeds. Food Chem. 6: 309-322.

[31] Urabe H, Sugawara Y, Ataka M, Rupprecht A 1998. Low-frequency Raman spectra of lysozyme crystals and oriented DNA films: dynamics of crystal water. Biophys J. 74: 1533-1540.

[32] Jen A, Merkle H P 2001. Diamonds in the rough: Protein crystals from a formulation perspective. Pharm Res. 18: 1483-1488.

Chapter 9

Calorimetric Study of Inulin as Cryo- and Lyoprotector of Bovine Plasma Proteins

Laura T. Rodriguez Furlán, Javier Lecot, Antonio Pérez Padilla,
Mercedes E. Campderrós and Noemi E. Zaritzky

Additional information is available at the end of the chapter

http://dx.doi.org/10.5772/54462

1. Introduction

Inulin is a generic term applied to heterogeneous blends of fructo-oligosaccharides [1] which are reserve carbohydrate sources present in many plant foods such as bananas, onions, garlic, leeks, artichokes and chicory, which represents the main commercial source. This polysaccharide has a wide range of both, nutritional and technological applications. Nutritionally, inulin is regarded as a soluble fiber which promotes the growth of intestinal bacteria, acting as a prebiotic. Also, is a non-digestible carbohydrate with minimal impact on blood sugar and unlike fructose, it is not insulemic and does not raise triglycerides being generally considered suitable for diabetics and potentially helpful in managing blood sugar-related illnesses [2-4]. Among the technological benefits, inulin is used as fat and sugar replacement, low caloric bulking agent, texturing and water-binding agent [5,6]. One general property of the saccharides is the stabilization of proteins by their incorporation into carbohydrate solutions before freeze-drying being this a known preservation procedure [7-10]. The previous incorporation of saccharide promotes the formation of amorphous, glassy systems, inhibits crystallization and influences the kinetics of deteriorative reactions upon storage by which its structured integrity is maintained [8,9,11,12]. To act successfully as a protectant, the saccharides should have a high glass transition temperature (T_g), a poor hygroscopicity, a low crystallization rate, containing no reducing groups. When freeze-drying is envisaged as a method of drying, a relatively high T'_g of the freeze concentrated fraction is preferable. Previous studies demonstrated that inulin meets these requirements being excellent protector of therapeutical proteins and viruses over the drying and storage processes [13,14].

The protein preserved by freeze-drying simplifies aseptic handling and enhances stability of protein products, with limited shelf lives in solution, by obtaining a dry powder without excessive heating. However, during the freeze-drying process the protein may lose its activity and must be protected from conformational changes or denaturation [11,15]. The stabilization of proteins conferred by saccharides during freeze- drying has been explained by several mechanisms. First, replacing the hydrogen bonding between water and protein stabilizes the protein during drying processes, and second, the formation of a glass matrix where the protein is encapsulated avoiding its unfolding and thus preserving its conformation during freeze-drying [8,12,16-18]. Therefore, through the correct selection of the saccharide it is possible to improve the stability of proteins through their encapsulation in a glassy matrix, where molecular mobility is quite limited so that the rates of diffusion-controlled reactions, like protein unfolding or protein aggregation, are reduced [16,19,20].

Information about the energy of a protein can be obtained by means of thermal denaturation studies, allowing the characterization of their behavior during freeze-drying cycle. Differential scanning calorimetry (DSC) is one of the most useful methods for assessing protein thermal behavior and to obtain thermodynamic parameters of folding-unfolding transitions [21].

During the freeze-drying of a protein solution with or without saccharides to protect the structure, the primary drying is the most time consuming stage of the process. It should be carry out at the maximum allowable temperature usually associated to the glass transition temperature of the maximally freeze concentrate solution (T'_g). Below this temperature a glassy state that behaves as an amorphous solid is obtained. If the temperature of the frozen system rises above the T'_g, the material becomes less viscous and freeze-drying may cause the loss of the porous structure and the product collapse [20,22,23]. In the freeze-dried sample, water is removed and the solute concentration in the matrix increases, obtaining a material with an amorphous structure that exhibits a glass-rubber transition at a specific temperature which is named as the glass transition temperature (T_g) [24-28]. It is noteworthy that amorphous materials are stable in the glassy state below T_g, when the temperature is higher the viscosity decreases and thus the rate of chemical reactions increases and crystallization events occur, increasing the rate of deterioration during storage [22,25,27-29]. Both transitions T'_g and T_g are important parameters in the development of the freeze-drying cycle because not only ensures the stability and quality of the product, but also allow to improve the efficiency of the manufacturing process [20,22,28,30].

A diagram of phases for the water-saccharide system is shown in Figure 1. The curve of the freezing temperature separates the zones corresponding to the liquid and the solid (ice) solution phases. In fact, this procedure is aimed at obtaining a glassy system at room temperature as indicated in D. To get to this state, the freeze-dried process indicated by the curve A-B-C-D-E is carried out. The curve for the glass transition temperature (T_g) is reached when the solution is overcooled (B-C) until the T'_g in point C, where the concentration of the vitrification agent (saccharide) is given by C'_g. Then the water is eliminated and the solute concentration increases (C-D-E), obtaining a solid with an amorphous structure that exhibits a glass transition temperature (T_g) [22,28].

Therefore, the determination of the freeze-drying cycle is important because of physical changes that occur in the solution during the process, its study can be applied to improve processability, quality, and stability of the product during storage [29].

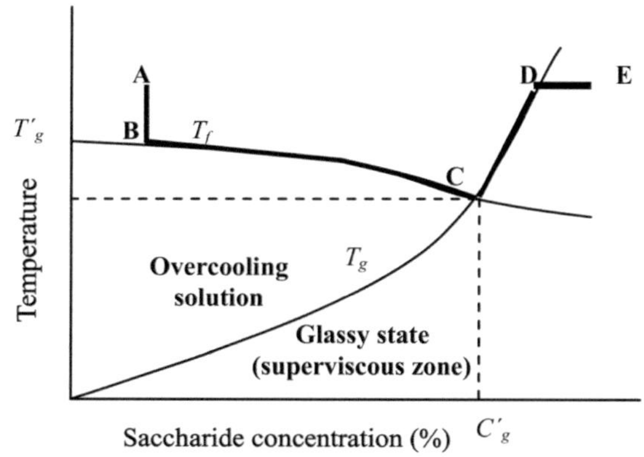

Figure 1. Phases diagram of the water–saccharide system. The curve A-B-C-D-E indicated the freeze-dried process. (T_f = freezing temperature; T_g= glass transition temperature; T'_g= glass transition temperature of the overcooling solution; C'_g = saccharide concentration)

Although many authors reported the use of saccharides as cryoprotectants of proteins and inulin as a good protector agent of some compounds, the present study is an attempt to evaluate inulin as cryoprotector of food proteins such as bovine plasma proteins, taking profit of the nutritional and technological benefit of the polysaccharide. Also there is a limited amount of data on glass transition temperatures for multicomponent mixtures and on the comparison of experimental and predicted values for such mixtures [28]. Then, the purposes of this study were *i)* to investigate the transition temperatures and the thermal denaturation of bovine plasma proteins stabilized with inulin in a glassy matrix in comparison to the effect of a monosaccharide (glucose) and a disaccharide (sucrose) at different concentrations using DSC, *ii)* to compare the quality, performance and storage conditions of these products.

The glass transition temperatures of the maximally concentrated frozen solutions (T'_g) were analyzed and compared to the experimental results by applying the predictive equations of Miller/Fox and Gordon/Taylor extended for multi-component systems. The glass transition (T_g) of the freeze dried multi-component mixtures, the onset crystallization temperature (T_c) of the solute at temperatures above T_g, in the freeze dried samples were determined. Furthermore, the kinetics of the denaturation and the thermal denaturation (T_d) of the freeze-dried samples, at different DSC scan rates, protein concentrations and pH, were analyzed and the thermodynamic compatibility of the different matrix components were determined. The enthalpy of change involved in the denaturation reactions of proteins (ΔH) was also determined. A kinetic model that describes bovine plasma proteins denaturation was proposed.

2. Materials and methods

2.1. Raw materials

The inulin used as cryoprotectant is mainly constituted by linear chains of fructose, with a glucose terminal unit, and has a molecular weight of 2400 g/mol. The commercial product was provided by Orafti Chile S.A. and was obtained from chicory. The other saccharides employed to compare their performance were: *i)* a monosaccharide, glucose (Parafarn, Argentine), with a purity of 99.99% and *ii)* a disaccharide, commercial sucrose (Ledesma S.A., Argentine).

The protein used in the study was spray dried bovine plasma (Yerubá S.A. Argentine). The molecular weights of the proteins were in the range of 15.000 to 80.000 Da. The composition was 76±5% proteins, <0.1% fat, 10% ash, 4% water, 1% low molecular weight compounds.

2.2. Preparation of Protein/carbohydrate samples: Concentration of bovine plasma proteins through ultrafiltration and freeze-drying treatments

The protein concentrate was obtained by means of a membrane process, which allowed protein concentration, eliminating insoluble macroscopic components, reducing the saline content [18]. The steps of the process were: i) the bovine plasma was dissolved in de-ionized water to a concentration of 3% w/v using a mixer at a low speed to avoid the formation of vortex and to minimize the appearance of foam; ii) the solution was passed through a porous support (Viledon FO 2431D, Germany) to remove macroscopic aggregates and reduce the saline content; iii) the feed solution (3 L) was thermostatized in a water bath and impelled with a centrifugal pump, first through a frontal flow stainless steel filter, with a pore size of 60 μm (Gora, Argentine) (this procedure of microfiltration (MF) reduces the amount of bacteria and spores and acts as cold pasteurization, moreover this stage protects the ultrafiltration (UF) membrane from fouling); and finally, iv) the UF was performed using Pellicon cassette module (Millipore, Bedford, MA, USA), containing modified polyethersulfone membranes with a molecular weight cut-off (MWCO) of 10 kDa, with a membrane area of 0.5 m^2. The concentration of proteins by UF was carried out by continuously removing the permeate stream until the desired concentration of 4% (w/v), was achieved. The experimental runs were performed at a transmembrane pressure (ΔP) of 1.5 bar, flow rate of (2.9 ± 0.05) L/min and a temperature of 10 °C. Additionally a discontinuous diafiltration (DD) process was applied to removal salts and other contaminant of low molecular weight. For this operation the starting solution was the UF concentrate, which was diluted to the initial volume (3 L) with de-ionized water in a single state and ultrafiltered to the desired concentration range.

The UF membrane undergoes a fouling process during protein permeation so a cleaning protocol may be applied. It was performed by applying a "Cleaning in Place" (CIP) procedure according to the manufacturer's instructions. At the end of each run, a cycle of water/ alkali (NaOH, pH=12.5 ± 0.5)/ water wash was applied to the membrane at (40 ± 2) °C and at a transmembrane pressure of 1 bar. Furthermore, a cleaning step using NaClO (commercial grade) 300 ppm was carried out at the same temperature and pressure to ensure sanitation and cleaning. Measurements of normalized water permeability were performed in order to verify recovery of flow through the membrane which ensures the recuperation of membrane permeability.

The bovine plasma protein (BPP) concentrate obtained by UF (concentration: 4 % w/v) was fractioned: A fraction as witness sample was reserved and the protective agent (glucose, sucrose, inulin) was added to the rest, in concentrations of 5%, 10% and 15% (w/v). A part of these solutions was reserved for DSC analysis to determine T'_g and the others were placed on stainless steel trays, frozen in a freezer at -40 °C and freeze-dried using a lyophilizer (Rificor S.A., Argentine) at 1 bar for 48 h. The samples temperature was controlled by a temperature sensor. The denatured protein content was determined before and after the freeze-drying.

2.3. Differential Scanning Calorimetry (DSC) measurements

Determination of T'_g in the protein solutions

The solutions containing plasma proteins–saccharides mixture were analyzed to determine T'_g at different pH values and saccharide concentrations by DSC with a Q100DTA Instrument (USA). The pH was adjusted using 0.1 N of NaOH and HCl. Protein concentrate solutions (average composition: saccharide 5% p/v - protein 4% p/v; saccharide 10% p/v - protein 4% p/v; saccharide 15% p/v - protein 4% p/v), (10 ± 2 mg) were weighed into aluminum DSC pans, hermetically sealed, and then loaded onto the DSC instrument at room temperature, using an empty pan as a reference. Samples Solutions were: (a) equilibrated at 20 °C and held for 1 min; (b) cooled at 2 °C/min until -80 °C for glucose, -60 °C for sucrose and -40 °C for inulin and held for 30 min; (c) warmed up to the annealing temperature (-50, -40 and -20 °C, for glucose, sucrose and inulin, respectively) by employing an annealing time of 30 min at heating rate of 2 °C/min [31]; (e) recooled at the same temperature of step (b) and held for 30 min; (f) warmed up to 0 °C at heating rate of 2 °C/min. The effectiveness of the procedure was verified corroborating the absence of ice devitrification in thermograms, that is to say the nonexistence of an exothermic peak previous to the ice melting.

Determination of T_g, T_c and T_d of proteins in the freeze–dried solids

Heat induced conformational changes on freeze-dried bovine plasma protein concentrate (BPP concentrate) in the amorphous carbohydrate matrix. The freeze-dried solids were analyzed to determine T_g, T_c and T_d at different pH values and saccharide concentrations by DSC with a Q100DTA Instrument (USA). The pH was adjusted using 0.1 N of NaOH and HCl. Protein concentrates (average composition: freeze-dried with saccharide 5% (p/v) = saccharide 35% p/p - protein 55% p/p; freeze-dried with saccharide 10% (p/v) = saccharide 64% p/p - protein 28% p/p; freeze-dried with saccharide 15% (p/v) = saccharide 79 % p/p - protein 14% p/p), (12.5 ± 2.5 mg) were weighed into aluminum DSC pans, hermetically sealed, and then loaded onto the DSC instrument at room temperature, using an empty pan as a reference.

Freeze–dried solids were equilibrated at 0 °C, held for 1 min and then warmed up to 200 °C at heating rate of 2 °C/min. To check the irreversibility of the reaction of heat-induced conformational changes, the samples after the end of the first heating stage described before, were re-scanned. For this, the protein-saccharide samples were cooled to 20 °C and stabilized during 5 min, and then warmed up to 200°C. Samples of freeze dried bovine plasma protein concentrate (BPP concentrates) in the amorphous carbohydrate matrix at pH

8, 6 and 4, at different heating rates of 2 and 5 °C/min in the temperature range 20–200 °C were analyzed. The pH was adjusted using 0.1 N of NaOH and HCl. Measurements were carried out on three separate samples (replicates). The following parameters were calculated at least in triplicate: T_d, at maximum heat flow, and ΔH, the enthalpy change involved in the overall heat-induced reactions within the protein molecules, that was determined by integrating the area beneath the enthalpy peak and above a straight baseline drawn in between the beginning and the end of the transition temperature range [32-34]; the T'_g and T_g were determined from the midpoint of the transition of the baseline shift on the amorphous sample.

In the freeze dried samples, at temperatures above T_g, the onset crystallization temperature (T_c) of the added solute was determined from the intersection of the baseline and the tangent of the exothermic peak. The enthalpy change involved in the overall heat-induced reactions within the protein molecules, ΔH_c, was determined by integrating the area beneath the exothermic peak and above a straight baseline drawn between the beginning and end of the transition temperature range [22,32,33].

2.4. Determination of native protein content

The native protein content is a measure of protein functionality preservation. It was determined after isoelectric precipitation of denatured/aggregated protein [18,35]. Dispersions of protein concentrate at 1% (w/v) were adjusted to pH value inferior of the pI of plasma proteins (~ 4.8) using 0.1 N of NaOH and HCl. An aliquot of the solution was centrifuged in a refrigerated ultracentrifuge (Beckman J2-HS) at 20,000 rpm 30 min at 5 ºC. Protein concentration in the supernatants was diluted in a dissociating buffer (EDTA 50 mM, urea 8 M, pH= 10) and determined by molecular absorptiometry at 280 nm. The results were reported as percentage of the total protein concentration [36]. The percentage of native protein content of suspensions at pH 4.8 was obtained as the ratio between soluble protein (*SP*) and total protein (*TP*) contents after aggregation of denatured protein (Eq. 1).

$$NP\% = \left(\frac{SP}{TP}\right) \text{x } 100 \tag{1}$$

2.5. Scanning electron microscopy

The microstructure of freeze-dried plasma concentrates with and without saccharides was analyzed by scanning electron microscopy (SEM) using an LEO1450VP equipment (Zeiss, Germany). Powder samples were mounted on double-sided carbon adhesive tape on aluminum stubs and gold-coated and processed in a standard sputter. The micrographs were obtained in high vacuum at 10 KeV.

2.6. Statistical analysis

The experimental data were statistically analyzed by the Tukey-Kramer multiple comparison test, in the cases where 2 or more comparisons were considered, assuming that a $P<0.05$ was statistically significant [37]. Statistical GraphPad InStat software (1998) was used.

3. Theoretical considerations

3.1. Equations for T'_g prediction

The Miller/Fox equation can be used for the determination of T'_g dependence with the composition in a multi-component system, assuming constant density of the solutions, independent of temperature [28,38,39]. For a ternary mixture (protein-saccharide-water), it can be written as:

$$\frac{1}{T_g} = \frac{m_1}{m_t T_{g1}(\rho_1 / \rho_t)} + \frac{m_2}{m_t T_{g2}(\rho_2 / \rho_t)} + \frac{m_3}{m_t T_{g3}(\rho_3 / \rho_t)} \tag{2}$$

where T_g, glass transition temperature; m, mass; ρ, density; the subscripts t, 1, 2, 3 mean: total and each pure component, respectively.

The Gordon and Taylor equation [40] predicts the plasticizing effect of water on the T_g for a multicomponent system. The equation has been used among others, for systems treated as binary mixtures, determining experimentally the glass transition of the respective solid [41,42]. Instead we proposed a system considering each individual component: bovine protein concentrate, saccharide and water, with each corresponding property [43]:

$$T_g = \frac{w_1 T_{g1} + k w_2 T_{g2} + k^2 w_3 T_{g3}}{w_1 + k w_2 + k^2 w_3} \tag{3}$$

where w_1, w_2, w_3, are the weight fraction of each component defined as (m_i / m_t), and k is an empirical constant proportional to the plasticizing effect of water. This parameter was calculated to fit experimental data from a nonlinear optimization procedure (Gauss Newton procedure) using the software Excel 2003 (Microsoft).

Eqs. (2) and (3) were used for the determination of T'_g of the frozen solutions.

3.2. Theory of protein unfolding

Unfolding of protein is suggested to involve at least two steps according to Lumry and Eyring model (1954). The first step is a reversible unfolding of the native protein (N). This is followed by an irreversible change of the denatured protein (D) into a final irreversible state (I) [44,45].

$$N \underset{k_{-1}}{\overset{k_1}{\leftrightarrow}} D \overset{k_2}{\rightarrow} I \tag{4}$$

A special case was when $k_2 >> k_{-1}$, where most of the D molecules will be converted to I as an alternative to refolding back to the native state. In this case, the denaturation process can be regarded as one-step process following first-order kinetics [44-46], (Eq.5).

$$N \xrightarrow{k} I \quad (5)$$

where the first-order rate constant k can be identified with k_1 of Eq. (4). The total absorbed heat now equals the enthalpy change from N to I; it was generally assumed that the enthalpy change from D to I was negligible compared to that from N to D [44].

Experimentally, the irreversibility of unfolding was verified in a rescan. For an irreversible process, in the DSC rescanned thermograms no transition could be observed.

4. Results and discussion

4.1. Effect of saccharides on glass transition of the freeze concentrated matrix

As was previously mentioned, to avoid collapse of the products during the freeze-dried process, a temperature below the glass transition temperature of the frozen concentrated solutions, must be attained. Inulin as protein protective agent was comparatively studied, employing mono and disaccharides. The thermograms of Figure 2 show the transition temperatures of the frozen solutions of bovine plasma with inulin compared to the other saccharides, obtained in a single scan.

The result indicated that at each saccharide concentration, T'_g was higher for inulin (Table 1), suggesting that it has a greater cryostabilizing effect on bovine plasma proteins than the other saccharides, improving product stability. It was also observed that T'_g increased with the molecular weight of the cryoprotectant that is: inulin > sucrose > glucose. The same tendency was reported previously by means of the evaluation of protein shelf life time [18]. By the other hand, it was reported that inulin exhibit better stabilizing properties than sucrose and trehalose in the prevention of the nonPEGlated lipoplexes aggregation [14]. Many studies concluded that transition temperatures increased with the saccharides molecular weight [22,23,27].For example T'_g of freeze–dried surimi depended strongly on the type and content of sugar and at each sugar level the T'_g was trehalose > sucrose > glucose > sorbitol [47].

Thermograms of bovine plasma solutions revealed the existence of two glass transitions (T'_{g1} and T'_{g2}) for glucose and sucrose as protective agents, evidenced as deviations in the base line (indicated by arrows in Figure 1). Similar results were found by Telis and Sobral [48] who worked with freeze–dried tomato. This may be because the presence of phases formed by different proportions of saccharide, water, and proteins present in the frozen solution [47-49]. Also it was observed that when the saccharide concentration increased, T'_{g1} and T'_{g2} increased and decreased respectively (Table 1). However a constant average value was maintained between both T'_g values for each sugar, being -51.2 ± 0.8 and -41.1 ± 0.1 for glucose and sucrose, respectively. Similar results were found in [47] on freeze–dried surimi product with trehalose. For inulin only one T'_g was found, which increased with the increase of saccharide concentration. From these results and considering that the water acts as plasticizer, i.e. decreases drastically T'_g of food polymers [26], it can be concluded that the conditions of the freeze-drying process, are linked directly to T'_g of the frozen solution.

Therefore, it is important to note that the higher value of T'_g observed in frozen solutions with inulin, allowed higher freezing temperatures during processing reducing production costs.

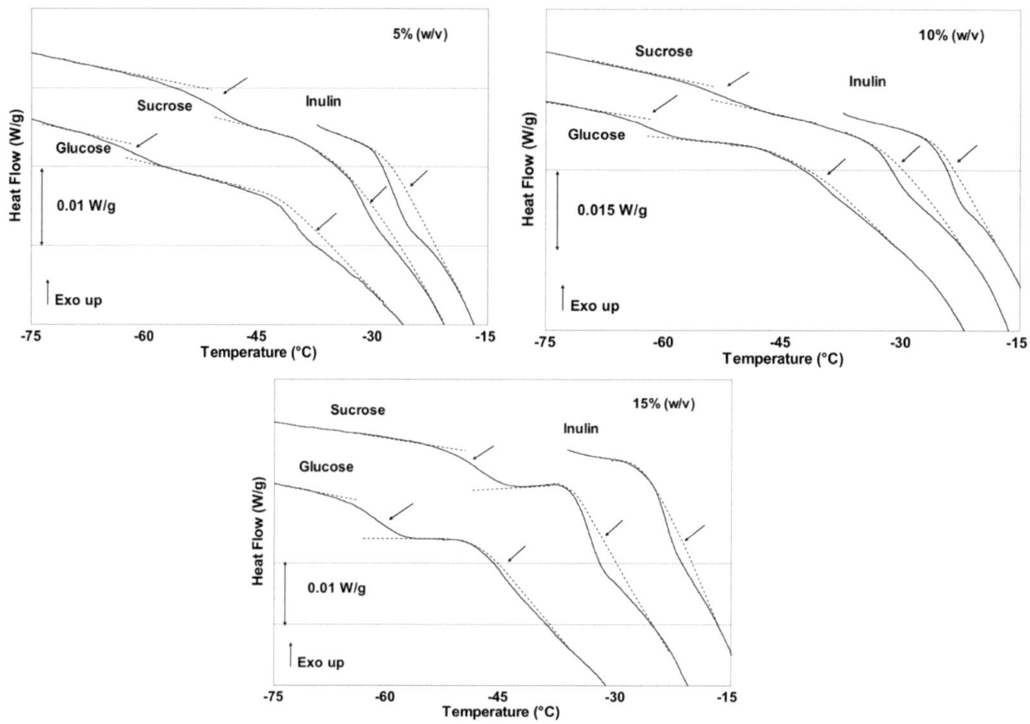

Figure 2. DSC thermograms for freeze bovine plasma protein-saccharide solutions. Down-arrows indicate T'_g. Scan rate = 2°C/min ; pH = 8.

Saccharide	Concentration (%, w/v)	T'_{g1} (°C)	T'_{g2} (°C)	T_g (°C)
Glucose	5	-62.50 ± 0.58^{a}	-39.24 ± 0.75^{a}	16.31 ± 0.38^{a}
	10	$-61.06 \pm 0.45^{a,b}$	-39.91 ± 0.83^{a}	41.52 ± 0.29^{b}
	15	-59.82 ± 0.68^{b}	-44.96 ± 0.49^{b}	60.31 ± 0.48^{c}
Sucrose	5	-51.48 ± 1.05^{c}	-31.15 ± 0.40^{c}	48.01 ± 0.56^{d}
	10	$-50.12 \pm 1.03^{c,d}$	-31.86 ± 0.60^{c}	52.48 ± 0.52^{e}
	15	-48.42 ± 0.98^{d}	-33.72 ± 0.45^{d}	64.28 ± 0.46^{f}
Inulin	5	-26.96 ± 0.68^{e}	-	48.85 ± 0.35^{d}
	10	-23.67 ± 0.55^{f}	-	66.18 ± 0.69^{g}
	15	-22.40 ± 0.45^{f}	-	69.25 ± 0.45^{h}

Table 1. Effect of type and concentration of cryoprotectant on glass transition temperature (T'_g) and lyoprotectant on glass transition temperature (T_g) of freeze bovine plasma proteins solutions (heating rate: 2 °C/min). Values represents the means ± standard deviation; n = 3. Values followed by different letters in the same column are significantly different from each other ($P < 0.05$).

The effect of water as a plasticizer of the mixture protein-saccharide was predicted by the Miller/Fox and Gordon–Taylor equations, the results, were compared with experimental values (Table 1). The data of T'_g of all pure components required for the Eq. (1) are listed in Table 2.

The densities of bovine plasma proteins, glucose, sucrose and inulin (at room temperature) were determined with a digital densimeter, and the results were: 0.4 ± 0.08 g/cm^3, 0.6 ± 0.05 g/cm^3, 0.8 ± 0.04 g/cm^3 and 0.3 ± 0.05 g/cm^3, respectively.

From literature the T_g of the water is -135 °C [41] and the T'_g of plasma protein is -11 ± 2 °C [22]. The T_g value of bovine plasma protein for Eq. (3), was 65 ± 3 °C. Entering this data into Eqs. (2) and (4), the predicted values of T'_g were obtained, which are listed in Table 3. The results showed that the glass transition property evaluated from the proposed models was in agreement with the experimental data with an average error of 4.86% for the Miller/ Fox equation and 0.09% for Gordon/Taylor equation. The value of k from the Gordon/Taylor equation is defined as the resistance to a T'_g decrease induced by the plasticizing effect of water [26,41,47]. The order found for k value of the saccharides was: inulin > sucrose > glucose. Although the highest value of k is for inulin, this saccharide has the highest T_g value, allowing a greater value T'_g and therefore generating a lower cost during processing, preventing also the collapse of the product at temperatures relatively higher during the freeze-drying.

Saccharide	T'_g (°C)		
	5 % (w/v)	10% (w/v)	15 % (w/v)
Glucose	-85	-79	-72
Sucrose	-59	-53	-46
Inulin	-17	-15	-13

Table 2. Data from references used in the calculation of T'_g by Miller/Fox and Gordon/Taylor modified equation [18].

	Glucose %(w/v)			Sucrose %(w/v)			Inulin %(w/v)		
	5	10	15	5	10	15	5	10	15
*ρ (g cm^{-3})	1.039	1.042	1.059	1.033	1.041	1.056	1.032	1.039	1.049
T_g' (°C) (Miller/Fox)	-63.99	-60.46	-56.1	-54.07	-51.24	-47.04	-30.43	-23.64	-19.51
Difference (%)	2.32	0.99	6.63	4.79	2.18	2.95	11.40	0.12	14.8
T_g' (°C) (Gordon/ Taylor modified)	-62.69	-61.26	-60.03	-51.38	-50.58	-48.47	-26.70	-24.23	-22.11
Difference (%)	0.30	0.33	0.35	0.19	0.91	0.10	0.97	2.31	1.31
k	3.5			4.1			4.5		

Table 3. Glass transition parameters for the multicomponent system: plasma bovine proteins-saccharides-water. *ρ: solution density (T=19.8°C)

4.2. Effect of saccharides on glass transition of the freeze-dried samples

The storage temperature of frozen or freeze-dried foods should be below the glass transition temperature as previously established [22,27,42,50]. Figure 3 shows the thermograms of the freeze-dried samples containing inulin compared with glucose and sucrose at different concentrations. The existence of these transitions evidenced the glassy state of the freeze–dried plasma protein/saccharides mixtures. Besides, Table 1 shows that T_g of the sample increases with increasing saccharide concentration. Similar results were found in the references [28,30]. This effect can be explained considering that sugar forms hydrogen–bridge bonds with proteins reducing the available volume for the interaction with water molecules, so water become less effective as plasticizer with an increase in saccharide content [51]. Also was observed that T_g of the freeze-dried samples increased with increasing of the molecular weight of the cryoprotectant. Processes of devitrification and hence product spoilage can occur if the temperature of storage is higher than the T_g of the sample. Therefore, the higher T_g value of inulin provides greater stability at higher temperatures, reducing the storage costs.

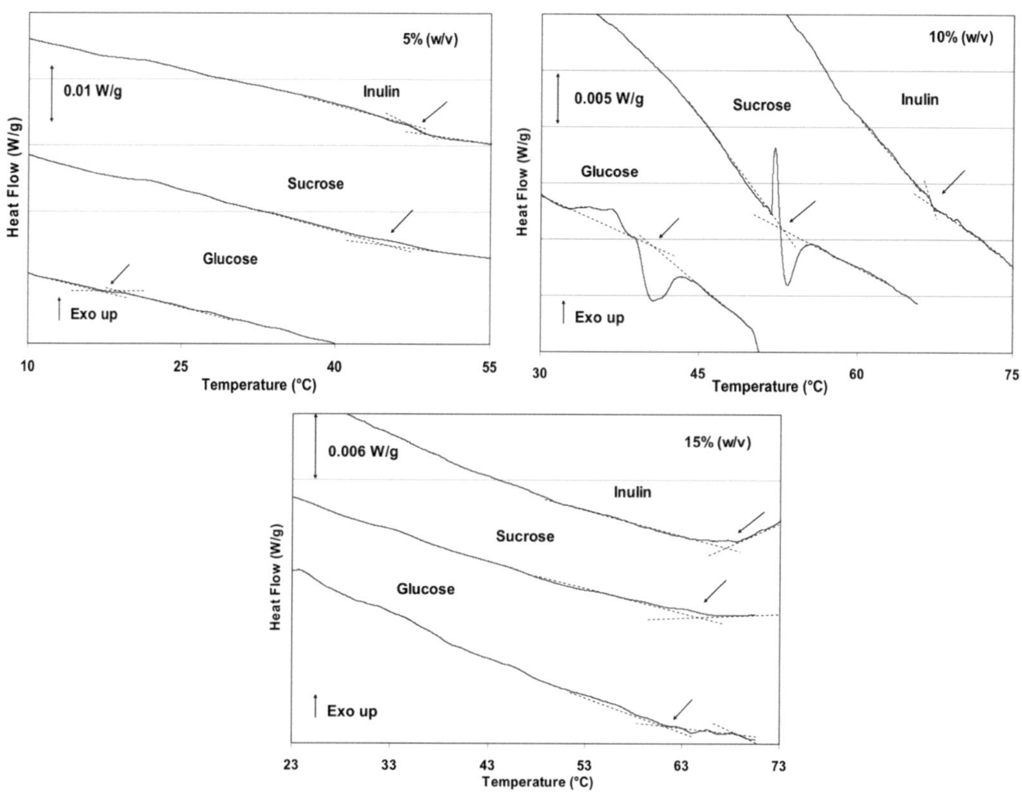

Figure 3. DSC thermograms for freeze–dried bovine plasma protein–saccharide mixtures. Down-arrows indicate T_g. Heating rate: 2°C min^{-1}; pH=8

4.3. Effect of saccharides on crystallization temperature of the freeze-dried samples

It is important to determine the crystallization temperature (T_c) of the freeze-dried samples since crystallization causes the most drastic changes on physical properties of food polymers and affects considerably food stability. The glass transition is often followed by crystallization of the solutes where the molecular mobility increases and the sample crystallizes increasing the rate of food spoilage [27,28,30].

Fig 4 shows the crystallization temperature (T_c) obtained from the intersection of the baseline and the tangent of the exothermic peak, and the crystallization enthalpy (ΔH_c) estimated as the area under the peak for the different protective agents at different concentrations. The crystallization temperature of freeze-dried samples was found to depend on the molecular weight and the saccharide concentration [27,30]. Therefore, the results showed that the presence of inulin at the same concentration than the other saccharides further increases the T_c value of freeze–dried solutions. Mixtures containing a saccharide concentration of 10 % (w/v) show an increase of ΔH_c, indicating a higher amorphous content. This behavior can be explained considering that a suitable proportion of saccharide and protein in the mixture allows a better interaction among these components [51,52,43].

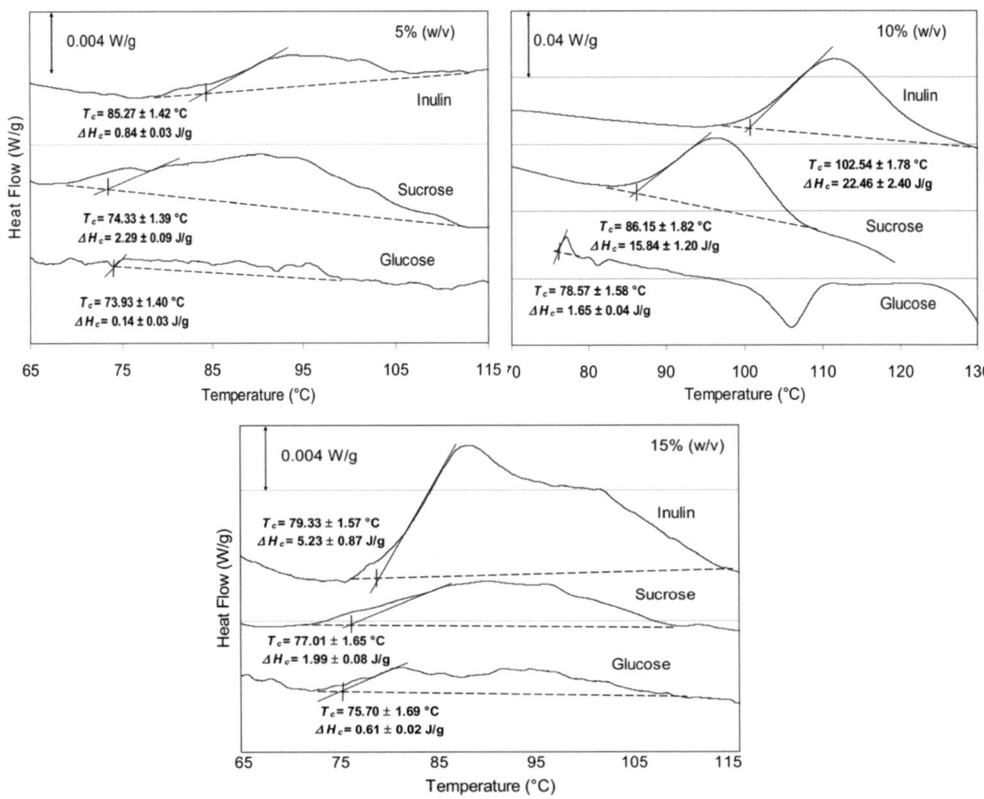

Figure 4. DSC thermogram for freeze-dried bovine plasma protein with the protective agents at different concentrations. The exothermic event indicates T_c. Heating rate: 2°C min^{-1}; pH=8.

4.4. Thermal denaturation of BPP in a matrix of saccharide

4.4.1. *Effect of saccharide type and concentration*

The thermal stability of BPP in a matrix of inulin compared with other saccharides was investigated using DSC. Table 4 shows the values of T_d obtained for BPP concentrate without protective agents and in different matrixes of glucose, sucrose and inulin at different concentrations. The value of T_d for BPP concentrate (88.19 ± 1.87 °C), was obtained from thermograms without protective agent and was similar to that reported in reference [53], for blood plasma. Comparing this value with the protein sample immersed in a matrix of saccharide, it was observed an increase in the value of T_d in all the cases, indicating a higher thermal resistance due to the stabilizing effect of saccharides. A similar behavior was observed in the DSC study of whey protein concentrates with the addition of honey [54]. Evaluating among the saccharides at the same concentration, it can be concluded that the higher the molecular weight of the carbohydrate, the higher was the T_d, thus inulin > sucrose > glucose. This behavior was in agreement with that reported in [55], in multi-block copolymers. With respect to the range of saccharide concentrations studied, optimum concentration was 10% (w/v), as it is shown in Table 4, in terms of the values of T_d and ΔH.

Saccharide	Concentration (%, w/v)	T_d (°C)	ΔH (J g^{-1})
Glucose		110.07 ± 1.22^{a}	0.84 ± 0.32^{a}
Sucrose	5	132.78 ± 2.12^{b}	$5.08 \pm 0.98^{b,c}$
Inulin		143.81 ± 0.89^{c}	$2.97 \pm 0.55^{a,d,c}$
Glucose		$107.27 \pm 0.85^{a,d}$	12.26 ± 0.92^{e}
Sucrose	10	144.95 ± 2.34^{c}	22.40 ± 1.23
Inulin		156.21 ± 1.12^{e}	12.22 ± 1.43^{e}
Glucose		104.91 ± 0.89^{d}	$3.77 \pm 0.98^{d,f}$
Sucrose	15	126.66 ± 1.54^{f}	7.01 ± 1.22^{b}
Inulin		132.57 ± 1.34^{b}	$5.78 \pm 0.76^{b,f,c}$

Table 4. Effect of saccharide concentration on the denaturation temperature of freeze dried BPP concentrate. Heating rate: 2°C min^{-1}. pH=8. Values followed by different letters in the same column are significantly different from each other ($P < 0.05$).

The functional structure of a protein in solution is determined by electrostatic forces, hydrogen bonds, Van der Waals interactions and hydrophobic interactions. All these interactions are influenced by water, becoming essential for the functional unfolding of most of the proteins. As water is eliminated during freeze-drying, peptide-peptide interactions prevail causing an alteration in the secondary, tertiary or quaternary structure of the protein, i.e. a conformational change of it. However, the presence of sugar displaces and supplants water forming hydrogen bonds with the dry protein which maintains its structured integrity into the glass matrix. In the case that the formation of the glass structure did not occur, the sugar would be excluded and it would not be available for the formation of hydrogen bonds to protect the dry protein from its unfolding or loss of conformation [13,14].

The protective effect of saccharides depends on its concentration, since as the concentration increases there are more possibilities of forming hydrogen bonds with the protein [11,18]. However, when concentrations were higher than 10 % (w/v), a lower protection was obtained. This result can be explained taking into account that at high concentrations, the saccharide starts to crystallize during freeze-drying, being prevented the formation of hydrogen bonds with the dry protein [12]. This behavior was confirmed by determination of the native proteins in the protein-saccharide matrixes employing eq. (1). The results are presented in Figure 5, which shows that there is a maximum at a concentration of 10% (w/v) for the different saccharides analyzed, indicating higher protein protection and stability.

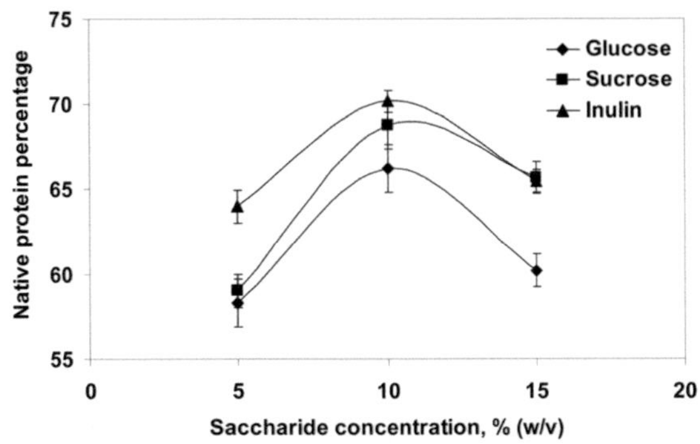

Figure 5. Native protein percentage of freeze dried BPP concentrate with different protective agents at different concentrations.

4.4.2. *Effect of pH*

To determine the application of these formulations is important to know the variation of T_d as a function of pH due to the wide range of environmental conditions existing in food. Table 5 shows the T_d values of BPP concentrate in a glassy matrix of saccharides at different pH values.

Saccharide	pH	T_d (°C)	ΔH (J g^{-1})
Glucose	8	107.27 ± 0.85^{a}	12.26 ± 0.82^{a}
Sucrose		144.95 ± 1.34^{b}	22.40 ± 0.97^{b}
Inulin		156.21 ± 1.12^{c}	12.22 ± 0.55^{a}
Glucose	6	102.94 ± 1.33^{d}	34.74 ± 0.92^{c}
Sucrose		134.56 ± 2.16^{e}	43.15 ± 1.23^{d}
Inulin		152.98 ± 1.52^{c,f}	42.95 ± 1.45^{d}
Glucose	4	101.74 ± 1.27^{d}	9.58 ± 0.98^{a,e}
Sucrose		107.67 ± 1.56^{a}	9.32 ± 0.72^{e}
Inulin		151.84 ± 1.89^{f}	9.35 ± 0.96^{e}

Table 5. Effect of pH and addition of saccharides on the denaturation temperature of BPP concentrate. Heating rate: 2°C/min. Values represents the means ± standard deviation; n = 3. Values followed by different letters in the same column are significantly different from each other (P< 0.05).

With increasing alkalinity of the medium there is an increase in the values of T_d for each saccharide (pH 8), indicating that BPP concentrate was more stable at higher pH. Similar results were found in previous works in porcine blood plasma proteins and whey protein concentrate [34,54]. Comparing between different saccharides at the same concentration, it can be seen that inulin presents a higher T_d in all the pH range. The maximum ΔH values were observed at pH 6 indicating a higher amount of native protein. Similar ΔH values at pH = 6 were reported by Dàvila in reference [34]. The lowest values of T_d and ΔH were found at pH 4, this may be to the proximity with the isoelectric point of proteins (pI: 4.8-5.8), thus decreasing the electrical net charge and facilitating aggregation reactions.

4.4.3. *Effect of scanning rate*

The protein-saccharide mixtures were studied at different scanning rates (2 °C/min and 5 °C/min). As an example Figure 6 shows the transition temperature and enthalpy for sucrose at 10 % (w/v).

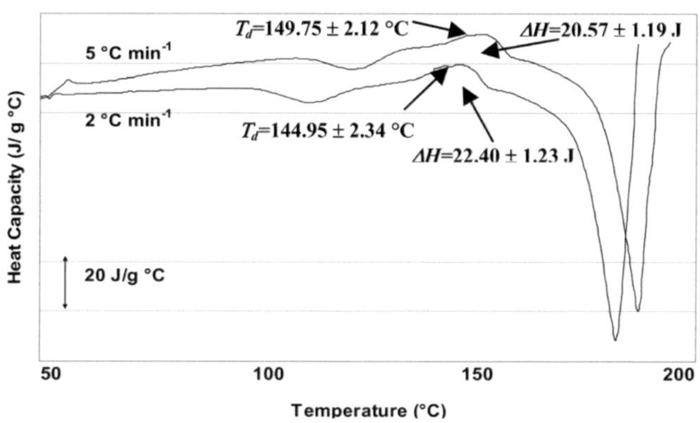

Figure 6. Effect of DSC heating rate on T_d values of freeze-dried BBP with sucrose 10%(w/v).

It was found for all the saccharides that T_d and ΔH are scanning rate dependent. T_d values increased 5 ± 2 °C in all the samples with increasing scanning rate, similar behavior was reported in references [56-58]. Furthermore, the ΔH decreased (~ 10%) with increasing scanning rate that was in agreement with the results reported in references [21,59]. Thus, the system was scanning rate dependent and so the thermal denaturation process was under kinetic control [33,44].

4.4.4. *Study of Irreversibility of the Thermal Denaturation of BPP*

The irreversibility of BPP denaturation was investigated by a multiple reheating experiment, according to the method proposed by by Idakieva and Michnik [45,60]. From the initial DSC scan, we have determined the values of the transition temperatures at 107°C, 145 °C and 156 °C for glucose, sucrose and inulin at 10% w/v, respectively (Table 5). DSC tests were carried out as successive scans, where the heating was carried out up to different final temperatures, with a cooling up to 20°C between scans (Figure 7).

For glucose, sucrose and inulin, the first heating was carried out up to 75°C, and 85°C (temperatures below the T_d for all the saccharides), respectively; no thermal effect was observed in the thermal denaturation peak during the reheating experiment. However, if the rescanning was stopped over their transition temperatures, the endothermic peak of T_d disappeared completely. Therefore, the endothermic peak of T_d disappeared completely upon rescanning the sample at temperatures above T_d; furthermore, as was previously described, the thermograms were scanning-rate dependent, suggesting both results that it was an irreversible event [61]. Similar behavior was also found for whey protein in an amorphous carbohydrate matrix [49], porcine blood plasma proteins [34] and BSA [33]. Irreversible denaturation of bovine plasma proteins might be due to processes such as aggregation, where hydrophobic interactions occur, and exposed thiol groups can form disulfide bonds, which result in an irreversible behavior [33]. Considering the Arrhenius law and the treatment developed in reference [43], the determination of the activation energy can be achieved from the experimental data. The obtained values were: 10443 J mol^{-1}, for BPP without protective agent; 27216 J mol-1, 32058 J mol^{-1} and 42099 J mol^{-1} for BPP with glucose, sucrose and inulin, respectively, all of them at 10% (w/v). The results showed that

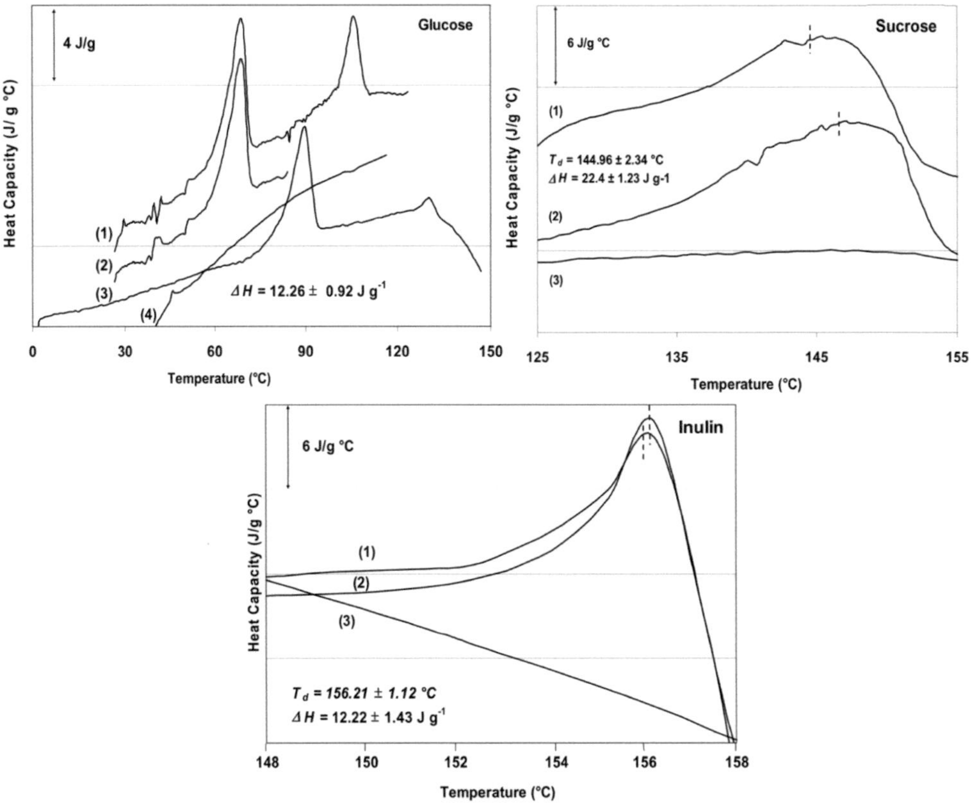

Figure 7. DSC thermograms of freeze dried BPP concentrate with saccharide at 10 % (w/v). DSC scans (2), (3), (4) represent thermograms from repeated heating and subsequent cooling. Scan (1) is a full scan to 120 °C (glucose), 155°C (sucrose) and 158 °C (inulin).

with the addition of protective agents the activation energy increased; besides with increasing molecular weight, the activation energy also increased. Therefore, the addition of saccharides, especially of inulin caused a decrease in the rate of degradation reactions, obtaining a higher stabilization upon storage [8,14,18].

4.4.5. Study of the blends morphology through SEM

Figure 8 sowed the SEM micrographs of blends of protein-saccharides.

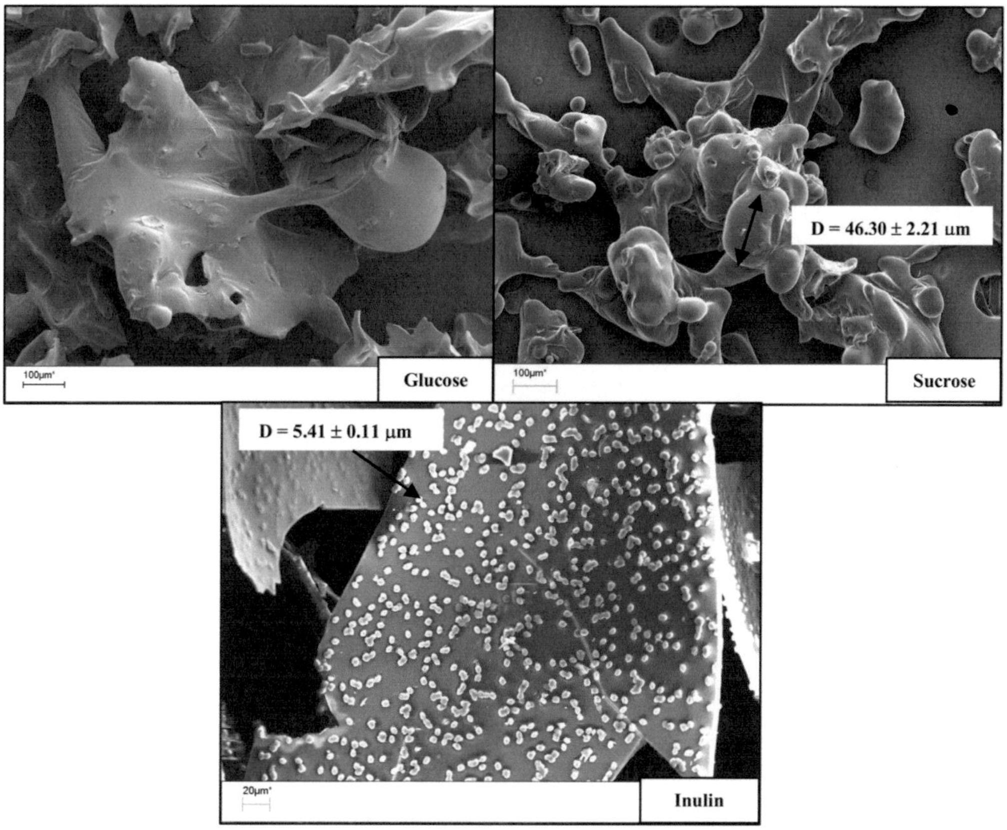

Figure 8. Scanning electron micrographs of the freeze-dried product with different saccharides, with a magnification of 200X for glucose and sucrose, 300X for inulin.

It was observed phases homogeneously distributed, indicating miscibility of the component in the matrix. The shapes were uniform, which was an attribute, linked with thermodynamic compatibility [62]. Based on the data previously obtained, comparing the transitions of the blends with respect to the value of the individual components, showed an increase in the value T_d. This increase in the T_d values can be attributed to greater miscibility of the components of the mixtures, confirming what was observed in the micrographs [61]. Therefore, these results are in agreement with the concept of miscibility, which is based on the variation of the thermal behavior with respect to the individual materials [63].

5. Conclusions

The thermodynamic properties of the solution and the freeze–dried bovine plasma proteins–saccharides mixtures were investigated in this study. The DSC thermograms demonstrated that the bovine plasma proteins– inulin mixtures have the highest glass transition temperature for the protein solution and also the highest glass transition and denaturation temperature for the freeze–dried powder, optimizing the freeze–drying process and also stabilizing and protecting the proteins during storage in conditions below the collapse temperature of the material. Thermograms revealed the existence of two glass transitions in solutions (T'_{g1} and T'_{g2}) for glucose and sucrose. With increasing saccharide content, the T'_{g1} and T'_{g2} of the samples increased and decreased, respectively. For inulin only one T'_g was found, which increased with saccharide concentration. Also was found that T'_g, T_g and T_c depended on the molecular weight of saccharides, increasing with the increasing of molecular weight, being inulin > sucrose > glucose. The proposed model allowed the prediction of transition temperature in a multicomponent mixture which is useful to design a freeze–drying cycle and storage stability of plasma protein concentrates. The addition of saccharides allowed the increase of the protein denaturation temperature and enthalpy, with an optimal saccharide concentration of 10% (w/v) and a pH range between 6 and 8. This change in the thermal properties shows a greater compatibility of the blends with 10% (w/v) saccharide, because this concentration causes the greatest changes in the values of T_d when compared with individual values of BPP. The results were corroborated by the SEM micrographs, showing homogeneously distributed phases, and denoting the highest miscibility between them. The temperature of thermal denaturation was scan rate dependent, and no thermal transition was detected in the re-scan experiments so it was concluded that the protein unfolding was irreversible and was adequately interpreted by the theoretical model employed.

Therefore, the results showed highest values of T_g and T_d in the freeze–dried samples of inulin proven that this compound is a better protein protective agent during storage than mono and disaccharides such as glucose and sucrose. In this way prevent the unfolding of bovine plasma proteins submitted to higher temperatures. Furthermore, the higher T'_g of frozen solutions of bovine proteins with inulin allows higher freezer temperatures during freeze–drying, reducing costs in a food elaboration. The finding about the inulin cryoprotant role of food proteins is relevant considering that it is a soluble fiber, categorized as a prebiotic, and being a valuable alternative as a functional ingredient for food formulation [64,65].

The findings regarding the protective effect of inulin on bovine plasma proteins, suggest that may be interesting the study of the behavior of formulated foods elaborated with the analyzed matrices (protein-saccharide-water) exposed to treatments such as cooling and freeze-drying.

Author details

Laura T. Rodriguez Furlán, Antonio Pérez Padilla and Mercedes E. Campderrós*
Research Institute of Chemical Technology (INTEQUI –CONICET-CCT San Luis), Faculty of Chemistry, Biochemistry and Pharmacy, UNSL, San Luis, Argentine

Javier Lecot and Noemi E. Zaritzky
Research and Development in Food Cryotechnology Centre (CIDCA- CONICET- CCT La Plata), Argentine

Noemi E. Zaritzky
Faculty of Engineering, UNLP, La Plata, Bs As, Argentine

6. References

[1] Niness K R (1999) Inulin and Oligofructose: What Are They? J. nutr. 129 (7): 1402-1406.
[2] Abrams S, Griffin I, Hawthorne K, Liang L, Gunn S, Darlington G, Ellis K A (2005) Combination of Prebiotic Short- and Long-chain Inulin-type Fructans Enhances Calcium Absorption and Bone Mineralization in Young Adolescents. Am. j. clin. nutr. 82: 471-476.
[3] Hempel S, Jacob A, Rohm H (2007) Influence of Inulin Modification and Flour Type on the Sensory Quality of Prebiotic Wafer Crackers. Eur. food res. technol. 224: 335-341.
[4] Nazzaro F, Fratianni F, Coppola R, Sada A, Pierangelo O (2009) Fermentative Ability of Alginate-prebiotic Encapsulated Lactobacillus Acidophilus and Survival under Simulated Gastrointestinal Conditions. J. funct. food 1(3): 319-323.
[5] Kip P, Meyer D, Jellema R H (2006) Inulins Improve Sensory and Textural Properties of Low-Fat Yoghurts. Int. dairy j. 16: 1098–1103.
[6] Ronkart S N, Paquot M, Fougnies C, Deroanne C, Blecker C S (2009) Effect of Water Uptake on Amorphous Inulin Properties. Food hydrocolloid 23: 922–927.
[7] Baeza R I, Pilosof A M R (2002) Calorimetric Studies of Thermal Denaturation of b-Lactoglobulin in the Presence of Polysaccharides. Lebensm.-wiss. technol. 35: 393–399.
[8] Buera P, Schebor C, Elizalde B (2005) Effects of Carbohydrate Crystallization on Stability of Dehydrated Foods and Ingredient Formulations. J. food eng. 67: 157-165.
[9] Claude J, Ubbink J (2006) Thermal Degradation of Carbohydrate Polymers in Amorphous States: A Physical Study Including Colorimetry. Food chem. 96: 402-410.
[10] Santivarangkna C, Higl B, Foerst P (2008) Protection Mechanisms of Sugars During Different Stages of Preparation Process of Dried Lactic Acid Starter Cultures. Food microbiol. 25: 429-441.
[11] Allison S D, Chang B, Randolph T W, Carpenter J F (1999) Hydrogen Bonding Between Sugar and Protein is Responsible for Inhibition of Dehydration-Induced Protein Unfolding. Biochem. biophys. 365: 289-298.
[12] Carpenter J F, Crowe L M, Crowe J H (1987) Stabilization of Phosphofructokinase with Sugars during Freeze-drying: Characterization of Enhanced Protection in the Presence of Divalent Cations. Biochim. biophys. acta 923(1): 109-115.

* Corresponding Author

[13] Hinrichs W L J, Prinsen M G, Frijlink H W (2001) Inulin Glasses for the Stabilization of Therapeutic Proteins. Int. j. pharmaceut. 215: 163–174.

[14] Hinrichs W L J, Sanders N N, De Smedt S C, Demeester J, Frijlink H W (2005) Inulin is a Promising Cryo- and Lyoprotectant for PEGylated Lipoplexes. J. control. release 103: 465-479.

[15] Rey Cabinet L, d'Etudes L, Switzerland J C (2004) Freeze Drying/lyophilization of Pharmaceutical and Biological Products. Maryland, U.S.A: Center for Biologics Evaluation and Research Food and Drug Administration.

[16] Liao Y H, Brown M B, Martin G P (2004) Investigation of the Stabilization of Freeze–dried Lysozyme and the Physical Properties of the Formulations. Eur j. pharm. Biopharm. 58: 15–24.

[17] Minson E I, Fennema O, Amundson C H (2006) Efficacy of Various Carbohydrates as Cryoprotectants for Casein in Skim Milk. J. food sci. 46(5): 1597-1602.

[18] Rodriguez Furlán L T, Pérez Padilla A, Campderrós M (2010) Inulin Like Lyoprotectant of Bovine Plasma Proteins Concentrated by Ultrafiltration. Food res. int. 43: 788-796.

[19] Costantino H R, Curley J G, Wu S, Hsu C C (1998) Water Sorption Behavior of Lyophilized Protein–sugar Systems and Implications for Solid-state Interactions. Int. j. pharm. 166: 211–221.

[20] Passot S, Fonseca F, Alarcon-Lorca M, Rolland D, Marin M (2005) Physical Characterization of Formulations for the Development of Two Stable Freeze–dried Proteins During Both Dried and Liquid Storage. Eur. j. pharm. biopharm. 60: 335–348.

[21] Guzzi R, Sportelli L, Sato K, Cannistraro S, Dennison C (2008) Thermal Unfolding Studies of a Phytocyanin. Biochim. biophys. acta 1784: 1997-2003.

[22] Chen T, Oakley D M (1995) Thermal Analysis of Proteins of Pharmaceutical Interest. Thermochim. acta 248: 229–244.

[23] Schenz T W (1995) Glass Transitions and Product Stability–an Overview. Food hydrocolloid 9(4): 307–315.

[24] Ahmed J, Prabhu S T, Raghavan G S V, Ngadi M. (2007) Physico-chemical, Rheological, Calorimetric and Dielectric Behavior of Selected Indian Honey. J. food eng. 79: 1207-1213.

[25] Gallegos Infante J A, Ochoa Martínez L A, Ortiz Corral C (2005) Glass Transition Temperature Behavior of a Model Blend of Carbohydrates. Cien. Tecnolog. Alimen. 5: 6–10.

[26] Noel T R, Parker R, Ring S G, Tatham A S (1995) The Glass-transition Behaviour of Wheat Gluten Proteins. Int j. boil. macromol. 17 (2): 81–85.

[27] Roos Y (1995) Characterization of Food Polymers using State Diagrams. J. food eng. 24: 339–360.

[28] Shah B N, Schall C A (2006) Measurement and Modeling of the Glass Transition Temperatures of Multi-component Solutions. Thermochim. acta 443: 78– 86.

[29] Katkov I I, Levine F (2004) Prediction of the Glass Transition Temperature of Water Solutions: Comparison of Different Models. Cryobiology 49: 62–82.

[30] Tattini Jr V, Parra D F, Polakiewicz B, Pitombo R N M (2005) Effect of Lyophilization on the Structure and Phase Changes of PEGylated-bovine Serum Albumin. Int. j. pharm. 304: 124–134.

[31] Sunooj K V, Radhakrishna K, George J, Bawa A S (2009) Factors Influencing the Calorimetric Determination of Glass Transition Temperature in Foods: a Case Study Using Chicken and Mutton. J. food eng. 91: 347–352.

[32] Akköse A, Aktas N (2008) Determination of Glass Transition Temperature of Beef and Effects of Various Cryoprotective Agents on Some Chemical Changes. Meat sci. 80: 875–878.

[33] Cao X, Li J, Yang X, Duan Y, Liu Y, Wang C (2008) Nonisothermal Kinetic Analysis of the Effect of Protein Concentration on BSA Aggregation at High Concentration by DSC. Thermochim. acta 467: 99-106.

[34] Dàvila E, Parés D, Cuvelier G, Relkin P (2007) Heat-induced Gelation of Porcine Blood Plasma Proteins as Affected by pH. Meat Sci. 76: 216-225.

[35] de Wit J N (1990) Thermal Stability and Functionality of Whey Proteins. J. dairy Sci. 73: 3602-3612.

[36] Giroux H J, Britten M (2004) Heat Treatment of Whey Proteins in the Presence of Anionic Surfactants. Food hydrocolloid 18: 685- 692.

[37] SAS USER GUIDE: Statistic. Versión (1989). SAS Inst. Inc., Cary, NC, USA.

[38] Fox T G (1956) Influence of Diluent and Copolymer Composition on the Glass Temperature of a Polymer System. B. am. phys. soc. 2(1): 123.

[39] Miller D P, de Pablo J J, Corti H (1997) Thermophysical Properties of Trehalose and its Concentrated Aqueous Solutions. Pharm res-dord 14(5): 578–590.

[40] Gordon M, Taylor J S (1952) Ideal Copolymers and the Second-order Transitions of Synthetic Rubbers. I. Non-crystalline Copolymers. J. appl. chem. 2: 493–500.

[41] Georget D M R, Smith A C, Waldron K W (1999) Thermal Transitions in Freeze– dried Carrot and its Cell Wall Components. Thermochim. acta 332: 203–210.

[42] Maitani Y, Aso Y, Yamada A, Yoshioka S (2008) Effect of Sugars on Storage Stability of Lyophilized Liposome/DNA Complexes with High Transfection Efficiency. Int. j. pharm. 356: 69–75.

[43] Rodriguez Furlán L T, Lecot J, Pérez Padilla A, Campderrós M, Zaritzky N (2011) Effect of Saccharides on Glass Transition Temperatures of Frozen and Freeze-dried Bovine Plasma Protein. J. food eng. 106: 74-79.

[44] Creveld L D, Meijberg W, Berendsen H J C, Pepermans H A M (2001) DSC Studies of Fusarium Solani Pisi Cutinase: Consequences for Stability in the Presence of Surfactants. Biophys. Chem. 92: 65-75.

[45] Idakieva K, Parvanova K, Todinova S (2005) Differential Scanning Calorimetry of the Irreversible Denaturation of Rapana Thomasiana (Marine Snail, Gastropod) Hemocyanin. Biochim. biophys. acta 1748: 50-56.

[46] Ramprakash J, Doseeva V, Galkin A, Krajewski W, Muthukumar L, Pullalarevu S, Demirkan E, Herzberg O, Moult J, Schwarz, F P (2008) Comparison of the Chemical and Thermal Denaturation of Proteins by a Two-state Transition Model. Anal. biochem. 374: 221-230.

[47] Ohkuma C, Kawaib K, Viriyarattanasaka C, Mahawanichc T, Tantratianc S, Takaia R, Suzuki T (2008) Glass Transition Properties of Frozen and Freeze–dried Surimi Products: Effects of Sugar and Moisture on the Glass Transition. Food hydrocolloid 22: 255–262.

[48] Telis V R N, Sobral P J A (2002) Glass Transitions for Freeze–dried and Air–dried Tomato. Food res. int. 35: 435–443.

[49] Sun W Q, Davidson P, Chan H S O (1998) Protein Stability in the Amorphous Carbohydrate Matrix: Relevance to Anhydrobiosis. Biochim. biophys. acta 1425: 245-254.

[50] Salman A D, Hounslow M J, Seville J P K (2006) Granulation. In: Sal, A. (Ed.), Handbook of Powder Technology, vol. 11. España: Elsevier.

[51] Gabbott P (2008) Principles and Applications of Thermal Analysis. Blackwell Publishing. Chapter 9.

[52] Dilworth S E, Buckton G, Gaisford S, Ramos R (2004) Approaches to Determine the Enthalpy of Crystallization, and Amorphous Content, of Lactose from Isothermal Calorimetric Data. Int. j. pharm. 284: 83–94.

[53] [53] Relkin P (1996) Thermal Unfolding of β-Lactoglobulin, α-Lactalbumin and Bovine Serum Albumin. A Thermodynamic Approach. Crit. rev. food sci. 36 (6): 565–601.

[54] Yamul D K, Lupano C E (2003) Properties of Gels from Whey Protein Concentrate and Honey at Different pHs. Food res. int. 36: 25-33.

[55] Penco M, Sartore L, Bignotti F, D'Antone S, Di Landro L. (2000) Thermal Properties of a New Class of Block Copolymers Based on Segments of Poly(D,L-lacticglycolic Acid) and Poly(e-caprolactone). Eur. polym. j. 36: 901-908.

[56] Kavitha M, Bobbili K B, Swamy M J (2010) Differential Scanning Calorimetric and Spectroscopic Studies on the Unfolding of Momordica Charantia Lectin. Similar Modes of Thermal and Chemical Denaturation. Biochimie 92: 58-64.

[57] Schubring R (1999) DSC Studies on Deep Frozen Fishery Products. Thermochim. acta 337: 89-95.

[58] Zamorano L S, Pina D G, Gavilanes F, Roig M G, Yu Sakharov I, Jadan A P, van Huystee, R B, Villar E, Shnyrov V L (2004) Two-state Irreversible Thermal Denaturation of Anionic Peanut (Arachis Hypogaea L.) Peroxidase. Thermochim. acta 417: 67-73.

[59] Vermeer A W P, Norde W (2000) The Thermal Stability of Immunoglobulin: Unfolding and Aggregation of a Multi-Domain Protein. Biophys. j. 78: 394-404.

[60] Michnik A, Drzazga Z, Kluczewska A, Michalik K (2005) Differential Scanning Microcalorimetry Study of the Thermal Denaturation of Haemoglobin. Biophys. chem. 118: 93-101.

[61] Rodriguez Furlán L T, Lecot J, Pérez Padilla A, Campderrós M E, Zaritzky N (2012) Stabilizing Effect of Saccharides on Bovine Plasma Protein: A Calorimetric Study". Meat sci. In press.

[62] Gallego K, López B L, Gartner C (2006) Estudio de Mezclas de Polímeros Reciclados para el Mejoramiento de sus Propiedades. Rev. Fac. Ing. 37: 59-70.

[63] Mousavioun P, Doherty W O S, George G (2010) Thermal Stability and Miscibility of Poly(hydroxybutyrate) and Soda Lignin Blends. Ind. crop. prod. 32(3): 656-661.

[64] Rodriguez Furlán L T, Pérez Padilla A, Campderrós M E (2010) Functional and Physical Properties of Bovine Plasma Proteins as a Function of Processing and pH, Application in a Food Formulation. Adv. j. food sci. tech. 2(5): 256-267.

[65] Rodriguez Furlán L T, Rinaldoni A N, Padilla A P, Campderrós M E (2011) Assessment of Functional Properties of Bovine Plasma Proteins Compared with other Proteins Concentrates, Application in a Hamburger Formulation. Am. j. food tech. 6 (9): 717-729.

Section 4

Thermal Analysis of Phase Transitions of Polymers and Paraffinic Wax

Chapter 10

Silver Particulate Films on Compatible Softened Polymer Composites

Pratima Parashar

Additional information is available at the end of the chapter

http://dx.doi.org/10.5772/54502

1. Introduction

Polymer/inorganic nanocomposites have been of great interest in recent years, not only for the novel properties of the nanocomposite materials but also for the continuously growing demand for the miniaturisation of electronics components, optical detectors, chemicals and biochemical sensors and devices.

Polymer matrices have been frequently used as particle stabilizers in chemical synthesis of metal colloids since these prevent agglomeration of the particles. Within the past decade, incorporating silver nanoparticle into a polymer matrix is more interesting because the resulting nanocomposites exhibit applications in catalysis, drug wound dressing and optical information storage.

It is difficult to disperse silver nanoparticle homogeneously into a polymer matrix by ex situ methods because of easy agglomeration of nanoparticles. At present, it is possible to obtain nanoparticles of different shape and size in nanostructured polymeric environment using various polymeric systems and different approaches. Numerous methods used toxic and potentially hazardous reactants. Increasing environmental concerns over synthesis route resulted in an attempt to adopt eco friendly methods.

One of the simplest techniques to form such particulate structures, which are generally known as island or discontinuous metal films, is through vacuum evaporation of metal on to a dielectric substrate by stopping the deposition at a very early stage. The temporal instability exhibited by island films even in vacuum is attributed to mobility of islands followed by coalescence [1]. Further, these films get oxidised when they are exposed to atmosphere. The oxidation of islands causes an irreversible increase in electrical resistance [2]. An interesting sub-surface particulate structure formation was reported when certain inorganic materials are deposited on to softened polymer substrates [3-6] and the morphology and formation of such structures depend on thermodynamic as well as deposition parameters [6, 5]. The use of softened polymer substrats provides the unique possibility of easily controlling the viscosity of the substrate to form a subsurface discontinuous silver particulate films. The morphology of sub-surface particulate structures also depends upon polymer metal interaction [7, 8]. The reported method is evaporation of silver on polymer substrate at high temperature and in vacuum of the order of 10^{-6} Torr. The ability to precisely tailor and optimize the nanocomposite structure creates opportunities for a wide range of applications.

2. Body

Pyridine-containing polymers have attracted interests in recent years because they can be used in various applications as water-soluble polymers and coordination reagents for transition metals, especially 4-vinylpyridine because of its more interesting properties resulting from higher accessibility of the nitrogen atom [9].

Deposition of silver on interacting polymers like Poly (2-vinylpyridine) and Poly (4-vinylpyridine) resulted in the formation of smaller particles (~ a few tens of nm) with smaller inter-particle separations whereas silver deposited on softened inert polymer like polystyrene (PS), irrespective of the deposited thickness is of highly agglomerated structures. Therefore, silver films on inert polymer lack in application due to room temperature resistances equalling that of the substrate. But, silver films on interacting polymers have room temperature resistance in the range of a few tens to a few hundred MΩ/sheet, which is desirable for device applications [8]. Both the interacting polymers are hygroscopic and costly. Therefore, blending an inert and stable polymer like PS with interacting polymers like P2VP and P4VP may provide a polymer matrix suitable for formation of subsurface silver films. Miscibility between the components polymers play a vital role in blending of polymers at the molecular levels. A compatible blend provides a firm basis for further application in devices. Earlier researchers [10-14] have suggested the improvement of miscibility of PS with P4VP by incorporating proton donors like poly (acrylic) acid and poly (p-vinyl phenol) or methacrylic acid into the chains of PS with P4VP in order to utilise its proton acceptor nature. Further, reversible addition-fragmentation chain transfer polymerization was developed by J.J. Yuan and et al [9] for the controlled preparation of PS/P4VP triblock copolymers as PS-b-P4VP-b-PS and P4VP-b-PS-b-P4VP.In order to retain the properties of both the polymers PS and P4VP, blending is carried out through solution casting and it is expected that combination of PS and P4VP should give rise to organised subsurface silver particulate structures with the advantages of both the polymers.

Polymer blending is a common way to develop new polymer materials with desirable combinations of properties. The main advantage of this method is to control the properties by varying the blend compositions [15]. A compatible blend is needed to have desirable combinations of properties of both the polymers. Compatibility of the two homopolymers is needed to an optimum extent for a blend to show superior properties. The compatibility signifies specific interaction such as dipole-dipole, ion-dipole and hydrogen bonding.

Various measurements like heat of mixing, viscometry, glass transition temperature, morphological studies by optical and electron microscopy, infrared spectroscopy and dynamic mechanical analysis, are used to study polymer compatibility. The compatibility of polymer composite is discussed using DSV, DSC, FTIR and SEM. Dilute solution viscometry is a simple and reliable method to investigate interactions of macromolecules in solution. It has been used as a complementary technique to prospect the effect of the position of nitrogen atom in the pyridine ring of P4VP on the interaction developed within PS/P4VP blends. This technique could not be applied to PS/P2VP blends because these blends show phase separation after twenty-four hour of preparation of solution. The criterion of single composition dependent glass transition is used to investigate the miscibility of polymer blends by DSC. Specific interactions most often liberate a heat of mixing and contribute towards the free energy of mixing. Fourier transform infrared spectroscopy is used to investigate specific interactions between the homopolymers in the blend compositions and compared to calorimetric results. SEM results confirm compatibility of blends at higher temperature.

Nanocomposites of metal nanoparticles in a polymer matrix have generated a great deal of interest which depends on the metal-polymer composition and their structure. Polymers are particularly attractive as the dielectric matrix in composites due to their versatile nature and can easily be processed into thin films. These nanocomposites exhibit a unique combination of desirable optical and electrical properties that are otherwise unattainable [16-20].All these properties depend on the size, size distribution and shape of the nanoparticles. The growth and arrangement of metal nanoparticles on various substrates are therefore key issues in all the fields of modern science and technology relating to nanoelectronics, photonics, catalysis and sensors [21].

Since past, polymer membranes have been studied as supporting materials for colloidal metals which are well known catalysis. Strictly speaking, such a membrane contained colloidal metal-rich and metal-poor phases and the localization of colloidal metals is governed by non–linear diffusion equations. Poly (styrene-b-2-vinyl pyridine) diblock copolymer forms micro phase separated film [22] and Ag ion added to such a film is localized in P2VP micro domains not in PS phase. Theoretical study in the self assembly of inorganic/block copolymers hybrids by Ginzburg and co-workers have predicted that affinity, size and amount of inorganic nanoparticles can be exploited to control the phase behaviour of inorganic/BCP hybrids [23].

Metal–polymer nanocomposite containing widely separated nanoparticles exhibit insulating behaviour. As the percentage of metal in composite increases, the nanoparticle separation decreases. At a certain thickness of silver on softened polymer substrate, nanoparticles are quite densely packed but separated by polymer gap such a film offer a host of unique property relevant to practical applications. These applications include high dielectric constant passives, electromagnetic interference shielding, sensors, and detector designed for a variety of specific purposes with high performances, sensitivity and flexibility [24]. Further, the morphology of the cluster films deposited on softened polymer substrates is dependent on the polymer-metal interaction. Gold deposited on polystyrene (PS) and subsequently annealed above its glass transition temperature results in a highly agglomerated film with large separation between the clusters, possibly due to inert nature of PS [1]. Silver deposited on PS also forms highly agglomerated subsurface particulate structure with large separations between the metal particles [25].Also, coalescence rate for gold particles in a poly(2-vinyl pyridine) matrix is much less than the coalescence rate for gold particles in a polystyrene matrix, indicating that homopolymer/metal interactions play an important role in the determination of the coalescence rate [7].Hence, this high coalescence rate in case of PS resulted in a highly agglomerated film even for a thickness of 300 nm. But, silver deposited on an interacting polymer like Poly (2-vinylpyridine) and poly (4-vinylpyridine) resulted in the formation of smaller particles (~ a few tens of nm) with smaller inter-particle separations [8].The differences in dispersion, size distribution and impregnation depth result from the differing natures of the polymer hosts and the processing conditions [26]. Therefore, it is desirable to restrict the nanoparticles to a small size regime along with a narrow size distribution by blending inert PS with interacting P2VP and P4VP.Therefore; it is interesting to blend PS with P2VP and P4VP in order to have desired morphology of silver particulate films on compatible polymer composite.

3. Experimental

Poly (4-vinyl pyridine) and Poly (2-vinylpyridine) used in this study, were procured from Sigma-Aldrich Chemicals Pvt. Ltd and Polystyrene from Alfa-Aesar (A Johnson Mathley company) respectively. The molecular weights of P4VP, P2VP and PS are 60,000, ~37,500 and 100,000, respectively. The structure of P4VP, P2VP and PS are (a), (b) and (c), respectively, as follows:

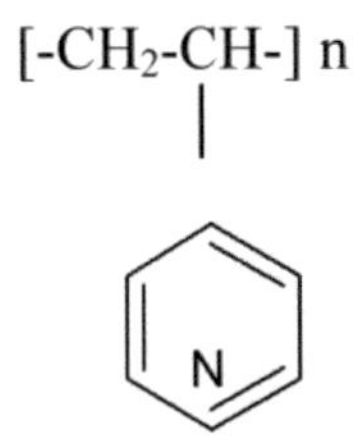

(a)

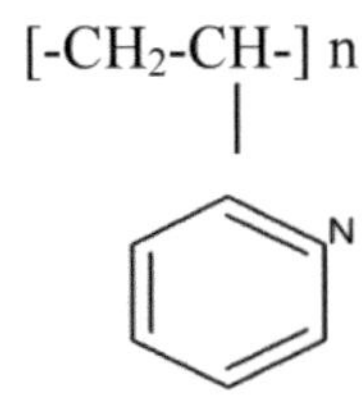

(b)

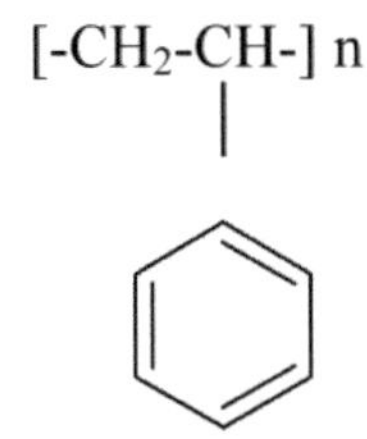

(c)

Polymer blends were prepared through solution blending by mixing in a common solvent, dimethylformamide (DMF). Blends of PS/P4VP with different compositions {PS (w)/P4VP (w) =0:100; 25:75, 50:50; 75:25; 100:0} were prepared. 1g of the total polymers at different ratios was dissolved in 20 ml of DMF at room temperature. Composite of PS/P2VP with different compositions {PS (w)/P2VP (w) = 0:100; 50:50; 100:0} were prepared. An amount of 0.5 g of the total polymers at different ratios, were dissolved in 5 ml of DMF at room temperature.

The stock solutions of PS, P4VP, and their different blend compositions were prepared in the common solvent DMF. Viscosity measurements were made using Ubbelohde Viscometer at 28^{0}C with an accuracy of ±0.2%.

For DSC study, the solvent is allowed to evaporate in a thermostat for 24 hours. The residuals of component polymers and their blend in powder form were then dried at 80^0 C for several days to ensure complete removal of any traces of residual solvent. The residuals of component polymers and their blends were found to be translucent. DSC measurements were carried out using a Shimadzu DSC-50. Small quantities of the samples, 8-10 mg were scanned at a heating rate of 5- 10 K/min^{-1} in the temperature range 28 to 220^{0}C under Nitrogen, N_2.

FTIR spectra of the blends were recorded using a Perkin-Elmer spectrometer (model 1000).

Thin films of homopolymers and their composite of approximately 5μm thickness were solution cast on a glass slide pre-coated with silver contacts with a gap of 1 cm X 1 c m for electrical studies. Silver films of various thicknesses were deposited on these substrates held at 457 K in a vacuum better than 8×10^{-6} Torr. The glass transition temperature of P4VP, P2VP and PS are 410, 357 and 373 K, respectively. Therefore, polymer substrates are softened at 457 K and sufficient fluidity is ensured. A chromel-alumel thermocouple was used to measure the substrate temperature by clamping it to the substrate surface holding the film. Source to substrate distance was maintained at 20 cm. A Telemark quartz crystal monitor (Model 850) was used to measure the deposition rate, as well as the overall film thickness. The deposition rate used was 0.4nm/s for all the films. Resistance measurements were carried out in-situ, using a Keithley electrometer model 617. The films were annealed at the deposition temperature for 1 hour before cooling them to room temperature. Stability of the films against exposure was studied by monitoring the film resistance during exposure to atmosphere by continuously leaking air into the vacuum chamber using a needle valve. The leak rate was such that the pressure increased by an order of magnitude in about a minute.

Optical absorption spectra of the silver particulate films were obtained on a Shimadzu UV-Vis-NIR spectrophotometer model SHIMADZU UV 3101 PC.

Scanning electron microscopy (SEM) measurements were carried out on Scanning electron microscope model JEOL JSM 5800 CV with image processing software.

4. Results and discussion

4.1. Viscosity measurements

The effectiveness of dilute-solution viscometry is based on the assumption that mutual interactions of macromolecules in solution have a great influence on the viscosity in TPS (two polymers in a solvent) [11]. Therefore, compatibility among the polymers depends on the fact that the repulsive interactions among polymer molecules cause their shrinkage, leading to a lowering of solution viscosity, while attractive interaction increases the viscosity.

The relative and reduced viscosities of homopolymer and their blends are found out from viscometric measurements. The intrinsic viscosity values of homopolymers and their blends were determined at 27⁰ C in DMF by extrapolation to zero concentration of the plots of reduced viscosity (η_{sp}/C) versus concentration as shown in figure 1. The plots are not perfect linear, but no crossover is seen. A sharp crossover in the plots of reduced viscosity versus concentration indicates incompatibility of blends [27]. Therefore, some order of compatibility is expected in the blends.

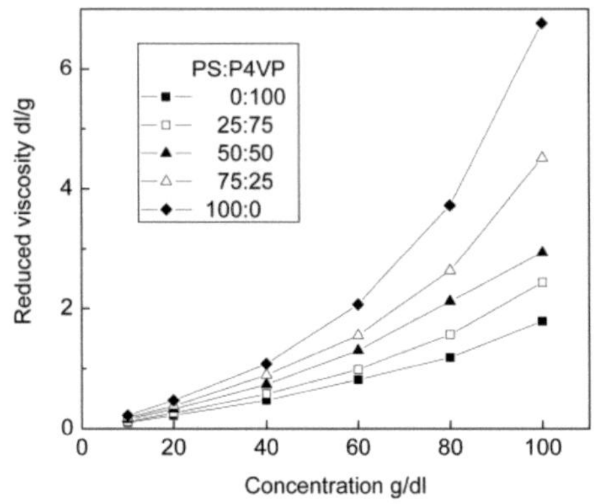

Figure 1. Reduced viscosity versus concentration composition of homopolymers in the blend for the PS/P4VP blends.

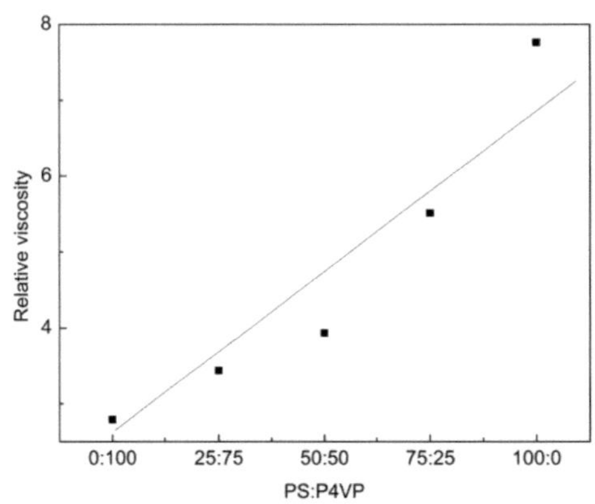

Figure 2. Relative viscosity versus original total concentration of 0.05g/ml.

Figure 2 shows a plot of relative viscosity versus blend composition at the original total concentration of 0.05g/ml. It is not found to be perfect linear for entire range. This indicates that the PS/P4VP blends are not hundred percent incompatible blend system [10, 28, 29].The concentration value is much lower than the critical concentration C″ estimated by C″ =1/ [η] [10].In the absence of specific interactions within the blend, polymer coils are independent if the solution concentration is below the critical concentration [10].The mixed solutions in DMF of PS and P4VP were clear indicating that no strong interactions are taking place between the blend components chains.

As proposed by Krigbaum and Wall [30], the specific viscosity η_{sp} of a solution polymer blends can be expressed as:

$$(\eta_{sp})_m = [\eta_1]\, C_1 + [\eta_2]\, C_2 + b_{11}C_1^2 + b_{22}C_2^2 + 2b_{12}C_1C_2 \quad (1)$$

Where $[\eta_1]$ and $[\eta_2]$ are the intrinsic viscosities of component polymers 1 and 2, C_1 and C_2 are the concentrations of polymers 1 and 2 in solution of polymer blend, b_{11} and b_{22} are specific interaction coefficients of polymers 1 and 2 in single polymer solutions and b_{12} is the interaction coefficient for the polymer blend of component polymers 1 and 2.The coefficient b_{11} is related to the constant k in the Huggins equation, when component polymer 1 is alone in the solution. This also applies to b_{22}.

$$\eta_{sp}/C = [\eta] + k\,[\eta]^2\, C \quad (2)$$

The relationship between b_{11} and k can be written as

$$b_{11} = k_1\,[\eta]^2 \quad (3)$$

Where k_1 is the Huggins constant for component polymer 1 in solution. The theoretical interaction coefficient between the two polymers, b_{12}^*, can be expressed as:

$$b_{12}^* = (b_{11}\, b_{22})^{1/2} \quad (4)$$

According to Krigbaum and Wall [30], information on the intermolecular interaction between polymer 1 and polymer 2 can be obtained by comparison of experimental b_{12} and theoretical b_{12}^*values. Hence, the miscibility of binary polymer blends can be characterized by the interaction parameter, Δb:

$$\Delta b = b_{12} - b_{12}^* \quad (5)$$

Negative values of Δb are found for solutions of incompatible polymer system, while positive values of Δb refer to attractive interaction in compatible systems. We can reduce the equation (1) to the following form when total concentration of the mixture (C) approaches zero.

$$(\eta_{spm}/C)_{c\to 0} = [\eta_1](C_1/C)_{c\to 0} + [\eta_2](C_2/C)_{c\to 0} \quad (6)$$

Polymer 1-polymer 2 interaction, Δb and theoretically $(\eta_{sp})_m$ can be calculated [15] as follows:

$$(\eta_{sp})_m / C_m = [\eta]_m + b_m C_m \quad (7)$$

Where C_m is the total concentration of polymers, C_1+C_2, and $[\eta]_m$ is the intrinsic viscosity of blend. It can be theoretically defined as;

$$[\eta]_m = [\eta]_1 X_1 + [\eta]_2 X_2 \quad (8)$$

Where X_1 and X_2 are weight fractions of polymer 1 and polymer 2, respectively. Interaction parameter, b_{12}, can be defined by the equation

$$b_m = X_1^2\, b_{11} + X_1X_2b_{12} + 2\, X_2^2\, b_{22} \quad (9)$$

Where b_m defines the global interaction between all polymeric species. b_{12} may be obtained experimentally by Eq. (7).

All the calculated and experimental values are summed up in Table 1. The experimental intrinsic viscosity values are compared with their weighed average values and are found to be lower than the theoretical values. Shih and Beatty [29] have studied immiscible systems by this method and found that the intrinsic viscosity always shows a negative deviation due to the repulsive interaction between the polymers. Hence, these blends were not thermodynamically compatible under equilibrium conditions.

The repulsive deviation causes a reduction in the hydrodynamic volume of the polymer molecules, and hence, the viscosity of the solution is reduced. It is found that Δb values are very much less than unity and negative for all the blends except for the blend 25:75, for which slightly positive value of Δb predicting some order of compatibility. Also, positive deviation in 25:75 can be attributed to increase in the proportion of the polar group, P4VP in the blend [11]. The difference between η_1 and η_2 are found to be large and therefore, a more effective parameter μ can be defined to predict about the compatibility [28].

$$\mu = \Delta b \,/\, (\eta_2 - \eta_1)^2 \tag{10}$$

Low values of μ observed in Table 1 may be due to weaker interaction between the polymers. The lower values of interaction parameters indicate that the PS and P4VP are not fully compatible, but physically miscible up to a certain extent.

Blend comp of PS/ P4VP	Intrinsic viscosity		Slope of red viscosity vs. concentration curve	Experimental Value, b_{12}	Theoretical value, b_{12}^*	Δb	μ
	Experimental(dl/g)	Theoretical(dl/g)					
0:100	0.167	0.167	0.018	-	-	-	-
25:75	0.284	0.402	0.024	0.053	0.034		
						0.019	0.021
50:50	0.325	0.638	0.03	0.017	0.034	-	-
						0.017	0.019
75:25	0.64	0.874	0.045	0.018	0.034	-	-
						0.016	0.017
100:0	1.11	1.11	0.067	-	-	-	-

Table 1. Intrinsic viscosity and interaction parametar of PS/P4VP blends.

4.2. Differential scanning calorimetery

DSC endothermograms for the homopolymers and their blends are shown in figure 3

All the blends exhibit a single T_g, intermediate between those of the parent polymers, PS and P4VP indicating the miscibility of these blends. The theoretical values of these can be predicted using Fox equation [31] and Gordon-Taylor equation [32].

$$1/T_g = X_1/T_{g1} + X_2/T_{g2} \tag{11}$$

$$T_g = (X_1 T_{g1} + k X_2 T_{g2}) / (X_1 + k X_2) \tag{12}$$

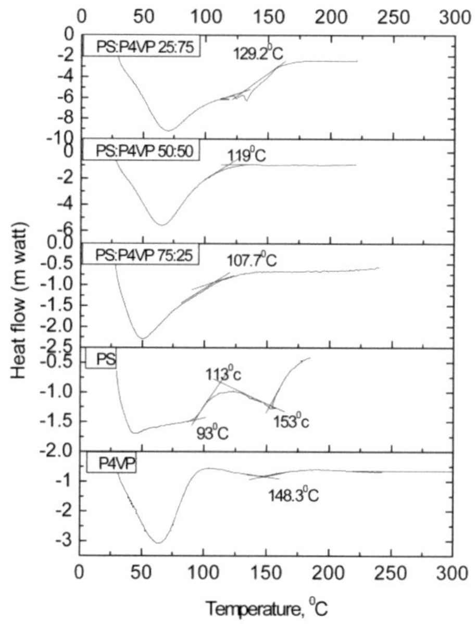

Figure 3. DSC thermograms of PS/P4VP blends

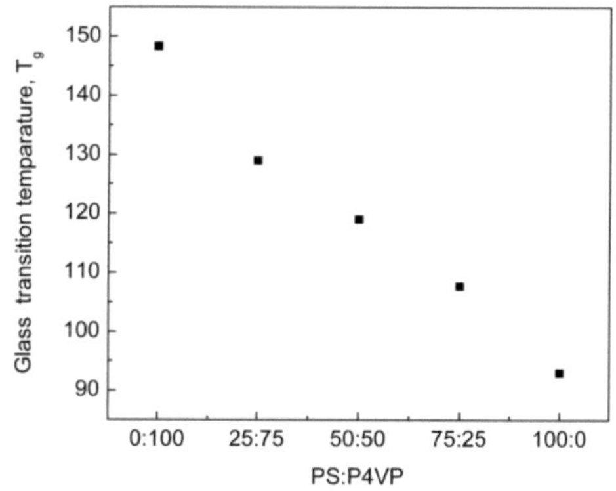

Figure 4. Glass transition temperature versus composition of PS/P4VP.

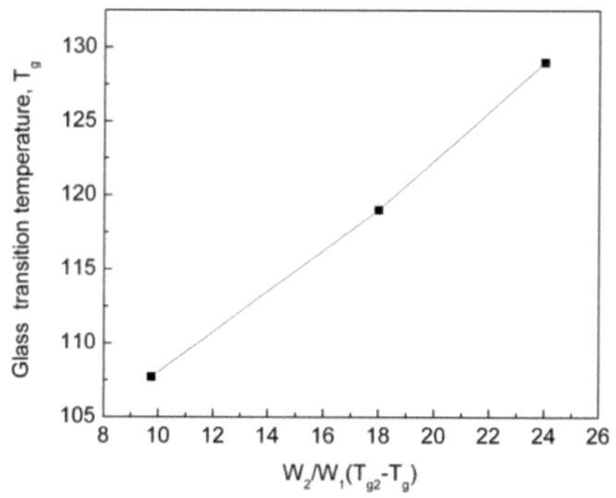

Figure 5. Verification of Gordon- Taylor equation for PS/P4VP blends.

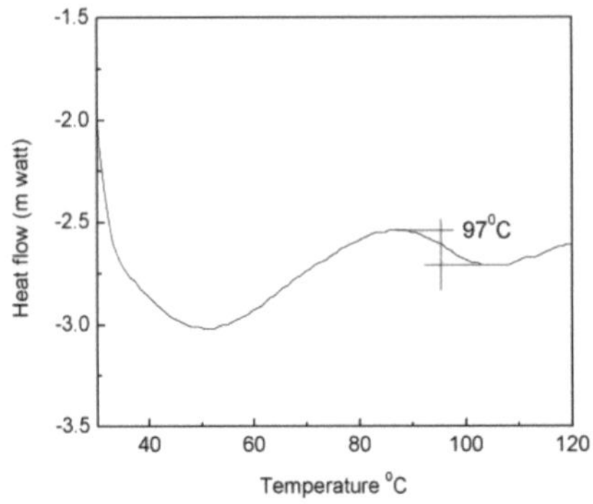

Figure 6. DSC thermogram of PS/P2VP, 50:50

Where X_1, X_2, T_{g1} and T_{g2} are the weight fractions and glass transition temperatures corresponding to polymer 1 and polymer 2, respectively. k is a constant which gives a semi-quantitative measure of degree of the interaction between the two polymers. All the experimental and calculated values of Tg, are shown in Table 2. Positive deviation observed from Fox equation is attributed to intermolecular interaction between the polymers. Figure 4 shows the plot of T_g, with blend composition. It is well established that when interactions between blend components are strong, such as those affected by Hydrogen bonding, the experimentally determined T_g of the blends are higher than those calculated from the additivity rule as a result of the reduction of polymer chains mobility in the blend [10]. In order to estimate the strength of the intermolecular interactions within the PS/P4VP blends, we used the Gorden-Taylor equation to verify through the linear fit in figure 5. Slope (k) of the straight line obtained is found to be 0.85, indicating interaction between the polymers [32]. The intercept is about 100.47^0 C which corresponds to T_g of pure PS.

Blend comp of PS/P4VP	Experimental T_g value (^{0}C)	Theoretical T_g value(^{0}C)
	DSC	Fox equation
0:100	148.3	-
25:75	129	125
50:50	119	116
75:25	107.7	107
100:0	93	-

Table 2. Experimental and theoretical glass transition temperatures of PS/P4VP blend.

Figure 6 shows DSC thermogram of the blend PS/P2VP, 50:50.This indicates a single T_g, about 370K intermediate between those of the parent polymers, PS and P2VP indicating the compatibility of the blend on melt mixing at higher temperature.

4.3. Fourier transform infrared spectroscopy

The proton donor PS copolymer, PSMAA (PS-methacrylic acid) with P4VP in solvent chloroform observed the specific interactions with the formation of hydrogen bonds, at a frequency, υ = 1607 cm^{-1} corresponding to the –COOH…N≤ hydrogen bond [11]. Therefore, the incorporation of MAA into PS results in an augmentation in miscibility with P4VP, in comparison with the PS/P4VP system, which is highly incompatible [11]. The present solution cast PS/P4VP blends, needs to be applied in thin film form at higher temperature around 200^0 C. Therefore, we need to ascertain about their compatibility at higher temperature. Figure 7 is the FTIR of the sample of the polymer blend PS/P4VP, 50:50 before and after DSC being carried out. This shows the absorption bands at 1598 and 1414 cm^{-1} corresponding to the pyridine ring of P4VP and at 822 cm^{-1} to the single substituted pyridine appeared in the spectra of PS/P4VP blend. Similarly, for PS, the absorption peaks at 1493cm^{-1} and 1448 cm^{-1}, which were characteristic of the phenyl ring and the peak at 697cm^{-1}, corresponding to the signals of the single substituted phenyl ring, appeared for the blend as well. Similar trends were observed by the triblockpolymers PS-P4VP-PS and P4VP-PS-P4VP, which were synthesized by chain transfer agent [33]. In case of PS-block-P4VP, the stretching bands overlap [14].The silver particulate film deposited on PS/P4VP (50:50) resulted in desired structure underlying the property of PS and P4VP [34].Figure 7 shows no shift in the frequency leads to the absence of hydrogen bond. Thus, the possibility of protonation of nitrogen of P4VP [14] is ruled out but some intermolecular interaction at higher temperature leads to single Tg composition.

Figure 8 shows FTIR of PS/P2VP, 50:50 after DSC which clearly indicates the absorption bands at 1594 and 1414 cm^{-1} corresponding to the pyridine ring of PS and at 822 cm^{-1} to the single substituted pyridine ring appeared in the spectra of PS/P2VP blend. Similarly, the absorption peaks at 1495cm^{-1} and 1448 cm^{-1}, which were characteristic of the phenyl ring and the peak at 697cm^{-1}, corresponding to the signals of the single substituted phenyl ring for P2VP spectra, also appeared in the spectra of PS/P2VP, 50:50 blend. No shift in the frequency ruled out the possibility of any hydrogen bond in PS/P2VP. But single T_g of the blend suggests some order of the compatibility at higher temperature.

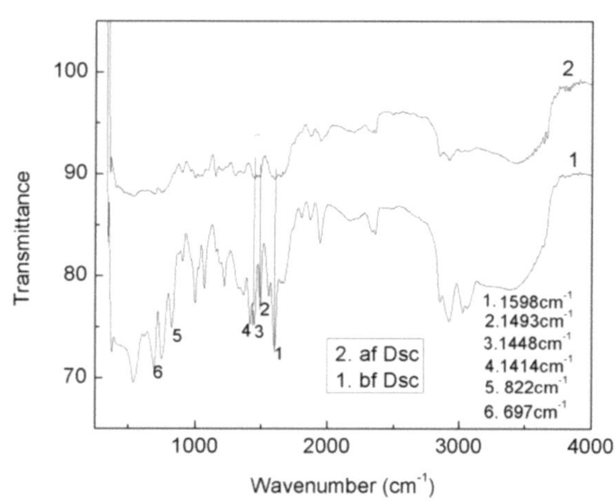

Figure 7. FTIR for PS/P4VP (50:50) before and after DSC.

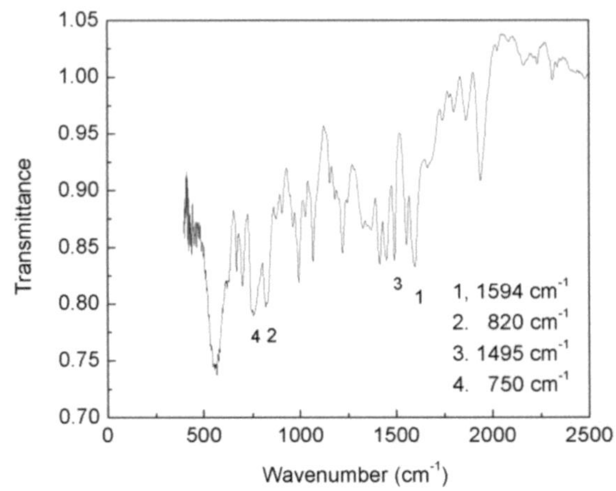

Figure 8. FTIR of PS/P2VP, 50:50 after DSC

4.4. Scanning electron microscopy

Fig. 9a and 9b show the SEM of PS: P4VP (50:50) samples after and before the DSC have been carried out. It is clear that after DSC the blend mixed better than the blend as obtained at room temperature by solution cast which eventually show phase separation after few days of preparation. Fig.9b clearly shows the dispersed phase of polymers whereas Fig. 9a indicates better compatibly of polymers. This can be attributed to mixing of homoplymers around 200^{0}C during the process of DSC. This is an indication of suitable compatibility of these blends at higher temperature.

Figs. 10a and 10b are the Scanning Electron Micrographs of PS/P2VP, 50:50 samples before and after the DSC have been carried out. It is clear that melt mixing at higher temperature gives more compatible blend than room temperature solution mixed blend. Thus, an order of compatibility is achieved in PS/P2VP, 50:50 blend as reported for PS/P4VP blends [35]. Therefore, we can expect formation of discontinuous silver subsurface film on the blend.

4.5. Electrical behaviour of discontinuous silver films on PS/P2VP and PS/P4VP

Figure 11&12 shows the variation of the logarithm of resistance against inverse of temperature for silver films of different thicknesses deposited on polymers and their composite at a temperature of 457 K, during cooling to room temperature. It is interesting to note that while some of the films show only negative temperature coefficient of resistance (TCR) some show almost zero TCR. Some of the films show negative TCR at higher temperatures and almost zero TCR at lower temperatures. The 50 nm thick silver films on pure PS and 75:25 blend of PS/P4VP show negative TCR. Silver on PS showed similar behaviour in our earlier studies resulting in room temperature resistance same as that of the substrate with the formation of large silver particles separated by large distances [25]. Blending the inert polymer PS with an interacting polymer like P4VP to the extent of 25% does not seem to alter the morphology of the particulate film as indicated by the electrical

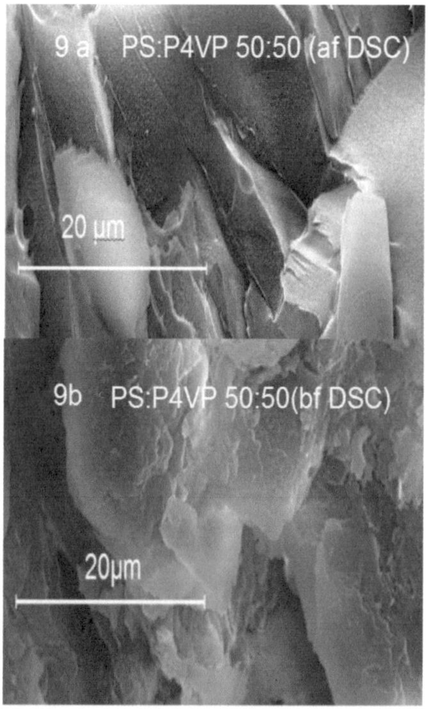

Figure 9. SEM photographs of PS/P4VP (50:50) a. after DSC and b. before DSC.

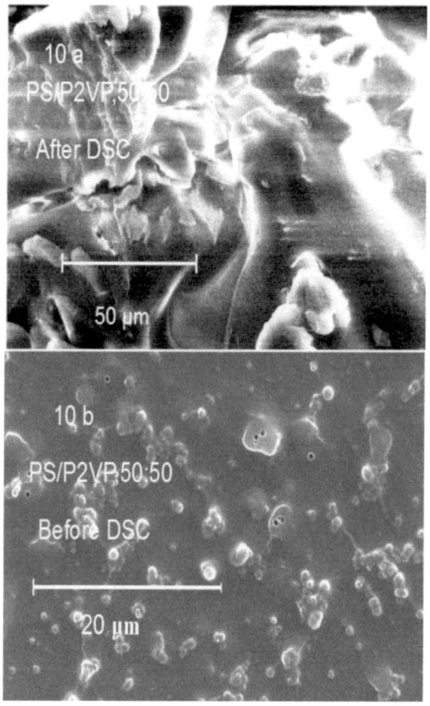

Figure 10. SEM photographs of PS/P2VP, (50:50) (a) after and (b) before DSC.

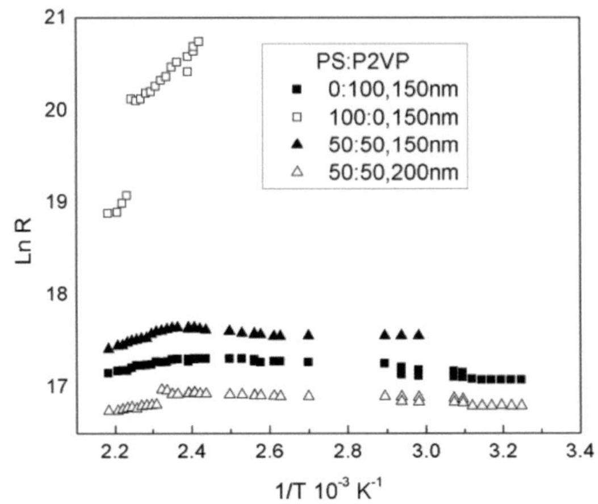

Figure 11. Variation of ln R with 1/T for silver films deposited on the composite PS/P2VP held at 457 K at a rate of 0.4 nm/s.

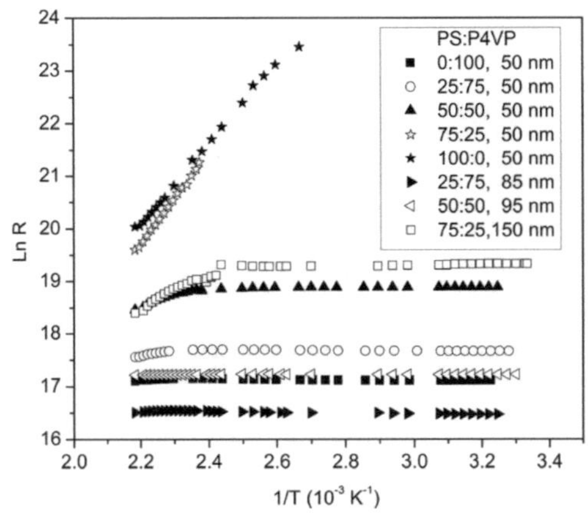

Figure 12. Variation of ln R with 1/T for silver films deposited on the composite PS/P4VP held at 457 K at a rate of 0.4 nm/s.

behaviour. When the P4VP content is increased to 50%, a negative TCR at high temperature followed by almost zero TCR at lower temperatures exhibited by the 50 nm thick film is similar to the behaviour observed earlier for the case of pure P2VP and P4VP [36, 37] indicating that the film consists of small particles separated by small distances. With further increase in P4VP content, the negative TCR part diminishes, giving rise to a near zero TCR .Similarly, the films on composites PS/P2VP (50:50) initially show negative TCR but zero TCR at lower temperature. Blending of PS with P2VP and P4VP seems to result in a positive effect on the composites. Therefore, negative TCR is totally vanishing and give rise to near zero TCR at room temperature for films deposited on the composites. The over all resistance of film deposited on composite decreases with increase in thickness of silver films deposited on the composite. Similar trend were reported that the electrical conductivity of composites is increased with high silver loading (30-80%) [38]. It is also interesting to note that even at 50% P4VP, with an increase of silver deposited, films show electrical characteristics as that of the films on pure P4VP [37]. Further, when 150 nm thick silver is deposited on a polymer blend with only 25% of P4VP, the films show desirable electrical characteristics in contrast with the very high room temperature resistance observed for films on pure PS even at 300 nm of silver [25].

Figure 13&14 show the variation of logarithm of resistance (ln R) with logarithm of pressure in torr [ln (pressure)]. It is seen that the resistances show large increase beyond a pressure of about 0.5 torr, for all the films. The variation in resistance is very small till that pressure. Similar characteristics were shown by silver films on softened P4VP [39] and P2VP [36]. It was shown through X-ray photoelectron spectroscopy (XPS) studies at various electrons take off angles that silver clusters are formed at a depth of a couple of nm from the polymer surface [36,37]. It is known that the formation of subsurface particulate structure formation is subject to certain thermodynamic [6] and deposition conditions [5]. While the thermodynamic conditions are met for the deposition of metals on most of the polymer substrates, deposition conditions used in the present study are similar to those used in our earlier studies. Therefore, it is reasonable to assume that the particles are formed just a couple of nm below the polymer surface. The behaviour of the particulate films upon exposure to atmosphere is attributed to oxidation of islands due to the inadequate polymer cover.

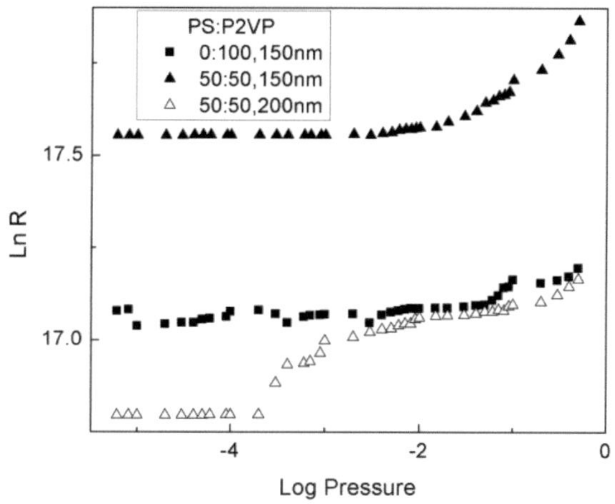

Figure 13. Variation of ln R with log (Pressure) for silver films deposited on composite PS/P2VP held at 457 K at a rate of 0.4 nm/s.

Figure 15 shows variation of room temperature resistance of silver particulate film of 150 nm thickness on PS/P2VP blends .It clearly indicates the decrease in resistance with increase in the amount of P2VP in the blend. Figure 16 shows variation of room temperature resistance of silver particulate film of various thicknesses on the blend PS/P2VP, 50:50.With the increase in the thickness of the silver particulate film the room temperature resistance of the film decreases [34]. Blending of P2VP with PS is expected to provide a polymer matrix where the size of silver clusters and inter-cluster separation can be modified because dispersion, size distribution and impregnation depth results from the natures of polymeric hosts [40].

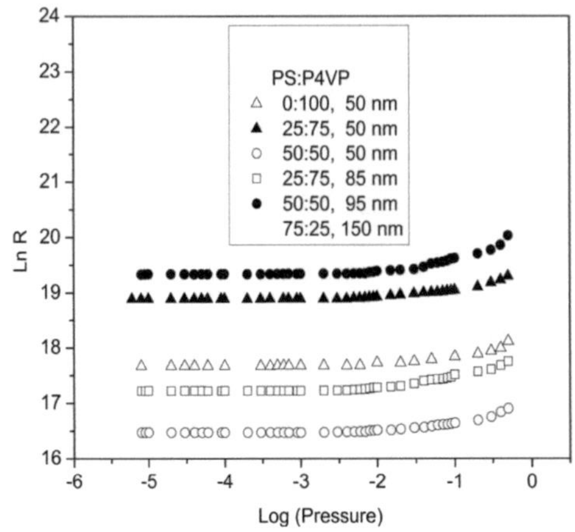

Figure 14. Variation of ln R with log (Pressure) for silver films deposited on composite PS/P4VP held at 457 K at a rate of 0.4 nm/s.

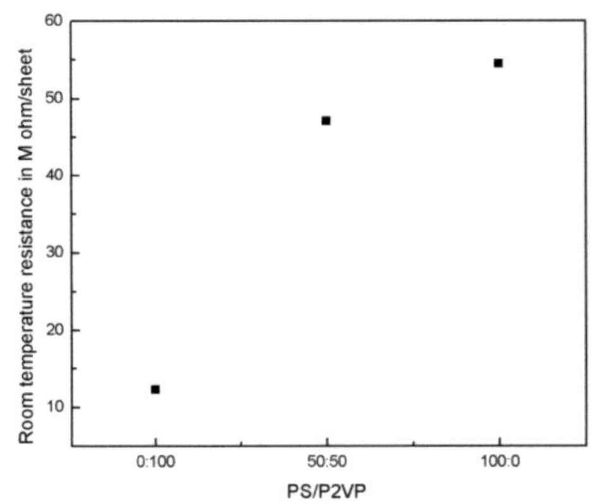

Figure 15. Variation of room temperature resistance of silver particulate film of 150 nm thickness verus composition of PS/P2VP.

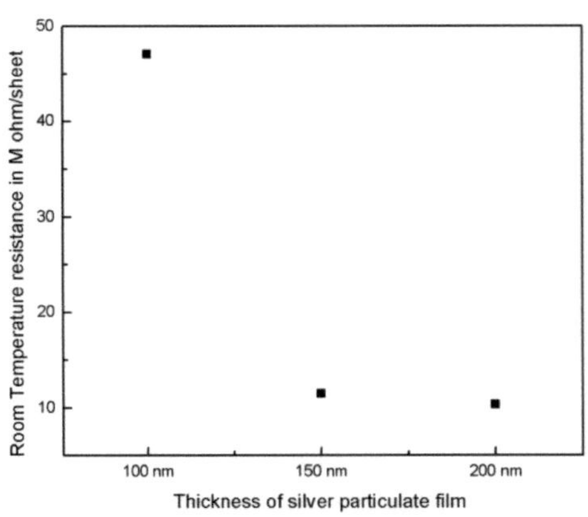

Figure 16. Variation of room temperature resistance of silver particulate film verus their thicknesses for PS/P2VP,50:50.

Figure 17 a,b,c shows the SEM of silver particulate films of thickness 200 nm on PS, P2VP and PS/P2VP, 50:50.It is evident from the figure that decrease in the particle size in the films of P2VP and PS/P2VP, 50/50 resulted in the close proximity of particles with reduction in the inter-particle separation. Also, the decrease in the size of silver cluster in the blend PS/P2VP, 50:50 improves the tunnelling effect as expected [34].Thus, the room temperature resistance of silver particulate film on the blend is now in the desirable range for applications.

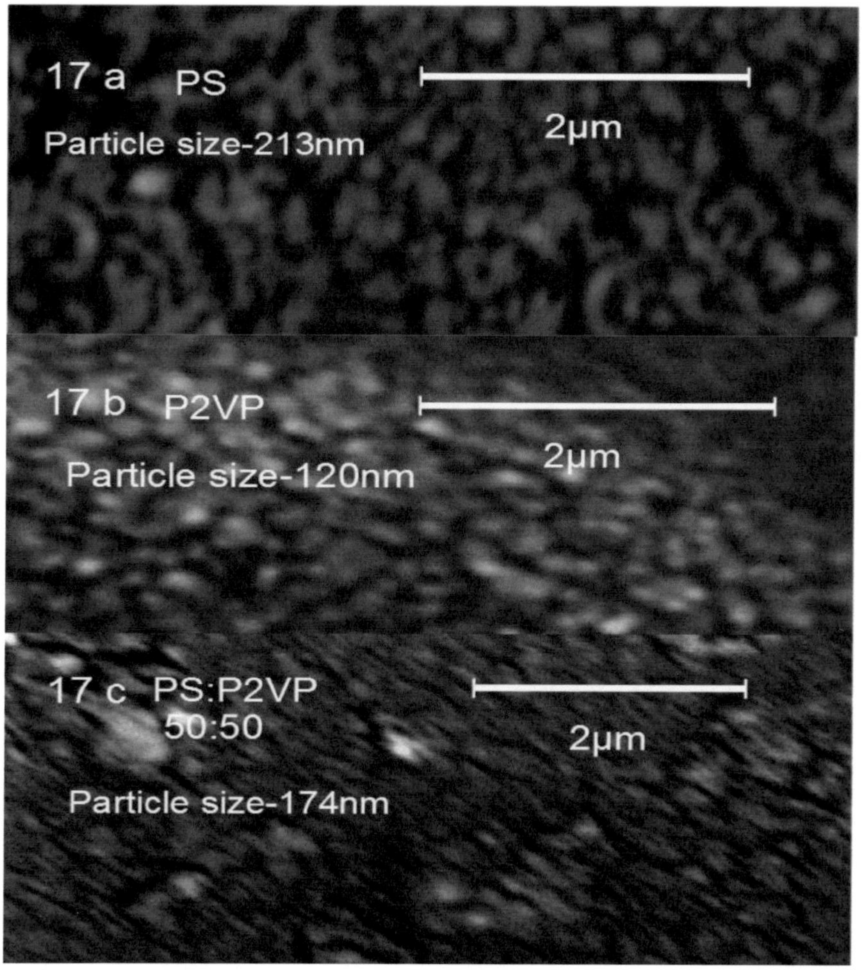

Figure 17. a,b,c: SEM of silver particulate films of thickness 200 nm on PS, P2VP and PS/P2VP, 50:50.

Figure 18 gives the variation of as deposited and room temperature resistances as a function of PS/P4VP composition for silver films of 50 nm thickness. It is seen that as the P4VP concentration increases there is a regular decrease of resistance at a fixed silver thickness. The plot of logarithm of these resistances with blend concentration gives linear fit as shown in figure19.Through this fit, one can estimate the resistance of the film at a particular blend and for the given conditions and thickness.

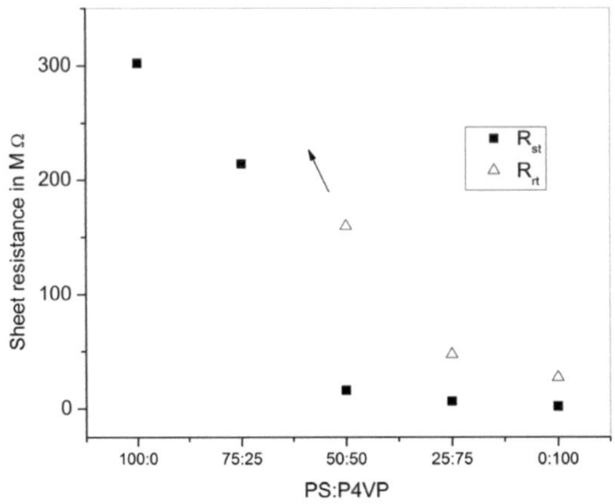

Figure 18. Variation of as-deposited and room temperature resistance with PS/P4VP blend composition for 50 nm silver films deposited at 457 K.

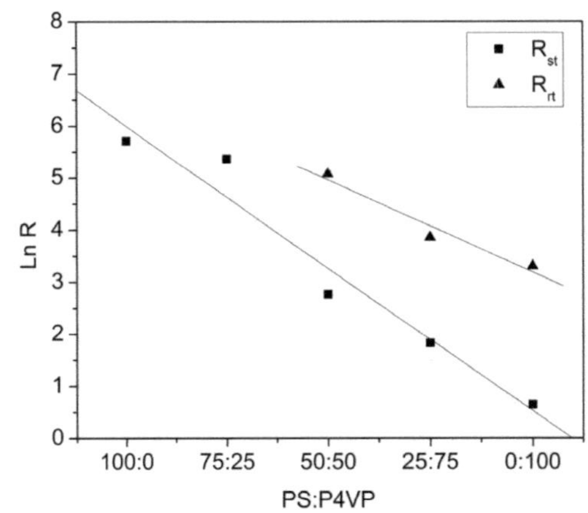

Figure 19. Variation of logarithm of resistances at 457 K and at room temperature with PS/P4VP blend composition for 50 nm silver films.

Table 3&4 gives the resistance data for the silver films of different thicknesses deposited on the PS/P2VP and P4VP, respectively.

Polymer PS:P2VP	Silver Film Thickness	Resistances (MΩ/sheet)				
		R_{ts}	$R_{1\ hra}$	R_{rt}	$R_{0.5T}$	R_{atm}
0:100	100 nm	12.28	32.2	32.68	34.22	63.4
0:100	150 nm	9.73	28.08	26.13	30.53	45.64
50:50	100 nm	47.07	111.57	114.76	489	>1000
50:50	150 nm	11.5	34.98	42.04	57.4	76.56
50:50	200 nm	10.35	29.9	30.1	38.41	68.25
100:0	150 nm	54.46	159.3	-	-	-

Table 3. Resistances for silver films deposited on PS/P2VP blends held at 457 K with a rate of 0.4 nm/s.

Polymer PS:P4VP	Silver film thickness	Resistances (MΩ/□)				
		R_{st}	$R_{1\,hra}$	R_{rt}	$R_{0.5t}$	R_{atm}
0:100	50 nm	1.9	26.8	27.1	32.8	72.3
25:75	50 nm	6.2	42.2	47.1	73.7	1042
25:75	85 nm	3.2	14.7	14.2	21.9	215
50:50	50 nm	15.9	119.5	159.4	241.3	2465
50:50	95 nm	2.9	29.9	30.1	52.2	418
75:25	50 nm	214	325	-	-	-
75:25	150 nm	14.8	98.2	248.9	501.3	5172
100:0	50 nm	302	491	-	-	-

Table 4. Resistances for silver films deposited on PS/P4VP blends held at 457 K with a rate of 0.4 nm/s.

4.6. Morphology of silver particulate films on PS/P2VP and PS/P4VP blends

4.6.1. Optical studies

Fig. 20a shows the optical absorption spectra recorded for 100 nm silver films deposited on polymeric blends of PS/P2VP held at 457 K at the deposition rate of 0.4 nm/s. It is well known that small silver particles embedded in polymer matrix exhibit plasmon resonance absorption and as a result absorption maxima occur in the visible-near infrared region and their spectral position depends on the particle size, shape, filling factor etc. in the polymer matrix. The surface plasmon resonance absorption for silver clusters in the polymer matrix generally occurs at a wavelength of ~ 430 nm [8]. It is well known that shift in the plasmon resonance peak towards higher wavelength occurs due to close proximity of the silver clusters [41-44].These nanoparticles exhibit unique optical properties originating from the characteristic surface plasmon by the collective motion of conduction electrons [43,44].Thus, the formation of silver nanoparticles can also be confirmed by UV/VIS absorption spectrum of composite films [44].Spectral position, half width and intensity of the plasmon resonance strongly depend on the particle size, shape and the dielectric properties of the particle material and the surrounding medium [45].Thus, the type of metal and the surrounding dielectric medium play a significant role in the excitation of particle plasmon resonance (PPR). The sensitivity of PPR frequency to small variations of these parameters can be exploited in various applications [46]. The differing natures of the polymeric hosts yield change in dispersion, size distribution and impregnation depth of silver clusters [26]. Therefore, silver particles embedded in PS/P2VP blends, a shift of the resonance position to higher wavelength (red shift) were found, which were correlated with changes of particle sizes and inter-separation in silver clusters. It is clearly seen (Fig.20a) that the plasmon resonance peak shifts towards the longer wavelength side for the PS/P2VP, 50:50 (435) as compared to pure polystyrene (429 nm). Also, there is increase in intensity of absorbing peaks which signify the decrease of particles size with the incorporation of P2VP into PS [44].P2VP exhibits two peaks (441,606 nm). It is interesting to note that PS/P2VP, 50:50 also shows an additional absorption band at higher wavelength (616 nm). The possible explanation is that silver nanoparticles are in a highly aggregated state leading to coupling of the plasmon vibrations between neighbouring particles [47].Similar results were found for silver particulate films of 150 and 200 nm films on PS/P2VP 50:50 blends. The shift in plasma resonance towards higher wavelength indicates close proximity and increase in particle size of silver nanoparticles with increasing thickness of silver particulate films [41].

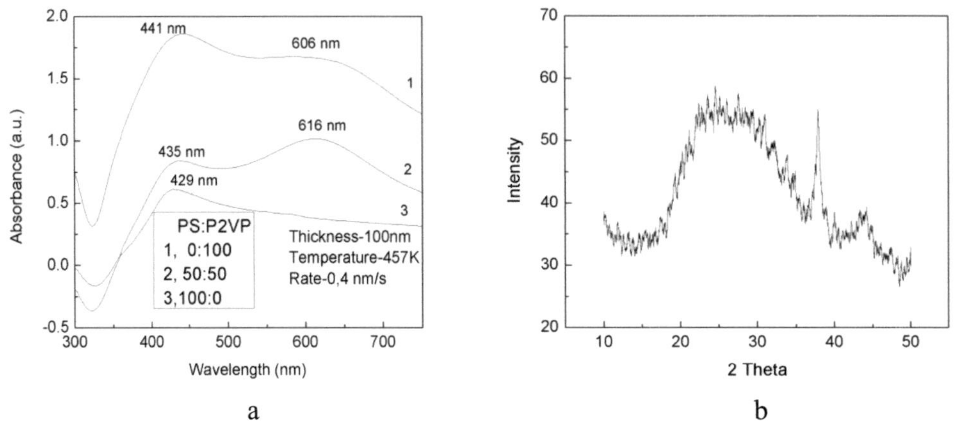

Figure 20. a: Optical absorption spectra for 100 nm silver particulate films deposited on PS/P2VP blends, **b**: XRD curve for silver particulate films of thickness 200 nm on PS/ P2VP, 50:50.

Fig.20 b shows the XRD pattern with the diffraction peak around 38^0 for the silver particulate films of thickness of 200 nm on PS/P2VP, 50:50. The broadening of the Bragg peaks indicates the formation nanoparticles. The particle sizes calculated from the Fig.20b is 40.6 nm for the silver particulate film of 200 nm on PS: P2VP, 50:50. The particle sizes estimated from XRD suggest that there is a small reduction in particle size due to blending of PS and P2VP. XRD of PS/P2VP, 50:50 for 200 nm has been carried out to have preliminary idea about average size of the silver clusters.

Fig. 21(a) shows the optical absorption spectra recorded for 50 nm silver films deposited on polymeric blends of PS/P4VP held at 457 K at the deposition rate of 0.4 nm/s. For silver particles embedded in PS/P4VP blends, a shift of the resonance position to higher wavelength (red shift) were found, which were correlated with changes of particle sizes and inter-separation in silver clusters. It is clearly seen (Fig.21a) that the plasmon resonance peak shifts towards the longer wavelength side in comparison to pure PS (485.5 nm).It is 598, 651.6 and 754.5 nm for PS: P4VP, 75:25, 50:50 and 25:75, respectively for 50 nm silver particulate films on them. Also, there is increase in intensity of absorbing peaks which can be attributed to the decrease of particles size with the incorporation of P4VP into PS [48].

Fig. 21 (b) shows the optical spectra recorded for the films of various thicknesses on PS/P4VP blends. The blend 50:50 shows the most promising result among all the silver particulate films on the blends. The intensity and shift of absorption peak is optimum (793 nm) for 95 nm film on PS/P4VP, 50:50.The silver particulate film of thickness 150 nm on PS/P4VP, 75:25 also shows a red shift (705 nm).This shift in the plasmon resonance peak towards higher wavelength can be attributed to close proximity of the silver clusters [42-45], perhaps this is the reason for the better electrical behaviour of the film of thickness 150 nm on PS/P4VP, 75:25 [34].

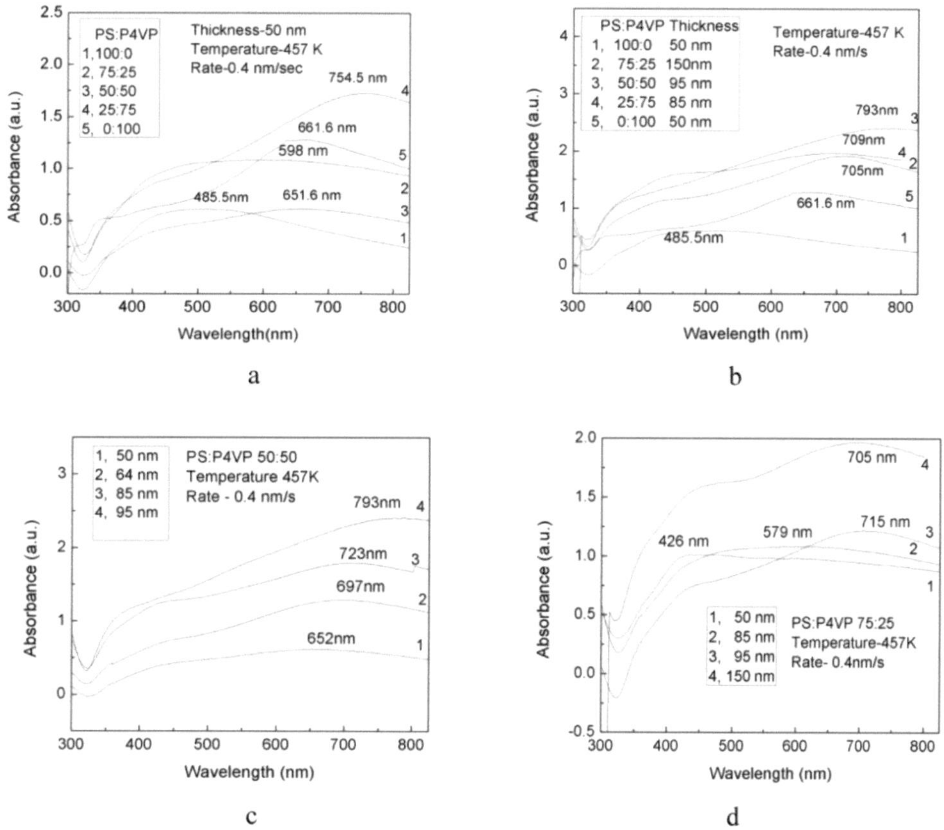

Figure 21. **a** Optical absorption spectra for 50 nm silver particulate films deposited on PS/P4VP blends, **b** Optical absorption spectra for the films of various thicknesses deposited on the PS /P4VP blends, **c** Optical absorption spectra for the films deposited on the blend PS/P4VP 50:50, **d** Optical absorption spectra for the film deposited on the blend PS/P4VP 75:25.

Fig. 21 (c) shows the optical spectra for films of varying thickness on PS/P4VP, 50:50.The shift in plasma resonance towards higher wavelength indicates close proximity of silver nanoparticles with increasing thickness of silver particulate films [15].

Fig. 21(d) shows the optical absorption spectra recorded for silver films on PS/P4VP, 75:25. It is clear that shift in surface plasma resonance is due to increase in particle size with the amount of silver deposited [8].The results are in agreement with the electrical properties of this blend which show increase in electrical conductivity on increasing silver loading [34].

As the fraction of the metal in a nanocomposite increases the nanoparticle separation decreases resulted in better electrical properties of nanocmposites [49]. Therefore, electrical behaviour of silver particulate films on PS/P4VP (50:50, 50 nm and 95 nm; 75:25, 50 nm and 150 nm) observed decrease in electrical resistance on increasing the thickness of film

[40].Thus, electrical studies of these blends suggest possibility of modification in morphology of silver particulate films on PS/P4VP as compared to films on PS.

Previous studies [8] of silver particulate films on PS for the 100 nm thickness exhibited minimum shift due to the presence of comparatively larger clusters with larger inter-cluster separations than on P2VP [8] for the same thickness. Also, silver particulate film on PS for the 50 nm thickness exhibited minimum shift due to the presence of comparatively larger clusters with larger inter-cluster separations than on P4VP [37] for the same thickness. Blending of P2VP with PS and P4VP with PS is expected to provide a polymer matrix where the size of silver clusters and inter-cluster separation can be modified because dispersion, size distribution and impregnation depth results from the natures of polymeric hosts [40].

Therefore, shift in the wavelength observed in optical spectra in the PS/P2VP, 50:50, and PS/P4VP, 75:25; 50:50; 25:75 can be attributed to modification in size distribution and better inter-cluster separations.

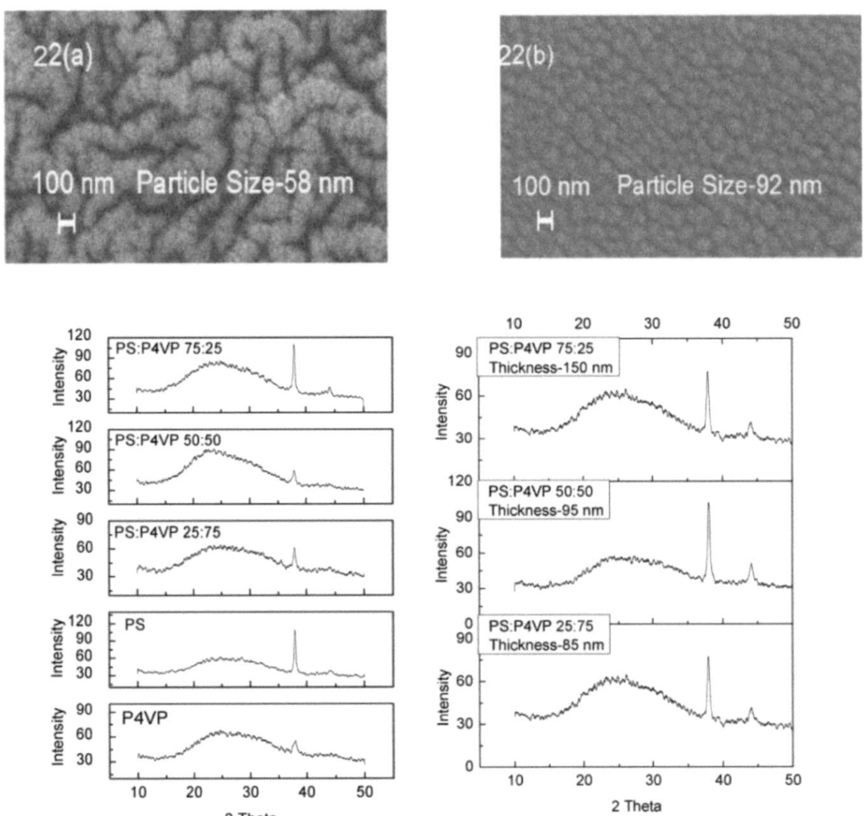

c d

Figure 22. **a,b**: SEM images of PS/P4VP, (a) 0:100 and (b) 100:0 for silver particulate films of thickness 50 nm. Acceleration voltage- 20 kV, Magnification 50 KX, **c**: XRD curves for silver particulate films of thickness 50 nm on P4VP, PS, PS/ P4VP, 25:75, 50:50, 75:25, respectively. **d**: XRD curves for silver particulate films on PS/P4VP, 75:25, 50:50 and 25:75 for 150, 95 and 85 nm thicknesses, respectively

Fig.22c shows the XRD pattern with the diffraction peak around 38^0 for the silver particulate films of thickness of 50 nm on P4VP, PS and PS/P4VP, 25:75, 50:50, and 75:25. The broadening of the Bragg peaks indicates the formation nanoparticles. The particle sizes calculated from the Fig.22c are 53, 51, 49, 46 and 30 nm for the blends (PS: P4VP) 100:0, 75:25, 50:50, 25:75 and 0:100, in that order. The particle sizes estimated from XRD suggest that there is a very small reduction in particle size due to blending of PS and P4VP.Figure 22d shows the XRD patterns for the silver particulate films on PS/P4VP, 75:25, 50:50 and 25:75 for 150, 95 and 85 nm thicknesses, respectively. The reflections at 38^0 and 44^0 correspond to metallic silver. The particle sizes calculated from the Fig.22d are 52.3, 51.6 and 53 nm for the blends (PS: P4VP) 75:25, 50:50, and 25:75, respectively for the diffraction angle 38^0.

4.6.2. Micro structural studies

Electrical properties of polymer/metal composite films are strongly linked to particles' nanostructure [40]. As the fraction of the metal in a nanocomposite increases the nanoparticle separation decreases resulted in better electrical properties of nanocmposites [49-50,52,53]. SEM of the silver particulate films (Fig.22a & 22b) on homopolymers (PS, P4VP) clearly shows the characteristic nature of these polymers. The size and inter-separation of silver clusters is less (average particle size-58 nm) in P4VP whereas size as well as inter-separation is wide in PS (average particle size-92 nm). As a result, PS do not show the desired electrical conductivity [51].Blending of P2VP and P4VP into PS modifies size, size distribution and inter-separation of silver particles deposited on their blends PS/P2VP, 50:50 and PS/P4VP, 50:50 and 75:25. Therefore, electrical behaviour of silver particulate films on PS/P2VP, 50:50 for 100, 150 and 200 nm observed decrease in electrical resistance on increasing the thickness of films [52]. Thus, electrical studies of these blends suggest possibility of modification in morphology of silver particulate films on PS/P2VP and PS/P4VP as compared to films on PS.

SEM of the silver particulate films of 200 nm on the homopolymers (PS, P2VP) and their blend PS/P2VP, 50:50 were shown in the Fig.23a. The acceleration voltage is 30 kV and magnification is 60 to 100 KX for all the SEM pictures. The particle sizes measured from respective SEM pictures are plotted as histogram. The corresponding histograms (Fig.23a) of silver particles of the films are shown side by side of the SEM pictures. The data fit into a log normal distribution for all the cases. Hence, the average size, $\bar{a}$ and geometric standard deviation, σ_a are determined from the log normal distribution of the curves. The average size, $\bar{a}$ and geometric standard deviation, σ_a are 205 nm and 4; 109 nm and 9, 129 nm and 3, respectively, for silver films on PS, P2VP and their blend (50:50). The size distribution and width of histograms as shown in the figure indicate that the particle size varies from 100 to 400 nm, 60 to 180 nm and 60 to 200nm for silver films on PS, P2VP and their blend (50:50).It is clear from the figure that the size and inter-separation of silver clusters is wide in PS whereas size as well as inter-separation is less in P2VP [8]. As a result, PS do not show the desired electrical conductivity [36].Blending of P2VP with PS modifies size, size distribution and inter-separation of silver particles on their blend PS/P2VP, 50:50 which results in improvement of tunnelling effect in the blend and the blend shows desired electrical behaviour [52]. This fact may be regarded as a consequence of the size as well as inter-separation evolution of nanoparticles during the ongoing deposition process.

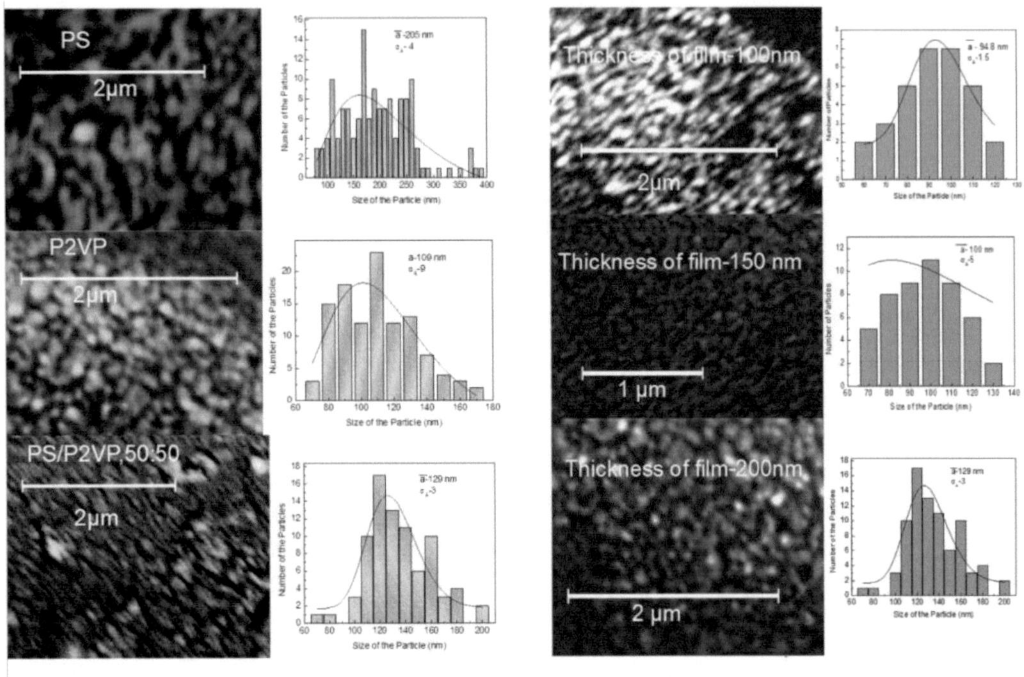

a

b

Figure 23. **a**: SEM of silver particulate films of 200 nm on PS, P2VP and PS/P2VP, 50:50 and their corresponding histograms. **b**: SEM of silver particulate films of 100, 150 and 200 nm on PS/P2VP, 50:50 and their corresponding histograms.

Fig.23b shows the particle size distribution for 100, 150 and 200 nm thickness films deposited on PS/P2VP, 50:50 and their corresponding histograms. The average size, ā and geometric standard deviation σ_a are 94.8 nm, 1.5 and 100 nm, 5 and 129 nm, 3 for the 100, 150 and 200 nm film, respectively. It is evident from the figure that size distribution varied from 55-125, 65-135 to 70-200 nm for the 100, 150 and 200 nm films, respectively. Such dispersion of silver nanoparticle within the PS/P2VP, 50:50 leads to better electrical behaviour [52].This electric behaviour is not observed even for 300 nm silver particulate films on PS [20].Hence, silver particulate films on PS/P2VP, 50:50 at low volume fraction of silver consist of isolated and widely dispersed nanoparticles. But systematic and controlled increase of silver volume in the blend matrixes has shown increase in the size of silver clusters [52-53].

Figs.24a to 27a show the SEM pictures of various thicknesses of silver films deposited on PS/P4VP blends. The acceleration voltage is 20 kV and magnification is 50 to 100 KX for all the SEM pictures. The particle sizes measured from respective SEM pictures are plotted as histogram in figs. 24b to 27 b. The corresponding histograms (Figs.24b to 27 b) of silver particles of the films are shown side by side of the SEM pictures. The data fit into a log normal distribution for all the cases. Hence, the average size, $\bar{a}$ and geometric standard deviation, σ_a are determined from the log normal distribution of the curves. The positive effect of blending P4VP with PS is clearly visible in these pictures. Figs. 24a, 25a show the particle size distribution for 50 and 95 nm thick silver films deposited on PS/P4VP, 50:50. The average size, $\bar{a}$ and geometric standard deviation, σ_a are 79.4 nm and ±1.2, respectively for the 50 nm film whereas the corresponding values for the 95 nm film are 95.4 nm and ±1.4. A closer look at the morphology of silver nanoparticles deposited on PS/P4VP, 50:50 in Figs.24a & 25a, clearly shows that particle size increases with the amount of silver deposited. The shape of the nanoparticles changes from near spherical particles to irregular ellipsoidal particles. The size distribution and width of histograms as shown in the figure indicate that the average size of the particle increases from 79.4 to 95.4 nm and the size distribution expands from 50-110 nm to 60-160 nm which results in improvement of tunnelling effect in PS/P4VP, 50:50 [40].And silver particulate film of thickness 95 nm show better electrical behaviour than silver particulate film of 50 nm on PS/P4VP, 50:50 [34]. It is evident that increase in size distribution decreases the inter-separation of silver clusters in this blend. This fact may be regarded as a consequence of the size as well as inter-separation evolution of nanoparticles during the ongoing deposition process.

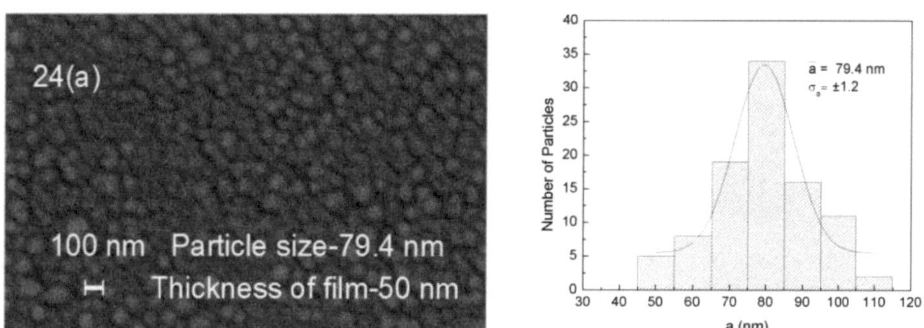

Figure 24. **a** SEM micrograph, Acceleration voltage-20 kV, Magnification 50 KX and **b** Corresponding histogram of 50 nm thick silver film on PS/P4VP, 50:50.

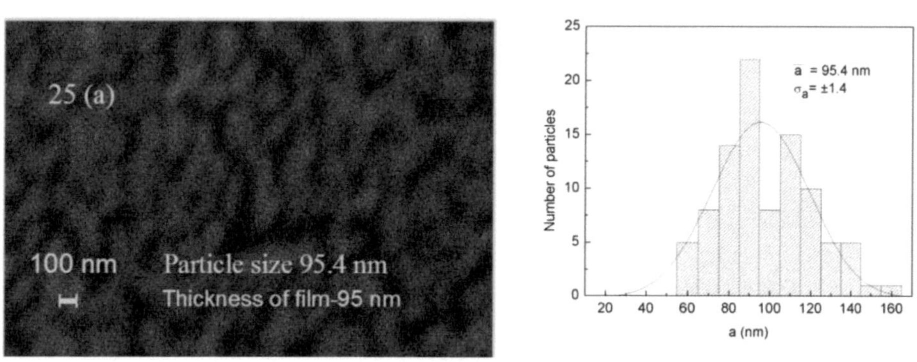

Figure 25. **a** SEM micrograph, Acceleration voltage-20 kV, Magnification 100 KX and **b** Corresponding histogram of 95 nm thick silver film on PS/P4VP, 50:50.

Figs.26a & 27a show the particle size distribution for 50 nm and 150 nm thickness films deposited on PS/P4VP, 75:25. The average size, ā and geometric standard deviation, σ_a are 88.6 nm and ±8.5 for the 50 nm film and 96.7 nm and ±4 for 150 nm film, respectively. It is evident from the figure that distribution of size increased from 50-120 nm to 60-140 nm .Such dispersion of silver nanoparticle within the PS/P4VP, 75:25 leads to better electrical behaviour [40].Such electric behaviour is not observed even for 300 nm silver particulate film on PS [36].Hence, silver particulate film of 50 nm thickness on PS/P4VP, 75/25 consist of isolated and widely dispersed nanoparticles. But systematic and controlled increase of silver in the PS/P4VP, 75:25 matrixes produced interesting result [40]. The silver particulate film of thickness 150 nm on PS/P4VP, 75:25 shows the room temperature resistance in few mega ohms a desirable range for applications.

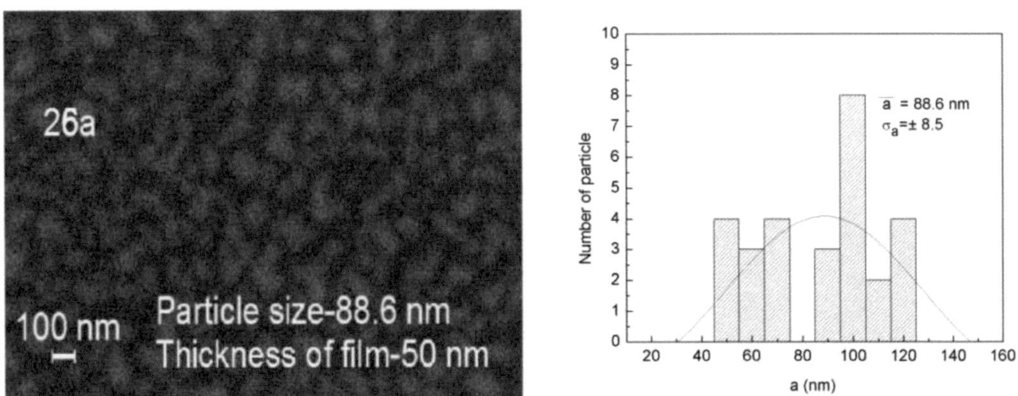

Figure 26. a SEM micrograph, Acceleration voltage-20 kV, Magnification 100 KX and **b** Corresponding histogram of 50 nm thick silver film on PS/P4VP, 75:25.

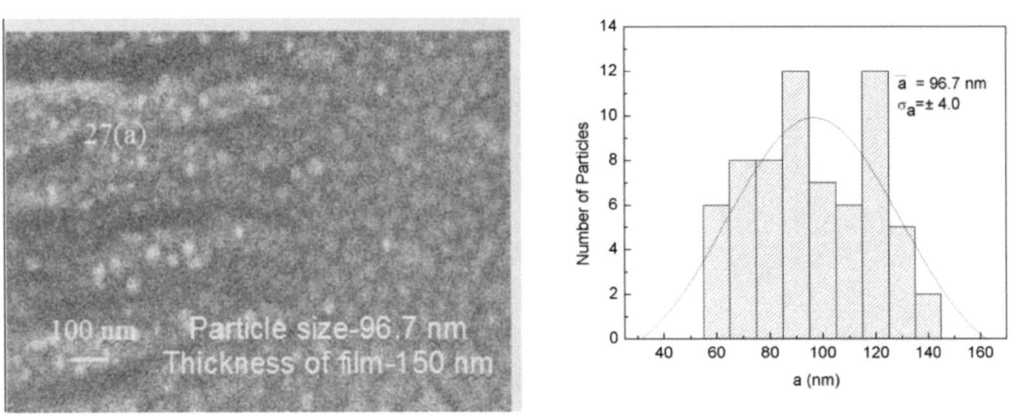

Figure 27. a SEM micrograph, Acceleration voltage-20 kV, Magnification 100 KX and **b** Corresponding histogram of 150 nm thick silver film on PS/P4VP, 75:25.

Table 5&6 has been compiled to show all the particles size of silver clusters embedded in PS/P2VP and PS/P4VP blends from XRD and SEM. The SEM has provided the morphology of silver clusters and their distribution. The XRD diffraction pattern represents the average throughout the film due to increased penetration and large beam size. The observed values of particle size from SEM are in the same range as the calculated values of the particle size from XRD. The difference in the values may be due to averaging over longer depths because of penetration of X-rays. Though, the trend of particle size measured from XRD and SEM are similar.

PS/P2VP	Thickness (nm)	Particle size (nm) XRD	SEM, $\bar{a}$	Standard deviation σ_a
0:100	200		109	9
50:50	100	-	94.8	1.5
	150	-	100	5
	200	40.6	129	3
100:0	200		205	4

Table 5. The average particle size for silver deposited on PS/P2VP blends. The deposition rate is 0.4nm/s and temperature is 457K.

PS/P4VP	Thickness (nm)	Particle size (nm) XRD	SEM, $\bar{a}$	Standard deviation σ_a
0:100	50	30	58	-
25:75	50	46.6	-	-
	85	53	-	-
50:50	50	49.1	79.4	1.2
	95	51.6	95.4	1.4
75:25	50	51.8	88.6	8.5
	150	52.3	96.7	4.0
100:0	50	53.3	92	-

Table 6. The average particle size for silver deposited on PS/P4VP blends. The deposition rate is 0.4nm/s and temperature is 457K.

5. Further research

The silver films deposited on polymer composites show an increase in resistance, when they are exposed to atmosphere. It may be possible to stabilise the resistance against exposure to atmosphere through the deposition of good quality inorganic passivators like alumina, zirconia etc, to make the films suitable for device applications. Also, silver films are more susceptible to oxygen in atmosphere than gold. Hence, deposition of gold nanopaticles on PS/P2VP and PS/P4VP composites may give more stabilised films which may be suitable films for sensitive applications.

6. Conclusion

The following conclusions may be drawn from the study on **Silver Particulate Films on Compatible Softened Polymer Composites**

Viscometry studies indicate a very small interaction parameter resulting in physically miscible blends of PS/ P4VP.DSC studies indicate a single T_g in all the cases indicating the formation of compatible blends. This may be due to some intermolecular interaction at higher temperature. Hence, the blends found some order of compatibility at higher temperatures. FTIR and SEM support the results of miscibility as well as DSC. The fairly compatible blend of PS/P2VP and PS/P4VP can be obtained on melt mixing at higher temperature. Deposition of silver on polymer blends coated substrate held at 457 K provides an approach to produce stable island films with reasonable control over their electrical resistance. Higher thickness films show almost zero TCR near room temperature, a desirable property for most of the devices. Low thickness films show a negative TCR, characteristic of island film. Silver particulate films deposited on composite blends show better electrical properties compared to pure PS.The deposition of silver particulate films by evaporation on PS/P2VP and PS/P4VP blends yields positive effect of blending PS with P2VP and P4VP.The size distribution and dispersion of silver nanoparticles is found to be dependent on the nature of the polymer host and thickness of particulate films. With the addition of P2VP, P4VP and amount of silver, morphology of the silver particulate films on PS/P2VP (50:50) and PS/P4VP (50:50, 75:25) could be modified to give the desired electrical results.

Author details

Pratima Parashar
Department of Materials Science, Mangalore University, Mangalagangotri, India
CET, IILM Academy of Higher Learning, Greater Noida, India

Acknowledgement

The author thanks DST for the XRD and NCL (Pune) for SEM facility. The author thanks DST for the funding through Women Scientist Scheme (WOS).

7. References

[1] Skofronick J G, Phillips W B (1967) Morphological Changes in Discontinuous Gold Films following Deposition. J. Appl. Phys. 38(12): 4791-4796.

[2] Fehlner F P (1967) Behavior of Ultrathin Zirconium Films upon Exposure to Oxygen. J. Appl. Phys. 38: 2223- 31.

[3] Kovacs G J, Vincent P S (1982) Formation and Thermodynamic Stability of a Novel Class of Useful Materials: Close-Packed Monolayers of Submicron Monodisperse Spheres Just below a Polymer Surface. J. Colloid. Interface. *Sci*. 90: 335-342.

[4] Kovacs G J, Vincentt P S (1983) Subsurface particulate film formation in softenable substrates present status and possible new applications. Thin Solid Films 100: 341-353.

[5] Kovacs G J, Vincentt P S, Trumblay C, Pundsak A L (1983) Vacuum deposition onto softenable substrates formation of novel subsurface structures.Thin Solid Films 101: 21-40.

[6] Kovacs G J, Vincentt P S (1984) Subsurface particulate film formation in softenable substrates present status and possible new applications.Thin Solid Films 111: 65-81.

[7] Kunz M S, Shull K R Kellock A J (1992) Morphologies of Discontinuous Gold Films on Amorphous Polymer Substrates. J. Appl. Phys. 72: 4458-4460.

[8] Mohan R K, Pattabi M (2001) Effect of polymer-metal particle interaction on the structure of particulate silver films formed on softened polymer substrates. J. New Mat. Electrochem. Systems 4: 11-15.

[9] Yuan JJ, Ma R, Gao Q, Wang YF, Cheng SY, Feng LX, Fan Z Q, Jiang L (2003) Synthesis and characterisation of Polystyrene/ Poly(4-vinylpyridine) Triblock Copolymers by Reversible Addition –Fragmentation Chain Transfer Polymerisation and Their Self-Assembled Aggregates in Water. J Appl Polymer Sci, 89:1017-1025.

[10] Bouslah N, Amrani F (2007) Miscibility and specific interactions in blends of poly [(styrene)-co-(cinnamic acid)] with poly (methyl methacrylate) and modified Poly(methyl methacrylate).Express polymer letters 1:44-50.

[11] Torrens F, Soria V, Codoner A, Abad C, Campos A (2006) Compatibility between polystyrene copolymers and polymers in solution via hydrogen bonding. Euro polymer Journal 42(10): 2807-2823.

[12] Bouslah N, Hammachin R, Amrani F (1999) Study of the compatibility of poly[styrene-*co*-(cinnamic acid)]/poly[(ethyl methacrylate)-*co*-(2-dimethylaminoethyl methacrylate)] blends. Macromolecular Chemistry and Physics 200(4):678-682.

[13] Jiao H, Goh S H, Valiyaveettil S (2001) Mesomorphic Interpolymer complexes and blends based on poly (4-vinylpyridine)- dodecylbenzenesulfonic acid complex and poly(acrylic acid) or poly(p-vinvylphenol. Macromolecules, 34:7162-7165.

[14] Kosonen H, Valkama S, Hartikainen J, Eerikainen H, Torkkel M, Jokela K, Serimaa R, Sundholm F, Brinke G, Ikkala O (2002) Mesomorphic Structure of Poly(styrene)-block-poly(4-vinylpyridine) with Oligo(ethylene oxide)sulfonic Acid Side Chains as a Model for Molecularly Reinforced Polymer Electrolyte. Macromolecules 35: 10149-10154.

[15] Wang X, Zuo J, Keil P, Grundmeier G (2007) Comparing the growth of PVD silver nanoparticles on ultra thin fluorocarbon plasma polymer films and self-assenbled fluoroalkyl silane monolayers. Nanotecnology 18:265303-265313.

[16] Heilmann A (2002) Polymer Films with Embedded Metal Nanoparticles, Berlin:Springler.

[17] Mayer A B R (2001) Colloidal metal nanoparticles dispersed in amphiphilic polymers. Polym. Adv. Technol. 12: 96-106.

[18] Beecroft L L (1997) Nanocomposite materials for optical applications. Chem.Mater. 9:1302-1317.

[19] Caseri W (2000) Nanocomposites of polymers and metals or semiconductors: Historical background and optical Properties Macromol.Rapid Commun. 21:705-722.

[20] Eilers H, Biswas A, Pounds T D, Norton M G (2006) Teflon AF/Ag nanocomposites with tailored optical properties. J.Mater. Res 21:2168-2171.

[21] Yonzon C R, Stuart D A, Zhang X, Mcfarland A D, Haynes CL, Duyne R P V (2005) Towards advanced chemical and biological nanosensors—An overview. Talanta 67:438-448.

[22] Saito R, Okamura S, Ishizu K (1992) Introduction of colloidal silver into a poly (2-vinyl pyridine) microdomain of microphase separated poly(styrene-*b*-2-vinyl pyridine) film. Polymer 33(5):1099-1101.

[23] Tao L, Ho R M, Ho J C (2009) Phase Behavior in Self-assembly of Inorganic/Poly(4-vinylpyridine)-b- Poly (ε-caprolactone) HybridMacromolecules 42:742-751.

[24] Pennelli G (2006) Lateral reduction of random percolative networks formed by nanocrystals: Possibilities for a new concept electronic device. Appl. Phys. Lett. 89:163513-163516.

[25] Rao K M, Pattabi M, Mayya K S, Sainkar S R, Murali Sastry M S (1997) Preparation and characterization of silver particulate films on softened polystyrene substrates. Thin Solid Films 310: 97

[26] Hassell T, Yoda S, Howdle S M., Brown P D (2006) Microstructural charecterisation of silver/polymer nanocomposites prepared using supercritical cabondioxide. J. of Phys: Conferrence Series 26:276-279.

[27] Dondos A S, Kondras P, Pierri P, Benoit H (1983) Hydrodynamic crossover in two-polymer mixtures from viscosity measurement. Macromolek Chem, 184(10):2153-2158

[28] Rao V, Ashokan P, Shridhar M H (1999) Studies on the compatibility and specific interaction in cellulose acetate hydrogen phthalate (CAP) and poly methyl methacrylate (PMMA) blend. Polymer 40:7167-7171.

[29] Shih K S, Beatty C L (1990) Blends of polycarbonate and poly(hexamethylene sebacate): IV. Polymer blend intrinsic viscosity behavior and its relationship to solid-state blend compatibility. Br Polym. J. 22 (1):11-17.

[30] Krigbaum W R, Wall F T (1950) Viscosities of binary polymeric mixtures. J Polym Sci. 5:505-514.

[31] Fox, T G (1956) Influence of Diluent and of Copolymer Composition on the Glass Temperature of a Polymer System..Bull Am Phys Soc. 1:123-125.

[32] Gordon M, Taylor J S (1952) Ideal copolymers and the second-order transitions of synthetic rubbers. I. Non-crystalline Copolymers. J Appl Chem 2:493-500

[33] Dong H.K, Won H.J, Sang C L, Ho C.K (1998) The compatibilizing effect of poly(styrene-*co*-4-vinylpyridine) copolymers on the polystyrene–polyethylene-based ionomer blends. J Appl. Polymer Sci. 69:807- 816.

[34] Parashar P, Pattabi M, Gurumurthy S C (2009) Electrical behaviour of discontinuous silver films deposited on softened polystyrene and poly (4-vinylpyridine) blends. J.Mater.Sci:Mater Electron 20: 1182-1185.

[35] Parashar P, Ramakrishna K, Ramaprasad A T (2011) A Study on Compatibility of Polymer Blends of Polystyrene/Poly(4-vinylpyridine). Journal of Appl Poly Sci 120 (3):1729-1735.
[36] Pattabi M, Rao K M, Sainker S R, Sastry M (1999) Structural studies on silver cluster films deposited on softened PVP substrate. Thin Solid Films 338:40-45.
[37] Rao K M, Pattabi M., Sainkar S. R, Lobo A., Kulkarni S K, Uchil J, Sastry M. S (1999) Preparation and characterisation of silver particulate structure deposited on softened poly(4-vinylpyridine) substrate. J. Phys. D: Appl.Phys. 32:2327-2336.
[38] Haoyan W, Eilers H (2008) Electrical Conductivity of Thin-Film Composites Containing Silver nanoparticles Embedded in a Dielectric Teflon AF Matrix. Thin Solid Films, 517: 575-581.
[39] Pattabi M , Rao K M (1998) Electrical behaviour of discontinuous silver films deposited on softened polyvinyl pyridine Substrate. J. Phys. D: Appl. Phys. 31: 19-23.
[40] Kiesow A, Morris J E, Radehaus C, Heilmann A (2003) Switching behavior of plasma polymer films containing silver nanoparticles. J.Appl. Phy. 94 (10): 6988-6990.
[41] Heilmann A, Kiesov A,Gruner M, Kreibig U (1999) Optical and electrical properties of embedded silver nanoparticles at low temperatures.Thin Solid Films 343-344:175-178.
[42] Fritzsche W, Porwol H, Wiegand A, Boronmann, Khler J M (1998) In-situ formation of Ag-containing nanoparticles in thin polymer film. Nanostrutured Materials, 10 (1) 89-97.
[43] Akamatsu K, Takei S, Mizuhata M, Kajinami A, Deki S, Fujii M, Hayashi S, Yamamoto K (2000) Preparation and characterization of polymer thin films containing silver and silver sulfide nanoparticles.Thin Solid Films 359:55-60.
[44] Carotenuto G (2001) Synthesis and characterization of poly (*N*-vinylpyrrolidone) filled by monodispersed silver clusters with controlled size. Appl. Organometal. Chem. 15(5):344-351.
[45] Kim J Y, Shin D H, Ihn K J, Suh K D (2003) Amphiphilic Polyurethane-co-polystyrene Network Films Containing Silver Nanoparticles. J Ind. Eng. Chem. 9(1):37-44.
[46] Heilmann A, Quinten M, Werner J (1998) Optical response of thin plasma-polymer films with non-spherical silver nanoparticles Eur.Phys. J.B, 3:455-461.
[47] Mandal S, Arumgam S K, Pasricha R, Sastry M (2005) Silver nanoparticles of variable morphology synthesized in aqueous foams as novel templates. Bull. Mater. Sci. 28(5):503-510.
[48] Kim JY, Shin DH, Ihn KJ (2005) Synthesis of CdS nanoparticles dispersed within amphiphilic poly(urethane acrylate-*co*-styrene) films. J.Appl.Polym Sci. 97(6):2357-2363.
[49] Biswas A, Bayer I S, Marken B, Pounds D, Norton M G (2007) Networks of ultra-fine Ag nanocrystals in a Teflon AF® matrix by vapour phase e-beam-assisted deposition Nanotecnology 18:305602-305608.
[50] Parashar P (2011) Morphology of Silver Particulate Films Deposited on Softened Polymer Blends of Polystyrene and Poly (4-vinylpyridine)J.Appl.Polymer Sci. 121 (2): 839-845.
[51] Takele H, Greve H, Pochstein, Zaporojtchenko V, Faupel F (2006) Plasmonic properties of Ag nanoclusters in various polymer matrices. Nanotechnology 17: 3499-3505.

[52] Parashar P (2011) Electrical behaviour of discontinuous silver films deposited on compatible Polystyrene/Poly (2- vinylpyridine) composite. J.Mater.Sci:Mater Electron, DOI 10.1007/s10854-011-0418-6.

[53] Parashar P (2011) Structural properties of silver particulate films deposited on softened polymer blends of polystyrene/poly (2-vinyl pyridine) J.Mater.Sci:Mater Electron, DOI 10.1007/s10854-011-0567-7

Chapter 11

Liquid-Solid Phase Equilibria of Paraffinic Systems by DSC Measurements

Luis Alberto Alcazar-Vara and Eduardo Buenrostro-Gonzalez

Additional information is available at the end of the chapter

http://dx.doi.org/10.5772/54575

1. Introduction

Several industrial sectors around the world deal with paraffinic wax in their processes or make use of it in their products. Hence, understanding physical properties of paraffins is of industrial importance. Some of these industrial sectors are: petroleum production, petroleum refining and products, chemical, energy and consumer products [1]. However, as it has been widely reported in literature [2-6], one of the most affected industrial sectors by the paraffin crystallization phenomena is the petroleum industry. Crude oils contain heavy paraffins that may form solid wax phases at low temperature in the pipelines and hydrocarbon production facilities. The problems caused by wax precipitation decreasing production rates and failure of facilities, are a major concern in the production and transportation of hydrocarbon fluids [7]. Paraffin waxes are mixtures of a wide range of high molecular weight alkanes that can crystallize from crude oils or solutions primarily due to temperature decreasing. They are rather non-polar molecules and their interactions are expected to be van der Waals or London dispersion type [4]. Paraffin waxes consist of branched (iso), cyclic and straight chain (normal) alkanes having chain lengths in excess of 17 carbon atoms (C_{17}) and potentially up to and over C_{100} [8]. However, despite the fact that crude oils are extremely complex systems containing a multitude of components, it is generally accepted that the crystallizing materials that form the deposits are primarily *n*-alkanes [9-10]. Therefore, in order to obtain a greater insight on the formation of wax deposits to prevent and solve these problems, it is necessary to get a deep knowledge of the mechanisms involved on the *n*-paraffins crystallization process.

Other industrial problems associated to the paraffin phase behavior have been reported in literature and summarized below. In diesel fuels production operations, fuel-filter plugging and other associated fuel handling problems can occur in cold weather due to paraffin crystallization. Moreover, fuels produced from Fischer–Tropsch syntheses that are currently being investigated for converting natural gas to liquids (fuels) can be particularly problematic due to amounts of higher molecular weight paraffin wax produced [1]. Phase equilibrium data of n-alkane systems with different solvents are of importance for the safe and efficient operation of chemical plants. They are necessary for high-pressure polymerization processes and for the design of oil-recovery processes. Besides its importance for technological processes such as crystallization and purification at high pressure, phase equilibrium properties provides a good tool for examining the thermodynamic nature of many systems [11].Recently, the use of phase change material (PCM) thermal energy storage has gained considerable attention because of its high storage density (amount of energy stored per unit mass), and a narrow temperature range for charging and discharging the storage. Paraffin waxes have been used as PCM for many applications because of their advantageous thermal performances and phase behavior [12]. Finally, the control of crystallization processes is a problem of quite general relevance, which appears in many practical fields such as pharmaceutical and specialty chemical industries [13-15]. "Crystal design or engineering" enables, in principle, a direct handling of the structure, size, and shape of crystals entering into the elaboration of materials. Classical means of controlling size, morphology, and polymorphic expression of crystals make use of parameters such as temperature, pH, supersaturation, and solvent quality [15].

Some experimental techniques reported in literature such as Microscopy and X-ray diffraction are powerful methods to determine the crystal structures but give limited insight into the crystallization process [16], while others methods used to get the liquid-solid equilibrium of paraffins have been used [17-18], but they are very complex due to they require the establishment of the equilibrium at each temperature of interest and the measurement of the composition of the phases present. Finally, visual methods have been also reported to measure solubility and phase behavior of paraffin waxes [1, 19]; however, these methods cannot be applied to test dark samples (e.g. black crude oils) [20]. Hence, for the study and measurement of paraffin crystallization process, Differential scanning calorimetry (DSC) is an experimental method widely used due to its simplicity, accuracy and fast response to monitor the phase transitions during cooling and heating that gives related thermodynamic quantities such as heat capacity and enthalpies of transition [3, 14, 16, 20-23]. DSC has been usually used for the determination of wax appearance and/or dissolution temperatures (WAT or WDT) in petroleum products [3, 20]. The WAT or cloud point is the singularly most important parameter relating to wax formation [4] and it is the temperature at which waxes first crystallize from solution during a cooling process. So that accurate WAT measurements by using reliable methods such as DSC are desirable since it represents a key factor to characterize the wax precipitation phenomena.

The objective of this chapter is to present the use of DSC technique on the measurement and characterization of the liquid-solid phase equilibria of paraffins. First, the details of the DSC method and the experimental conditions used to get the key properties to characterize the liquid-solid phase equilibria of paraffins are described. Then, experimental studies about the effect of the chemical nature of solvent and asphaltenes on liquid-solid phase behavior of paraffinic model systems; are presented and discussed in these sections in order to show specific applications on the use of the Calorimetry to carry out relevant studies of phase equilibria properties. Finally, this chapter presents the characterization of the wax precipitation phenomena by using DSC measurements in crude oils that present solids deposition problems during their production and transporting, where the results obtained by using DSC technique are compared with those obtained with other techniques such as rheometry, spectroscopy and densitometry; in order to show advantages and disadvantages of the use of DSC method to measure liquid-solid phase equilibria of wax in crude oils.

2. DSC methodology applied to measure liquid-solid phase equilibria of paraffins

As it was mentioned above, there are many experimental works in literature [3, 14, 16, 20-23, 28-30] reporting the use of DSC to study the paraffins crystallization process. In this section, it is described the DSC methodology to characterize the liquid-solid phase equilibria of paraffins in model systems and crude oil samples. The objective of this section is to provide the details of the DSC technique and the experimental conditions used to get the key properties that characterize the liquid-solid phase equilibria of paraffins.

The measurement principle of differential scanning calorimetry (DSC) is based on the measurement of the difference in the heat flows to the sample crucible and reference crucible. These heat flows are directly proportional to the temperature difference between the furnace and crucible, but inversely proportional to the thermal resistance of the system. In Figure 1 is shown the measuring cell, furnace and liquid nitrogen cooling chamber of the Shimadzu DSC-60A differential scanning calorimeter used in the experiments to be presented in this work.

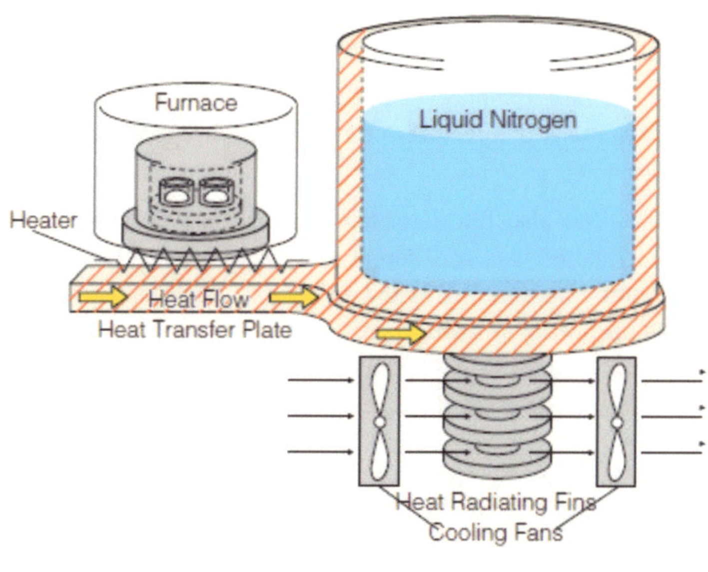

Figure 1. DSC measuring cell and temperature control system (Source: Shimadzu).

The method to obtain the liquid-solid phase equilibrium properties from DSC experiments is explained below. A calibration procedure of the DSC equipment should be performed before carrying out the experiments by using Indium or series of high purity normal paraffins as standard [22]. Each sample (between 10 and 20 mg) is first heated until reaching a temperature higher than expected crystallization onset temperature (WAT) but without reaching the boiling point of the sample. Then, the sample is held isothermally for 1 min., and then cooled to the desired temperature at a pre-defined rate. The cooling/heating rate can be variable; in general low heating/cooling rates would be desirable from an equilibrium point of view [20]. However, by using low cooling rates higher WAT are obtained with a loss of sensitivity to identify the DSC peak onsets, whereas high cooling rates depress measured WAT due to supercooling effects [20]. Differences about ± 1°C on DSC WAT measurements have been observed when using 1, 5 and 10 °C/min as cooling rates in single paraffin solutions [16]; whereas for crude oil mixtures and by using low cooling rates of 0.1 to 1 °C/min, differences about ± 1-2°C were reported [31] Therefore, in the experiments presented in the following sections, we employ a heating/cooling rate of 5 °C/min because it provided sufficient experimental speed and sensitivity to identify onsets of the exo and endothermic peaks. In order to delete any thermal history effects, two heating/cooling cycles are employed, so that crystallization and melting properties are obtained from the second cycle. The crystallization onset temperature (WAT) is determined as the onset of the exothermal peak during the cooling process corresponding to the liquid–solid transition. Under heating conditions, the melting temperature is recorded as the onset of the endothermal peak, whereas the wax disappearance temperature (WDT), temperature at which the last precipitated paraffin re-dissolves in the oil or solution, can be recorded as the endset of the solid–liquid endotherm. Finally, due to that the total energy released during cooling or heating process is proportional to the area between the base line and the exothermal peak or endothermal peak, respectively, the enthalpies of crystallization and melting of the waxy model systems are calculated from the integration of heat flow curve.

In Figure 2 is shown the determination of the equilibrium temperatures (WAT and melting temperature) as well as the enthalpies from the DSC thermograms according to the method explained above.

The DSC technique allows also the determination of the wax precipitation or solubility curve (amount of precipitated wax at different temperatures) as it has been reported [18, 28, 32-33]. It is carried out by assuming that the amount or fraction of precipitated wax in the total wax content is proportional to the percent of accumulated heat released in the total heat released (Crystallization enthalpy), thus the amount of precipitated wax at different temperatures can be determined by dividing the accumulated heat released by the heat of crystallization. This procedure is depicted in Figure 3, where the accumulated heat released for the exothermic peak related to the crystallization of the system 6 wt % of C_{36} in *n*-decane is plotted as an example [28].

Finally, by using DSC data, we can determine the degree of crystallinity for pure solutes in solvent systems or mixtures by using the following equation [28, 34]:

$$\textit{Percent crystallinity} = [\Delta H_m / \Delta H_{m^\circ}] \times 100 \qquad (1)$$

where ΔH_m is the melting enthalpy of the mixture measured by DSC and ΔH_{m° is the melting enthalpy of the 100% crystalline solute.

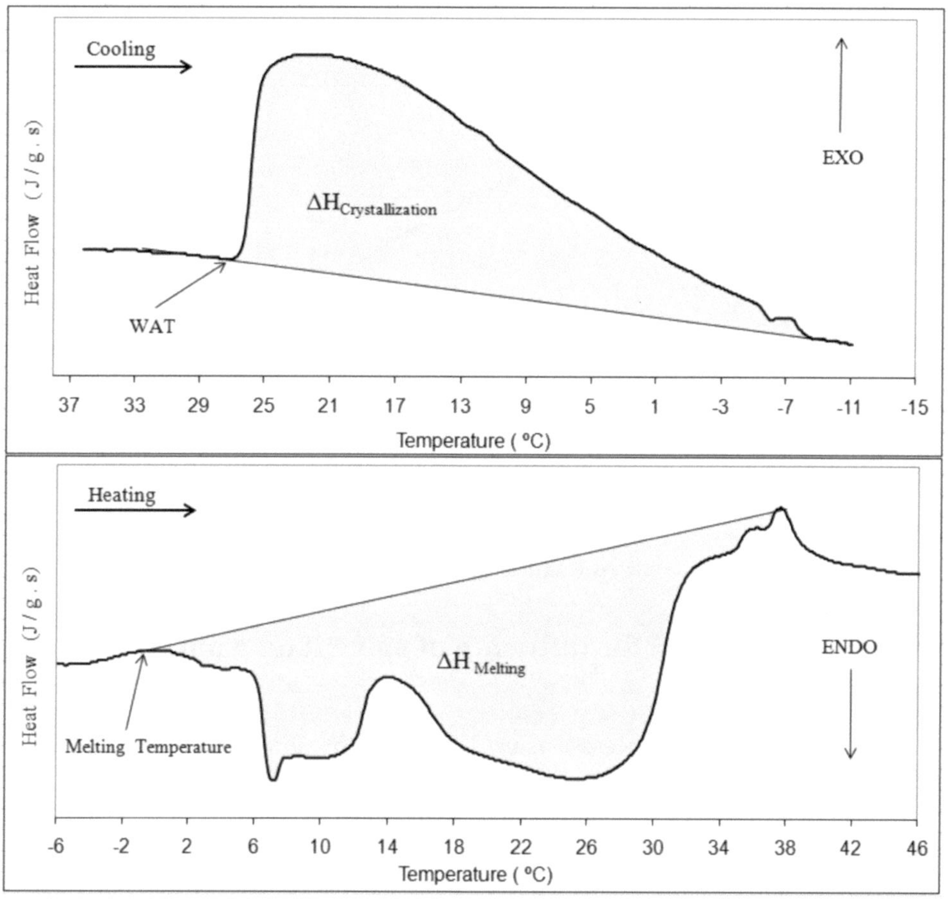

Figure 2. Example of DSC measurements on a paraffinic model system.

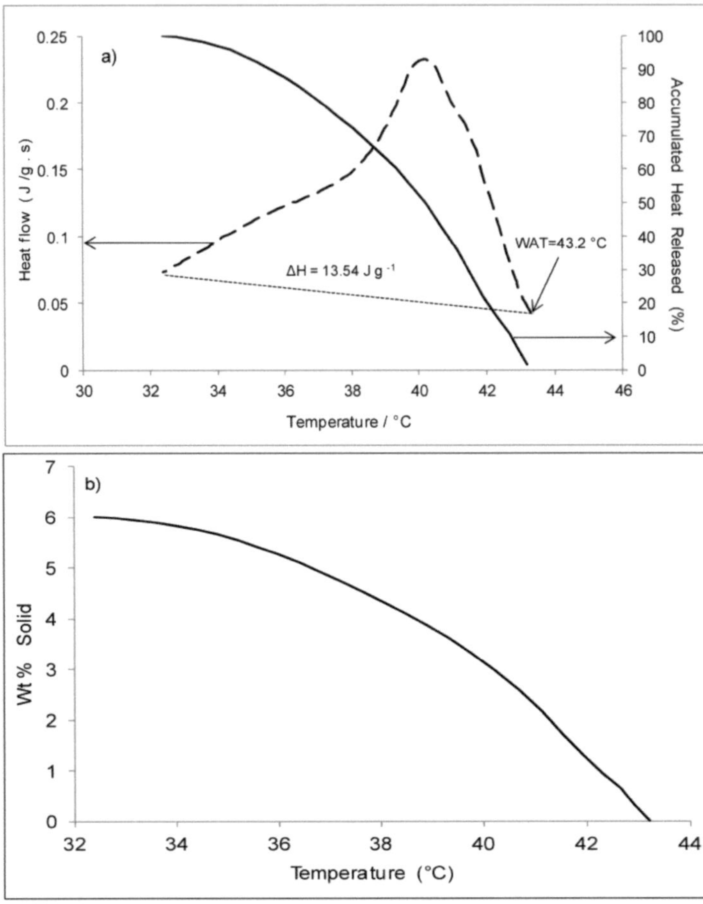

Figure 3. Determination of wax solubility curve of the system 6 wt.% of C_{36} in *n*-decane: a) accumulated heat released for the DSC exothermic peak and b) wax solubility curve obtained [28].

3. Experimental study of the influence of solvent on paraffin crystallization

The paraffin crystallization process can be influenced by many factors such as paraffin composition, solvent nature, polidispersity, rate of cooling, pressure, kinetics and presence of impurities [1,9, 14, 28, 30, 35-36], so that a better knowledge of factors affecting wax solubility will also improve the understanding of the wax precipitation phenomena in the petroleum industry described above. The studies about the effect of solvent on solubility of waxes reported in literature [1] have shown that waxes do not exhibit ideal solution behavior when crystallizing and that their solubility in a solvent increases as both the solvent molecular size and solvent solubility parameter decrease. The influence of the shape and size of the solvent on solute–solvent interaction and on the n-alkanes solubility has been also described in literature, hence it has been reported that globular or spherical solvents destroy the conformational order in liquid long-chain hydrocarbons [28, 37]. Aromatic solvents have been reported as a help in both inhibiting wax crystal formation and decreasing the amount of the wax deposited [38]. The experimental studies reported in literature have been carried out evaluating the effect of the solvent on cloud point or wax dissolution temperatures. Nevertheless, the wax gelation and deposition processes are actually originated due to the amount of paraffin crystals formed during cooling below WAT. This makes important to evaluate the influence of solvent on the amount of crystallized paraffin at temperatures below WAT. Therefore, in this section is presented an application of the Differential Scanning Calorimetry (DSC) to study the liquid-solid phase behavior of a high molecular weight *n*-paraffin: hexatriacontane ($C_{36}H_{72}$), in presence of solvents of different chemical nature in order to get a better understanding of the interactions solute-solvent on the paraffin crystallization mechanism.

Figure 4 shows the DSC thermograms of the crystallization and melting behavior of 6 wt. % of hexatriacontane (C_{36}) in different pure and mixed solvents systems. As can be seen, a single and well defined peak is observed during cooling and heating processes, related to the crystallization and melting of the monodisperse sample of the heavy paraffin C_{36} in different solvent systems. However, the endothermic peaks seem to be broader than the exothermic, so while the identification of the crystallization onset temperatures was straightforward; the melting onset temperatures were difficult to identify.

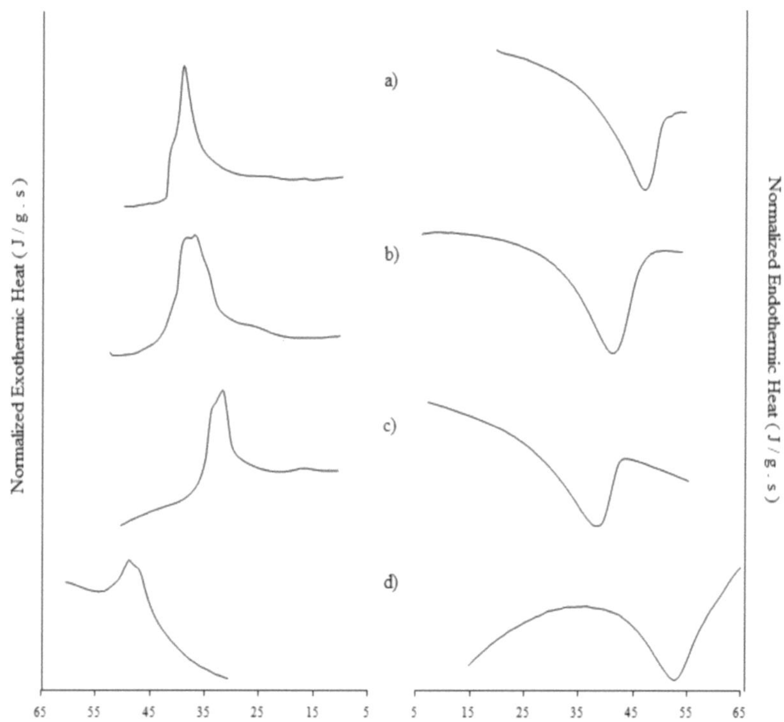

Figure 4. DSC exothermic and endothermic peaks of 6 wt% of C_{36} in different simple and mixed solvents systems: a) 94% of n-decane, b) 47% n-decane + 47% 1-phenyldodecane, c) 47% n-decane + 47% xylene and d) 94 % of squalane [28].

Crystallization and melting properties of the model systems investigated are shown in Table 1. The influence of the solvent aromaticity over the solution of C_{36} in *n*-decane was studied by adding mono-aromatic solvents: xylene and 1-phenyldodecane. The data show clearly the effect of the solvent chemistry on those properties. Lower values of crystallization and melting enthalpies are obtained in presence of aromatic solvents. The magnitude of the enthalpies decrease is related to the aromaticity of the solvent mixture, where greater aromaticity causes a greater diminishing of crystallization and melting enthalpies. Hence, the aromatic single rings interspersed among hexatriacontane molecules hinder their interactions, preventing an efficient ordering during cooling, then the paraffin crystal networks of a solid phase formed in such circumstance result significantly less ordered, as indicated by the lower values of the crystallinity index calculated from DSC data (see Table 2) for the model systems with aromatic solvents.

Solvent System	WAT (°C)	Enthalpy of Crystallization (J/g)	Melting Temperature (°C)	Enthalpy of Melting (J/g)	WDT (°C)	Solvent system aromaticity [a]
94 % n-decane	43.2	13.54	30.52	12.48	45.6	0
47% n-decane + 47% xylene	37.5	8.05	11.96	9.31	37.92	0.43
47% n-decane + 47% 1-phenyldodecane	47.5	10.66	20.25	10.09	41.82	0.12
94 % squalane	54.5	6.36	37.1	12.56	52.89	0

[a] Calculated as the aromaticity factor of aromatic solvent multiplied by its molar fraction in the mixture, where aromaticity factor of the xylene and 1-phenyldodecane are 0.75 and 0.333 respectively.

Table 1. Crystallization and melting properties of the system 6 % of C_{36} in different solvents systems [28]

Solvent System	Onset Crystallization Temperature (°C)	Endset Crystallization Temperature (°C)	ΔT[a] (°C)	Δt[b] (min)	DSC Crystallinity [c] (%)
94 % n-decane	43.2	32.4	10.08	2.16	7.21
47% n-decane + 47% xylene	37.5	23.5	14	2.8	5.38
47% n-decane + 47% 1-phenyldodecane	47.5	22.5	25	5	5.83
94 % squalane	54.5	41.08	13.42	2.68	7.26

[a] ΔT = Onset - Endset
[b] For a cooling rate = 5 °C/min
[c] For a melting enthalpy of hexatriacontane = 172.9 J/g [39]

Table 2. DSC crystallization data of the system 6 % of C_{36} in different solvent systems [28]

A way to explain the depressing effect of aromatic solvents over the enthalpies of crystallization and melting is through the entropy change that suffers the system during liquid - solid phase transition. It can be assumed that magnitude of the entropy change in the crystallization and melting processes for a paraffinic – aromatic mixture is lower than that for a 100% aliphatic system, like the C_{36} - n-decane system, because of the poor crystal network ordering of the solid phase formed in presence of aromatic solvents, which is reflected in the lower values of their crystallinity index. On another hand, a crystal network with greater disorder introduced by the aromatic molecules of the solvent is weaker, so its melting temperature tends to be lower. The WAT is also lowered due to the presence of aromatic solvents; however, in the case of the 1-phenyldodecane, a less aromatic solvent than xylene and with a greater molecular weight than decane, its aliphatic chain of 12 carbons has a significant "ordering" effect in the paraffinic crystal network that surpasses the effect of its aromatic rings regards to the depressing of the WAT and melting temperature observed with the xylene.

In contrast, by changing the solvent system from n-decane to squalane (a C_{24} aliphatic chain with six methyl branches), the WAT is significantly increased to 54.5 °C with a dramatic depression of the enthalpy of crystallization. This behavior is influenced by the size and structure of the solvent, in a similar way to that observed for the 1-phenyldodecane – n-decane solvent system. The methyl branches of the C_{24} iso-paraffin inhibits the efficient ordering of C_{36} molecules during crystallization process, forming a less ordered solid phase than that formed in C_{36}-decane system, which is reflected in the low enthalpy of crystallization measured (6.36 J/g), on the other hand, the highest WAT observed is a consequence of the greater size of squalane with respect to decane. In fact, as has been reported, solubility decreases (greater WAT) as the solvent size increases due to the inability of the bigger solvent molecule to effectively contact and solvate the solute [1]. Furthermore, despite the disorder in the crystalline arrangement caused by the six methyl branches of the squalane, the effect of its size (24 carbon length) results in the highest values of both temperature and enthalpy of melting.

Table 1 shows also the WDT of C_{36} in different solvents during melting process. Under ideal conditions, the values of WAT and WDT should be similar; however, as has been reported [22] differences between both values can be attributed to the experimental uncertainty and kinetic effects (e.g. supercooling). Our DSC results showed differences in the range between 0.88-5.38 °C which could be attributed to the heating/cooling rate used of 5 °C/min. The effect of solvent chemistry on WDT of C_{36} was similar to that observed for the melting temperature discussed above.

The results presented before showed a significant influence of solvent chemistry on crystallization and melting of C_{36}; however, in order to get a better understanding of the paraffin crystallization process in presence of solvents of different chemical structure at temperatures below WAT, solubility curves were obtained by using the DSC data as can be seen in Figure 5. As expected, the crystallization process of C_{36} in squalane starts before respect to the other systems, as a consequence of solvent chain length, followed by the other mixtures according to their respective WATs.

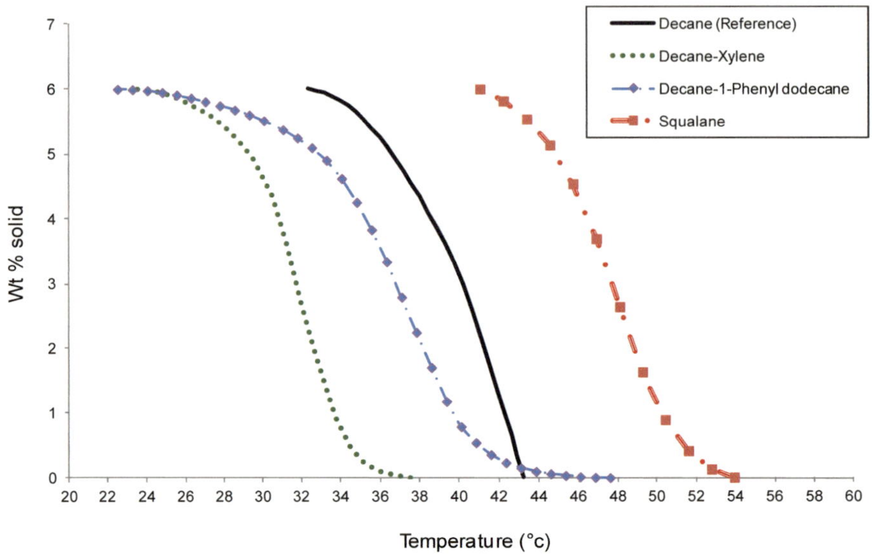

Figure 5. DSC Solubility curves of 6 wt % of C_{36} in different simple and mixed solvents systems [28].

The chemical structure of the solvent affects significantly the crystallization rate of the Hexatriacontane C_{36}. Table 2 shows that the crystallization process is slower for the systems with aromatic solvents. An evident consequence of this is the fact that lower amounts of solids are formed in these systems with respect to the decane system below its WAT as can be observed in Figure 5; this can be attributed to the aromatic ring interfering with the normal crystal growth, retarding the conformational ordering of C_{36} molecules in the solid phase created.

The degree of crystallinity obtained by DSC for the solid phase formed in presence of squalane, shown in Table 2, points out the effect of competition between the size and branching of the aliphatic solvents in the crystallinity of the solid phases formed. Due to the greater size of squalane with regard to the *n*-decane, it would be expected a greater crystallinity index for the solid phase formed in presence of squalane; however its ramifications limit the possibility of achieving an efficient conformational ordering of the crystal network in the solid phase, which results in a crystallinity index value similar to that obtained for the *n*-decane system.

The results presented in this section showed that the solvent aromaticity was a key factor that results in inhibition of the paraffin crystallization process, decreasing WAT, by promoting the creation of a solid phase partially disordered due to the presence of aromatic single rings interspersed among paraffin molecules hindering their efficient ordering during the cooling process.

4. DSC study of the effect of asphaltenes on liquid-solid phase equilibria

Asphaltenes are the most heavy and polar fraction in the crude oil. The asphaltene fraction is formed by many series of relatively large molecules containing aromatic rings, several heteroaromatic and napthenic ring plus relatively short paraffinic branches [8]. Some authors have proposed that asphaltenes form aggregates with a core formed by aromatic regions, and aliphatic chains on the periphery interacting with the surrounding oil [40-41]. The aliphatic portions of the asphaltene permit an interaction of asphaltenes with waxes. A handful of studies have evaluated the asphaltene-wax interactions and their effect on wax crystallization and gelation properties [28-30, 42-45]. The study of the influence of the asphaltenes and their chemical nature on the wax crystallization phenomena of paraffinic model systems; is presented in this section. DSC measurements allowed getting crystallization properties of paraffinic model systems in presence of asphaltenes of different origin with different molecular structure.

The effect of asphaltenes was studied on liquid-solid equilibrium of the binary system: tetracosane ($C_{24}H_{50}$) - octacosane ($C_{28}H_{58}$). The composition used in these model systems was 15 and 10 wt% of C_{24} and C_{28} respectively in the following solvent systems: a) 75% of decane, b) mixture of 37.5 % of decane + 37.5 % of xylene and c) mixture of 37.5 % of decane + 37.0 % of xylene + 0.5 % of asphaltenes. The asphaltene samples used in this study, labeled as AsphPC and AsphIri, were extracted from two Mexican crude oils of the southern region. These asphaltenes of different chemical nature were characterized by using elemental analysis, vapor pressure osmometry, ^{1}H and ^{13}C NMR spectroscopy in order to get their molecular parameters. Details of the experimental techniques used for the characterization of these asphaltenes and their effect on phase-equilibrium and rheological properties of waxy model systems have been reported recently in literature [28, 30]. Some of the molecular parameters of these asphaltenes are shown in Table 3. As can be seen, the AsphPC asphaltenes are more aromatic than AsphIri asphaltenes, and its aromatic core is also bigger and more condensed, whereas the aromatic core of AsphIri asphaltenes is richer in alkyl substituents comprising methyl groups, alkyl chains and naphthenic rings [30]. DSC exothermic peaks of the model systems are plotted in Figure 6 and their crystallization properties are shown in Table 4.

Symbol	Definition	Asphaltene sample	
		AsphIri	*AsphPC*
R_A	Aromatic rings	7.09	20.69
f_a	Aromaticity factor	0.50	0.67
ϕ	Condensation index	0.53	0.71
n	Average number of carbon atoms per alkyl substituents	5.99	5.47
σ	Aromatic substitution index	0.55	0.37
n_{ac}	Average length of the alkyl chains	11.84	11.12

Table 3. Molecular parameters of AsphPC and AsphIri asphaltenes [30]

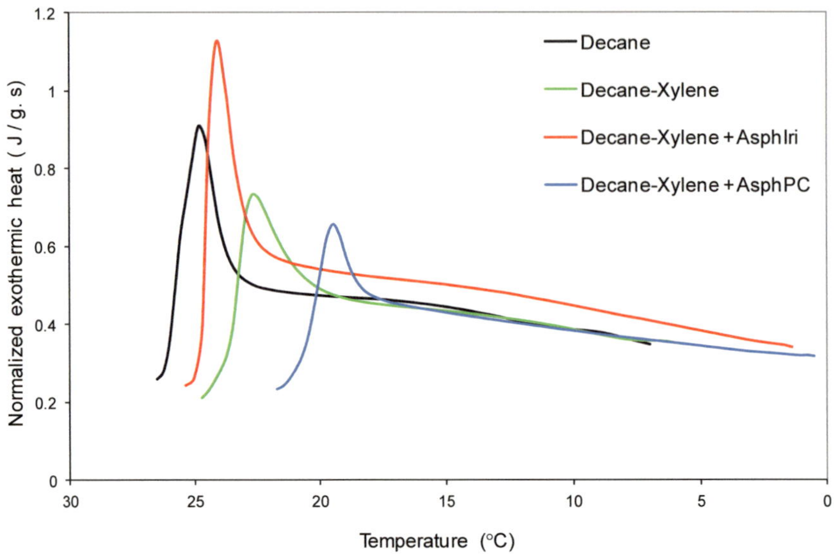

Figure 6. DSC exothermic peaks of the binary system C_{24}-C_{28} in different solvent systems and with asphaltenes of different chemical nature [28].

Solvent System	WAT (°C)	Enthalpy of Crystallization (J/g)	Endset (°C)	ΔT^a (°C)	Δt^b (min)
Decane	26.5	37.09	7	19.5	3.9
Decane-Xylene	24.81	36.51	6.09	18.72	3.74
Decane-Xylene + AsphIri	25.66	58.42	1	24.66	4.93
Decane-Xylene + AsphPC	21.77	33.01	0.53	21.24	4.24

$^a\Delta T$ = WAT - Endset; ^{b} For a cooling rate = 5 °C / min

Table 4. DSC crystallization data of the binary system C_{24}-C_{28} in different solvent systems and with asphaltenes [28]

As can be observed, the presence of an aromatic solvent as the xylene decreases slightly both WAT and crystallization enthalpy due to the disorder-effects generated by the aromaticity as was discussed before. However, crystallization properties were notably affected by the presence of asphaltenes where their chemical nature played an important role. It has been reported that flocculated asphaltenes providing nucleation sites for waxes increase WAT [44], but also it has been reported [30] that asphaltenes decreased very slightly the WAT. According with other studies [45] the effect of the asphaltenes on the WAT depends of the aggregation state of the asphaltenes.

In these model systems, the results obtained showed a slight increasing of WAT due to presence of the aliphatic asphaltenes (AsphIri) and a decreasing in presence of the more aromatic asphaltenes (AsphPC) with respect to the C_{24}-C_{28} in decane-xylene model system without asphaltenes. For another hand, the crystallization heat increases significantly in presence of AsphIri asphaltenes, whereas diminishes moderately in presence of AsphPC asphaltenes. These results make evident the effect of the chemical nature of different asphaltenes over the crystallization behavior of paraffinic systems. A greater abundance of aliphatic chains in AsphIri asphaltenes permits a better interaction with the paraffins of the C_{24}-C_{28} system promoting co-crystallization phenomena, where the asphaltenes are partially integrated to the crystal network and probably acting as nucleation sites, causing a slight increasing of WAT. Moreover, the partial immobilization of the paraffins engaged in the interactions with the asphaltene alkyl chains may promote a *"quasicrystallization"* phenomenon of the paraffins in the asphaltene network, such interaction results in exothermic effects as has been reported in literature [46, 47], which explains the significant increasing of crystallization heat of the model system with AsphIri asphaltenes as observed in Table 4. On the other hand, the most aromatic asphaltenes (AsphPC) with a bigger and more condensed aromatic core and with a smaller amount of aliphatic substituents inhibit in some extent the paraffin-asphaltene interactions so that they cannot be incorporated to the paraffin crystal structure hindering nucleation process and crystal network growth, which results in a WAT decreasing with a lower crystallization enthalpy due to the formation of a disordered solid phase.

The wax solubility curves of these systems are plotted in Figure 7, as can be seen the effect of solvent and asphaltenes of different chemical nature is evident. Regarding to the C_{24} - C_{28} mixture in *n*-decane, the presence of xylene reduces moderately both the amount of solid formed and the WAT, although the temperature interval of their solubility curves are very similar. However, the effect of asphaltenes on wax solubility curve is very significant considering its low concentration in the model system (0.5%), particularly in the case of the more aromatic AsphPC asphaltene. As in the case of C_{36} in different solvents, the rate of wax precipitation for the C_{24}-C_{28} mixture in xylene is significantly affected by the presence of a small amount of highly aromatic compounds such as the asphaltenes. The data in Table 4 shows also that the presence of asphaltenes in the paraffinic system increases around 25% the time required to precipitate the total of paraffins (Δt) for a cooling rate of 5°C/ min.

These results showed that asphaltenes are practically acting in this system as inhibitors of the paraffins precipitation. When the asphaltenes have a structure highly condensed with a certain degree of aromatic substitution, that allows some kind of interaction with the paraffins, the inhibition effect is greater due the steric interference and the disorder generated in the paraffinic network which difficult the molecular recognition among paraffin molecules avoiding the growth of stable crystalline networks and therefore the formation of the solid phase. Otherwise a less condensed aromatic structure with a greater substitution degree have a better interaction with paraffins and thus, it could play a role as nucleation site increasing both WAT and the amount of solid phase formed, at least in a certain temperature range as can be observed in Figure 7, but even in such case the disruptive effect that introduces the aromatic core of the asphaltenes prevails inhibiting the crystallization process, as it was observed for the model system with the AsphIri asphaltenes. These phenomena are sketched in Figure 8.

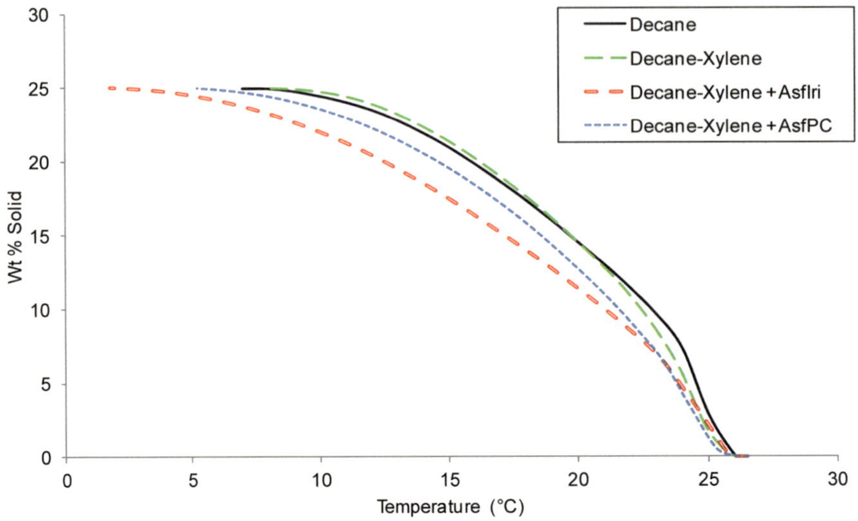

Figure 7. Wax solubility curves of the binary system C_{24}-C_{28} in different solvent systems and with asphaltenes of different chemical nature [28].

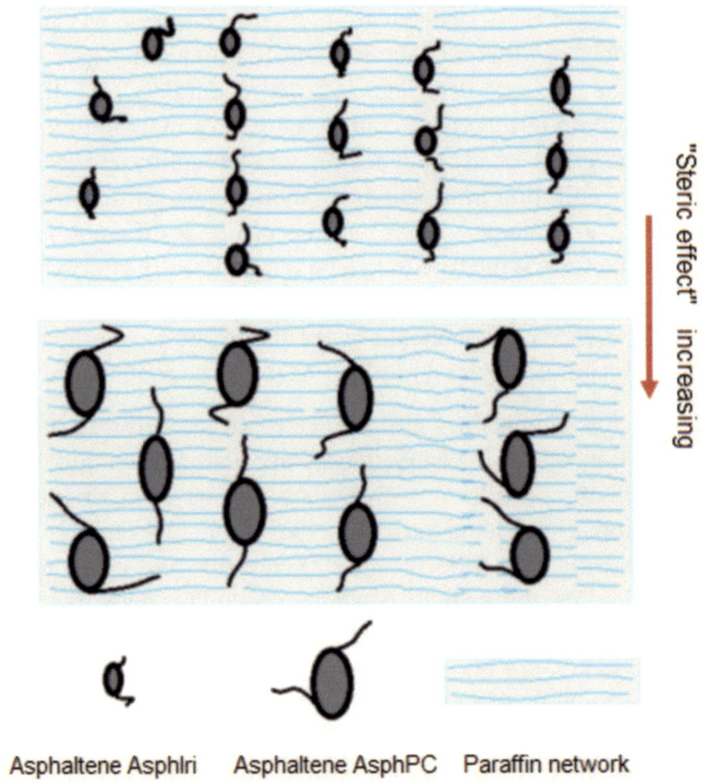

Figure 8. Schematic representation of the "Steric effect" of asphaltenes of different chemical nature on paraffin crystallization.

5. Wax precipitation study in crude oils by DSC measurements

With the ongoing trend in deep water developments, flow assurance has become a major technical and economic issue. The avoidance or remediation of wax deposition is one key aspect of flow assurance [3-4, 6-7]. In order to develop solutions to the wax deposition problem is necessary to get a deep understanding of the crystallization phenomena in which the crude oil composition, particularly the content of high molecular weight paraffins and asphaltenes have a significant impact [29]. Comparison of experimental methods for measurement of wax precipitation in crude oils have been reported in literature [5, 21, 24-27, 29], where wax detection limits vary depending on the measurement technique, oil composition, thermal history, time of measurement and fluid properties related to crystal nucleation and growth [27]. In this section is presented a characterization of the wax precipitation phenomena in crude oils by using DSC measurements. Wax appearance temperature is measured by using DSC and these results are compared with those obtained by using other techniques such as rheometry, spectroscopy and densitometry in order to show advantages and disadvantages of the use of DSC method to measure liquid-solid phase equilibria of waxes in crude oils. The importance to get the wax melting temperature and crystallinity degree by DSC is analyzed also as key parameters to evaluate the propensity of crude oils to present wax precipitation problems during crude oil production and transporting. Three Mexican crude oils of the southern region labeled as RDO1, RDO2 and J32 are studied in this work. Crude oils RDO1 and RDO2 present wax precipitation and deposition problems during their production and transportation, whereas crude oil J32 presents a severe asphaltene precipitation and deposition problem along the well during primary production. The cloud point temperatures of the crude oils and the crystallization and melting properties of their isolated waxes were determined by using the DSC method according to the procedures and conditions described and presented in section II of this chapter. However, in these experiments for waxes, under heating conditions, the wax melting point was recorded as their endothermic peak temperature.

In Figure 9 are shown the DSC thermograms of crude oils during cooling process. As it has been reported [22], during cooling, a decrease of the solvating power of the oil matrix results in precipitation of solid particles of wax.

The thermograms of crude oils RDO1 and RDO2 show one well defined exothermal peak from which the WAT can be easily determined 19.2 and 18.5°C respectively. In contrast the thermogram of crude oil J32 presents two exothermal peaks, first one well defined, a liquid-solid phase transition whose onset corresponding to a WAT of 28°C and a second one broad not well defined, with an onset around 20.8°C.

Other methods were employed to determine the WAT of the crude oil samples, these experimental techniques were fourier transform infrared spectroscopy (FT-IR), rheometry and densitometry. Figure 10 shows the WAT determination by these methods described briefly below. FT-IR method is based on the fact that the absorbance between the wave numbers 735 and 715 cm^{-1} attributed to the rocking vibrations of the long chain methylene (LCM) groups (the major component of the solid wax formed in crude oils), has been found to increase with a decrease in temperature due to the formation of a solid phase [48]. Below the WAT, the higher absorptive of the solid phase made up of LCM grups, contributes strongly to the total absorbance, which gives rise a change in the slope of the plot of the area of the absorbance peak (735 to 715 cm^{-1}) of each FT-IR spectrum as a function of the temperature. The temperature where the change in slope occurs is recorded as the WAT as observed in Figure 10 a). Rheometric WAT measurements were carried on the basis that petroleum fluids exhibit non-Newtonian behavior below WAT and Newtonian behavior above the WAT which follows the Arrhenius temperature dependence:

$$\mu = A\, e^{\,Ea\,/\,RT} \tag{2}$$

where μ is the Newtonian dynamic viscosity, *A* is the Arrhenius pre-exponential factor, *Ea* is the activation energy of viscous flow, *R* is the universal gas constant and *T* is the absolute temperature. The formation and growing of solid wax crystals dispersed in the crude oil medium causes a viscosity increasing during a cooling process [35]. In this way, from viscosity-temperature curves, WAT is recorded as the temperature of the deviation of the Arrhenius law as it is shown in Figure 10 b). Finally, the WAT determination by using densitometry method is carried out by identifying the temperature at which a sharp change in the slope of density – temperature curve obtained during a cooling process that become evident the onset of wax crystallization as can be observed in Figure 10 c).

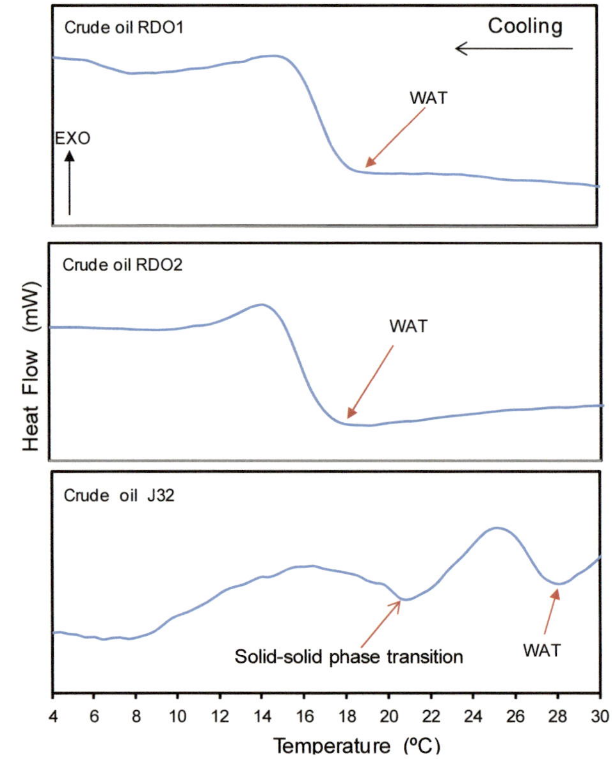

Figure 9. Exothermic peaks from DSC Thermograms of crude oils during cooling [29].

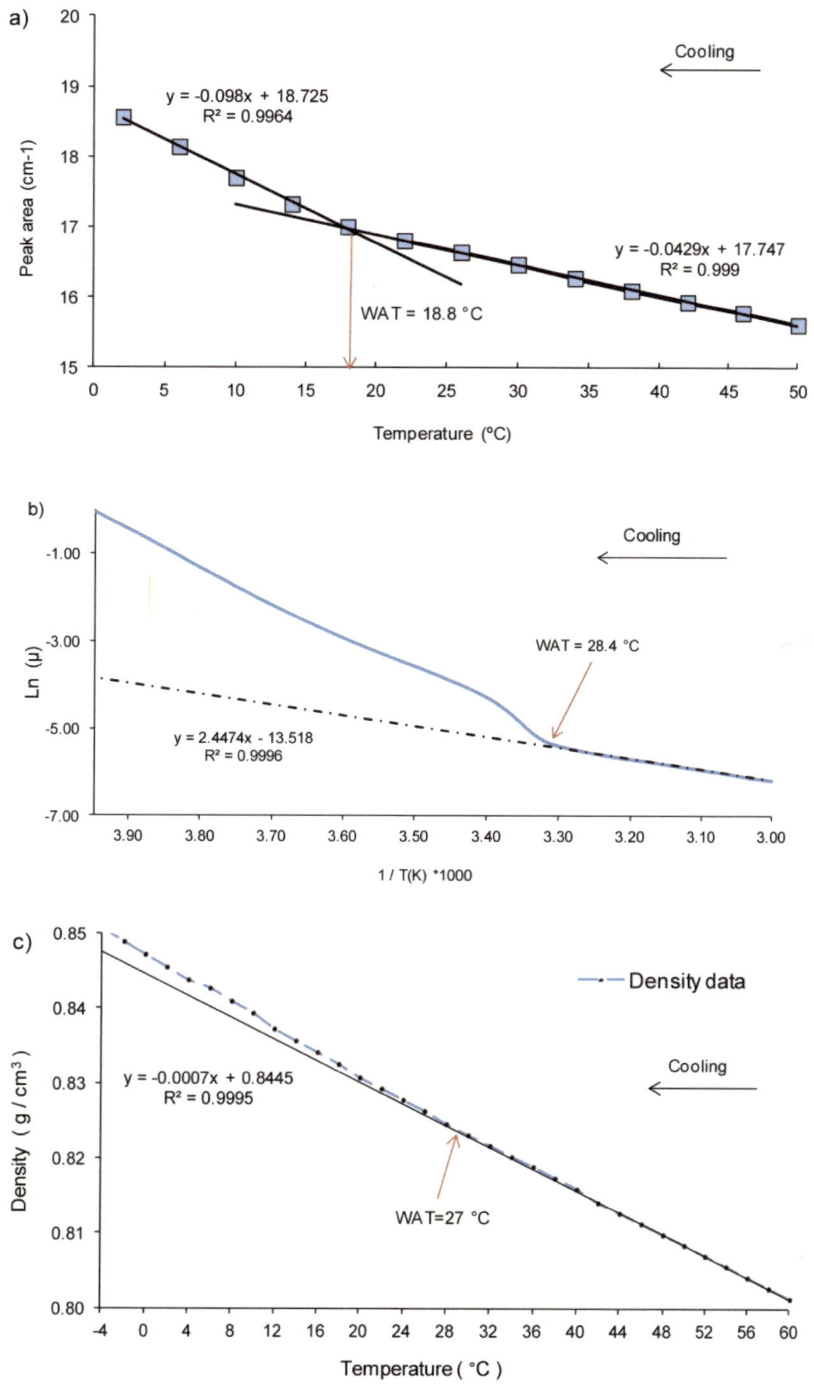

Figure 10. WAT determination by using different experimental methods: a) FT-IR WAT determination for crude oil RDO1, b) Rheometric WAT measurement for crude oil RDO2 from Arrhenius plot and c) Density-Temperature profile for crude oil RDO1 [29].

Table 5 shows the WAT measurements by using the different techniques for the crude oils investigated in this work. According to these results, good agreement is obtained between DSC and FT-IR method in the WAT measurement, whereas Rheometry method apparently overestimated cloud point for crude oils RDO1 and RDO2 and underestimate it in the case of crude oil J32. WAT detection difficulties related with effects of crude oil composition were found when using Densitometry method where the uncertainty associated with the WAT measurement was high.

Crude oil	WAT (°C)				Average (°C)	Deviation (%)
	DSC	Rheometry	Densitometry	FT-IR		
RDO1	19.2	32.1	27	18.8	24.2	± 26.5
RDO2	18.5	28.4	21	16.5	21.1	± 24.6
J32	28	16.3	*ND	30.7	25	± 30.6

*Not detectable

Table 5. WAT measurements of crude oils [29]

The good agreement between DSC and FT-IR methods can be explained in terms of their high sensitivity to the energetic variations related to the phase transition phenomena of paraffins. The Rheometry and Densitometry methods, however, need that a critical amount of solid wax come out of solution to produce a detectable change in the rheological properties or density of the crude oil sample and thus to identify the WAT. On the other hand, the WAT values of the crude oils RDO, measured by the rheometric technique were higher than DSC values ; hence liquid-liquid demixing effects not detected by DSC and FT-IR methods during cooling can have a significant impact on rheological behavior of crude oil samples resulting in an overestimation of WAT by the rheometric technique.

Waxes isolated from crude oils were also analyzed by DSC experiments in order to get their crystallization and melting properties. Figure 11 shows the exothermic and endothermic peaks obtained from DSC thermograms of the crude oil waxes. As can be observed, during cooling and heating, a single and well-resolved peak was obtained for each of the RDO1 and RDO2 waxes, whereas exo and endothermic peaks of Wax J32 were broader and partially well defined between 60 and 6 °C.

The crystallization and melting properties of waxes characterized by DSC are shown in Table 6. It can be seen that wax J32 has the highest crystallization temperature (60.2 °C) followed by wax RDO2 (52.3 °C) and wax RDO1 (50.1 °C). In the case of crystallization enthalpies, the lowest was obtained for wax J32 and the highest for wax RDO1. As expected and due to the crystallizing materials that form the waxy deposits are primarily *n*-alkanes, there is a correlation between wax crystallization temperature and crude oil WAT (see Figure 12).

The wax melting temperature is an important parameter that can be used to define the temperature at which pipe walls or storage facilities may need to be heated in order to remove solid deposits [49]. The lowest melting temperature found in wax J32 (35.9 °C) helps to minimize the propensity of crude oil J32 to present wax precipitation problems during production and transportation, despite of its highest both crystallization temperature whereas crude oils RDO1 y RDO2, whose wax fractions have the highest melting temperatures present wax precipitation and deposition problems in the well head and downstream. A correlation between temperatures and enthalpies of melting was observed in waxes analyzed by DSC, wax RDO1 with highest melting temperature (45.8 °C) has also the highest melting enthalpy (112.89 J/g). This correlation found in this work is in agreement with previous results reported in literature [50]. Finally in Figure 13 is plotted the relationship between wax melting temperature and wax crystallinity, where a good correlation can be observed. In fact and as expected, the lowest degree of crystallinity of wax J32 indicates that its crystal structure is weaker and less stable, with a higher disorder degree, and thus making it easier to melt as pointed out its lower melting temperature.

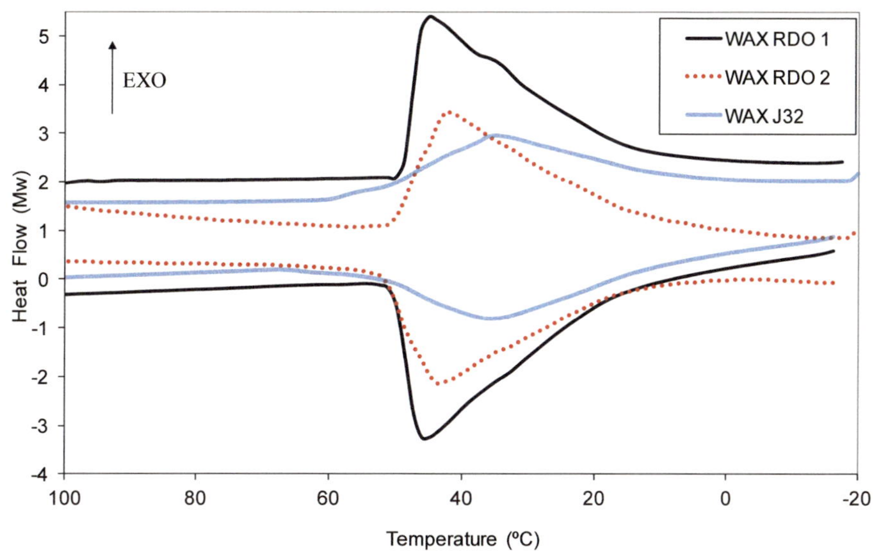

Figure 11. Exo and endothermic peaks from DSC thermograms of waxes [29].

Property	Wax		
	RDO1	**RDO2**	**J32**
Crystallization Temperature (°C)	50.1	52.3	60.2
Crystallization Enthalpy (J/g)	110.8	107.11	82.29
Melting Temperature (°C)	45.8	43.6	35.9
Enthalpy of melting (J/g)	112.89	110.56	102.13
DSC Crystallinity (%)	70.29	68.84	62.46

Table 6. Crystallization and melting properties of waxes [29]

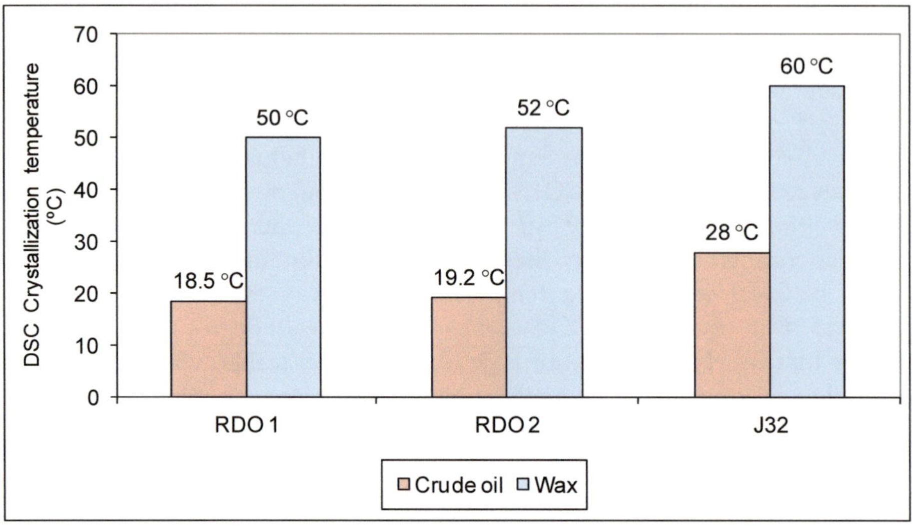

Figure 12. Corrrelation between DSC crystallization temperatures of crude oils and their wax fraction

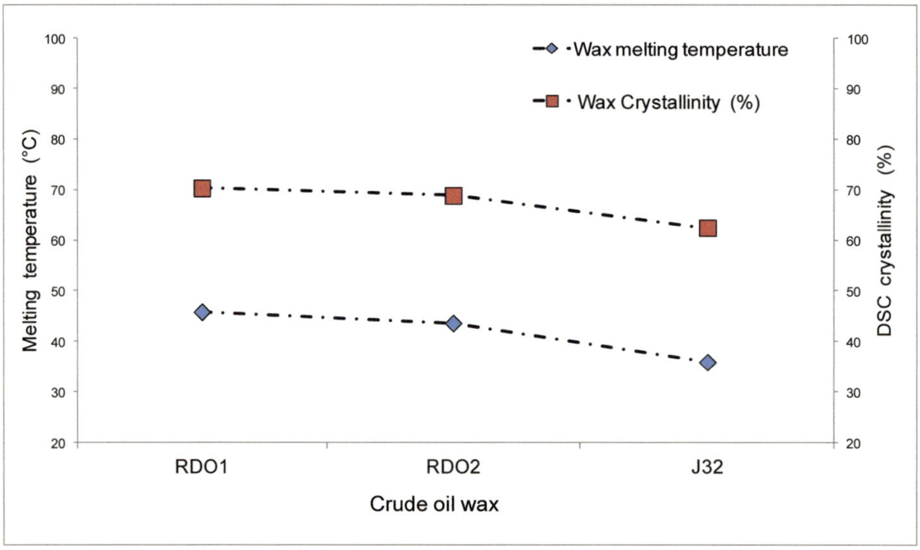

Figure 13. Corrrelation between wax melting temperature and crystallinity.

As conclusions of this section is established that from the methods presented here for the WAT determination, DSC technique is recommended for the identification of phase transitions in crude oils due to its high sensitivity and reliability. Crude oils RDO1 and RDO2, whose wax fractions have the highest melting temperatures and crystallinity degree present wax precipitation and deposition problems in the well head and downstream, whereas the crude oil J32 that has the wax fraction with the lowest melting temperature and crystallinity degree do not present wax deposition problems during production. DSC measurements allowed identifying key parameters useful to evaluate the wax precipitation in crude oil samples.

6. Conclusions

In this chapter we have showed the usefulness of the DSC technique on the characterization of the liquid-solid phase equilibrium of paraffins. It was shown that DSC is an experimental method widely used due to its simplicity, accuracy and fast response to monitor phase transitions. Good capabilities and advantages of the DSC method were shown in order to carry out relevant studies on thermodynamic properties of paraffins. In this chapter it was shown that DSC measurements allowed elucidating the interactions of paraffins with solvents and asphaltenes of different chemical nature during the crystallization and melting phenomena. DSC technique was very useful on the characterization of the wax precipitation phenomena in crude oils due to its high sensitivity and reliability for the identification of phase transitions and crystallinity measurement, which allows evaluating tendency of crude oils to present wax precipitation problems in production and transporting operations. Finally, in order to develop solutions to the wax formation problems presented in some industrial sectors such as petroleum industry is necessary to get a deep understanding of the paraffin crystallization phenomena where the DSC method can give us valuable information as was shown in this chapter.

Author details

Luis Alberto Alcazar-Vara and Eduardo Buenrostro-Gonzalez
Instituto Mexicano del Petróleo, Programa Académico de Posgrado. Eje Central Lázaro Cárdenas, México, D.F.

Acknowledgement

The authors express gratitude to the Instituto Mexicano del Petróleo (IMP) for both providing facilities and granting permission to publish results. L.A.A.V thanks CONACYT and the Programa Académico de Posgrado of IMP for the economic support granted during his Ph.D. studies. The authors thank to Mr. J.A. Garcia-Martinez from IMP for his excellent assistance on asphaltenes characterization.

7. References

[1] Jennings DW, Weispfennig K (2005) Experimental solubility data of various n-alkane waxes: Effects of alkane chain length, alkane odd versus even carbon number structures and solvent chemistry on solubility. Fluid Phase Equilibria. 227:27–35.

[2] Burger ED, Perkins TK, Striegler JH (1981) Studies of Wax Deposition in the Trans. Alaska Pipeline. J. Petroleum Tech. 33:1075-86.

[3] Rønningsen HP, Bjørndal B, Hansen AB, Pedersen WB (1991) Wax precipitation from North Sea oils. 1. Crystallization and dissolution temperature, and Newtonian and non-Newtonian flow properties. Energy & Fuels. 5:895–908.

[4] Leontaritis KJ (1995) The Asphaltene and Wax Deposition Envelopes. The Symposium on Thermodynamics of Heavy Oils and Asphaltenes, Area 16C of Fuels and Petrochemical Division, AIChE Spring National Meeting and Petroleum Exposition, Houston, Texas, March 19-23.

[5] Hammami A, Raines MA (1999) Paraffin Deposition from Crude Oils: Comparison of Laboratory Results to Field data. SPE Journal. 4(1):9-18.

[6] Alboudwarej H, Huo Z, Kempton E (2006) Flow-Assurance Aspects of Subsea Systems Design for Production of Waxy Crude Oils. SPE Annual Technical Conference and Exhibition in San Antonio, Texas, USA. 24-27 September.

[7] Coutinho JAP, Edmonds B, Moorwood T, Szczepanski R, Zhang X (2006) Reliable wax predictions for flow assurance. Energy & Fuels. 20:1081–1088.

[8] Speight JG (1999) The Chemistry and Technology of Petroleum (3rd ed.), Marcel-Dekker, New York.

[9] Garcia MD, Orea M, Carbognani L, Urbina A (2001) The Effect of Paraffinic Fractions on Crude Oil Wax Crystallization. Petroleum Science and Technology. 19(12):189–196.

[10] Garcia MD, Urbina A (2003) Effect of Crude Oil Composition and Blending on Flowing Properties. Petroleum Science and Technology. 21(5-6):863–878.

[11] Domanska U, Morawski P (2005) Influence of Size and Shape Effects on the High-Pressure Solubility of n-Alkanes: Experimental Data, Correlation and Prediction. Journal of Chemical Thermodynamics. 37:1276-1287.

[12] He B, Martin V, Setterwall F (2004) Phase Transition Temperature Ranges and Storage Density of Paraffin Wax Phase Change Materials. Energy. 29: 1785–1804.

[13] Giron D (2002) Applications of Thermal Analysis and Coupled Techniques in Pharmaceutical Industry. Journal of thermal analysis and Calorimetry. 68: 335-357.

[14] Senra M, Panacharoensawad E, Kraiwattanawong K, Singh P, Fogler HS (2008) Role of n-Alkane Polydispersity on the Crystallization of n-Alkanes from Solution. Energy & Fuels. 22:545–555.

[15] Marie E, Chevalier Y, Eydoux F, Germanaud L, Flores P (2005) Control of n-Alkanes Crystallization by Ethylene-Vinyl Acetate Copolymers. Journal of Colloid and Interface Science. 290(2):406-418.

[16] Guo X, Pethica BA, Huang JS, Prud'homme RK (2004) Crystallization of Long-Chain n-Paraffins from Solutions and Melts As Observed by Differential Scanning Calorimetry. Macromolecules. 37:5638–5645.

[17] Van Winkle TL, Affens WA, Beal EJ, Mushrush GW, Hazlett RN, DeGuzman J (1987) Determination of Liquid and Solid Phase Composition in Partially Frozen Middle Distillate Fuels. Fuel. 66(7):890-896.

[18] Coutinho JAP, Ruffier-Meray V (1997) Experimental Measurements and Thermodynamic Modeling of Paraffinic Wax Formation in Undercooled Solutions. Ind. Eng. Chem. Res. 36:4977– 4983.

[19] Bhat NV, Mehrotra AK (2004) Measurement and Prediction of the Phase Behavior of Wax-Solvent Mixtures: Significance of Wax Disappearance Temperature. Ind. Eng. Chem. Res. 43(12): 3451-3461.

[20] Hammami A, Ratulowski J, Coutinho JAP (2003) Cloud Points: Can We Measure or Model them? .Petroleum Science and Technology. 21(3&4):345-358.
[21] Hammami A, Raines MA (1999) Paraffin Deposition from Crude Oils: Comparison of Laboratory Results to Field data. SPE J. 4(1): 9-18
[22] Hansen AB, Larsen E, Pedersen WB, Nielsen AB (1991) Wax precipitation from North Sea crude oils. 3. Precipitation and dissolution of wax studied by differential scanning calorimetry. Energy & Fuels. 5:914–923.
[23] Guo X, Pethica BA, Huang JS, Adamson DH, Prud'homme RK (2006) Effect of Cooling Rate on Crystallization of Model Waxy Oils with Microcrystalline Poly(ethylene butane). Energy & Fuels. 20:250–256.
[24] Monger-McClure TG, Tackett JE, Merrill LS (1999) Comparisons of Cloud Point Measurement and Paraffin Prediction Methods .SPE Production & Facilities. 14(1):4–10.
[25] Lira-Galeana C, Hammami A (2000) Wax Precipitation from Petroleum Fluids: A Review. in: Asphaltenes and Asphalts 2. Yen TF, Chilingarian G eds. Elsevier Science Publishers: Holland. pp. 557-608.
[26] Leontaritis KJ, Leontaritis JD (2003) Cloud Point and Wax Deposition Measurement Techniques. SPE Paper No. 80267, SPE International Symposium on Oilfield Chemistry, Houston, Texas. February 5-8.
[27] Coutinho JAP, Daridon JL (2005) The Limitations of the Cloud Point Measurement Techniques and the Influence of the Oil Composition on its Detection. Petroleum Science and Technology. 23: 1113-1128.
[28] Alcazar-Vara LA, Buenrostro-Gonzalez E (2012) Experimental Study of the Influence of Solvent and Asphaltenes on Liquid-Solid Phase Behavior of Paraffinic Model Systems by using DSC and FT-IR Techniques. Journal of Thermal Analysis and Calorimetry. 107(3): 1321-1329.
[29] Alcazar-Vara LA, Buenrostro-Gonzalez E (2011) Characterization of the Wax Precipitation in Mexican Crude Oils. Fuel Processing Technology. 92(1): 2366–2374.
[30] Alcazar-Vara LA, Garcia-Martinez JA, Buenrostro-Gonzalez E (2012) Effect of Asphaltenes on Equilibrium and Rheological Properties of Waxy Model Systems. Fuel. 93: 200–212.
[31] Svetlichnyy DS, Didukh AG, Aldyarov AT, Kim DA, Nawrocki M, Baktygali AA (2011) Study of Heat Treatment and Cooling Rate of Oil Mixtures Transported by "Kumkol-Karakoin-Barsengir-Atasu" Pipeline. Online journal Oil and Gas Business. Available: http://www.ogbus.ru/eng/authors/Svetlichnyy. Accessed 2012 April 13.
[32] Han S, Huang Z, Senra M, Hoffmann R, Fogler HS (2010) Method to Determine the Wax Solubility Curve in Crude Oil from Centrifugation and High Temperature Gas Chromatography Measurements. Energy & Fuels. 24:1753-1761.
[33] Coto B, Martos C, Peña JL, Espada JJ, Robustillo MD (2008) A New Method for the Determination of Wax Precipitation from non-Diluted Crude Oils by Fractional Precipitation. Fuel. 87: 2090-2094.
[34] Conti DS, Yoshida MI, Pezzin SH, Coelho LAF (2006) Miscibility and Crystallinity of Poly(3-hydroxybutyrate)/Poly(3-hydroxybutyrate-co-3-hydroxyvalerate) Blends. Thermochimica Acta. 450:61–66.

[35] Paso K, Senra M, Yi Y, Sastry AM, Fogler HS (2005) Paraffin Polydispersity Facilitates Mechanical Gelation. Ind. Eng. Chem. Res. 44:7242-7254.

[36] Vieira LC, Buchuid MB, Lucas EF (2010) Effect of Pressure on the Crystallization of Crude Oil Waxes. II. Evaluation of Crude Oils and Condensate. Energy & Fuels. 24:2213-2220.

[37] Domanska U, Morawski P (2005) Influence of Size and Shape Effects on the High-Pressure Solubility of n-Alkanes: Experimental Data, Correlation and Prediction. Journal of Chemical Thermodynamics. 37:1276-1287.

[38] Rakotosaona R, Bouroukba M, Petitjean D, Dirand M (2008) Solubility of a Petroleum Wax with an Aromatic Hydrocarbon in a Solvent. Energy & Fuels. 22:784–789.

[39] Dirand M, Bouroukba M, Chevallier V, Petitjean D, Behar E, Ruffier-Meray V (2002) Normal Alkanes, Multialkane Synthetic Model Mixtures, and Real Petroleum Waxes: Crystallographic Structures. Thermodynamic Properties and Crystallization. J. Chem. Eng. Data. 47:115–143.

[40] Mullins OC, Betancourt SS, Cribbs ME, Dubost FX, Creek JL, Andrews AB, Venkataramanan L (2007) The Colloidal Structure of Crude Oil and the Structure of Oil Reservoirs. Energy & Fuels. 21(5):2785-2794.

[41] Carbognani L, Rogel E (2003) Solid Petroleum Asphaltenes Seem Surrounded by Alkyl layers. Petroleum Science and Technology. 21(3-4):537-556.

[42] Garcia MD, Carbognani L (2001) Asphaltene-Paraffin Structural Interactions. Effect on Crude Oil Stability. Energy & Fuels. 15(5):1021-1027.

[43] Venkatesan R, Ostlund JA, Chawla H, Wattana P, Nyden M, Fogler HS (2003) The Effect of Asphaltenes on the Gelation of Waxy Oils. Energy & Fuels. 17(6):1630-1640.

[44] Kriz P, Andersen SI (2005) Effect of Asphaltenes on Crude Oil Wax Crystallization. Energy & Fuels. 19(3):948-953.

[45] Tinsley JF, Jahnke JP, Dettman HD, Prud'home RK (2009) Waxy Gels with Asphaltenes 1: Characterization of Precipitation, Gelation, Yield Stress, and Morphology. Energy & Fuels. 23(4):2056-2064.

[46] Mahmoud R, Gierycz P, Solimando R, Rogalski M (2005) Calorimetric Probing of n-Alkane-Petroleum Asphaltene Interactions. Energy & Fuels. 19:2474 –2479.

[47] Stachowiak C, Viguie J-R, Grolier J-P, Rogalski M (2005) Effect of n-Alkanes on Asphaltene Structuring in Petroleum Oils. Langmuir. 21:4824-4829.

[48] Roehner RM, Hanson FV (2001) Determination of Wax Precipitation Temperature and Amount of Precipitated Solid Wax versus Temperature for Crude Oils using FT-IR Spectroscopy. Energy & Fuels. 15:756–763.

[49] Flow assurance design guideline (2001) Deepstar IV project.

[50] Alghanduri LM, Elgarni MM, Daridon JL, Coutinho JAP (2010) Characterization of Libyan Waxy Crude Oils. Energy & Fuels. 24:3101–3107.

Chapter 12

Thermal Analysis of Phase Transitions and Crystallization in Polymeric Fibers

W. Steinmann, S. Walter, M. Beckers, G. Seide and T. Gries

Additional information is available at the end of the chapter

http://dx.doi.org/10.5772/54063

1. Introduction

Each year about 50 Million tons polymer is processed to fibers worldwide [1]. Polymeric fibers are manufactured into all sorts of daily as well as industrial goods [2, 3]. The most prominent materials are thermoplastic among which poly(ethylene terephtalat) (PET), polyamides (PA) and polypropylene (PP) make up the largest fraction [4]. Other thermoplastic polymers such as poly(vinylidene fluoride) (PVDF) belong to niche markets with highly specialized applications [5-7].

The most distinctive property of synthetic fiber materials which separates them from other polymeric products is the strong anisotropic material structure. Geometrically this is characterized by a rather high aspect ratio of diameter to length which can reach several magnitudes of order in filament fibers. An exemplary PET textile multi-filament bundle of 300 single filaments in one flat yarn weighs around 100 g per 10.000 meters length (100 dtex). This yields in a single filament diameter of roughly 3 μm. A common industrial bobbin holds up to 25 kg of a virtually endless length of yarn, which in this case would be 2.5 million meters. The predominant cross section is circular in shape. Nonetheless, depending on the fiber application other cross-sections are possible and also common [3,4].

On a structural level the strong anisotropic character of a thermoplastic fiber is mainly caused and influenced by the production process. Hence, the spinning process of thermoplastic fibers shall be explained briefly. Next to the direct spinning process in which the polymer is directly processed to fibers right after the polymerization process, the most common process is the extruder based fiber production. The polymer granules are heated and transferred into a molten state inside the extruder [8]. The melt is then conveyed into a gear pump which ensures a constant flow of mass. This constant polymer flow then is being pressed through filtration layers and finally extruded through capillaries. Following the extrusion the polymer is drawn down vertically and solidifies while cooling from extrusion temperature down to the ambient air temperature. Usually the fibers are drawn down mechanically by rollers [4,8]. Figure 1 illustrates a classical melt spinning line. The polymer granules are feed through a hopper (A) into an extruder (B). The molten polymer is transported through heated pipes (C) to a gear pump (D). The gear pump feeds the spin pack (E) in which several layers of filtration are placed. The polymer is then extruded through capillaries and exits the spinpack into the quenching zone (F) where a laminar air flow ensures constant cooling conditions. After solidification and before touching the first roller or godet (H) the filaments are usually coated with a spin finish (G).

The most important process parameters of the melt spinning process are: the extrusion temperature $T_{Extrusion}$; the mass flow through each single capillary $m_{Throughput}$; the density of the melt ϱ_{Melt}; the cross-sectional area of the capillary $A_{Capillary}$; the viscosity η of the melt at the local temperature and the draw down speed v. Of course, there are numerous other parameters that affect the process such as the surface quality of the capillary walls, the form of the capillary rim, the ambient air profile consisting of flow direction, speed and temperature and others. For a basic comprehension these are neglected at this point. There is large number of extensive publications on these aspects available such as [9-13].

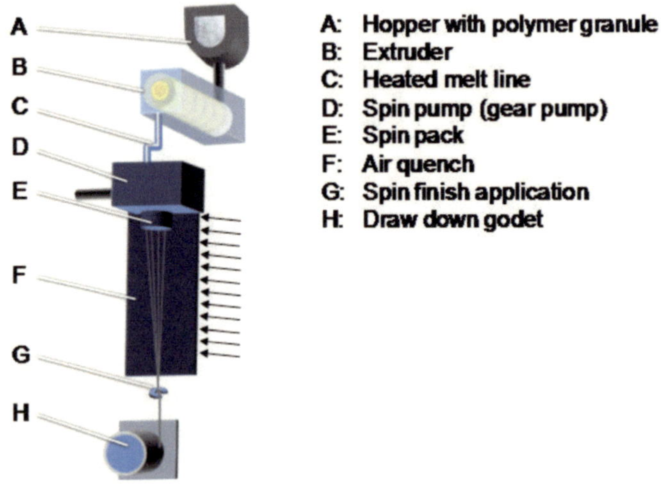

Figure 1. Schematic overview of a conventional melt spinning line

The cooling of the material from melt to ambient temperature takes place under tremendous stretching stress, which is characterized by the ratio between draw down and melt extrusion speed. This is commonly referred to as melt draw ratio (MDR). This melt draw ratio can vary between small one digit figures for rather thick filaments, e.g. fishing line applications, up to values well beyond 100 for fine filaments with diameters in the range of 1 to 50 µm. In most applications the fibers undergo a consecutive stretching or drawing stage after solidification. Thus this is called solid-state-drawing (SSD). Herein the filament is usually run between two rollers whereas the second roller is run at a higher speed than the first one. The speed ratio of the two rollers is referred to as the solid state draw ratio (SSDR) [2,11,12]. Usually, the solid-state-drawing is usually performed under elevated temperature levels.

This is usually realized by heated rollers, so that the fibers heat up before and after drawing when in contact with the surface of the rollers. Other principles can facilitate chamber ovens, contact heating plates or overheated steam. The process parameters of the drawing state also have a significant influence on the material structure, the orientation state and also the relaxation state. For example do fibers which are drawn and not properly heat-set a high degree of shrinkage which is usually unacceptable for most applications [2,4].

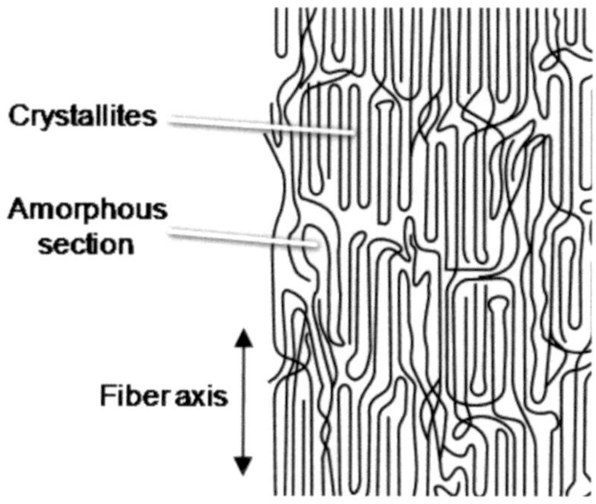

Figure 2. Model for the morphology in a polymeric fiber in allusion to [14]

In both cases the melt drawing as well as the solid state drawing the macromolecules are oriented along the fiber axis (see figure 2). This results in a unique morphology and crystalline structure which can be found only in fibrous materials in this highly oriented state [9,11]. Although there remains some controversy and vivid discussion about the resulting crystalline structure, scientist and researchers agree in the fact that the highly elongated material exhibits deformed crystallites as well as a highly ordered amorphous phase in the non crystalline regions.

For the description of the crystalline structure, most commonly the Stacked-Lamellae and the Shih-Kebab models are quoted [14]. This is usually based on small angle x-ray scattering data, which however do not provide a real image of the structure [15]. A closed theory about the development of the various morphologies is not available. For the thermo-dynamical description of the crystallization the Gibbs free energy is used [16]:

$$\Delta G_m = \Delta H - T \cdot \Delta S \tag{1}$$

Elongated states cannot be discussed within the equilibrium thermodynamics, because of the missing isotropic character of the morphology. Nonetheless, an adoption of the equilibrium theories through consideration of anisotropic influences in analogy to the electro-magnetic field theory is possible [16,17]. In literature [16] within the discussion of entropy elasticity it is described that elongation of a melt through an outside force results in an increase of the free energy of the material. Thus a deformation will result in an increased Gibbs energy ΔG_m in spinning and thus will significantly influence the crystallization of the polymer [16].

For the ideal Gaussian polymer chain this energy contribution corresponds with a reduction of entropy $\delta S_{\Delta L}$ in the elongated state [11]:

$$\Delta G_{m,\Delta L} = \Delta H - T \cdot (\Delta S - \delta S_{\Delta L}) \tag{2}$$

This change in entropy causes an increase of the crystallization temperature as well as an enhancement of crystallization rates. For polymorphic materials simple extensions of this theory are given in [18,19]. The different morphological phases are described through varying energy levels which depend on temperature and state of elongation and strain [19,20]. Depending on the process parameters and the material properties these phenomena are more or less prominent and detectable. Since the degree of crystallinity, the degree of orientation within the molecular structure and tendency of a material to crystallize when stored at temperatures above the glass transition temperature, thermal analysis is one key analytical method to investigate fiber materials, processes and fiber product properties.

2. Experimental method

The following text deals with the experimental investigation of polymeric materials. In this context the method of differential scanning calorimetry (DSC) is described and it will be pointed out what the general procedure is like and which experimental parameters have to be considered.

Differential scanning calorimetry follows the principle of the measurement of heat flow differences. By performing DSC a sample whose temperature is increased gradually and then subsequently cooled down is investigated and finally compared to a reference probe. Therefore it is possible to determine enthalpies and melting points of an arbitrary polymeric fiber. In this context a variation of the involved parameters offers a possibility to draw conclusions about underlying properties such as equilibrium values but concerning the execution of the experiment all of the possibly modifiable parameters have to be regarded carefully to perform DSC correctly. In order to perform DSC a furnace which can be heated up and cooled down homogenously is required. Inside this oven there are two mountings for the samples and each mounting is equipped with a high-sensitive temperature sensor [21,23,24]. The general set up is depicted in figure 3.

One mounting (left mounting in figure 3) is for the crucible which contains the prepared sample. The lid of the crucible has at least one hole to allow an exchange with the surrounding atmosphere. Furthermore, pressure build-up in the crucibles is prevented if parts of the sample vaporize.The other mounting (right mounting in figure 3) is for an empty crucible which functions as a reference. Due to the usage of such a reference only effects caused by the sample itself are observable in the final thermogram. The oven is purged with a gas (sample gas), so that transitions and chemical reactions in different atmospheres can be examined. To avoid oxidation processes a protective gas (e.g. N_2 or Ar) can be used to create an atmosphere around the sample during the process of DSC. Otherwise, air or oxygen can be selected. Furthermore, the space around the oven is purged with a protective gas (N_2) to avoid ice formation at low temperatures [21-24].

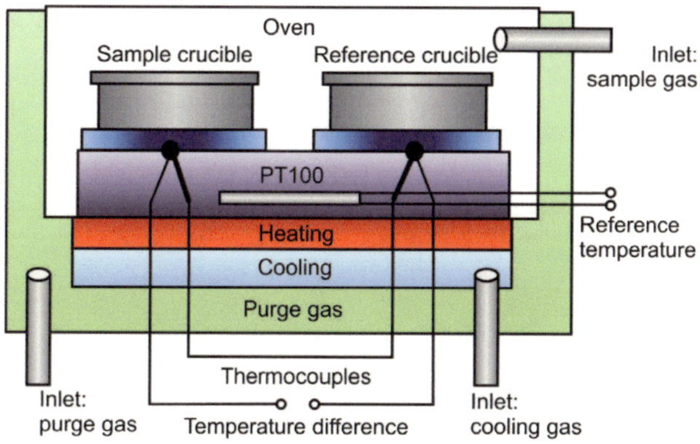

Figure 3. Schematic set up of differential scanning calorimetry

In a special form of differential scanning calorimetry a temperature-modulation is used. Usually one pre-defined frequency is used for temperature modulation [23,24]. The special TOPEM® (Mettler Toledo brand name) technique developed by Mettler Toledo allows the frequency-independent separation of reversing and non-reversing components of the heat flow by analyzing the impulse response of the sample to a pulse of stochastically varied length. Therefore a separation of overlapping effects is possible. Due to this separation extra insight can be gained [22].

3. Sample preparation

The process of sample preparation is of significant importance for the success of experimental investigation of polymeric fibers and has to be handled with great care. Additionally, different aspects of preparation have to be considered simultaneously in order to provide valid and reliable experimental results.

Before beginning with the experimental procedure itself, several aspects have to be dealt with. In general, both granules and fibers are treated in the same manner: The samples have to be reduced to small pieces so that they fit into the crucibles. In this context it is inevitably necessary to consider that the preparation method directly influences the results which are provided by differential scanning calorimetry. Therefore optimum conditions and parameters have to be found in order to determine certain effects (such as glass transition, crystallization or melting). Otherwise these effects still would be observable but not as good as if the optimum conditions were adjusted. Figure 4 conveys an idea which steps are necessary for sample preparation.

The sample has to be reduced to small pieces in order to perform the experiments. Therefore it is possible to alter the form and size of these pieces which yields different results. For example, the reduction to smaller pieces results in different observations concerning the gained thermogram. Usage of a sample with reduced size leads to a decrease of the peak and a lowered melting point. Therefore one can conclude that the mechanical aspect of preparation cannot be neglected and has to be treated carefully. Figures 5 a) and b) deliver an impression which mechanical appearance of the sample has to be chosen in the best case.

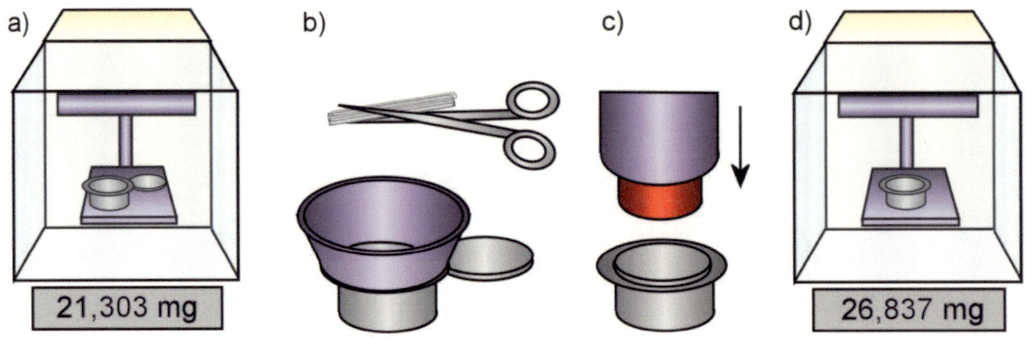

Figure 4. Preparation of fiber samples for DSC:
a) Determination of the empty weight of the crucible
b) Reduction of the fiber to small pieces
c) Insertion of the crucible into cavity
d) Determination of the total weight.

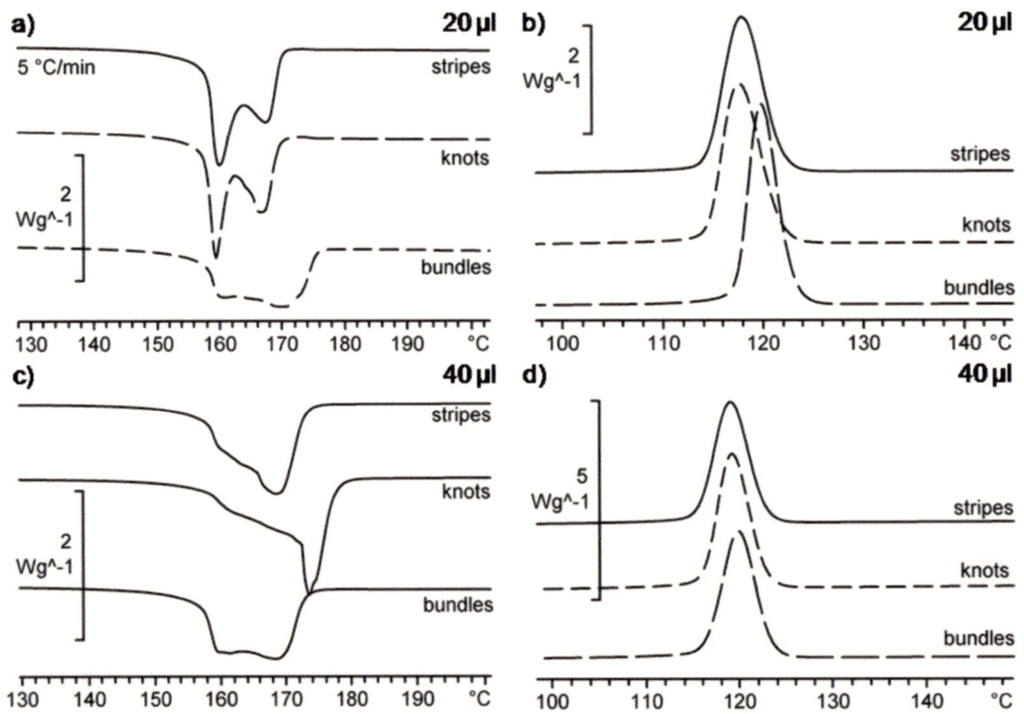

Figure 5. Effect of preparation conditions and crucible size
a) Heating process with crucible volume of 20 µl,
b) Cooling process with crucible volume of 20 µl,
c) Heating process with crucible volume of 40 µl,
d) Cooling process with crucible volume of 40 µl.

Obviously, best results will be achieved if the sample is manufactured into short fiber chops (stripes). Especially knots cause irregularities as an unsteady graph in the thermogram. The explanation for this phenomenon is that stripes offer an increased contact surface whereas knots melt discontinuously. Additionally, this result is independent on the volume of the crucible. In figures 5 c) and d) a similar thermogram is presented but the volume of the crucible is doubled (from 20 μL to 40 μL) and what is apparent here is the fact that again the short fiber chops provide the steadiest graph. Additionally, one can conclude that only the usage of the crucible with 20 μL volume delivers reliable results due to the fact that for the 40 μL crucible the graph inside the thermogram is rather uneven. This effect is caused by the better heat contact in the smaller crucibles, since the lid is pressed to the bottom of the crucible during the preparation process.

Concerning the weighted portion another remarkable effect can be observed: The peak height increases whereas the peak width drastically increases (see figure 6). Furthermore the onset temperature also increases logarithmically as a function of the sample mass. Nevertheless this result is rather obvious due to the fact that for the melting process of a sample with increased weight a higher amount of energy is required than for a sample with less weight. Concerning the interpretation of the results an advantage of less sample weight is that the peak sharpness is increased and therefore overlapping effects can be observed easier. In this context it is necessary to keep in mind that it is possible that thermal events that cause only little effects might be missed. These observations are depicted in the following diagram (figure 6), where the double peak during melting of polypropylene gets more separated for lower sample masses.

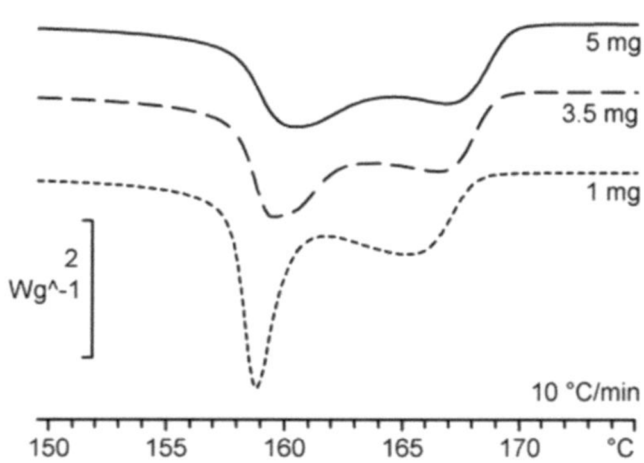

Figure 6. Normalized thermograms for different sample weight of polypropylene fibers

4. Influence of experimental parameters

To illustrate the effects caused by the variation of different parameters, several thermograms are shown in the following paragraph. All these thermograms refer to the investigation of fibers and as an example polypropylene (PP) was used. Firstly, the heating rated is varied as it is depicted in figure 7 a).

Analysis of the thermogram presented above conveys that the alteration of the heating rate causes a strong effect on the results delivered by differential scanning calorimetry. With all other parameters remaining constant the peak height and width increases with increasing heating rate. Moreover, a double peak is of importance. Phase transitions only occur or can be separated from the melting process if lower heating rates are used. For high heating rates they cannot be observed.

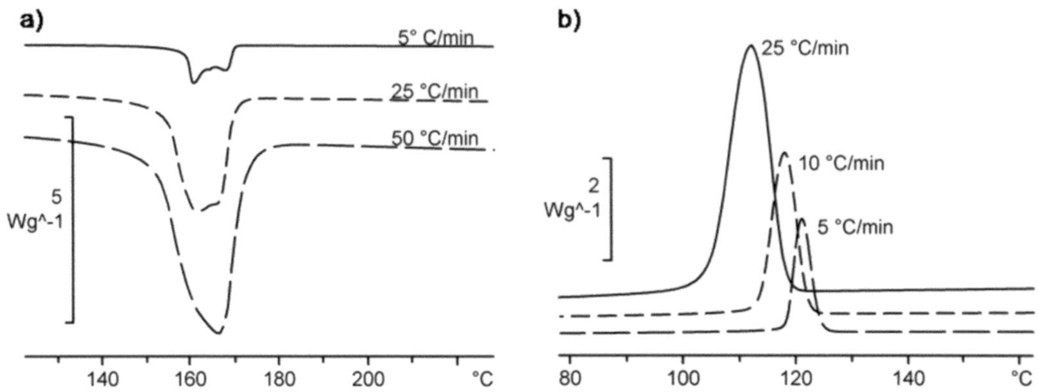

Figure 7. a) normalized thermograms for different heating rates for polypropylene fibers
b) normalized thermograms for different cooling rates of polypropylene granule

Secondly, the variation of the cooling rate is also important as it is depicted in figure 7 b). Beside other observations it is very obvious that the crystallization peak is shifted in positive temperature direction with decreasing cooling rates and the absolute height of the peak is also decreasing. Apparently the correct choice of the cooling rate is as important as it is for the heating rate for the analysis of crystallization processes. By taking into account different cooling rates and extrapolating the peak temperature to isothermal conditions (not measurable in DSC), the true crystallization point can be evaluated.

4.1. Fibers from commodity polymers

All results provided by the last paragraph yield that the alteration of heating and cooling rate causes a strong effect on the results of DSC. Due to this fact it is vital for the success of the analysis to consider the effect of the altering of especially those parameters mentioned above on the properties of a polymeric material which are in the center of interest. Among others the properties crystallization, melting and glass transition shall usually be investigated. In order to support the successful performance of DSC several hints and recommendations for various polymers will be presented in this paragraph. Typical commodity polymers are polypropylene (PP) [25], polyamide (PA6) [26] and polyethylene terephthalate (PET) [27]. In the following paragraph the effects of variations of heating and cooling rate on these polymers will be examined and presented.

Starting with PP (LyondellBasell Moplen HP561R [25]), in the following thermogram the results of DSC with different heating rates are depicted (figure 8 a)).

With increasing heating rate an increase of the peak height and width is noticeable. Additionally, the experiment's velocity is decreasing and the melting process starts earlier.

Noticeable is also the double peak when low heating rates were used. During the melting process of fibers it is likely that a phase transition from α- to γ-phase takes place [28]. The obversation of this transition is depending on the choice of the heating rate because it is possible that the material melts directly or the effect is superimposed by others. Due to the fact that a characteristic amount of energy is necessary, it is possible to observe this transition as a peak in the thermogram if the heating rate is chosen correctly. After complete transition to γ-phase the sample will melt completely and another peak is observable in the thermogram. Now the distance between these mentioned peaks is depending on the heating rate and if this rate is adjusted inappropriately both effects are no longer separate from each other. Another possibility is that no phase transition occurs because the temperature is risen quickly enough to start the melting process directly.

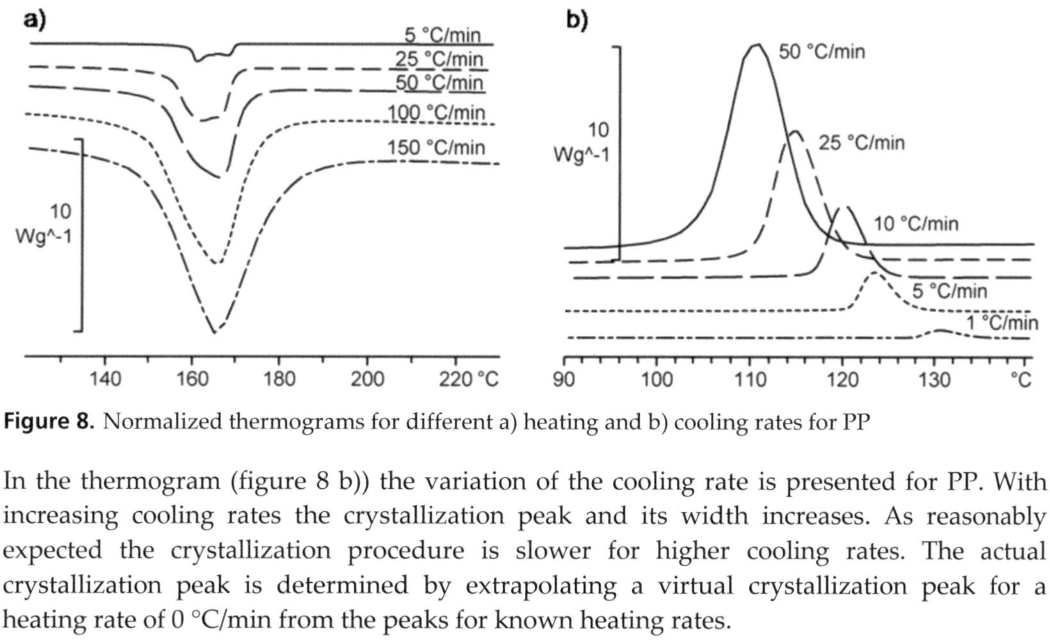

Figure 8. Normalized thermograms for different a) heating and b) cooling rates for PP

In the thermogram (figure 8 b)) the variation of the cooling rate is presented for PP. With increasing cooling rates the crystallization peak and its width increases. As reasonably expected the crystallization procedure is slower for higher cooling rates. The actual crystallization peak is determined by extrapolating a virtual crystallization peak for a heating rate of 0 °C/min from the peaks for known heating rates.

As the previous results were gained from investigations of PP similar experiments were performed for PA6 (BASF Ultramid B24N03 [26]). Beginning with a variation of the heating rate as it is depicted in figure 9 a).

Similar to PP the peak width and height is increased for higher heating rates. In this case another remarkable effect can be observed:

With heating rates higher than 25 °C/min a glass transition is starting below 50°C. This indicates that if heating rates below 25 °C/min were chosen this kinetic transition causes a too small effect and is therefore not visible. If the heating rate is then increased further another peak becomes observable. What is visible here is a phase transition from γ- to α-phase between 120 °C and 160°C [29]. For the observation of these two effects, it is therefore necessary to choose a heating rate above 25 °C/min.

In analogy the cooling rate is varied as it is depicted in figure 9 b). Similar to PP, the crystallization peak shifts to lower temperatures for higher heating rates and true crystallization point can be gained by an extrapolation of the peak temperature. The glass transition point cannot be observed since the actual state of the polymer chains is frozen in during cooling and changes in heat capacity will only be visible by heating up the sample again. Furthermore, the phase transition is also not visible since the material crystallizes in the α phase.

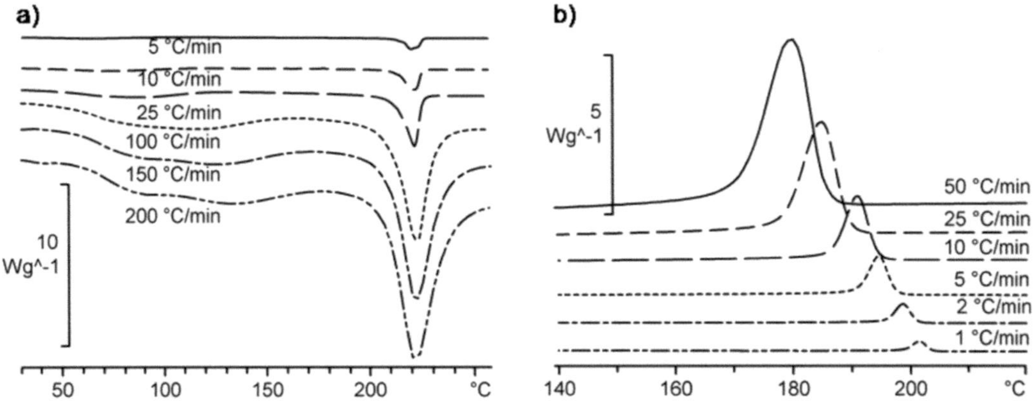

Figure 9. Normalized thermograms for different a) heating and b) cooling rates for PA6

Finally, DSC is performed for PET (Invista Polyester Chips 4048 [27]) and again heating and cooling rates are varied. Firstly, the heating rate is altered (figure 10 a)).

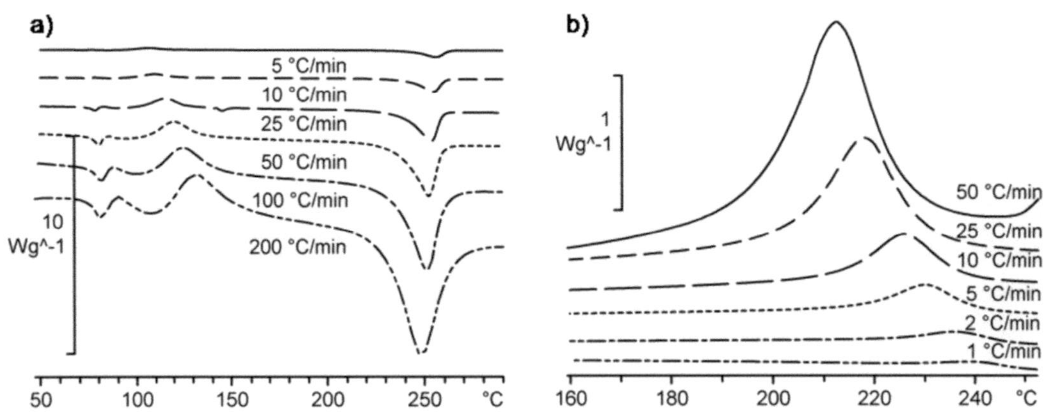

Figure 10. Normalized thermograms for different a) heating and b) cooling rates for PET

Again, the peak width and height increases with higher heating rates. But the glass transition is in this case even more remarkable: Beginning with a heating rate of 25 °C/min this transition becomes visible and is much more pronounced than it was in the investigation of PA6, since the mobilization of polymer chains during the transition has a larger influence on the heat capacity. Furthermore, a relaxation process can be observed shortly after glass transition, indicating that the elongated state of the amorphous polymers chains was frozen in during the rapid cooling in the spinning process. At high heating rates another effect can be observed: after the glass transition and relaxation second crystallization occurs, which pronounces the quenching of the material during the spinning process. After heating over the glass transition point, the mobility of the amorphous or mesomorphous polymer chains increase so that the already present crystallites can grow.

Secondly, the cooling rate is changed for PET sample fibers (figure 10 b)). The analysis of this diagram yields similar results as for the investigation of PP and PA6. With increased cooling rates higher and broader crystallization peaks become visible. Additionally, these peaks are shifted to lower temperatures with increased cooling rates.

4.2. Recommendations

Using the results from the previous paragraph, one can conclude that the appropriate choice of heating and cooling rates is essential for the experiment's success. Therefore, the following table contains valuable information concerning the right choice of these rates in order to gain maximum benefit from performing differential scanning calorimetry of commodity polymers. In this context it is necessary to consider that with too high heating rates some effects may superimpose and therefore might not be visible.

Polymer	Process	Effect	Temperatures	Rates
PP	Heating	Phase transition ($\alpha \rightarrow \gamma$)	156 – 162 °C	< 25 °C/min
PP	Heating	Melting	162 – 170 °C	all
PP	Cooling	Crystallization	100 – 125 °C	all
PA6	Heating	Glass transition	50 – 60 °C	> 25 °C/min
PA6	Heating	Phase transition ($\gamma \rightarrow \alpha$)	100 – 170 °C	> 10 °C/min
PA6	Heating	Melting	200 – 240 °C	all
PA6	Cooling	Crystallization	160 – 200 °C	all
PET	Heating	Glass transition with relaxation	70 – 90 °C	> 25 °C/min
PET	Heating	Crystallization	90 – 160 °C	> 10 °C/min
PET	Heating	Melting	220 – 270 °C	all
PET	Cooling	Crystallization	190 °C – 240 °C	all

Table 1. Recommendations for the right choice of heating and cooling rates

5. Poly(vinylidene fluoride) fibers

Poly(vinylidene fluoride) is a thermoplastic, semicrystalline fluoropolymer with the monomer unit [CF_2-CH_2]. Due to the high fluorine content, it exhibits excellent chemical stability [30]. Furthermore, the polar side groups are responsible for the piezoelectric, pyroelectric and ferroelectric properties of the material, which are only present in one crystalline phase of the polymorphic material, the so called β phase [31,32]. In total, four different crystalline phases can occur [33]. An overview of the crystalline phases, together with the conditions for their formation, is given in table 2.

Crystalline phase	Molecular conformation	Crystalline unit cell (a-b-projection)	Polarity	Conditions for formation
α	TGTG′	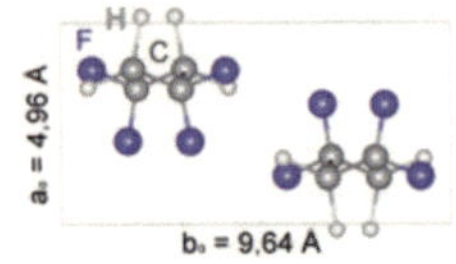	non-polar	Cooled from melt, cast from solution
β	TT	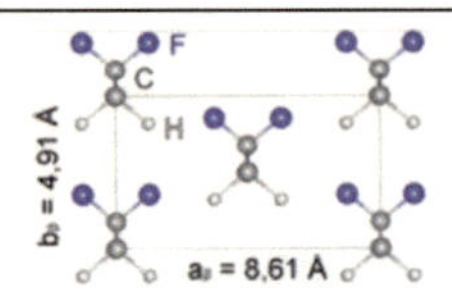	polar (high)	Mechanical stress, high electric field
γ	TTTGTTTG′	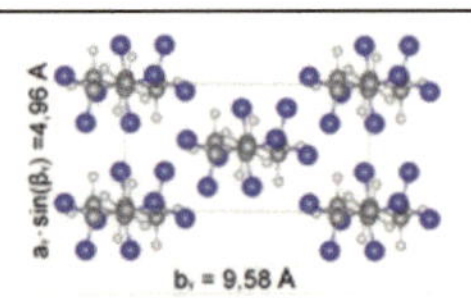	non-polar	Heat treatment, cast from solution
δ	TGTG′	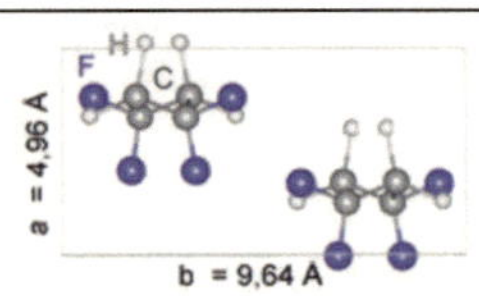	polar (low)	Electric field

Table 2. Crystalline phases of poly(vinylidene fluoride) [34-37]

If the material is present in the β phase, it can be used to create sensors or actuators, which are commonly used in the form of films in microphones, hydrophones or headphones. Here, the necessary process conditions (drawing of the films or high electric fields) for the β phase formation are well known [31,38].

In the case of fibers, the material could potentially be used as sensor or actuators. Possible applications include direction sensitive and spatially resolved strain measurement, which is useful for health monitoring in medical / smart textiles or structural health monitoring in fiber reinforced composites. However, suitable process conditions for fiber spinning, drawing and further processing steps have to be found, whereas β phase crystallites have to be formed and not be destroyed along the process chain [39-42]. Therefore, methods of thermal analysis were developed to identify the presence of the β phase, which are validated by additional X-ray diffraction measurements (WAXD) and dynamic mechanical analysis (DMA) [40,42]. In the following section of this chapter, these methods and their validation will be demonstrated, and they will be applied to gather information about phase transitions during melt spinning, drawing and heat treatment. Therefore, thermal analysis is a powerful tool for process analysis and the development of a process chain for the creation of piezoelectric sensor fibers.

6. Experimental details

Conventional and temperature modulated DSC are carried out on a DSC 1 from Mettler Toledo, Greifensee, Switzerland, equipped with a FRS5 sensor having 56 thermocouples. During the experiments, heating and cooling rates were varied between 1 °C/min and 20 °C/min in a temperature range between -90 °C and 250 °C, which corresponds to 50 °C below the glass transition and 70 °C above the melting point. As checked by thermogravimetric analysis, no weight loss and therefore no polymer degradation occurs in this temperature region during the relevant residence times. For the temperature modulated DSC analysis, TOPEM® technique by Mettler Toledo was used, where the constant heating rate was modulated with heat pulses of 0.5 °C height and stochastically varied length between 15 and 30 s (corresponds to 33.3 and 16.6 MHz). Experimental conditions for DSC analysis are summarized in table 3.

Parameter	Value	Unit
Starting temperature	-90	°C
End temperature	250	°C
Heating / cooling rate	1 / 2 / 5 / 10 / 20	°C/min
Constant heating / cooling rate (TOPEM®)	0.5 / 1 / 2	°C/min
Heat pulse height (TOPEM®)	0.5	°C
Heat pulse length (TOPEM®)	15 - 30	s
Crucible size (granule)	40	µl
Crucible size (fiber)	20	µl
Purge gas volume rate	50	ml/min

Table 3. Parameters for DSC experiments

Results of thermal analysis are compared to the polymer structure and polymer chain orientation, which are determined by wide angle x-ray diffraction (WAXD). Here, a 2D image plate system IPDS II from STOE & Cie GmbH, Darmstadt, Germany, is used for simultaneous analysis of structure and texture. Since the three most important crystalline forms of PVDF α, β and γ have unique diffraction patterns, the method can be easily used to identify the phases. By additional experiments with a heating chamber and heating rates similar to DSC, the underlying phase transitions can be assigned directly to their thermal effect. Since β phase formation can be achieved by mechanical stress, thermal properties of the material are further correlated to dynamic mechanical analysis (DMA), which is carried out on a DMA/SDTA861e by Mettler Toledo. Here, mechanical relaxation processes determined from the phase shift tan(δ) between storage and loss modulus are correlated to their contributions to heat flow.

7. Properties of raw material

In this chapter, some information about the thermal properties and the crystallization behavior of the raw material (Solvay Solef® PVDF 1006 [43]) will be given, which can be later on compared to changes in the fiber material. A typical DSC thermogram (with heating and cooling rate of 10 °C/min) of PVDF quiescently cooled from the melt (with a cooling rate of 1 °C/min) can be found in figure 11. As indicated in the figure, the melting and crystallization peak can be identified clearly.

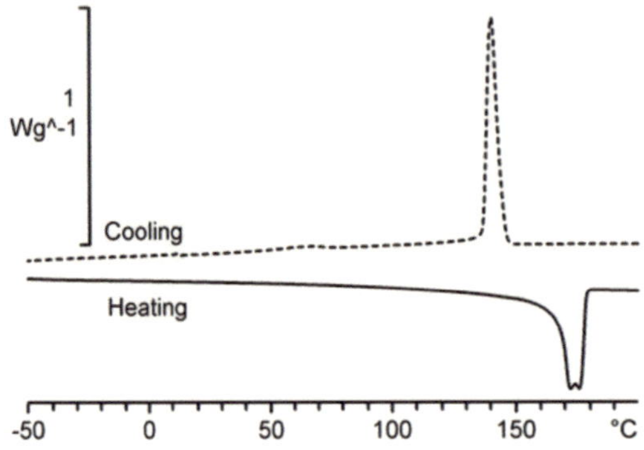

Figure 11. DSC thermogram (heating and cooling) for PVDF bulk material

The endothermal peak correlates to the melting of the material, which takes place in the temperature region between 165 and 180 °C. During the cooling phase, the crystallization (exothermal peak) takes place between 150 and 140 °C. When compared to X-ray data and polarizing microscopy, this behavior can be assigned to a non-textured α phase (figure 12 a), orientation factor f = 0) with spherulitic morphology (figure 12 b)). The material properties are summarized in table 4.

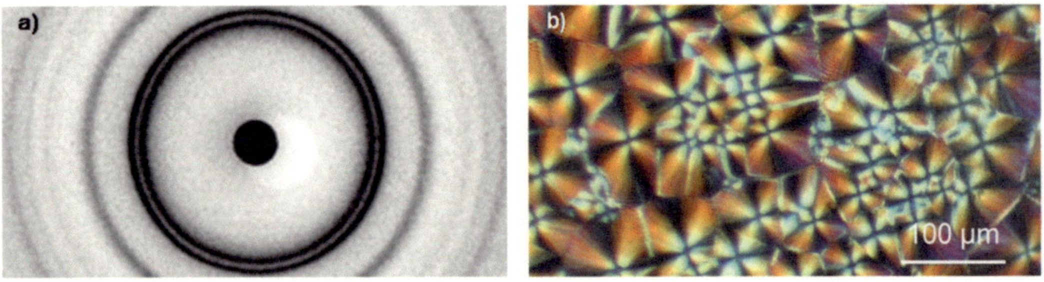

Figure 12. a) Diffraction pattern of quiescently cooled PVDF sample,
b) Polarizing microscopy image of quiescently cooled PVDF sample.

8. Influence of the spinning process

PVDF fibers (multifilaments) are produced in an industrially relevant high speed spinning process. The winding speed is varied between 100 and 2.500 m/min, which correlates to a melt draw ratio of 40 respectively 100. Due to this high draw ratios, polymer chains are oriented and orientation induced crystallization takes place. However, mechanical stress is relatively low during the melt drawing, so that the material crystallizes in a textured α phase (orientation factor f ≈ 1). This information can be extracted from the X-ray data (figure 13 a)).

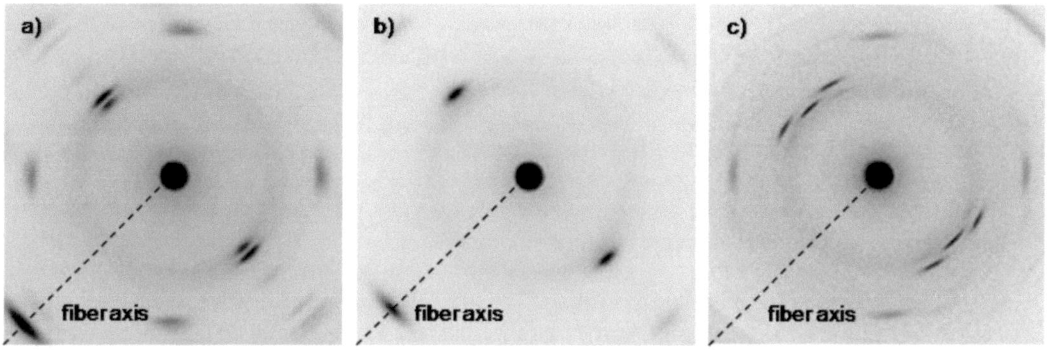

Figure 13. a) Diffraction pattern of melt spun PVDF fiber (winding speed 2.500 m/min),
b) Diffraction pattern of a drawn PVDF fiber (DR = 1.5, drawing T_D = 140 °C),
c) Diffraction pattern of a thermally treated PVDF fiber (T = 165 °C).

Depending on the heating rate, the melting behavior changes in the fiber material. After the melting peak, a second peak occurs (figure 14 a)). By further analysis (section "phase transitions during heat treatment"), this peak correlates to the melting of the γ phase, which is converted from α phase during the heating process. This peak gets larger for lower heating rates, since the process of γ phase conversion has more time to take place. Further evidence for changes in this kinetics can be found in the shape of the melting peak, whereas the first part of the peak becomes larger for lower heating rates. Furthermore, γ phase formation is more promoted in fibers produced with higher winding speeds (figure 14 b)). Here, also the same effect on the melting peak (shift to the first part) can be found.

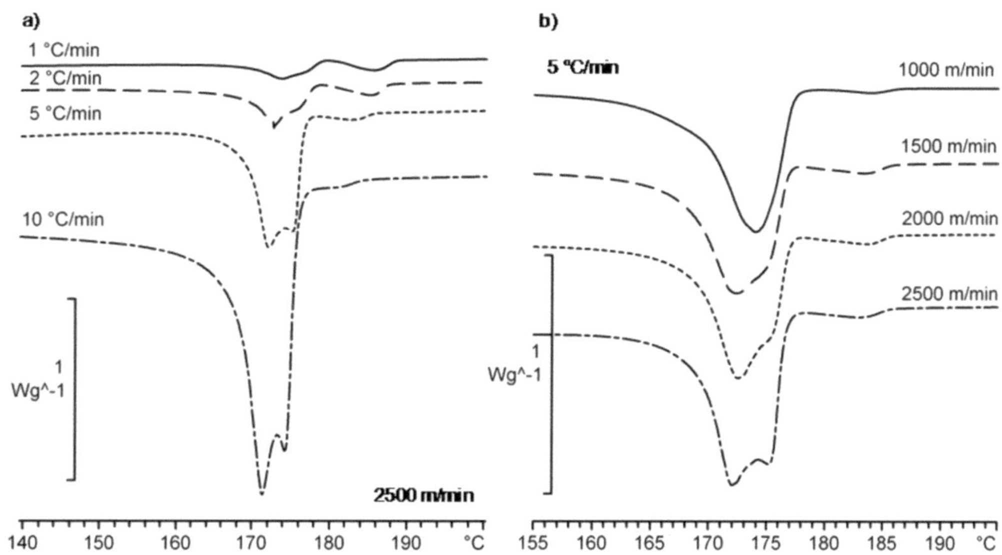

Figure 14. Melting behavior of melt spun PVDF fibers:
a) as a function of the heating rate,
b) as a function of the take up velocity.

Typical values for thermal and structural properties for a melt spun fiber can be found in table 4. Even though the material is cooled down with about 1.000.000 °C/min during fiber spinning, the melting enthalpy is exactly in the same range compared to the slowly cooled sample. This emphasizes the strong enhancement of crystallization rates due to the uniaxial orientation in the spinning process.

9. Influence of the drawing process

In the drawing process, a plastic deformation of the material in the solid state takes place, whereas the drawing ratio DR and drawing temperature T_D can be varied. Starting from the standard process (DR = 1.4, T_D = 140 °C), drawing ratio and temperature are varied separately between 1.0 and 1.6 as well as 40 °C and 160 °C respectively. The effect of both parameters on the melting behavior is displayed in figure 15.

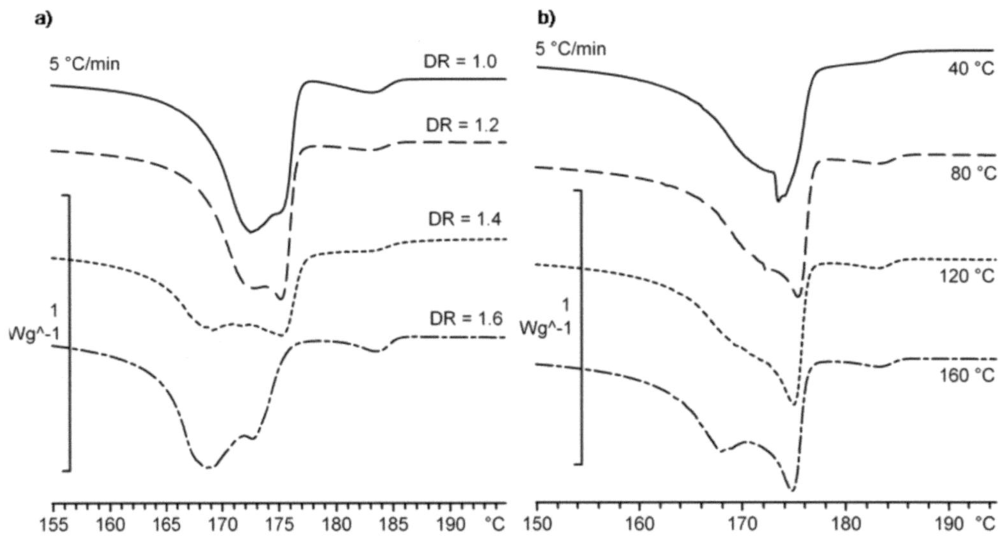

Figure 15. a) Changes in the melting peak as a function of draw ratio,
b) Changes in the melting peak as a function of drawing temperature.

A main effect of the drawing process is a modification of the melting peak, whereas peak temperature shifts to lower temperatures with higher drawing ratios (168 °C for the highest draw ratio compared to 173 °C for an as-spun fiber). In the X-ray diffraction pattern (figure 13 b)), the formation of β phase can be identified, so that the modification of the melting peak correlates to this crystalline phase. Like in the undrawn fiber, the peak coming from γ phase melting can also be found. The same effect can be found by increasing the drawing temperature. The underlying phase transitions during the melting process will be described in detail in the next section. At lower temperature regions, further peaks can be found (figure 16) depending on the drawing temperature. For low temperatures, a signature close after the glass transition point (-35 °C) can be found. A second peak can be found at higher temperatures, which is present in all fibers. For undrawn fibers, the peak temperature is around 55 °C and for drawn fibers about 5 °C higher.

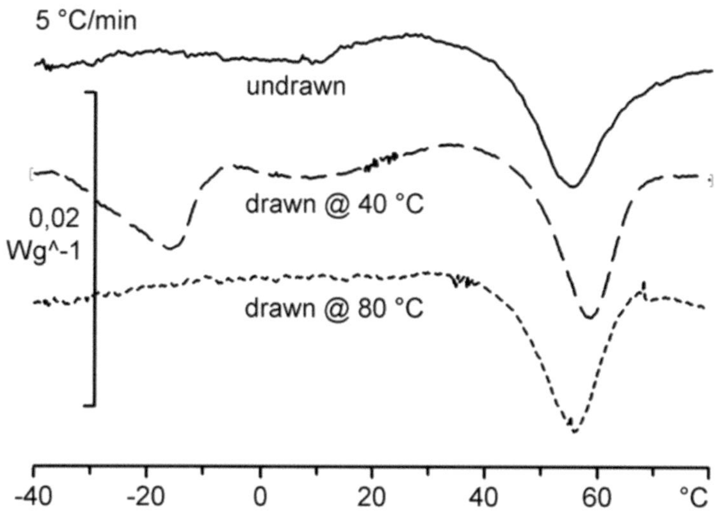

Figure 16. Relaxation processes at different drawing temperatures (draw ratio DR = 1.4)

Both peaks correlate a relaxation process, which can be found in DMA measurements, which are displayed in figure 17. Here, glass transitions can be identified easily. Another relaxation process (known as α_c relaxation in other types of polymers [44]) can be found at 55 °C (α phase) and at 60 °C (β phase), so there is clear evidence for a heat contribution to DSC measurements of this relaxation process. A reason for the occurrence of the first peak in coldly drawn fibers is a drawing temperature below the relaxation temperature, so that more energy is stored in the material and released when heated above glass transition temperature.

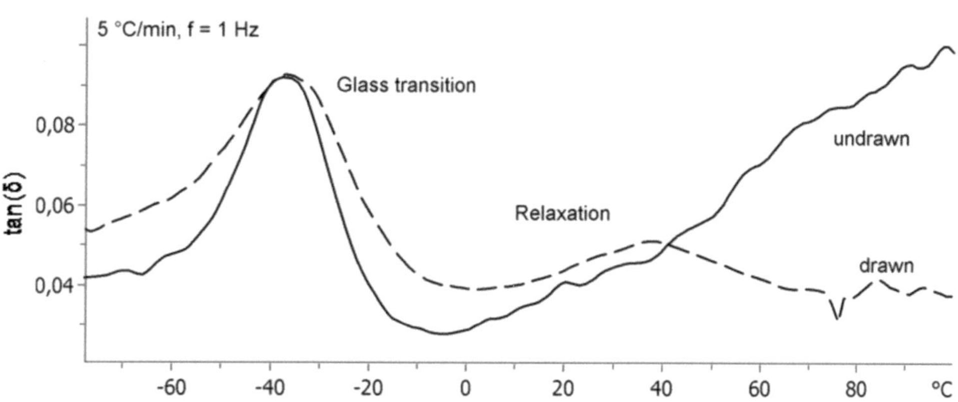

Figure 17. Phase shift tan(δ) in dynamic mechanical response of the material

When compared to X-ray diffraction measurements, β phase amount is drastically increased without affecting overall crystallinity if the fibers are drawn above the relaxation temperature. Therefore, the relaxation process takes place in the α phase and promotes β phase formation. The properties of all fibers are summarized in the table 4.

Parameter	Value	Value	Value	Unit
Type of material	Granules	Melt spun fiber	Drawn fiber	-
Relaxation temperature range	-	36 - 65	40 - 71	°C
Relaxation temperature peak	-	55	60	°C
Relaxation enthalpy	-	1.6	1.3	J/g
Melting range	168 - 180	168 - 185	165 - 185	°C
Melting peak temperature	174	173	168	°C
Melting enthalpy	49.6	48.2	45.8	J/g
Crystallization range	140 - 150	-*	-*	°C
Crystallization peak temperature	145	-*	-*	°C
Crystallization enthalpy	50.0	-*	-*	J/g
Crystalline phase	α	α	β	-
Morphology	spherulitic / non-textured	textured	textured	-

Table 4. Thermal and structural properties of PVDF (*: not detectable since determined by process)

10. Phase transitions during heat treatment

Heat treatment is important in the further processing of the fibers. If they are to be used as piezoelectric sensors, a polarization process has to take place at elevated temperatures without having a reconversion of β phase to α phase. To identify such a transition and explain phase transformations during the melting procedure of the material, temperature modulated measurements can be used. However, the occurrence of different crystalline phases has to be validated X-ray measurements of heated fibers. The results for temperature modulated measurements (TOPEM ®) are displayed in figure 18 for a undrawn and a highly drawn fiber, whereas heat flow is separated into reversing and non-reversing parts.

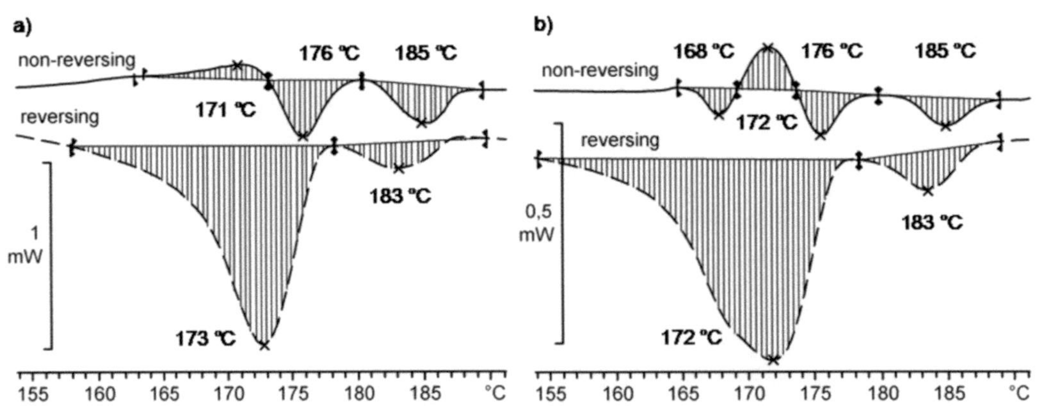

Figure 18. TOPEM ® results with constant heating rate of 2 °C/min (melting peak):
a) PVDF fiber containing α phase,
b) PVDF fiber containing β phase.

For an interpretation of the results, the highly oriented crystalline structures have to be taken into account. They were formed far from equilibrium and therefore must contribute to the non-reversing heat flow. On the other side, energy stored in the solidified amorphous parts due to short-range bonding to neighbor polymer chains contributes to the reversing heat flow. However, also crystalline phases contribute to the reversing heat flow if time scale of the phase transitions is larger than the pulse length applied to the sample. In total, 3 (α phase fiber) or 4 (β phase fiber) transformations in the crystalline regions take place. The same amount of phase transitions can also be found in X-ray measurements. The additional peak in the β phase fibers is caused by a reconversion to the α phase. After this transition, the changes in the crystalline structure are the same for both fibers, starting from α phase conversion to γ phase, going on with the melting of the remaining α crystallites and finally ending with the melting of the γ phase. Since the melting of the α phase takes place shortly after the conversion to γ phase, the amount of γ phase (compare to figure 14) strongly depends on the heating rate. All phase transitions during heat treatment are summarized in table 5.

Transition	Temperature range (peak)	Heat flow
$\beta \rightarrow \alpha$	165 °C – 170 °C (168 °C)	endothermal
$\alpha \rightarrow \gamma$	166 °C – 173 °C (171 °C)	exothermal
$\alpha \rightarrow$ melt	173 °C – 180 °C (176 °C)	endothermal
$\gamma \rightarrow$ melt	180 °C – 190 °C (185 °C)	endothermal

Table 5. Summary of phase transitions in PVDF fibers during heat treatment

11. Overview over phase transitions

For the selection of the right process settings for the production of piezoelectric fibers, it is necessary to get an overview of all phase transitions which can occur in the processes (melt spinning, solid state drawing and heat treatment/polarization). All these transitions were described in the previous sections and are summarized in figure 19.

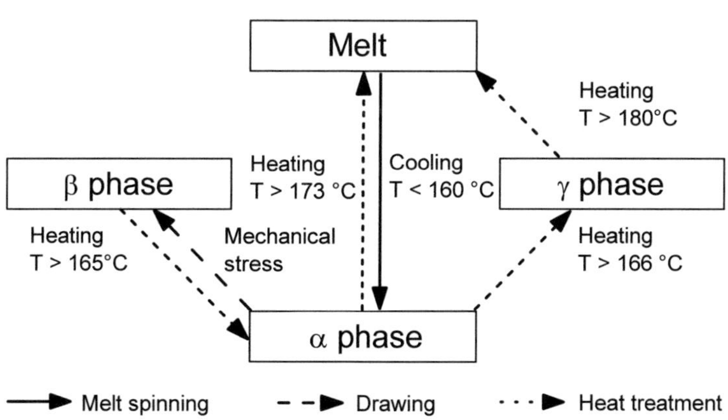

Figure 19. Overview of phase transitions in poly(vinylidene fluoride) fibers

After melt spinning, the fibers should be drawn at high ratios (close to maximum elongation) to form as much β phase fraction as possible. Drawing temperature should be at least above 55 °C to enhance β phase formation and prevent relaxation processes. For further polarization processes, temperature should be kept below 160 °C to keep the material in the β phase. For the process development, calorimetric measurements are of great value to identify the phase transitions, so that they can be realized in the process by choosing the right parameters. Furthermore, thermal analysis is of great value for further quality control, since the present crystalline phases can be detected by a simple method (compared to techniques like X-ray diffraction).

12. Carbon nanotube composite fibers

Carbon nanotubes (CNT) are allotropes of pure carbon in a form of a cylindrical structure. Depending on the number of graphite sheets forming the tube, they are divided into single wall (SW-CNT) and multi wall carbon nanotubes (MW-CNT). Beside excellent mechanical properties, CNT offer high electrical and thermal conductivity. When added to a polymer matrix, their properties are partially transferred to the polymer nanocomposite material [45]. In the case of mechanical reinforcement, an increase of Young's modulus can be observed in many polymer matrices [45,46]. One of the most interesting aspects is the formation of electrical conductive paths, so called percolative networks, in otherwise insulating materials [45,47]. Compared to other conductive fillers like carbon black (CB), the amount of CNT for reaching electrical conductivity is extremely low, whereas percolation thresholds below 0.1 % were observed [48].

However, not only the desired material properties in solid state change by the additivation of CNT. The nanoparticles interact with the polymer matrix and influence the rheological properties of the polymer melt, but also the crystallization behavior [46]. Especially in the melt spinning process, process settings have to be adjusted to allow the production of CNT modified nanocomposite fibers [49-53]. Therefore, the nancomposites have to be analyzed by DSC to understand the effects of CNT on the different polymer matrices, so that the right process parameters for fiber production can be found. Furthermore, thermal analysis provides useful information about the functional properties of the material, since crystallization conditions have a large influence on the electrical conductivity.

13. Experimental conditions

For each polymer, conventional DSC was carried out at least 50 °C below the glass transition (except of polyethylene, where glass transition is too low) and 50 °C above the melting point. Heating rate was varied between 2 °C and 20 °C, whereas results with 5 °C/min are shown in this chapter, since all effects to be demonstrated can be found at this heating rate.

Electrical properties of the composite materials where checked with a LCR meter, whereas the materials were tested in a frequency range between 1 Hz and 100 kHz. By checking AC conductivity, effects of partially not connected conductive networks can be found.

Furthermore, the values determined at low frequencies can be considered as the DC conductivity of the material.

14. Compounding in different polymer matrices

Depending on the type of polymer, CNT can influence the crystallization process in two different ways. Both effects will be described in this section with the help of the examples polypropylene (PP, Basell Moplen HP561R [25]) and polyamide 6 (PA6, BASF B24N03 [26]).

In PP (compare figure 20 a)), crystallization peak shifts to higher temperatures with increasing amount of CNT. This effect has two reasons. First, the high thermal conductivity of CNT allows the latent heat to be transferred faster from the polymer to the surrounding medium, to that sample temperature is closer to the reference temperature compared to unmodified samples. However, this effect shifts the peak only about 0.5 °C per w% CNT added to the polymer. The more dominant effect is the acting of CNT as foreign substance in the polymer, on which crystal nuclei can be formed at higher temperatures. After this nucleation, crystallites can grow in the usual way. By adding only 1 w% of CNT, this effect shifts the melting peak by 5 °C to higher temperatures.

The effect in PA6 is different (see figure 20 b)). The form of the crystallization peak changes to that a double peak can be found. Here, the part of the peak at lower temperatures (which is also shifted compared to unmodified material), has the same origin like in PP. Here also higher crystallization temperature is caused by enhanced thermal conductivity and crystal nucleation on the CNT. The part of the peak at higher temperatures is caused by another effect, the direct crystal growth on the CNT. The particles act as nuclei for the polymer chains and the crystallites can grow at higher temperatures, since no undercooling is needed for crystal growth compared to nucleation.

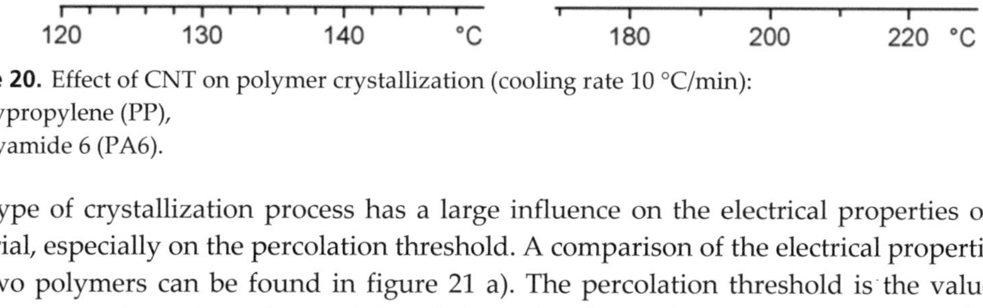

Figure 20. Effect of CNT on polymer crystallization (cooling rate 10 °C/min):
a) Polypropylene (PP),
b) Polyamide 6 (PA6).

The type of crystallization process has a large influence on the electrical properties of the material, especially on the percolation threshold. A comparison of the electrical properties of the two polymers can be found in figure 21 a). The percolation threshold is the value for CNT concentration, where the specific resistivity drastically decreases over several orders of magnitude. Above the threshold, resistivity only decreases slowly due to the higher amount of conductive filler. For polypropylene, the threshold can be found at 3 w% compared to 7 w% for polyamide 6.

For polypropylene, CNT act as nucleation seeds, but the material has no chemical affinity to the nanoparticles. Therefore CNT try to form aggregates in the polymer matrix, which can build conductive paths through the whole fiber. The region between the aggregates is then filled by PP crystallites during the crystallization process. Because of the separation of the different regions, only lower amount of CNT is needed to form conductive paths. Therefore, the dynamic percolation threshold defined by the crystallization process is significantly lower than it would be for a uniform distribution of the particles in the fiber. This behavior can be observed with transmission electron microscopy (figure 21 b)).

For polyamide 6, where polymer crystallites directly grow on the particles, a chemical affinity between the components is given and single CNT are separated by the polymer matrix (see figure 21 c)). Therefore, CNT are well distributed in the polymer and the percolation threshold is defined by the geometry of the nanoparticles (length and diameter) as well as their orientation to the fiber matrix. Single CNT in PA6 can be oriented better compared to the aggregates in PP, so that the resistivity of a fiber is highly sensitive to the process parameters in the spinning and drawing process.

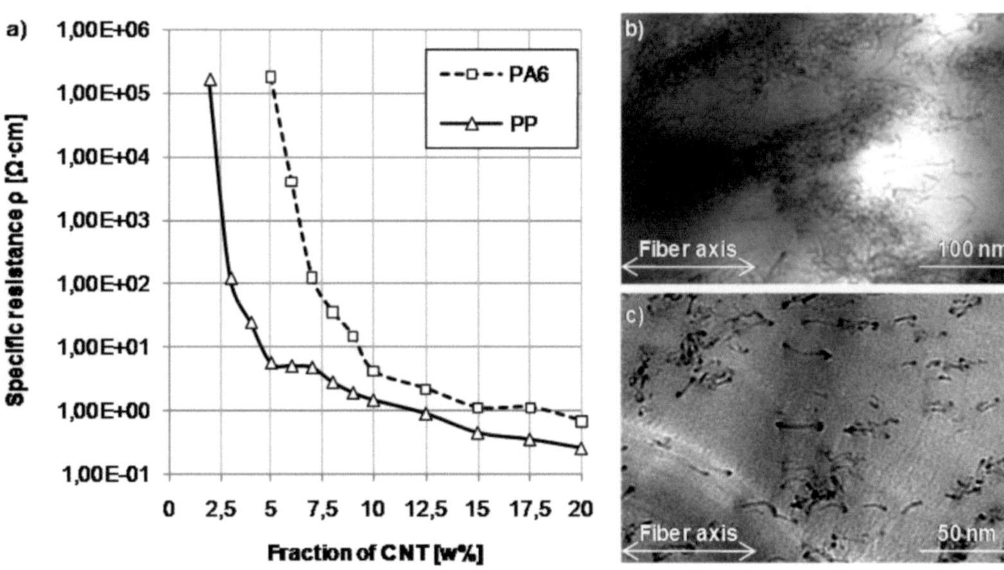

Figure 21. a) Electrical conductivity as a function of CNT weight fraction in different polymer matrices (PP and PA6),
b) Transmission electron microscopy image of 10 w% CNT in PP,
c) Transmission electron microscopy image of 5 w% CNT in PA6.

15. Influence of the spinning process

As described above, CNT have a major influence on the crystallization behavior in the different polymers. Even though the material is cooled down much faster during melt spinning and orientation induced crystallization is dominating the structure formation in the material, crystallization phenomena are drastically altered due to the presence of the nanoparticles. An example for the changes in the crystallization of polypropylene can be found in figure 22. If the material is heated above the relaxation point (approx. 50 °C), recrystallization phenomena can be found in the unmodified fibers. If CNT are doped to the PP matrix, no recrystallization peak can be found. Since the melting enthalpy is constant for different CNT fractions compared to the raw material (after recrystallization), it can be concluded that crystallization rates are enhanced by the presence of the particles. This observation is correlating with the fact, that the material cannot be drawn at high ratios in the molten state if CNT are present.

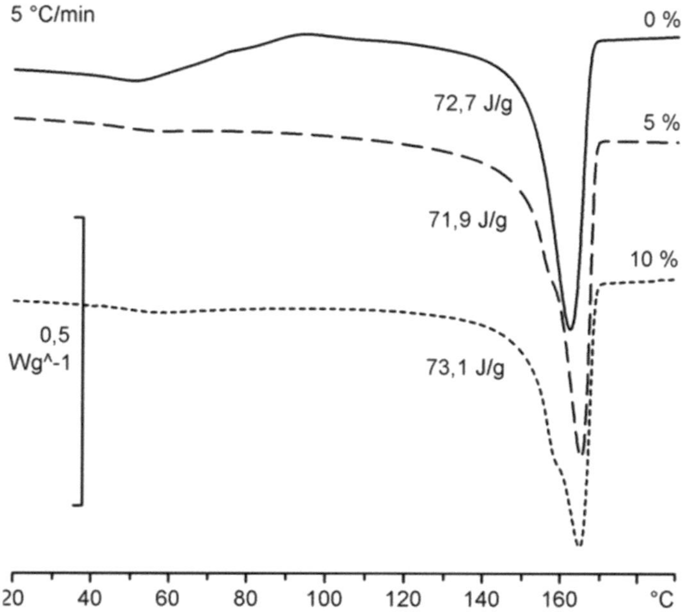

Figure 22. Changes in the melting behavior as a function of CNT concentration in polypropylene fibers spun with a winding speed of 50 m/min

In polyamide 6, the thermal property affected mostly is the glass transition. This effect is displayed in figure 23. By adding higher amounts of CNT, the glass transition temperature shifts to higher values. Thus a T_G of 60 °C is reached for a concentration of 10 w% compared to 52 °C for the same type of unmodified material. Since polymer crystallites grow on the CNT surface, the material is the polymer matrix has a better chemical bonding to the nanoparticles and the polymer chain mobility is lowered due to their presence. Therefore, more energy is needed to mobilize the polymer chains resulting in a higher glass transition temperature.

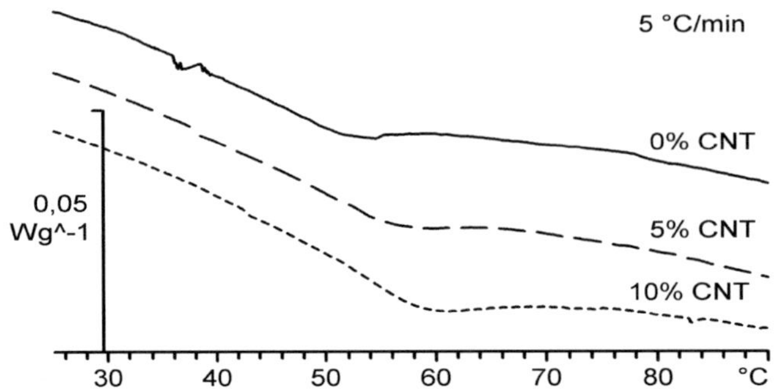

Figure 23. Changes in the glass transition as a function of CNT concentration in polyamide 6 fibers spun with a winding speed of 50 m/min

16. Effects in blend systems

In blend systems, two or more polymers are mixed together without forming a mutual polymer chain. If both components are not compatible with each other, they are separated in the fiber material, but influence each other [54]. Depending on the chemical structure of the polymers, CNT are attracted more by one of the components and aggregate in this material. This effect will be explained on a mixture of polypropylene (PP, LyondellBasell Moplen HP561R [25]) and low-density polyethylene (LDPE, LyondellBasell Lupolen 1800S [55]). The crystallization and melting behavior can be found in figure 24. In the crystallization process, the PP peak shifts by the additivation of CNT, which indicates the nucleation of PP

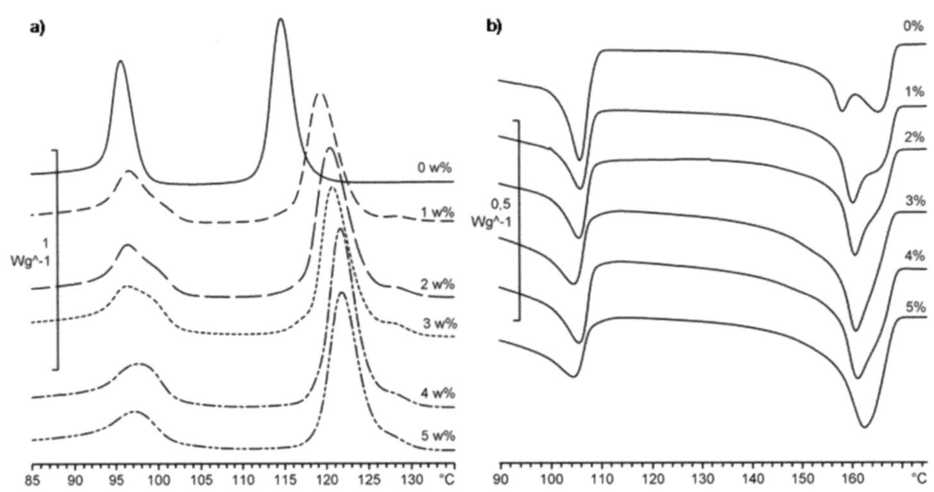

Figure 24. Effect in blend systems of polypropylene / polyethylene (1:1, heating/coolingrate 5 °C/min) as a function of CNT concentration:
a) Crystallization phenomena during cooling of bulk material,
b) Melting phenomena during heating of fibers.

on the CNT and therefore the presence of nanotubes in this polymer. For LDPE, there is no temperature shift of the crystallization peak. However, the peak gets broader. This indicates major changes in the phase separation behavior, but without changing the crystallization process by the nanotubes. Possible reasons for the broadening are more diverse sizes of regions of LDPE in PP, which tend to crystallize at different temperatures. Further evidence of the aggregation of CNT in PP can be found in the melting process. While the form of melting peak of LDPE is not changed at all, the PP changes from a double peak to a single peak. Since the occurrence of the double peak indicates a phase transformation before the actual melting procedure, it can be concluded that the present crystalline phase (α phase) is stabilized by the presence of CNT and the nanotubes are situated in the PP phase of the blend system.

17. Impact on the development of electrically conductive fibers

As stated above, the choice of the base polymer as well as the melt spinning of the fibers has a large influence on the electrical conductivity. The base materials can be classified into two groups. In polymers like polypropylene, CNT act as nucleation seeds. The particles can aggregate and easily form conductive paths, so that a low percolation threshold can be observed. This class of materials is very useful for the production of electrically conductive fibers with high spinning speeds, since electrical conductivity is less affected by the spinning process and possible mechanical treatment during the use of the fibers. For polymers like polyamide 6, crystallites grow on the surface of CNT, so that particles are separated and the percolation threshold is higher. Since single CNT can be oriented easily, the spinning and drawing process has a larger influence on the conductivity. This makes the processing of the materials more instable. However, they can potentially be used to create sensors, whereas mechanical stress and deformation can be detected by changes in the resistivity.

18. Conclusion

Taking into account the unique morphology of polymeric fibers, special analysis methods are needed for the development of new materials and processes. Due to high deformation of the molten polymer in the spinning process as well as the solid-state deformation during drawing, the polymer chains are highly oriented along the fiber axis. Differential scanning calorimetry, especially with temperature modulation, is one of the most important tools for fiber development. It can be used for a fundamental study of new materials and their behavior during spinning and processing as well as quality control of commidity polymers.

For the analysis of commodity polymers, the effect of sample preparation and experimental parameters is demonstrated. For the preparation of samples, small crucibles with fibers cut into small pieces are useful to measure thermograms with clearly visible effects. The choice of parameters has a large influence on the thermal effects observable in the results. Depending on the heating and cooling rate, effects like glass transition, structural phase transitions, melting and crystallization can be revealed from the thermograms. For the observation of these effects in the most common fiber polymers (polypropylene, polyamide 6 and poly(ethylene terephthalate)), recommendations for experimental parameters are given and the temperature ranges for these effects are indicated.

The development of new spinning processes is demonstrated for poly(vinylidene fluoride), whereas DSC can be used to detect the formation of the piezoelectric β phase. With the help of temperature modulated DSC compared to X-ray diffraction and dynamic mechanical analysis, a process chain for the generation of piezoelectric sensor fibers with possible spinning and drawing parameters was developed. Since the formation of the piezoelectric crystallites can be detected by DSC, it is powerful tool for future development of piezoelectric sensors and actuators.

Electrically conductive nanocomposites based on carbon nanotubes are used as an example for the development of new materials. The presence of nanoparticles influences the crystallization process dependent on the chemical structure of the polymer matrix, whereas two different mechanisms can be detected. The resulting structure then determines the percolation threshold for electrical conductivity. If polymer crystallites can directly grow on the CNT, the particles are separated and percolation threshold is only determined by the particle geometry and orientation. In the other case, the affinity of the polymer chains to the CNT is lower, so that they act as nucleation seeds. CNT can then aggregate and form conductive paths. Due to the higher separation of the two components, percolation threshold is lower.

Author details

W. Steinmann*, S. Walter, M. Beckers, G. Seide and T. Gries
Institut für Textiltechnik (ITA) der RWTH Aachen University,
Aachen, Germany

Acknowledgement

Special thanks to the Deutsche Forschungsgemeinschaft (German research foundation, DFG) for funding the project GR 1311/10-2.

19. References

[1] Engelhardt A (2011) Global synthetic industrial filament yarn and fiber markets. Chemical Fibers International. j. 61: 122.
[2] Fourné F (1995) Synthetische Fasern: Herstellung, Maschinen und Apparate, Eigenschaften; Handbuch für Anlagenplanung, Maschinenkonstruktion und Betrieb. München: Hanser. 880 p.
[3] Gries T, Sattler H (2005) Chemiefasern. In: Winnacker K, Küchler L, editors. Chemische Technik. Weinheim: Wiley
[4] Wulfhorst B, Gries T, Veit D (2006) Textile Technology. München: Hanser. 320 p.

* Corresponding Author

[5] Walter S, Steinmann W, Gries T, Seide G, Schenuit H, Roth G (2010) Production of textile fabrics from PVDF multifilament yarns with textile titer. Technical Textiles. j. 53: E95-E97.

[6] Walter S, Steinmann W, Gries T, Roth G, Seide G, Schenuit H. Tools for savers of life : production of warp-knitted fabrics from PVDF multifilament yarns of textile fineness. Kettenwirk-Praxis. j. 45: 34-36.

[7] Walter S, Steinmann W, Seide G, Gries T, Roth G (2012) Development of innovative fibre materials for technical applications : fine polyvinylidene fluoride filaments and fabrics. Filtration. j. 12: 60-64.

[8] Michaeli W (2010) Einführung in die Kunststoffverarbeitung. München: Hanser. 256 p.

[9] Ziabicki A (1976) Fundamentals of fiber formation. London: Wiley. 504 p.

[10] Ziabicki A, Kawai H (1985) High-Speed Fiber Spinning – Science and Engineering Aspects. London: Wiley. 586 p.

[11] Salem D (2000) Structure formation in polymeric fibers. München: Hanser. 578 p.

[12] Beyreuther R, Brünig H (2007) Dynamics of fibre formation and processing. Berlin: Springer. 365 p.

[13] Nakajima T (1994) Advanced fiber spinning technology. Cambridge: Woodhead. 276 p.

[14] Prevorsek D, Oswald H (1990) Melt-spinning of PET and nylon fibers. In: Schultz J, Fakirov F, editors. Solid state behavior of linear polyesters and polyamides. Englewood Cliffs: Prentice Hall.

[15] Chu B, Hsiao B (2001) Small-Angle X-ray Scattering of Polymers. Chemical Reiews. j. 101:1727-1761.

[16] Mandelkern L (2002) Crystallization of Polymers – Volume 1 equilibrium concepts. Cambridge: University Press. 448 p.

[17] Fließbach T (2006) Statistische Physik. München: Elsevier. 408 p.

[18] Strobl G, Cho T (2007) Growth kinetics of polymer crystals in bulk. European Physics Journal. j. 23:55-65.

[19] Banik N, Boyle F, Sluckin T, Taylor P (1979) Theory of structural phase transitions in poly(vinylidene fluoride). Physical Review Letters. j. 43: 456-460.

[20] Strobl G (2007) A multiphase model describing polymer crystallization and melting. Lecture Notes in Physics. j. 714: 481-502.

[21] Frick A, Stern C (2006) DSC-Prüfung in der Anwendung. München: Hanser. 164 p.

[22] Schawe J, Hütter T, Heitz C, Alig I, Lellinger D (2006) Stochastic temperature modulation: A new technique in temperature-modulated DSC. Thermochimica Acta. j. 446:147-155.

[23] Cheng S (2002) Handbook of Thermal Analysis and Calorimetry, Volume 3: Applications to Polymers and Plastics. Amsterdam: Elsevier. 828 p.

[24] Hatakeyama T, Quinn F (2000) Thermal Analysis: Fundamentals and Applications to Polymer Science. London: Wiley.

[25] LyondellBasell (2009) Product Date and Technical Information Moplen HP561R. Rotterdam: LyondellBasell.

[26] BASF Corporation (2011) Datasheet Ultramid B24N03. Ludwigshafen: BASF Corporation.

[27] Invista Resins & Fibers GmbH (2009) Product Specification Polyester Chips 4048. Gersthofen: Invista Resins & Fibers GmbH.

[28] Hsiao B (2011) Polymorphism, Preferred Orientation and Morphology of Propylene-Based Random Copolymer Subjected to External Force Fields (dissertation). State University of New York at Stony Brook.

[29] Liu Y, Cui L, Guan F, Gao Y, Hedin NE, Zhu L, Fong H (2007) Crystalline Morphology and Polymorphic Phase Transitions in Electrospun Nylon 6 Nanofibers. Macromolecules. j. 40:6283-6290.

[30] Drobny J (2001) Technology of fluoropolymers. Boca Raton: CRC Press. 227 p.

[31] Nalwa H (1995) Ferroelectric Polymers – Chemistry, Physics and Applications. New York: Marcel Dekker. 912 p.

[32] Herbert J (1982) Ferroelectric transdurcers and sensors. New York: Gordon and Breach. 464 p.

[33] Lovinger A (1983) Poly(vinylidene fluoride). Developments in Crystalline Polymers. j. 1:195.

[34] Takahashi Y, Matsubara Y, Tadokoro H (1983) Crystal structure of form II of poly(vinylidene fluoride). Macromolecules. j. 16:1588-1792.

[35] Hasegawa R, Takahashi Y, Chatani Y, Tadokoro H (1972) Crystal structures of three crystalline forms of poly(vinylidene fluoride). Polymer Journal. j. 3:600-610.

[36] Weinhold S, Litt M, Lando J (1980) The crystal structure of the gamma phase of poly(vinylidene fluoride). Macromolecules. j. 13: 1178-1183.

[37] Takahashi Y, Tadokoro H (1980) Crystal structure of form III of poly(vinylidene fluoride)

[38] Wang T, Herberg J, Glass A (1988) The Application of Ferroelectric polymers. Glasgow: Blackie and Son. 304 p.

[39] Du C, Zhu B, Xu Y (2007) Effect of stretching on crystalline phase structure and morphology of hard elastic PVDF fibers. Journal of Applied Polymer Science. j. 104:2254-2259.

[40] Steinmann W, Walter S, Seide G, Gries T, Roth G, Schubnell M (2011) Structure, properties, and phase transitions of melt-spun poly(vinylidene fluoride) fibers. Journal of Applied Polymer Science. j. 120:21-35.

[41] Lund A, Hagström B (2010) Melt-spinning of Poly(vinylidene fluoride) Fibers and Influence of Spinning Parameters on β-phase crystallinity. Journal of Applied Polymer Science. j. 116:2685-2693.

[42] Walter S, Steinmann W, Schütte J, Seide G, Gries T, Roth G, Wierach P, Sinapius M (2011) Characterisation of piezoelectric PVDF monofilaments. Materials Technology. j. 26:140-145.

[43] Solvay Solexis (2003) SOLEF 1006 PVDF Homopolymer (data sheet). Brussels: Solvay.

[44] Hougham G (1999) Fluoropolymers – Volume 2 Properties. New York: Kluwer. 408 p.

[45] Grady B (2011) Carbon Nanotube Polymer Composites. Hoboken: Wiley. 352 p.

[46] Lee S, Kim M, Kim S, Youn J (2008) Rheological and electrical properties of polypropylene/MWCNT composites prepared with MWCNT masterbatch chips. European Polymer Journal. j. 44:1620-1630.

[47] Wescott J, Kung P, Maiti A (2006) Conductivity of carbon nanotube polymer composites. Applied Physics Letters. j. 90.

[48] Alig I, Pötschke P, Pegel S, Dudkin S, Lellinger D (2008) Plastic composites containing carbon nanotubes: Optimisation of processing conditions and properties. Rubber Fibre Plastics. j. 3:92-95.

[49] Steinmann W, Walter S, Seide G, Gries T (2011) Melt spinning of electrically conductive bicomponent fibers. In: Adolphe D, Schacher L, editors: 11th World Textile Conference AUTEX 2011, 8-10 June 2011, Mulhouse, France. Book of Proceedings, Volume 2. Mulhouse : Ecole Nationale Supérieure d'Ingenieurs Sud-Alsace. pp. 716-721.

[50] Wulfhorst J, Steinmann W, Walter S, Seide G, Heidelmann M, Weirich T, Gries T (2011) Nanoadditivation of meltspun filament yarns. In: ICONTEX 2011 International Congress of Innovative Textiles, 20-22 October 2011. Çorlu/Tekirdağ : Namik Kemal University. pp. 39-39.

[51] Wulfhorst J, Steinmann W, Walter S, Seide G, Gries T (2011) Antibacterial behaviour and electrical conductivity of textiles by melt spinning of yarns with incorporated nanoparticles. In: Lahlou M, Koncar V, editors: Book of Abstracts / 3rd Edition of the International Conference on Intelligent Textiles and Mass Customisation ITMC'2011, October 27, 28 & 29, 2011, Casablance & Marrakesh, Morocco. Casablanca: ESITH . p. 33.

[52] Skrifvars M, Soroudi A (2009) Melt Spinning of Carbon Nanotube Modified Polypropylene for Electrically Conducting Nanocomposite Fibers. Solid State Phenomena. j. 151:43-47.

[53] Liu K, Sun Y, Lin X, Zhou R, Wang J, Fan S, Jiang K (2010) Scratch-resistant, highly conductive and high-strength carbon nanotubes-based composite yarn. ACS Nano. j. 10:5827-5834.

[54] Kyrylyuk A, Hermant M, Schilling T, Klumperman B, Koning C, van der Schoot P (2011) rolling electrical percolation in multicomponent carbon nanotube dispersions. Nature Nanotechnology. j. 6:364-369.

[55] LyondellBasell (2011) Product Date and Technical Information Moplen HP561R. Rotterdam: LyondellBasell.

Section 5

Indirect Calorimetry to Measure Energy Expenditure

Chapter 13

Energy Expenditure Measured by Indirect Calorimetry in Obesity

Eliane Lopes Rosado, Vanessa Chaia Kaippert and Roberta Santiago de Brito

Additional information is available at the end of the chapter

http://dx.doi.org/10.5772/55605

1. Introduction

Obesity is a multifactorial disease characterized by excessive deposition of fat in adipose tissue, which may be due to excessive energy intake, and or changes in body energy expenditure, resulting in positive energy balance [1].

Mika Horie et al [2] demonstrated that obese women had higher total energy expenditure (TEE), compared with normal weight. However, this increase may be due to increased basal metabolic rate (BMR) due to higher fat-free mass (FFM) and energy demand during physical activity. However, Melo et al [3] found that obese individuals are economical, the metabolic point of view. Therefore, the energy expenditure (EE) per kilogram of body weight at the give time is lower in obese individuals.

The low metabolic rate, expressed relative to FFM seems to be a risk factor for weight gain [4]. In a prospective study in Pima Indians, Ravussin et al. [5] showed that both the low resting metabolic rate (RMR) and low TEE increased risk of weight gain. The basal energy expenditure (BEE) and resting (REE) can be obtained through BMR and RMR, respectively, multiplied by 24 hours (1440 minutes).

There are several methods for the assessment of EE with different levels of precision, including indirect calorimetry, which measures the metabolic rate by the determination of oxygen consumption (O_2) (with a spirometer), the production of carbon dioxide (CO_2) and excretion of urinary nitrogen, for a given period of time [6]. This technique relies on the fact that all the O_2 consumed and CO_2 produced is due to the oxidation of the three major energy substrates, which are fats, carbohydrates and proteins [7].

Recognizing the need to estimate EE in institutions that have no indirect calorimetry, researchers have proposed the use of specific equations, developed from calorimetry studies in groups of individuals with similar clinical characteristics [8]. Although the estimate of EE is the most common method, the predictive equations might generate errors [9]. Shetty [10] considers that the equations used to estimate the BEE in normal weight adults have reasonable precision (coefficient of variation 8%).

In clinical practice it is impracticable to measure the calorimetric methods for EE, so the international use of the equations was recommended, modified from a compilation of data carried out by Schofield [11]. Studies conducted in different ethnic groups found that these equations provide high BEE estimates, particularly for residents in the tropics [12-14]. Wahrlich and Anjos [14] confer these differences to the fact that equations have been developed mostly from population samples of North America and Europe which show differences in body composition, and live in different environmental conditions.

It is known that in populations with severe obesity is actually more difficult to fit the equations, because there is the difficulty in choosing the weight to be applied in the equation, which may influence a lot the results [15]. The use of current weight leads to the overestimation of the results independent of the equation to be applied, and the use of ideal or adjusted weight can result in the underestimation of energy needs [16].

Considering that obesity is a chronic disease of epidemiological importance, nutritional intervention studies have been developed in order to propose strategies for the prevention and treatment of this disease. The chronic imbalance between energy intake and EE results in positive energy balance and body weight gain. One way of evaluating the influence of dietary components in the EE, it through the measurement of energy metabolism by indirect calorimetry.

Obesity is also considered multifactorial, with genetic and environmental causes. In this sense, also studied the influence of candidate genes to obesity in metabolic variables, and indirect calorimetry is used in these studies.

The purpose of this chapter is to assess the importance of indirect calorimetry in the assessment of EE in obese individuals, both in study of nutrition intervention and influence of genes in EE, and in the validation and adequacy of existing prediction equations, which were not created for this population.

2. Indirect calorimetry

Indirect calorimetry remains a gold standard in measuring EE in the clinical settings. Indirect calorimetry offers a scientifically-based approach to customize a patient's energy needs and nutrient delivery to maximize the benefits of nutrition therapy. With recent advances in technology, indirect calorimeters are easier to operate, more portable, and affordable. Increased utilization of indirect calorimetry would facilitate individualized patient care and should lead to improved treatment outcomes [17].

Indirect calorimetry is considered a standard method, after validation by comparison with the direct calorimetry [18], however, its use is restricted to research due to the demanding cost and time for its conclusion [14], requiring the use of prediction equations in clinical practice.

According to Green (1994) [19], this technique is based on the principles that there are no considerable reserves of O_2 in the body, the O_2 uptake reflects the oxidation of nutrients, that all the chemical energy in the body comes from the oxidation of carbohydrates, fats and proteins, and that the ratio of O_2 consumption and CO_2 produced for the oxidation of these macronutrients are fixed. After determining the concentration of O_2 inspired and CO_2 expired, the calculation of RMR can be done with the equation of Weir (1949) [20, 21].

Given the difficulties associated with urine collection 24 hours, and found little difference between the results using the complete formula and simplified (2%), many authors have chosen to use the equation of Weir (1949) disregarding urinary losses of nitrogen [22].

The amount of O_2 used for oxidation and CO_2 production will depend on the substrate being oxidized. The respiratory quotient (RQ = Volume CO_2/Volume O_2) varies with the nutrients are being oxidized [14]. The table below describes the values of RQ complexes corresponding to the use of energy substrates [22].

RQ	Interpretation
0.67	Ethanol oxidation
0.71	Fat oxidation
0.82	Protein oxidation
0.85	Oxidation of a mixed diet
1.0	Carboydrate oxidation
1.0 – 1.2	Lipogenesis

Adapted of Materese (1997) [22].

Table 1. Values for interpretation of respiratory quotient (RQ) according to substrate oxidation.

The RQ is divided into non-protein RQ, which reflects the participation of carbohydrates and fats, and protein RQ, which represents the use of proteins. The rate of protein oxidation is obtained by determining the amount of nitrogen excreted in the urine during the test [7, 8].

The evaluation of RMR should only be initiated after a period of rest to minimize possible effects of recent physical activities such as dressing, driving or walking. In practice, it is recommended around twenty minutes of rest, for longer periods can cause the individual to sleep or get impatient [14].

During the measurement of RMR, the ambient temperature should be maintained in the neutral thermal zone, ie between 25 and 26°C, by Henry & Emery (1986) [14]. In order to evaluate the effects of temperature on EE, Dauncey (1981) [23] subjected women to direct calorimetry for thirty (30) hours at 22°C and 28°C. The evaluation of the RMR morning for 30 minutes, after twelve hours of fasting, demonstrated that the production of heat at the lowest temperature (22°C) was significantly greater (mean of 11 ± 3.2%) compared to the higher temperature (28ºC). Wahrlich & Anjos (2001) [14] point out that at ambient temperatures below the neutral zone, the use of protective clothing may be sufficient to prevent the increase in EE caused by the cold.

Compher et al. (2006) [24] conducted a systematic literature review to determine the optimal conditions for obtaining reliable measures of RMR by indirect calorimetry. Based on this survey, the following information was highlighted by the authors: 1) food, ethanol, caffeine and nicotine affect RMR for a variable number of hours after consumption, so its use should be controlled prior to measurement, 2) daily activities increased the RMR, however, a short rest period (20 minutes) before the test is sufficient, 3) moderate or vigorous physical activities presented most impact on the metabolism and therefore should be avoided in hours prior to measurement of RMR; 4) the measure with a duration of ten minutes with the first five minutes disregarded, and the remaining five minutes with a coefficient of variation of up to 10% guarantee accurate measurements of RMR, 5) the trial site should be physically comfortable and should be evaluated for ten to twenty minutes of rest before the start of measurement.

For research on assessment of RMR recommended: fasting for at least 6 hours, abstaining from caffeine during the night, nicotine and alcohol for at least 2 hours, of moderate physical activity for at least 2 hours, and vigorous physical activity for 14 hours [24].

In relation to the proper position for holding the indirect calorimetry, the authors emphasize that the most important is to ensure that each individual is physically comfortable during the test, and that measures are always taken in the same position. However, they warn that some positions require increased muscle tone and, therefore, could influence the measurement of RMR, and for research that has attempted to evaluate the RMR should keep the individual in the supine posture with or slightly higher [24].

The time required for obtaining an accurate measurement of RMR is only about five to ten minutes, discarding the first five minutes, provided that no changes occur in over 10% VO_2 and VCO_2 and 5% in RQ. Accordingly, one measure would be sufficient to describe the RMR for twenty-four hours. However, if you cannot guarantee the stability of the readings, the combination of two or three repetitions increase the precision of the measurements for the extrapolation to twenty-four hours [8, 24].

For studies involving analysis of the thermic effect of food (TEF), the indirect calorimetry should be conducted for a period of 6 hours, as measured by shorter periods are not able to fully assess the TEF [24].

With regard to environmental characteristics, it is recommended that the room temperature is comfortable, that is, between 20 and 25°C, the environment is quiet, with soft lighting and humidity control [24].

In females, should be avoided that the energy metabolism assessments are performed during the luteal phase because this phase of the menstrual cycle are described changes such as water retention, increased of body weight and energy demand, changes in lipid profile and metabolism of some nutrients (vitamin D, calcium, magnesium and iron), emotional hypersensitivity, aches and changes in eating behavior [25].

So, indirect calorimetry is a simple and affordable tool for measuring EE and for quantifying the utilization of macronutrients. Its use is becoming increasingly widespread, but it is necessary to know its methodological features and its theoretical and practical limitations. Indirect calorimetry measures the rate of REE, the major component of the TEE. Coupling the measurement of body composition to that of REE expands the diagnostic potential of indirect calorimetry. Once the lean and fat compartments have been measured, it is possible to establish on the basis of REE whether an individual is hyper- or hypometabolic. The clinical applications are practically unlimited [26].

Although the basic principles of indirect calorimetry are well established, it is important to recognize that there are several potential pitfalls in the methodology and data interpretation that must be appreciated to properly understand and apply the results derived from this technique. One must recognize that the fundamental measurement provided by indirect calorimetry is the net disappearance rate of a substrate regardless of the metabolic interconversions that the substrate may undergo before its disappearance from its metabolic pool. Under most circumstances, direct oxidation represents the major route by which a substrate disappears from its metabolic pool, and the two terms are often used interchangeably. However, under conditions when rates of gluconeogenesis, ketogenesis, or lipogenesis are elevated, the presumed equivalence between oxidation and disappearance may no longer apply, even though the actual measurements derived from indirect calorimetry remain valid. When indirect calorimetry is combined with other in vivo metabolic techniques (e.g., the insulin clamp or radioisotope turnover methods) it can provide a powerful tool for noninvasively examining complex metabolic processes [27].

Indirect calorimetry can also evaluate BEE. The main difference between REE and BEE is that REE is measured after the individual dislocation to the exam site, necessitating the prior resting period of 30 minutes to neutralize the effects of the physical activity performed [28]. Study found that REE is 10-15% higher than the BEE [29].

3. Utilization of indirect calorimetry in obesity

3.1. Evaluation of prediction equations

Kross et al [30] evaluated the accuracy of multiple regression equations to estimate REE in critically ill patients, especially for obese patients. A total of 927 patients were identified, including 401 obese patients. There were bias and poor agreement between measured REE and REE predicted by the Harris-Benedict, Owen, American College of Chest Physicians, and Mifflin equations ($p > 0.05$). There was poor agreement between measured and predicted REE by the Ireton-Jones equation, stratifying by sex. Ireton-Jones was the only equation that was unbiased for men and those in weight categories 1 and 2. In all cases except Ireton-Jones, predictive equations underestimated measured REE. The authors concluded that none of these equations accurately estimated measured REE in this group of mechanically ventilated patients, most underestimating energy needs. The authors concluded that is necessary to develop predictive equations for adequate assessment of energy needs.

Ullah et al [31] compared measured REE using the bedside with indirect calorimetry commonly used prediction equations, considering that the accuracy of prediction equations for estimating REE in morbidly obese patients is unclear. A total of 31 morbidly obese patients (46 kg/m^2) were studied. Pre-operative REE with indirect calorimetry was measured and compared with estimated REE using the Harris-Benedict and Schofield equations. All patients subsequently underwent a Roux-en-Y gastric bypass and were repeated measurements at six weeks and three months following surgery. The equations overestimated REE by 10% and 7%, by Harris-Benedict and Schofield equations, respectively. After weight loss the difference between the estimated and measured REE reduced to 1.3%. The accuracy improved after surgery induced weight loss, confirming their validity for the normal weight population. The study demonstrated that indirect calorimetry should be used in morbid obesity.

Cross-sectional study developed by our research group (unpublished data) with 92 women (35.60 ± 6.66 years) with excess body weight (34.41 ± 4.71 kg/m^2), Brazilian and Spanish. This study assessed the women in a metabolic unit, after fasting for 12 hours without performing strenuous physical activity in the last 24 hours and with minimal effort. The evaluation was performed using the open-circuit respiratory hood with indirect calorimetry (Deltatrac Metabolic Monitor-R3D) [6]. For the calculation of EE, it was used the values of the following volumes; inspired O_2 (VO_2), expired CO_2 (VCO_2) (ml / min) and urinary nitrogen [6-28], obtained by the calorimeter. In Brazilian women, it was found that the estimates obtained by Harris-Benedict, Shofield, FAO / WHO /ONU and Henry & Rees did not differ from REE of indirect calorimetry, which presented higher values than the equations proposed by Owen, Mifflin-St Jeor and Oxford. In Spanish women, also the equations proposed by Owen, Mifflin-St Jeor and Oxford presented EE lower than the indirect calorimetry, while the other equations did not differ from the indirect calorimetry. Both are women, Brasilian and Spanish, the best equations were FAO / WHO / ONU, Harris-Benedict Shofield and Henry & Rees.

Study aimed to validate the published predictive equations for REE in 76 normal weight (44.8 kg, 19.0 kg/m^2) and 52 obese (64.0 kg, 25.9 kg/m^2) Korean children and adolescents in the 7-18 years old age group. The open-circuit indirect calorimetry using a ventilated hood system was used to measure REE. Sixteen REE predictive equations were included, which were based on weight and/or height of children and adolescents, or which were commonly used in clinical settings despite its use based on adults. For the obese group, the Molnar, Mifflin, Liu, and Harris-Benedict equations provided the accurate predictions of > 70% (87%, 79% 77%, and 73%, respectively). On the other hand, for non-obese group, only the Molnar equation had a high level of accuracy (bias of 0.6%, RMSPE of 90.4 kcal/d, and accurate prediction of 72%). The accurate prediction of the Schofield (W/WH), WHO (W/WH), and Henry (W/WH) equations was less than 60% for all groups [32].

Alves et al. (2009) [33] compared the RMR obtained by indirect calorimetry with predict equations (Harris-Benedict (HB) and Ireton-Jones (IJ)) in 44 patients with excess body weight. The nearest RMR in fasting was obtained with the equation HB using the current body weight (1.873 ± 484 kcal / day and 1798 ± 495 kcal / day for HB and indirect calorimetry, respectively). However, the authors emphasize the need to employ the indirect calorimetry for the determination of EE of obese, because despite the similarity found between the absolute REE measured by indirect calorimetry and the prediction equation, there are significant ranges of variability, suggesting that the ideal method and more accurate to obtain the actual REE in this population is the indirect calorimetry.

3.2. Evaluation of the effect of nutritional interventions and obesity candidate genes in energy expenditure

Study with 60 obese women (34.59 ± 7.56 years) was conducted in order to evaluate the influence of fat diet and peroxisome proliferator-activated (PPARγ2) and β2-adrenergic receptor genes on energy metabolism. It was found that polymorphism in PPARgamma2 resulted in increased in fat oxidation, regardless of genotype of β2-adrenergic receptor gene. Polyunsaturated fatty acids (PUFA) intake can assist in weight loss, but the genotype of the genes assessed determines the type of fat that should be ingested [34].

The same research group developed another study with sixty obese women (30–46 years) which were divided into two groups depending on the genotype of PPARγ2 (Pro12Pro and Pro-12Ala/Ala12Ala). At baseline and after two nutritional (short- or long-term) interventions, measurement of anthropometrical and body composition (bioelectrical impedance) variables, dietary assessments, energy metabolism (indirect calorimetry) measurements as well as biochemical and molecular (PPARγ2 genotype) analyses were performed. All women received a high-fat test meal to determine the post-prandial metabolism (short term) and an energy-restricted diet for 10 weeks to determine the effect of diet in long term. The Pro12Ala polymorphism in the PPARγ2 gene influenced energy metabolism in the assayed short- and long-term situations since the response to both nutritional interventions differed according to the genotype. The results suggest that fat oxidation and EE may be lower in Pro12Pro carriers compared to Pro12Ala/Ala12Ala genotypes, while in obese women with Pro12Ala/Ala12Ala polymorphisms in the PPARγ2 gene fat oxidation was negatively correlated with the monounsaturated fatty acids (MUFA) and PUFA (%) intake [35].

The difference in structure of fatty acids, including the chain length, degree of unsaturation and the position of the double bond, can affect the rate of oxidation of fatty acids. Medium chain saturated fatty acids (MCSFA) are more easily oxidized than the long chain saturated fatty acids (LCSFA), while unsaturated fatty acids (UFA) are more easily oxidized compared to saturated chain acids (SFA) with the same chain length [36].

Using indirect calorimetry studies also indicate that the PUFA shows a higher oxidation compared to SFA, both in men and in obese normal [37, 38]. Piers et al (2002) [39] found that changes in the type of fat dietary may have a beneficial effect on reducing body weight in men who consume high fat content, since the oxidation rate postprandial (assessed by Indirect calorimetry) of nutrient is increased after a high MUFA meal, compared with the SFA.

Casas-Agustench et al (2009) [40] aimed to compare the acute effects of three fatty meals with different fat quality on postprandial thermogenesis and substrate oxidation. Evaluated twenty-nine healthy men aged between 18 and 30 years in randomized crossover trial, comparing the thermogenic effects of three isocaloric meals: high in PUFA from walnuts, high in MUFA from olive oil, and high in SFA from fat-rich dairy products. Indirect calorimetry was used to determine RMR, RQ, 5-H postprandial EE and substrate oxidation. Five hours postprandial thermogenesis was higher by 28% after the high PUFA meal ($p = 0.039$) and by 23% higher after the high MUFA meal ($p = 0.035$), compared with the high SFA meal. Increased fat oxidation rates no significantly after the two meals rich in UFA and decreased non significantly after the high SFA meal. Postprandial RQ, carbohydrate and protein oxidation measures were similar among meals. The authors concluded that fat quality determined the thermogenic response to a fatty meal but clear effects on substrate oxidation.

Another study was conducted to evaluate whether postprandial abnormalities of EE and / or lipid oxidation are present in healthy, normal-weight individuals with a strong family history of obesity and thus at high risk to become obese. They conducted a case-control study. A total of 16 healthy young men participated in the study. Eight individuals had both parents overweight (father's and mother's body mass index > 25 kg / m^2) and eight had both parents with normal body weight (father's and mother's body mass index < 25 kg / m^2). The group of individuals with overweight parents was similar to that with normal-weight parents (control group) in terms of body mass index and FFM. EE was measured by indirect calorimetry, and blood samples were taken for the evaluation of metabolic variables in the fasting state and every hour for 8 h after a standard fat-rich meal (protein 15%, 34% carbohydrate, 51 fat %, 4090 kJ). Fasting and postprandial EE, and fasting fat and carbohydrate oxidation were both in similar groups. On the contrary, postprandial carbohydrate oxidation (incremental area under curve) was significantly higher and that of fat oxidation lower in the group of individuals with overweight parents. They concluded that normal-weight individuals with a strong family history of obesity present a reduced fat oxidation in the postprandial period. These metabolic characteristics may be considered the early predictors of weight gain and are genetically determined probably [41].

Differences in meal-induced thermogenesis and macronutrient oxidation between lean ($n = 19$) and obese ($n = 22$) women after consumption of two different isocaloric meals, one rich in carbohydrate (CHO) and one rich in fat were examined. Women were studied on two occasions, one week apart. In one visit they consumed a CHO-rich meal and in the other visit a fat-rich meal. The two meals were isocaloric and were given in random order. REE and macronutrient oxidation rates were measured and calculated in the fasting state and every hour for 3 h after meal consumption. Meal-induced thermogenesis was not different between lean and obese subjects after the CHO-rich ($p = 0.89$) or fat-rich ($p = 0.32$) meal, but it was significantly higher after the CHO-rich compared with the fat-rich meal in the lean and the obese individuals ($p < 0.05$). Protein oxidation rate increased slightly but significantly after the test meals in both groups ($p < 0.01$). Fat oxidation rate decreased after consumption of the CHO-rich meal ($p < 0.001$), whereas it increased after consumption of the fat-rich meal in both groups ($p < 0.01$). CHO oxidation rate increased in both groups after consumption of the CHO-rich meal ($p < 0.001$). Oxidation rates of protein, fat, and CHO during the experiment were not significantly different between lean and obese participants. In conclusion, it was verified that meal-induced thermogenesis and macronutrient oxidation rates were not significantly different between lean and obese women after consumption of a CHO-rich or a fat-rich meal [42].

In a parallel-arm, long-term feeding trial, 24 lean and 24 overweight participants received a daily peanut oil load in a milk shake equivalent to 30% of their REE for eight weeks to evaluate the effects of peanut oil intake on appetite, EE (indirect calorimetry at baseline and week 8), body composition, and lipid profile. Energy intake increased significantly in the overweight but not in the lean participants. A statistically significant body weight gain (median 2.35 kg) was also observed among the overweight subjects, although this corresponded to only 43% of the theoretical weight gain. In the overweight participants, the REE was significantly increased by 5% over the intervention, but no significant difference was observed in the lean subjects. As expected, REE was significantly higher in the overweight than in the lean participants. No marked differences of appetite were observed over time in either group or between overweight and lean participants. These data indicate that ingestion of peanut oil elicits a weaker compensatory dietary response among overweight compared with lean individuals. Body weight increased, albeit less than theoretically predicted [43].

The effects of a moderate-fat diet, high in MUFAs, and a low-fat (LF) diet on EE and macronutrient oxidation before and after a 6-mo controlled dietary intervention were compared. Twenty-seven overweight (body mass index 28.1 $\pm$ 0.4 kg/m^2) nondiabetic subjects (18–36 years) followed an 8-wk low-calorie diet and a 2-wk weight-stabilizing diet and then were randomly assigned to a MUFA ($n = 12$) or LF ($n = 15$) diet for 6 mo. Substrate oxidation and 24-h EE were measured by whole-body indirect calorimetry. The first measurement (0 mo) was taken during the weight-stabilizing diet, and the second measurement was taken after the 6-mo intervention. A tendency was seen toward a lower 24-h EE with the MUFA than with the LF diet ($p = 0.0675$), but this trend did not remain after adjustment for the initial loses of fat mass and FFM ($p = 0.2963$). Meal-induced thermogenesis was significantly ($p < 0.05$) lower with the MUFA than with the LF diet. Despite a slightly lower meal-induced thermogenesis, the MUFA diet had an effect on 24-h EE that was not significantly different from that of the LF diet after a 6-mo controlled dietary intervention [44].

Study with 24 healthy, overweight men (body mass index between 25 and 31 kg/m^2) compared the effects of diets rich in medium-chain triglycerides (MCTs) or long-chain triglycerides (LCTs) on body composition, EE, substrate oxidation, subjective appetite, and *ad libitum* energy intake. At baseline and after four weeks of each dietary intervention, EE was measured using indirect calorimetry. Average EE was 0.04 $\pm$ 0.02 kcal/min greater ($p < 0.05$) on day 2 and 0.03 $\pm$ 0.02 kcal/min (not significant) on day 28 with functional oil (64.7% MCT oil) compared with olive oil consumption. Similarly, average fat oxidation was greater ($p = 0.052$) with functional oil compared with olive oil intake on day 2 but not day 28.

Consumption of a diet rich in MCTs results in greater loss of adipose tissue compared with LCTs, perhaps due to increased EE and fat oxidation observed with MCT intake [45].

A controlled randomized dietary trial was conducted with 26 overweight or moderately obese men and women (body mass index 28-33 kg/m^2) to test the hypothesis that n-3-polyunsaturated fatty acids (n-3-PUFA) lower body weight and fat mass by reducing appetite and *ad libitum* food intake and/or by increasing EE. Diets were administered in an isocaloric fashion for 2 weeks followed by 12 weeks of *ad libitum* intake. The n-3-PUFA and control diets were identical in all regards except for the fatty acid composition. Both groups lost similar amounts of weight when these diets were consumed *ad libitum* for 12 weeks [mean (SD): -3.5 (3.7) kg in the control group vs. -2.8 (3.7) kg in the n-3-PUFA group, $F(1,24) = 13.425$, $p = 0.001$ for time effect; $F(1,24) = 0.385$, $p = 0.541$ for time × group interaction]. No differences were founds between the n-3-PUFA and control groups with regard to appetite as measured by visual analogue scale, *ad libitum* food intake or, REE as measured by indirect calorimetry, diurnal plasma leptin concentrations, or fasting ghrelin concentrations. These results suggest that dietary n-3-PUFA do not play an important role in the regulation of food intake, EE, or body weight in humans [46].

4. Conclusion

Indirect calorimetry is useful technique in the metabolic evaluation of obese individuals. Despite some methodological limitations, is still the best way to estimate this variable in this population, which is useful both in studies of dietary intervention, intended to propose new strategies for prevention and treatment of obesity, and for validation of predictive equations for energy expenditure in this population.

Abbreviations

BEE - basal energy expenditure
CHO – carbohydrate
EE - energy expenditure
FFM - fat-free mass
LCSFA - long chain saturated fatty acids
LCT - long-chain triglycerides
LF – low-fat
MCSFA - Medium chain saturated fatty acids
MCT - medium-chain triglycerides
MUFA - monounsaturated fatty acids
PPARγ2 - peroxisome proliferator-activated
PUFA - polyunsaturated fatty acids
REE – resting energy expenditure
RMR - resting metabolic rate
RQ - respiratory quotient
SFA – saturated fatty acids

TEE - energy expenditure
TEF - thermic effect of food
UFA - unsaturated fatty acids
VO_2 - inspired O_2
VCO_2 - expired CO_2

Author details

Eliane Lopes Rosado, Vanessa Chaia Kaippert and Roberta Santiago de Brito
Nutrição e Dietética Departament, Federal University of Rio de Janeiro, Rio de Janeiro, Brasil

Acknowledgement

This work was supported by Conselho Nacional de Pesquisa (CNPq) and Fundação de Amparo à Pesquisa do Estado de Minas Gerais (FAPEMIG).

We thank the Federal University of Viçosa and Navarra University.

5. References

[1] Flatt J-P (2007) Differences in Basal Energy Expenditure and Obesity. Obesity. 15(11):2546-8.
[2] Mika Horie L, González MC, Raslan M, torrinhas R, Rodrigues NL, Verotti CC, Cecconello I, Heymsfield SB, Waitzberg DL (2009) Resting energy expenditure in white and non-white severely obese women. Nutr Hosp. 24(6):676-81.
[3] Melo CM, Tirapegui J, Ribeiro SML (2008) Gasto energético corporal: conceitos, formas de avaliação e sua relação com a obesidade. Arq Bras Endocrinol Metab. 52(3):452-464.
[4] Astrup A, Buemann B, Toubro S, Ranneries C, Raben A (1996) Low resting metabolic rate in subjects predisposed to obesity: a role for thyroid status. Am J Clin Nutr. 63(6):879-83.
[5] Ravussin E, Lillioja S, Knowler WC, Christin L, Freymond D, Abbott WG, Boyce V, Howard BV, Bogardus C (1988) Reduced rate of energy expenditure as a risk factor for body-weight gain. N Engl J Med 1988;318(8):467-72.
[6] Ferrannini E (1988) The theoretical bases of indirect calorimetry: a review. Metabolism. 37(3):287-301.
[7] Jéquier E, Acheson K, Schutz Y (1987) Assessment of energy expenditure and fuel utilization in man. Ann Rev Nutr. 7:187-208.
[8] Diener JRC (1997) Calorimetria indireta. Rev Ass Med Brasil. 43(3):245-53.
[9] Frankenfield D, Roth-Yousey L, Compher C (2005) Comparison of Predictive Equations for Resting Metabolic Rate in Healthy Nonobese and Obese Adults: A Systematic Review. J Am Diet Assoc. 105:775-89.
[10] Shetty P (2005) Energy requirements of adults. Public Health Nutrition. 8(7A):994–1009.
[11] Schofield WN (1985) Predicting basal metabolic rate, new standards and review of previous work. Hum Nutr Clin Nutr. 39 suppl 1:5-41.

[12] Piers LS, Diffey B, Soares MJ, Frandsen SL, McCormack LM, Lutschini MJ, O`Dea K (1997) The validity of predicting the basal metabolic rate of young Australian men and women. Eur J Clin Nutr. 51:333-7.
[13] Cruz CM, Silva AF, Anjos LA (1999) A taxa metabólica basal é superestimada pelas equações preditivas em universitárias do Rio de Janeiro, Brasil. Arch Latinoam Nutr. 49(3):232-7.
[14] Wahrlich V, Anjos LA (2001) Aspectos históricos e metodológicos da medição e estimativa da taxa metabólica basal: uma revisão da literatura. Cad Saúde Pública. 17(4):801-817.
[15] Rodrigues AE, Mancini MC, Dalcanale L, Melo ME, Cercato C, Halpern A (2010) Padronização do gasto metabólico de repouso e proposta de nova equação para uma população feminina brasileira. Arq Bras Endocrinol Metab. 54(5):470-476.
[16] Weg MWV, Watson JM, Klesges RC, Clemens LHE, Slawson DL, MCClanahan BS (2004) Development and cross-validation of a prediction equation for estimating resting energy expenditure in healthy African-American and European-American women. Eur J Clin Nutr. 58:474-80.
[17] Haugen HA, Chan LN, Li F [2007] Indirect calorimetry: a practical guide for clinicians. Nutr Clin Pract. 22(4):377-88.
[18] Benedetti FJ, Bosa VL, Mocelin HT, Paludo J, Mello ED, Fischer GB (2011) Gasto energético em adolescentes asmáticos com excesso de peso: calorimetria indireta e equações de predição. Rev Nutr. 24(1): 31-40.
[19] Green JH (1994) Assessment of energy requirements. In: Heatley RV, Green JH, Losowsky MS, editors. Consensus in Clinical Nutrition. Cambridge: Cambridge University Press. pp. 22-37.
[20] Weir JB (1949) New methods for calculating metabolic rate with special reference to protein metabolism. J Physiol. 109:1-9.
[21] Delany JP, Lovejoy JC (1996) Energy expenditure. Endocrinol Metab Clin North Am. 25(4): 831-846.
[22] Matarese LE (1997) Indirect calorimetry: technical aspects. J Am Diet Assoc. 97:S154-S160.
[23] Dauncey MJ (1981) Influence of mild cold on 24 h energy expenditure, resting metabolism and diet induced thermogenesis. Br J Nutr. 45:257-267.
[24] Compher C, Frankenfield D, Keim N, Roth-Yousey L (2006) Best practice methods to apply to measurement of resting metabolic rate in adults: a systematic review. J Am Diet Assoc. 106(6):881-903.
[25] Sampaio HAC (2002) Aspectos nutricionais relacionados ao ciclo menstrual. Rev Nutr. 15(3):309-317.
[26] Battezzati A, Viganò R [2001] Indirect calorimetry and nutritional problems in clinical practice. Acta Diabetol. 38(1):1-5.
[27] Simonson DC, DeFronzo RA [1990] Indirect calorimetry: methodological and interpretative problems. Am J Physiol. 258(3 Pt 1):E399-412.
[28] Kamimura MA, Avesani CA, Draibe SA, Cuppari L (2008) Gasto energético de repouso em pacientes com doença renal crônica. Rev Nutr. 21(1):75-84.

[29] Poehlman ET, Horton ES (2003) Necessidades Energéticas: Avaliação e Necessidades em Humanos. In: Shils ME, Olson JÁ, Shike M, Ross AC. Tratado de Nutrição Moderna na Saúde e na Doença. São Paulo: Ed Manole, 2003. Vol 1. 9º edição.

[30] Kross EK, Sena M, Schmidt K, Stapleton RD (2012) A comparison of predictive equations of energy expenditure and measured energy expenditure in critically ill patients. J Crit Care. 14; [Epub ahead of print].

[31] Ullah S, Arsalani-Zadeh R, Macfie J [2012] Accuracy of prediction equations for calculating resting energy expenditure in morbidly obese patients. Ann R Coll Surg Engl. 94(2):129-32.

[32] Kim MH, Kim JH, Kim EK (2012) Accuracy of predictive equations for resting energy expenditure (REE) in non-obese and obese Korean children and adolescents. Nutr Res Pract. 6(1):51-60.

[33] Alves VGF, Rocha EEM, Gonzalez MC, Fonseca RBV, Silva MHN, Chiesa CA (2009) Assessment of resting energy expenditure of obese patients: comparison of indirect calorimetry with formulae. Clin Nutr. 28:299-204.

[34] Rosado EL, Bressan J, Hernández JAM, Martins MF, Cecon PR (2006) Efecto de la dieta e de los genes PPARγ2 y β2-adrenérgico en el metabolismo energético y en la composición corporal de mujeres obesas. Nutr Hosp. 21(3):317-331.

[35] Rosado EL, Bressan J, Hernández JAM, Lopes IM (2010) Interactions of the PPARγ2 polymorphism with fat intake affecting energy metabolism and nutritional outcomes in obese women. Ann Nutr Metab. 57(3-4):242-250.

[36] DeLany JP, Windhauser MM, Champagne CM, Bray GA (2000) Differential oxidation of individual dietary fatty acids in humans. Am J Clin Nutr. 72(4):905-911.

[37] Jones PJ, Schoeller DA (1988) Polyunsaturated:saturated ratio of diet fat influences energy substrate utilization in the human. Metabolism. 37:145–151.

[38] Jones PJ, Ridgen JE, Phang PT, Birmingham CL (1992) Influence of dietary fat polyunsaturated to saturated ratio on energy substrate utilization in obesity. Metabolism. 41:396– 401.

[39] Piers LS, Walker KZ, Stoney RM, Soares MJ, O'Dea K (2002) The influence of the type of dietary fat on postprandial fat oxidation rates: monounsaturated (olive oil) vs saturated fat (cream). Int J Obes. 26:814–821.

[40] Casas-Agustench P, López-Uriarte, Bullo M, Ros E, Gómez-Flores A, Salas-Salvadó J (2009) Acute effects of three high-fat meals with different fat saturations on energy expenditure, substrate oxidation and satiety. Clin Nutr. 28:39-45.

[41] Giacco R, Clemente G, busiello L, Lasorella G, Rivieccio AM, Rivellese AA, Riccardi G (2004) Insulin sensitivity is increased and fat oxidation after a high-fat meal is reduced in normal-weight healthy men with strong familial predisposition to overweight. Int J Obes. 28(2):342-8.

[42] Tentolouris N, Alexiadou K, Kokkinos A, Koukou E, Perrea D, Kyriaki D, Katsilambros N (2011) Meal-induced thermogenesis and macronutrient oxidation in lean and obese women after consumption of carbohydrate-rich and fat-rich meals. Nutrition. 27(3):310-315.

[43] Coelho SB, Sales RL, Iyer SS, Bressan J, Costa NMB, Lokko P, Mattes R (2006) Effects of peanut oil load on energy expenditure, body composition, lipid profile, and appetite in lean and overweight adults. Nutrition. 22:585-592.

[44] Rasmussen LG, Larsen TM, Mortensen PK, Due A, Astrup A (2007) Effect on 24-h energy expenditure of a moderate-fat diet high in monounsaturated fatty acids compared with that of a low-fat, carbohydrate-rich diet: a 6-mo controlled dietary intervention trial. Am J Clin Nutr. 85:1014-1022.

[45] St-Onge M-P, Ross R, Parsons WD, Jones PJH (2003) Medium-chain triglycerides increase energy expenditure and decrease adiposity in overweight men. Obes Res. 11(3):395-402.

[46] Kratz M, Callahan HS, Yang PY, Matthys CC, Weigle DS (2009) Dietary n-3-polyunsaturated fatty acids and energy balance in overweight or moderately obese men and women: a randomized controlled trial. Nutr Metab. 6:24.

Section 6

Applications of Calorimetry into Propellants, Alloys, Mixed Oxides and Lipids

Thermal Decomposition Kinetics of Aged Solid Propellant Based on Ammonium Perchlorate – AP/HTPB Binder

R. F. B. Gonçalves, J. A. F. F. Rocco and K. Iha

Additional information is available at the end of the chapter

http://dx.doi.org/10.5772/52109

1. Introduction

Despite the widespread use and long investigative history of ammonium perchlorate(AP)-fuel mixtures, it still can be said that AP alone and AP/HTPB (hydroxyl-terminated-polybutadiene) composites remain among the most confounding materials in the research setting [1]. Since the physical structure of composite propellants like the AP/HTPB composite is heterogeneous, the combustion wave structure appears to be also heterogeneous. During the combustion, at the burning surface, the decomposed gases from the ammonium perchlorate particles and fuel binder (HTPB) are interdiffused and produce diffusion flame streams. Due this, the flame structure of AP composite propellants is complex and locally three-dimensional in shape.

Ammonium perchlorate (NH_4ClO_4) is a powerful oxidizer salt largely used in solid propellant formulations for application in airspace and defense materials industries. It is obtained by reaction between ammonia and perchloric acid, or by double decomposition between an ammonium salt and sodium perchlorate, and crystallizes with romboedric structure in room temperature and pressure, with relative density of 1.95 [2] Similarly to most ammonium salts, AP thermal decomposition occurs before its fusion. When submitted to a low heating rate, decomposes releasing gases chlorine, nitrogen and oxygen and water in the vapor state; while with a high heating rate stimulus there are instant reactions with high energy release.

During the combustion process of AP crystals at high pressures, is possible to observe the formation of a tiny layer of ammonium perchlorate in liquid phase at the grain surface [3], followed by a region where it is presented in gaseous phase.

According to Beckstead and Puduppakkam [4], the combustion of a monopropellant can be divided in three regions (condensed, liquid-gas two-phase region and gas region). The two-phase region consists of liquid and gaseous species resulting from the melting and/or decomposition of the solid phase. The precise division between the two-phase and gas-phase region (i.e. the 'burning surface') is not well defined due to chemical reactions, bubbles, and condensed material being convected away from the surface. In the gas phase region of a monopropellant, the flame is essentially premixed. The species emanating from the surface react with each other and/or decompose to form other species. A wide variety of reactions involving many species occur in the gas flame until equilibrium is reached in the final flame zone.

Thermal decomposition of AP, as its combustion processes, have been experimentally studied and reported in the literature. The thermal decomposition of AP may be observed by differential thermal analysis (DTA) and thermal gravimetry (TG), in the figure below [5]. A heating rate of 0.33 K/s was used on the analysis.

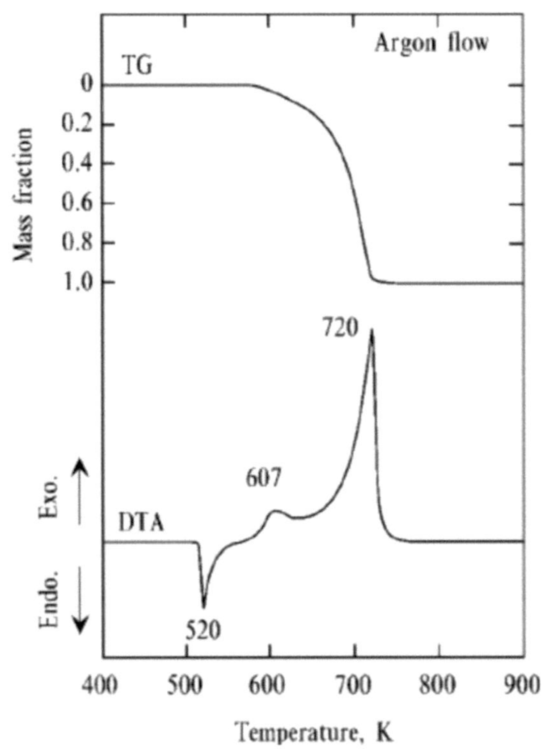

Figure 1. TG and DTA of AP decomposition [5]

The phase transition from orthorhombic to cubic crystal lattice (ΔH = -85 kJ/kg) is represented by the endothermic peak on 520 K. The exothermic events on 607 K and 720 K are due to the proper decomposition of the AP crystal in ammonia and perchloric acid, followed by the formation of chloridric acid and oxygen (decomposition of $HClO_4$), according to the reactions below.

$$NH_4ClO_4 \rightarrow NH_3 + HClO_4$$
$$HClO_4 \downarrow HCl + 2\,O_2$$

These oxygen molecules will be used as oxidizer in binders combustion, when the AP is used in a composite propellant or even when is burning by itself.

The combustion mechanism of AP has been studied and modified. The table below shows the elementary reactions which take part on the combustion process. This mechanism was proposed by Gross [6], according to literature data. It is very interesting the analysis of the combustion for its close relation to the thermal decomposition. When high pressure or high temperatures are used, the material suffers combustion instead of thermal decomposition, i.e. there's a higher velocity of decomposition and higher energy release, but the process is usually incomplete.

Reaction	A	b	Ea
$HClO_4=ClO_3+OH$	1.00E+14	0.0	3.91E+04
$HClO_4+HNO=ClO_3+H_2O+NO$	1.50E+13	0.0	6.00E+03
$ClO_3=ClO+O_2$	1.70E+13	0.5	0,00E+00
$Cl_2+O_2+M=ClO_2+Cl+M$	6.00E+08	0.0	1.12E+04
$ClO+NO=Cl+NO_2$	6.78E+12	0.0	3.11E+02
$ClO+ClOH=Cl_2+HO_2$	1.00E+11	0.0	1.00E+04
$ClOH+OH=ClO+H_2O$	1.80E+13	0.0	0,00E+00
$HCl+OH=Cl+H_2O$	5.00E+11	0.0	7.50E+02
$Cl_2+H=HCl+Cl$	8.40E+13	0.0	1.15E+03
$ClO+NH_3=ClOH+NH_2$	6.00E+11	0.5	6.40E+03
$NH_3+Cl=NH_2+HCl$	4.50E+11	0.5	1.00E+02
$NH_3+OH=NH_2+H2O$	5.00E+07	1.6	9.55E+02
$NH_2+O_2=HNO+OH$	3.00E+09	0.0	0,00E+00
$NH_2+NO=H_2O+N_2$	6.20E+15	-1.3	0,00E+00
$HNO+OH=NO+H_2O$	1.30E+07	1.9	-9.50E+02
$HNO+O_2=NO_2+OH$	1.50E+13	0.0	1.00E+04
$HNO+H=H_2+NO$	4.50E+11	0.7	6.60E+02
$NO+H+M=HNO+M$	8.90E+19	-1.3	7.40E+02
$HO_2+N_2=HNO+NO$	2.70E+10	0.5	4.18E+04
$NO+HO_2=NO_2+OH$	2.11E+12	0.0	4.80E+02
$H+NO_2=NO+OH$	3.47E+14	0.0	1.48E+03
$H_2+OH=H_2O+H$	2.16E+08	1.5	3.43E+03

Table 1. AP combustion mechanism

$k = A\,T^{b}\exp(-E/RT)$. Units: A (mol-cm-s-K), E (J/mol).

Based on combustion mechanisms, the burning process may be simulated and analyzed by some specific softwares. In a previous work [7], these simulations were done, considering a perfectly stirred reactor, different internal pressures and a specific temperature profile. The combustion simulation results may be observed in the figure below, which show the behavior of AP combustion with different internal pressures of the combustion chamber.

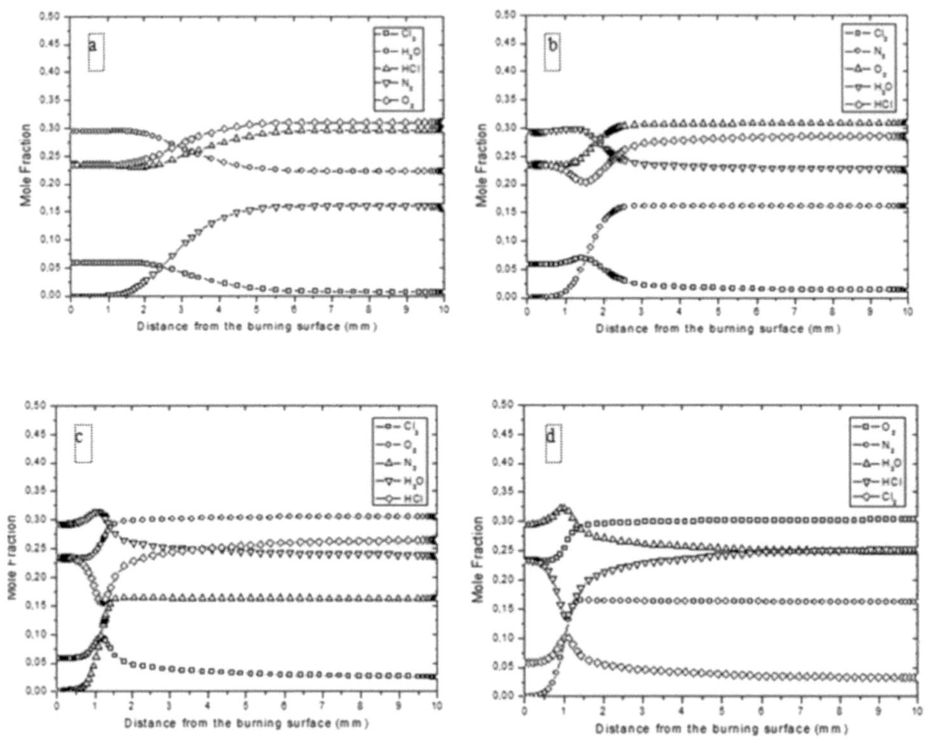

Figure 2. AP combustion at a) 1 atm; b) 5 atm; c) 30 atm and d) 60 atm.

The “elbows” appear due to the increase of the occurrence of intermediate reactions in the flame zone. This phenomenon generates a great variation on the mole fractions of intermediates (as the high temperature enhance the speed of the slower reactions, generating more radicals), which modify the concentration of the main species (specially in the flame zone), so the different slope is observed. As the pressure in the combustion chamber increases, there is an approximation of the flame to the material's surface and accentuation of the "elbows" presented on the flame region, indicating the influence of the speed increase of elementary reactions in the decomposition process of the material in study. This gain in chemical speed reactions may be converted in gain in thrust of rocket motors and specific impulse of solid propellant grains.

When a composite propellant is used, like AP-HTPB, the combustion process depends on the diffusion of the gases generated on the initial decomposition of the oxidizer, which surrounds the binder molecules at the burning surface. The combustion mechanism has higher complexity as new components are added, because there are the elementary reactions for each component decomposition and their interactions in chamber, as well as the formation and decomposition of new intermediary species, especially in the flame region. The proposed mechanism for AP-HTPB combustion may be observed in Table 2 below.

Reaction	A	b	Ea
$Cl_2+O_2+M=ClO_2+Cl+M$	6.00E+08	0	1.12E+04
$ClO+NO=Cl+NO_2$	6.78E+12	0	3.11E+02
$HCl+OH=Cl+H_2O$	5.00E+11	0	7.50E+02
$Cl_2+H=HCl+Cl$	8.40E+13	0	1.15E+03
$NH_3+Cl=NH_2+HCl$	4.50E+11	0.5	1.00E+02
$NH_3+OH=NH_2+H_2O$	5.00E+07	1.6	9.55E+02
$NH_2+O_2=HNO+OH$	3.00E+09	0	0.00E+00
$NH_2+NO=H_2O+N_2$	6.20E+15	-1.3	0.00E+00
$HNO+OH=NO+H_2O$	1.30E+07	1.9	-9.50E+02
$HNO+O_2=NO_2+OH$	1.50E+13	0	1.00E+04
$HNO+H=H_2+NO$	4.50E+11	0.7	6.60E+02
$NO+H+M=HNO+M$	8.90E+19	-1.3	7.40E+02
$HO_2+N_2=HNO+NO$	2.70E+10	0.5	4.18E+04
$NO+HO_2=NO_2+OH$	2.11E+12	0	4.80E+02
$H+NO_2=NO+OH$	3.47E+14	0	1.48E+03
$H_2+OH=H_2O+H$	2.16E+08	1.5	3.43E+03
$CH_4+Cl=CH_3+HCl$	2.50E+13	0	3.83E+03
$CH_4+H=CH_3+H_2$	6.60E+08	1.6	1.08E+04
$CH_4+OH=CH_3+H_2O$	1.00E+08	1.6	3.12E+03
$CH_3+H+M=CH_4+M$	1.27E+16	-0.6	3.83E+02
$CO+OH=CO_2+H$	4.76E+07	1.2	7.00E+01
$CO+ClO=CO_2+Cl$	3.00E+12	0	1.00E+03
$CO+ClO_2=CO_2+ClO$	1.00E+10	0	0.00E+00
$H+O_2=O+OH$	8.30E+13	0	1.44E+04
$CH_2+H_2=CH_3+H$	5.00E+05	2	7.23E+03
$CH_2+H+M=CH_3+M$	2.50E+16	-0.8	0.00E+00
$CH_4+O=CH_3+OH$	1.02E+09	1.5	6.00E+02
$OH+CH_3=CH_2+H_2O$	5.60E+07	1.6	5.42E+03
$C_2H_4+O_2=2CO+2H_2$	1.80E+14	0	3.55E+04
$NH_2+NO_2=2HNO$	1.40E+12	0	0.00E+00
$NH_2+ClO=HNO+HCl$	2.50E+12	0	0.00E+00
$O_2+HNO=NO+HO_2$	1.00E+13	0	1.30E+04
$H+Cl+M=HCl+M$	5.30E+21	-2	-2.00E+03
$Cl+Cl+M=Cl_2+M$	3.34E+14	0	-1.80E+03
$Cl+HO_2=ClO+OH$	2.47E+13	0	8.94E+02
$ClO+O=Cl+O_2$	6.60E+13	0	4.40E+02
$H+HCl=Cl+H_2$	7.94E+12	0	3.40E+03
$HCl+O=Cl+OH$	2.30E+11	0.6	9.00E+02

Cl_2+O=Cl+ClO	2.51E+12	0	2.72E+03
N_2O+M=N_2+O+M	6.20E+14	0	5.61E+04
N_2O+OH=N_2+HO_2	2.00E+12	0	2.11E+04
N_2O+O=NO+NO	2.90E+13	0	2.32E+04
N_2O+O=N_2+O_2	1.40E+12	0	1.08E+04
N_2O+H=N_2+OH	4.40E+14	0	1.89E+04
2H+M<=>H_2+M	1.00E+18	-1	0.00E+00
2H+H_2<=>2H_2	9.00E+16	-0.6	0.00E+00
2H+H_2O<=>H_2+H_2O	6.00E+19	-1.3	0.00E+00
2H+CO_2<=>H_2+CO_2	5.50E+20	-2	0.00E+00
ClO_2+NO=ClO+NO_2	1.00E+11	0	0.00E+00
Cl+ClO_2=ClO+ClO	5.00E+13	0	6.00E+03
ClO+ClO=Cl_2+O_2	1.00E+11	0	0.00E+00
Cl+HO_2=HCl+O_2	1.80E+13	0	0.00E+00
Cl+O_2+M=ClO_2+M	8.00E+06	0	5.20E+03
NO_2+O=NO+O_2	1.00E+13	0	6.00E+02
HNO+HNO=H_2O+N_2O	3.95E+12	0	5.00E+03
NO_2+NO_2=NO+NO+O_2	1.00E+14	0	2.50E+04
Cl+N_2O=ClO+N_2	1.20E+14	0	3.35E+04
OH+OH=H_2O+O	6.00E+08	1.3	0.00E+00
NH_2+NO_2=H_2O+N_2O	4.50E+11	0	0.00E+00
HNO+NH_2=NH_3+NO	5.00E+11	0.5	1.00E+03
ClO+HNO=HCl+NO_2	3.00E+12	0	0.00E+00
HCl+HO_2=ClO+H2O	3.00E+12	0	0.00E+00
NH_2+NO=H+N_2+OH	6.30E+19	-2.5	1.90E+03
NH_2+OH=H_2O+NH	4.00E+06	2	1.00E+03
NH_2+NH_2=NH+NH_3	5.00E+13	0	1.00E+04
NH+NO=N_2+OH	1.00E+13	0	0.00E+00
NH+NO=H+N_2+O	2.30E+13	0	0.00E+00
Cl+NH_2=HCl+NH	5.00E+10	0.5	0.00E+00
ClO_2+NH=ClO+HNO	1.00E+14	0	0.00E+00
N+NO_2=NO+NO	1.00E+14	0	0.00E+00
N+N_2O=N_2+NO	5.00E+13	0	0.00E+00
NH+OH=H_2O+N	5.00E+11	0.5	2.00E+03
NH+OH=H_2+NO	1.60E+12	0.6	1.50E+03
NH+NH_2=N+NH_3	1.00E+13	0	2.00E+03
HO_2+CH_3<=>O_2+CH_4	1.00E+12	0	0.00E+00
CH_2+CH_4<=>2CH_3	2.46E+06	2	8.27E+03

Table 2. AP-HTPB combustion mechanism[a]

$$k = A\,T^{b} \exp(-E/RT).$$ Units: A (mol-cm-s-K), E (J/mol).

M: any metal surface or metallic additive used only as support or catalyst

[a] Kinetic data composed of [8]

Similarly, the combustion process of ammonium perchlorate formulated with hydroxyl terminated polybutadiene was simulated in a perfect stirred reactor (with 70/30 proportion), with variations in the internal chamber pressure (Figure 3 below).

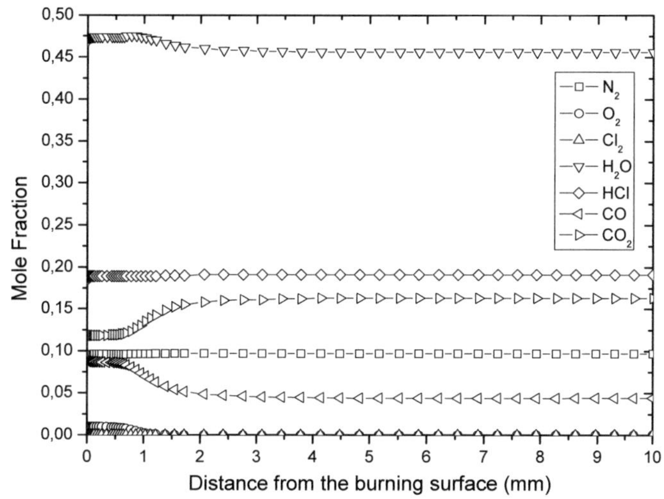

Figure 3. AP/HTPB combustion [7]

The combustion process of AP/HTPB has presented invariable with pressure. This behavior should be attributed to the homogeneous dispersion of AP admist the binder, in the solid phase, and to the lack of this species in relation to the binder (generating lower concentrations of O_2 than necessary). Also, in the gas phases, it is assumed that all of the liquid AP and HTPB present on the condensed phase decompose to form gaseous species; evaporation is not included.

In this simulation, the oxygen molar fraction suffers a decrease (and cancels), according to the reactions with HTPB decomposition products, for the formation of carbon monoxide and dioxide. Also, it is interesting to highlight that in this case the carbon monoxide molar fraction suffers a decrease, because the restriction of oxidizer species makes that the oxygen presented in CO to be also used as oxidizing source, viewing the reactive behavior of this specie. In this simulation, the molar fractions of CO and CO_2 are not null initially, because given the system temperature, HTPB suffers an initial decomposition that should not be discarded, generating both carbon oxides.

There is always the premise in all simulations and all studies that the materials are in a perfect state, flawless. Unfortunately, this is not the reality in most industries or laboratories, when there's low turnover. Therefore, the materials may suffer many different changes in their structure or properties. The main one is the aging process.

The aging process is one of the most significant factors responsible for changes in the activation energy of solid propellants (usually reduction). This phenomenon can be defined as the growth of cross bonds in the polyurethane chain, altering the mechanical properties of traction resistance and elongation, in comparison of the properties just after the fabrication [9]. This aging process can be responsible for the appearance of failures and cracks in the grains, which compromise the propellant performance.

The AP/HTPB composite decomposition and the combustion mechanism have been extensively investigated in the last decades and the appearance of advanced methods of diagnostics, like flash pyrolysis, thermogravimetry and differential scanning calorimetry, led to the ressurgence of the interest. These methods are widely used for the investigation of thermal decomposition of organic materials [10], polymers [11,12], composites [13] and explosives [14].

Kissinger [15] and Ozawa [16] and Flynn [17] demonstrated that differential scanning calorimetry (DSC) technique, based on the linear relation between peak temperature and heating rate, can be used to determine the kinetics parameters of a thermal decomposition (activation energy, rate constant). The Ozawa method is one of the most popular methods for estimating activation energies by linear heating rate and it is the so-called isoconversional method. Thermal analysis cannot be used to elucidate the complete mechanism of a thermal degradation but the dynamic analysis has been frequently used to study the overall thermal degradation kinetics of polymers and composites because it gives reliable information on the frequency factor(A), the activation energy (E) and the overall reaction order [18].

In the present work, the differential scanning calorimetry (DSC) technique and the Ozawa dynamic method were used to determine the kinetic parameters of the aged and non-aged solid propellant, AP/HTPB, thermal decomposition. The Kissinger method for obtaining the activation energy value was also employed for a comparison purpose.

2. Experimental

2.1. Materials and apparatus

AP was obtained from Avibras Indústria Aeroespacial S.A.; HTPB from Petroflex Industry S.A., a subsidiary of Petrobras – Petróleo do Brasil S.A.; IPDI from Merck; DOA from Elekeiroz S.A.. The composite propellant was produced in a batch process of 5 kg mass (pilot plant) using a planetary mixer under vacuum atmosphere during 2 hours. All raw materials are incorporated in HTPB polyol, starting with AP that was classified to a medium size of 300 micrometers. When all ingredients are added to the HTPB polyol, the IPDI curing agent can be mixed to the liquid propellant. The propellant curing process was conducted in a temperature of 60 Celsius during a 120 hs period time.

The synthetic aging process was conducted by exposing the cured propellant formulation to a temperature of 338 K for 300 days in a muffle (FNT-F3-T 6600W) that was monitored day by day during this period.

The polyurethane network was obtained by curing HTPB polymer samples with IPDI (isophore diisocianate) at an [NCO]/[OH] equivalent ratio of 0.95, at 338 K for 120 h. The NCO/OH ratio is defined as the equivalent ratio between the materials containing NCO (IPDI) groups and those containing OH groups (HTPB) and it affects the mechanical properties of cured composite propellant [13,14]. The chemical composition of the propellant was (weight) binder 22% and others 78%. The synthetic aging process was conducted by exposing the propellant formulation to a temperature of 338 K for 300 days.

DSC curves were obtained on a model DSC50 Shimadzu in the temperature range of 298-773 K, under dynamic nitrogen atmosphere (ca. 50 mL/min). Sample masses were about 1.5 mg, and each sample was heated in hermetically sealed aluminum pans. Seven different heating rates were used for the non-aged samples: 10.0, 15.0, 20.0, 30.0, 35.0, 40.0 and 45.0 K min^{-1}; for the aged samples, three different heat rates were used: 30, 35 and 40 K min^{-1}. DSC system was calibrated with indium (m.p.= 429.6 K; ΔH_{fus}=28.54 Jg^{-1}) and zinc (m.p.= 692.6 K).

2.2. Kinetic approach

The method used in the analysis of composite samples was based on DSC experiments in which the temperatures of the extrapolated onset of the thermal decomposition process and the temperatures of maximum heat flow were determined from the resulting measured curves for exothermic reactions. DSC curves at different heating rates, β, for non-aged and aged composite samples are shown in Figs. 4 and 5, respectively.

In order to determine the kinetic parameters of the degradation step Ozawa and Kissinger's methods were applied. They were both derived from the basic kinetic equations for heterogeneous chemical reactions and therefore have a wide application, as it is not necessary to know the reaction order [19] or the conversional function to determine the kinetic parameters. The activation energy determined by applying these methods is the sum of activation energies of chemical reactions and physical processes in thermal decomposition and therefore it is called apparent.

The temperatures of exothermic peaks, T_P, can be used to calculate the kinetic parameters by the Ozawa method [16,17]. These parameters are the activation energy, Ea, and the pre-exponential factor, A, relatives to the decomposition process.

A linear relationship between the heating rate (log β) and the reciprocal of the absolute temperature, T_P^{-1}, may be found and the following linear equation can be established:

$$\log ß = a.T_P^{-1} + b \quad (1)$$

where a and b are the parameters of the linear equation: a is -0.4567E/R (slope) and b is a constant (linear coefficient). R is the gas constant.

Assuming that the rate constant follows the Arrhenius law and that the exothermic reaction can be considered as a single step process, the conversion at the maximum conversion rate is invariant with the heating rate when this is linear. Having in account such assumptions, eq.

(1) may be applied to the exothermic peak maximum temperature considering different heating rates [15,19]. Thus carrying out several experiments at different heating rates a plot of log ß vs $1/T_P$ may be done and the activation energy can be estimated directly from the slope of the curve using the following equation derived [14] from the eq.(1):

$$Ea = -\ 2.19\ R\ [d \log ß/\ d\ T_P^{-1}] \tag{2}$$

where - d log ß/ d T_P^{-1} = parameter a (eq.1).

With the same above assumptions, the Kissinger method[8] may be used to calculate the activation energy and the pre-exponential factor from the maximum rate condition which will occur at the maximum exothermic peak temperature, T_P.

The Kissinger method is based on the plot of ln (β/T_P^2) vs. $1/T_P$. Activation energy is calculated from the slope of the curve using the following equation:

$$Ea = R\ d[\ln \beta/T_P^2]\ /\ [d\ (1/T_P)] \tag{3}$$

Once time E is known the values of pre-exponential factor, A, are calculated with the equation:

$$A = (\beta\ E \exp Ea/RT_P)\ /\ RT_P^2 \tag{4}$$

The temperature dependence of the specific rate constant k is described by the Arrhenius equation:

$$k = A \exp\ (\ -Ea/RT_P) \tag{5}$$

The kinetic Shimadzu software, based on the Ozawa method, feed with the exothermic peak temperatures and the heating rate data, gives the Arrhenius kinetic parameters (Ea, A) relative to the thermal decomposition of composite and, consequently, with the eq. (5) the overall rate constant can be calculated.

There's also a possibility of using Flynn and Wall methodology with TGA analysis (constant heating rate TGA) [17], once it requires less experimental time, although this method is limited to single-step decompositions and first order kinetics. The approaches are the following:

This first approach requires at least three determinations at different heating rates (Fig. 4 below), following the Arrhenius equation:

$\left(\frac{d\alpha}{dt}\right) = A.exp\left(\frac{-E\alpha}{RT}\right)\left(1-\alpha\right)^n$, where α represents the fraction of decomposition and n is the reaction order.

Re-arranging, the equation turns to:

$E\alpha = \left(\frac{-R}{c}\right)\frac{d\ln\beta}{d\left(\frac{1}{T}\right)}$, where c is a constant for n=1.

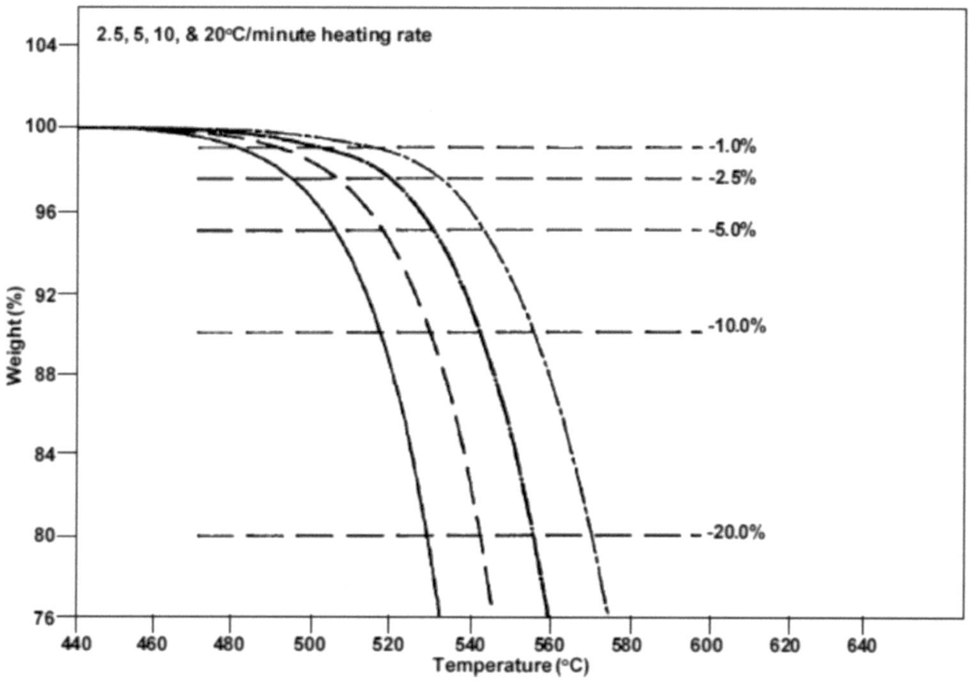

Figure 4. Constant heating rate TGA plots – Flynn & Wall method [20]

From this curve is possible to construct a ln β vs 1/T plot. The slope of this new curve is used to calculate the activation energy.

3. Results and discussion

The activation energy and kinetic parameters of thermal decomposition of propellant samples were calculated by Ozawa method using DSC curves at different heating rates: 10.0, 15.0, 20.0, 30.0, 35.0, 40.0 and 45.0 K min^{-1} for the non-aged ones and 30.0, 35.0 and 40.0 K min^{-1} for the aged ones.

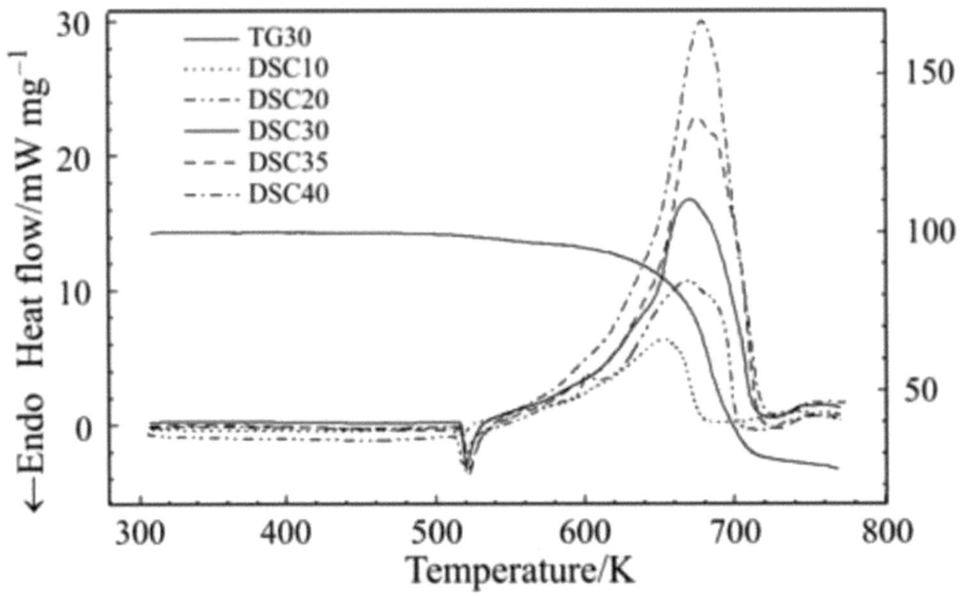

Figure 5. DSC curves of thermal decomposition of non-aged composite samples, AP/HTPB, at the heating rates: 10, 20, 30, 35 and 40 K min^{-1} and TG curve with a heating rate of 30 K min^{-1}

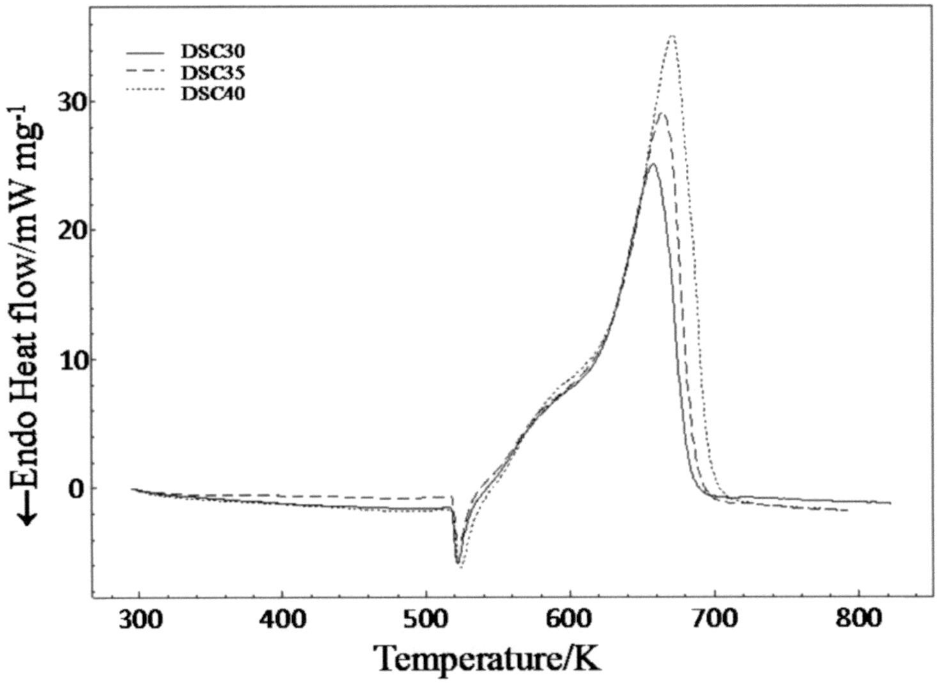

Figure 6. DSC curves of thermal decomposition of aged composite samples, AP/HTPB, at the heating rates: 30, 35 and 40 K min^{-1}

The exothermic events have different maximum temperatures in both cases; the higher the heating rate, higher is the maximum temperature of the peak.

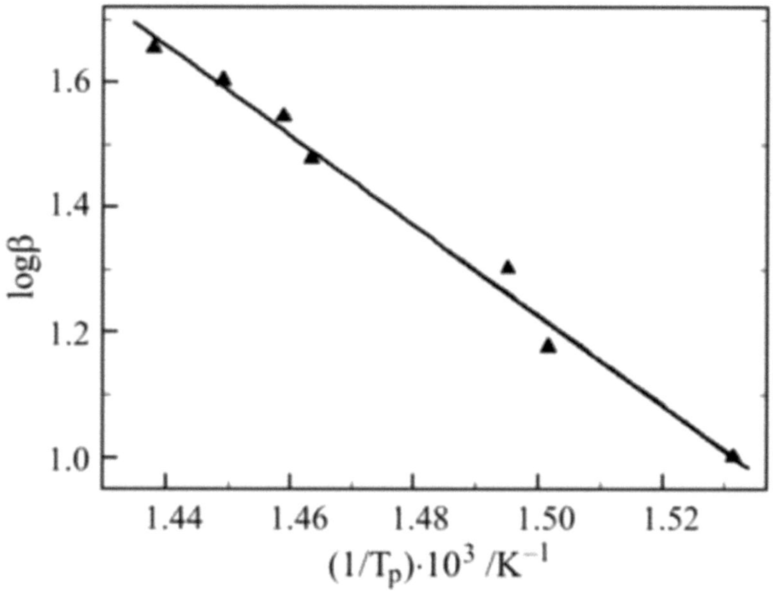

Figure 7. Ozawa plot for AP/HTPB samples at 10, 15, 20, 30, 35, 40 and 45 K min^{-1}

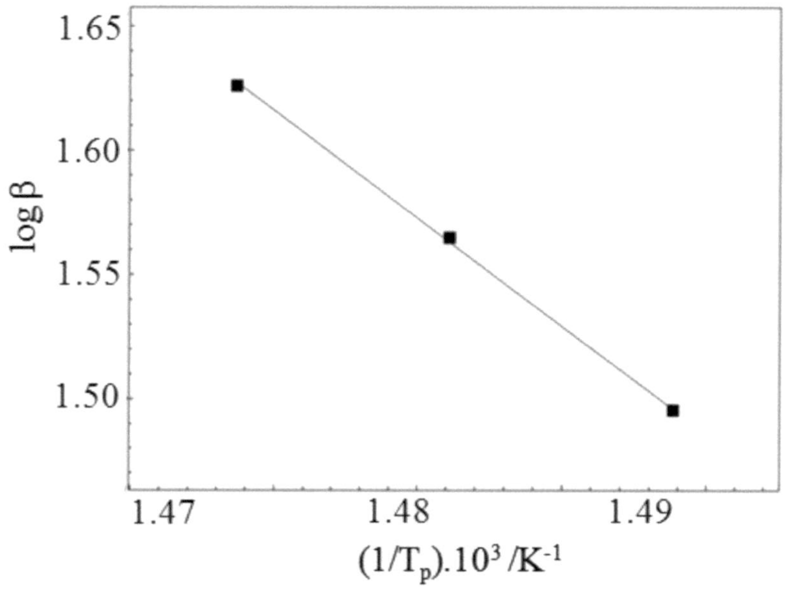

Figure 8. Ozawa plot for aged AP/HTPB samples at 30, 35 and 40 K min^{-1}

The DSC curves are presented in Figs. 5 and 6. The DSC curves show that the first stage is endothermic and the second stage is exothermic. The endothermic event is quite similar for the different heating rates used and it shows the same peak temperature. This event occurs around 520 K and it was not considered because it represents a phase transition of ammonium perchlorate (AP) from the orthorhombic to the cubic form [21,22]. Together with DSC curves obtained for different heating rates (non-aged samples), Fig. 1, the TG curve for 30.0 K min^{-1} was included to show that in the region corresponding to the endothermic peak (DSC curves) there is no any weight loss or, at least, it is imperceptible and, the same behavior was observed in all of other TG curves for different heating rates.

Figures 7 and 8 show the plot of log β vs the reciprocal of the absolute temperature relative to each maximum of the exothermic stage. The values of the activation energy were found to be 134.5 kJ mol^{-1} (non-aged samples) and 79.0 kJ mol^{-1} (aged samples). Sell et al.[23] using thermogravimetry at heating rates between 0.5 and 10 K min^{-1} studied the decomposition kinetics of the AP/HTPB propellant samples with isoconversional method and the calculated activation energies are between 100 and 230 kJ mol^{-1}.

From the slope of Kissinger plot (ln (β/T_P^2) vs. $1/T_P$) and eq. (3) the activation energy was also calculated and is 126.2 kJ mol^{-1} for the non-aged samples, therefore quite similar to that obtained using the Ozawa method.

The thermal decomposition of solid composite propellant is a multistep process and the reaction mechanism changes with the temperature and, consequently, the activation energy varies with the extent of the reaction. DSC data are used to estimate the activation energies of thermal decomposition of propellant samples because the global decomposition reaction is taken in account. Implicit in any discussion about the decomposition is the fact that the overall process is complex, and any derived rate parameters do not correspond to an elementary single step. TG/DTG results are in agreement with this assumption.

The pre-exponential factor was found to be 2.04 10^{10} min^{-1} (non-aged samples) and $1.29.10^6$ min^{-1} (aged samples) and the reaction orders for the global composite decomposition were estimated in 0.7 (non-aged) and 0.6 (aged) by the kinetic Shimadzu software based in the Ozawa method. This value is quite different from the Arrhenius assumption where the reaction order is always considered as 1.0. For practical purposes the Arrhenius parameters, like the corrected reaction order, can be used to estimate the overall rate constant (k) for thermal decomposition using the eq. (5).

The differences found in the kinetics parameters between the original and the aged samples, specially the activation energy (Ea), confirm the practical observation that energetic materials like the composites used in solid propellant rocket motors require less energy to start the combustion process as they age. Besides, considering the heating rate of 40 K min^{-1} for the original and for the aged samples, a reduction in the enthalpy of the decomposition's exothermic phase was observed (2.56 to 1.15 J g^{-1}).

Cohen [24] studied the kinetics of the surface pyrolysis of HTPB and, assuming zero-order kinetics, they found the activation energy of 71 kJ mol^{-1}. Comparison between the activation energies for the propellant decomposition and the activation energies for decomposition of individual ammonium perchlorate (AP) or/and HTPB binder suggests that the overall kinetics of the mass loss is determined by the reaction between the binder and the decomposition products of AP [24].

Ammonium perchlorate is widely used as an oxidizer in energetic composites and it is one of the most important raw materials in propellant formulations where it represents at least 80 % of total mass of composite solid propellants, so its contribution on the thermal decomposition behavior of propellant samples is always very important. The addition of burning rates catalysts like Fe_2O_3 on the propellant formulation alters the thermal decomposition behavior of AP, and consequently the thermal decomposition behavior of the propellant. Shin-Ming [22] showed that the presence of these catalysts compounds reduce the maximum decomposition reaction temperature in AP samples.

Another important aspect of DSC curves is the correlation of maximum temperature of exothermic peak obtained for each heating rate applied to the composite sample during the experiments. This correlation can be used to determine the burning rate characteristics of a composite solid propellant with a specific formulation. The burning rate characteristics are an important ballistic parameter of the energetic composite like solid propellant. Xiao-Bin [25] showed that the burning rates of propellants were very closely related to the exothermic peak temperature of ammonium nitrate (AN) that is used as an oxidizer in smokeless propellant formulation.

In the present work, the DSC curves at different heating rates were obtained for original and synthetically aged samples that have the same raw materials and with the same manufacture process. These conditions are necessary because differences in the raw materials, as ammonium perchlorate (AP) particle size, can affect the thermal decomposition behavior of the composite. In other words, the decomposition mechanism of AP powder of fine particle size differs that of AP of larger particle size.

4. Conclusions

For energetic materials like composite solid propellant, it is critical to use the minimum sample size and low heating rates to avoid the risks to potential damage of the DSC cell resulting in DSC curves with a lot of interferences caused by the detonation behavior of composite samples. In opposition to this criteria, in this study, high heating rates were used (10.0 to 45.0 K min^{-1}), but to compensate this condition very low sample sizes were used ($\approx$ 1.5 mg). Despite these heating rates are not close to the rocket motor chamber conditions (heating rates estimated as 10^6 K s^{-1}) the slower heating rates used in this work allow one to get a better insight into the reaction kinetics mechanisms.

The Ozawa and Kissinger methods demonstrated that differential scanning calorimetry technique, based on the linear relation between peak temperature and heating rate, can be used to determine the kinetics parameters of thermal decomposition reaction of energetic materials giving reproducible results.

The DSC curves do not show any interference and the kinetic data obtained using the maximum temperatures (reciprocal, in K^{-1}) and the respective heating rates are very close to the results found in the literature, at very lower heating rates [26-29].

Author details

R. F. B. Gonçalves, J. A. F. F. Rocco and K. Iha
Instituto Tecnológico de Aeronáutica, CTA, São José dos Campos, S.P., Brasil

Acknowledgement

The authors gratefully acknowledge financial support from FAPESP (Fundação de Amparo à Pesquisa do Estado de São Paulo) and CNPq (Conselho Nacional de Desenvolvimento Científico e Tecnológico) for the fundings and investiments.

5. References

[1] Brill T B, Budenz B T (2000) Flash Pyrolysis of Ammonium Perchlorate-Hydroxyl-Terminated-Polybutadiene Mixtures Including Selected Additives, Progress in Astronautics and Aeronautics, AIAA, Vol.185, pp. 3.

[2] Beckstead M W, Puduppakkam K, Thakre P, Yang V (2007) Progress in Energy and Combustion Science, 33, 497.

[3] Boggs T L (1970) AIAA Journal, 8, 5, 867.

[4] Beckstead M W, Puduppakkam K V (2004) Modeling and Simulation of Combustion of Solid Propellant Ingredients Using Detailed Chemical Kinetics, 40th AIAA/ASME/SAE/ASEE Joint Propulsion Conference and Exhibit.

[5] Kubota N (2002) Propellants and Explosives, Wiley-VCH.

[6] Gross M L (2007) Two-dimensional modeling of AP/HTPB utilizing a vorticity formulation and one-dimensional modeling of AP and ADN.

[7] Gonçalves R F B, Rocco J A F F, Machado F B C, Iha K (2012) Ammonium perchlorate and ammonium perchlorate-hydroxyl terminated polybutadiene simulated combustion, Journal of Aerospace Technology and Management, v 4, n1.

[8] Korobeinichev O, Ermolin N, Chernov A, Emel'yanov I (1990) AIAA/SAE/ASME/ASEE 26th Joint Propulsion Conference.

[9] Celina M, Minier L, Assink, R (2002) Development and application tool characterize the oxidative degradation of AP/HTPB/Al propellants in a propellant reability study, Thermochimica Acta, Vol. 384, pp. 343, 349.

[10] Mathot V B F (2001) New routes for thermal analysis and calorimetry as applied to polymeric systems, J. Therm. Anal. Cal., Vol. 64, pp. 15, 35.
[11] Maijling J, Simon P, Khunová V (2002) Optical Transmittance Thermal Analysis of the Poly(Ethylene Terephtalate) Foils, J. Therm. Anal. Cal., Vol. 67, pp. 201, 206.
[12] de Klerk W P C, Schrader, M A, van der Steen A C (1999) Compatibility Testing of Energetic Materials, Which Technique?, J. Therm. Anal. Cal., Vol. 56, pp. 1123, 1131.
[13] Stankovic M, Kapor V, Petrovic S (1999) The Thermal Decomposition of Triple-Base Propellants, J. Therm. Anal. Cal., Vol. 56, pp. 1383-1388.
[14] Jones D E G, Feng H T, Augsten R A, Fouchard R C (1999) Thermal Analysis Studies on Isopropylnitrate, J. Therm. Anal. Cal., Vol. 55, pp, 9, 19.
[15] Kissinger H E (1957) Reaction Kinetics in Differential Thermal Analysis, Anal. Chem., Vol. 29, pp. 1702.
[16] Ozawa T, Isozaki H, Negishi A (1970) A new type of quantitative differential analysis, Thermochimica Acta., Vol. 1, No. 6, pp. 545, 553.
[17] Flynn J H (1966) A quick, direct method for the determination of activation energy from thermogravimetric data, Thermochim. Acta, Vol. 4, pp. 323.
[18] Park J W, Lee H P, Kim H T, Yoo K O (2000) A kinetic analysis of thermal degradation of polymers using a dynamic method, Polym. Degrad.Stabil., Vol. 67, pp. 535.
[19] Ozawa T (2001) Temperature control modes in thermal analysis, J. Therm. Anal. Cal., Vol. 64, 2001, pp. 109, 126.
[20] Sauerbrunn S, Gill P, Decomposition Kinetics using TGA, TA Instruments.
[21] Na-Lu L, Tsao-Fa Y (1991) The thermal behavior of porous residual ammonium percholate, Thermochim. Acta, Vol. 186, pp. 53.
[22] Shin-Ming S, Sun-I C, Bor-Horng W (1993) The thermal decomposition of ammonium perchlorate (AP) containing a burning-rate modifier, Thermochim. Acta, Vol. 223, pp. 135.
[23] Sell T, Vyazovkin S, Wight C A (1999) Thermal decomposition kinetics of PBAN-Binder and composite solid rocket propellants, Combust. Flame, Vol. 119, pp. 174.
[24] Cohen N S, Fleming R W, Derr R L (1974) Role of binders in solid propellant combustion, AIAA Journal, Vol. 6, pp. 212.
[25] Xiao-Bin Z, Lin-Fa H, Xiao-Ping Z (2000) Thermal decomposition and combustion of GAP/NA/Nitrate Ester propellants, Progress in Astronautics and Aeronautics, AIAA, Vol. 185, pp. 413.
[26] Du T (1989) Thermal decomposition studies of solid propellant binder HTPB, Thermochim. Acta, Vol. 138, pp. 189.
[27] Rocco J A F F, Lima J E S, Frutuoso A G, Iha K, Ionashiro M, Matos J R, Suárez-Iha M E V (2004) Thermal degradation of a composite solid propellant examined by DSC – Kinetic study, J. Therm. Anal. Cal., Vol. 75, pp. 551, 557.

[28] Andrade J, Frutuoso A G, Iha K, Rocco J A F F, Bezerra E M, Matos J R, Suárez-Iha M E V (2008) Estudo da decomposição térmica de propelente sólido compósito de baixa emissão de fumaça, Quim. Nova, Vol. 31, No. 2, pp. 301-305.
[29] Andrade J, Iha K, Rocco J A F F, Bezerra E M (2007) Análise térmica aplicada ao estudo de materiais energéticos, Quim. Nova, Vol. 30, No. 4, pp. 952, 956.

Chapter 15

Numerical Solutions for Structural Relaxation of Amorphous Alloys Studied by Activation Energy Spectrum Model

Kazu-masa Yamada

Additional information is available at the end of the chapter

http://dx.doi.org/10.5772/53723

1. Introduction

What is the physical process which is producing the general characteristic of glass? It is said that the transformation of the liquid-system to glass one is a final theme in physics through into the twenty-first century. Furthermore, the development of amorphous-material devices and specimen modification methods is closely related with the obviousness of high thermal stability and stability of relaxation.

The aim of this research is to clarify numerical solutions for nano-structure relaxation processes focusing on the activation energy in transition metal (ie Cu, Fe) based amorphous alloys. Activation energy for structural relaxation process in a metal type amorphous ternary and quaternary alloys, with cross sections of typically 0.03 mm x 2.0 mm, prepared by chill-block melt spinning has been investigated by Differential Scanning Calorimetry (DSC) with a cyclically heating technique [1,2,3]. Activation energies for structural relaxation with a spatial quantity in amorphous materials have been discussed by use of a total relaxed ratio function that depends on annealing temperature and time. In the present work in amorphous ternary and quaternary alloys, the distributions for the Activation Energy Spectrum (AES) with derivative-type relaxed ratio function were observed. Another result has been also established that the "reversible" AES model energy distribution though the cyclically nano-structural relaxations were in good agreement with the presented experimental results of transition metal based amorphous alloys.

There has been recently considerable that the glassy alloys are representative of the bulk formed ultra-fine structure [1]. Particularly Cu has been shown to be good base element for bulk glass-forming alloy with fully glassy sections recently by use of die injection casting [2,3]. Binary Cu - (Zr or Hf) alloys have been found to form an amorphous phase over a wide composition range. However, addition of Ti in both these binary systems greatly increased the glass forming ability (GFA), with the critical diameter for fully amorphous rods being at least 4 mm for $Cu_{60}Zr_{30}Ti_{10}$, $Cu_{60}Hf_{20}Ti_{20}$ and $Cu_{55}Hf_{25}Ti_{20}$ [2,3]. Meanwhile the understanding of the structural relaxation process is essential in the development of stability for amorphous alloys, as well as in establishing stable working temperature to avoid the degradation of strength. Therefore, high thermal stability of quasi-stable amorphous materials for Cu based alloys. The atomic mechanism of diffusion in amorphous alloys is still poorly understood as compared to that in crystalline alloys. However, measurements of diffusivity in amorphous alloys have been limited so far because of the experimental difficulties of measuring the very small diffusion coefficients, usually less than 10^{-17} m^2s^{-1}, which are typical of amorphous alloys below their crystallization temperatures [4,5].

In the present work, using Differential Scanning Calorimetry (DSC) thermal analysis has been made to determine the activation processes [6,7,8], and to evaluate whether it represents the thermodynamically stable form of CuHfTi and CuHfTi-B glass-forming amorphous alloys.

2. Experimental procedure

Cu-based alloy ingots of composition $Cu_{60}Hf_{20}Ti_{20}$, $(Cu_{60}Hf_{22}Ti_{18})_{0.99}B_1$ and $(Cu_{60}Hf_{22}Ti_{18})_{0.97}B_3$ were prepared by arc-melting mixtures in an argon gas atmosphere purified with a Ti getter. The alloy compositions represent the nominal values but the weight losses in melting were negligible. The alloy ingots were inverted on the hearth and re-melted several times, to ensure compositional homogeneity. Ribbon samples of each alloy, with cross sections of typically 0.03 mm * 2.0 mm, were produced by chill-block melt spinning in a sealed inactive gas atmosphere. The amorphous state of the specimen of the ribbon samples was confirmed by X-ray diffraction.

The endothermic/exothermic heats for relaxation process were measured by differential scanning calorimetry (DSC) of DSC3100s of MacScience Co., Ltd (Bruker Japan Co., Ltd.) at a constant heating rate of 1.00 K/s. And the ordered specimens were prepared by annealing used in the electric furnace of the DSC. Pre-annealing and following long-time main annealing by a DSC furnace are at a 700 K for 1800 s and at a 580 K for 6000 s, respectively. The maximum temperature of 700 K is enough to suppress the crystallization and to measure optimistically the structural relaxation for these three kinds of specimen [9].

2.1. Theory with DSC annealing process

Due to insufficient data of thermal stability of amorphous alloys, the following points are left as future problems. Even bulk glass-forming alloy, also amorphous alloys is a non-equilibrium state. The certain overall atoms in an amorphous alloy are in non-stable state rather than in the stable crystalline state. Therefore, not only crystallization over a certain wide temperature range but also re-arrangement of atoms occurs. The structural relaxation process is one of the essential phenomena in some non-equilibrium materials. Thereby, to study the structural relaxation is important to investigate the constitutional property of the amorphous alloys. Furthermore, the structural relaxation is closely connected with the stability of specific examples related to the bulk glass-forming amorphous alloy. It is also necessary to know this property from a viewpoint of the application development.

Consider the population of an assembly of reaction centre for structural relaxation, that is to say isolated double wells potential model (or so called Two Level System, TLS) as shown in Fig. 1[9]. A relaxation centre which is isolated and in a particular structural configuration, permits an atom to be either in a higher energy position at state 0 or in a lower energy position at state 1 in Fig. 1. The axis of abscissas is the configuration variable for relaxation processes, and the position 1/2 on the axis in Fig.1 is the saddle point for the energy wall between the position 0 and 1.

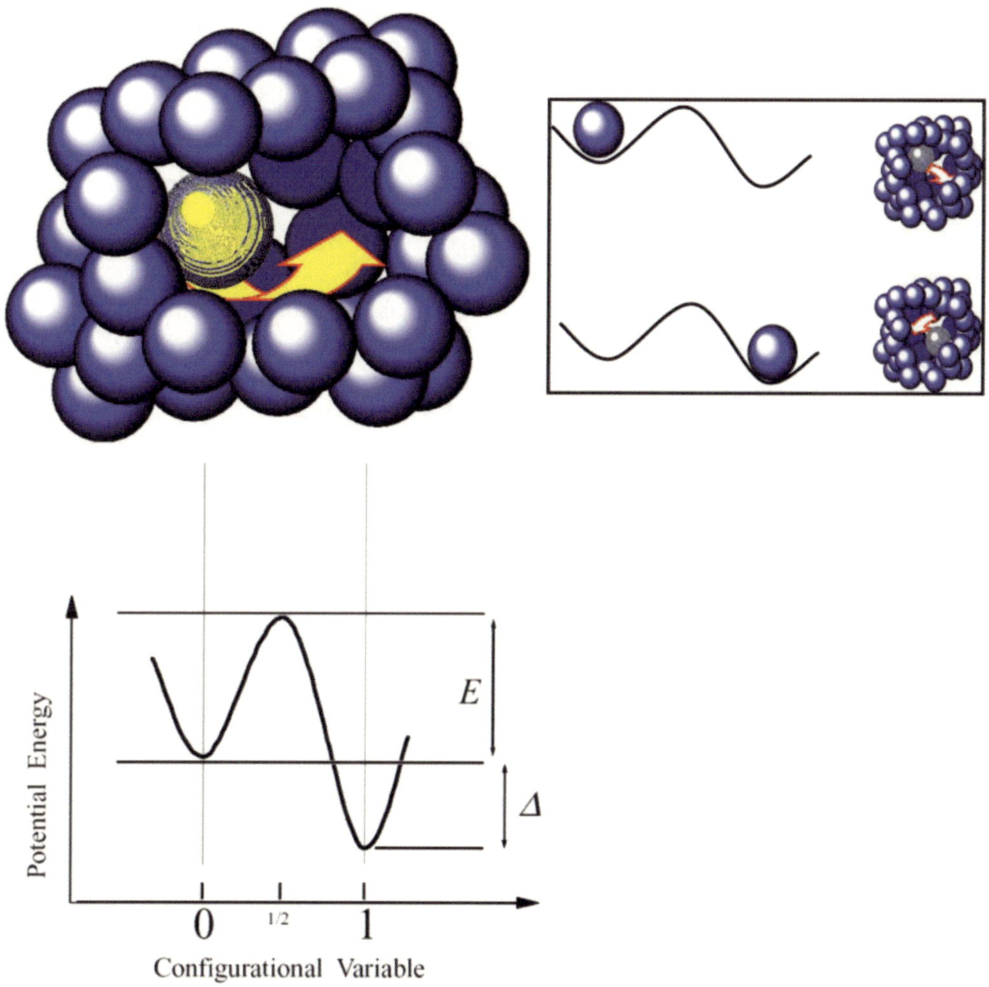

Figure 1. Schematic illustrations of relaxation centre and energy levels for the corresponding two level system

On the population of an assembly of this model, activation energy spectrum (AES) in structural relaxation processes, J. A. Leake, J. E. Evetts and M. R. J. Gibbs [10,11] describe the phenomena of physical available and variable property with good agreement between the theory and the experimentation. The theory assumes exponent factor nearly equal one-dimension for chemical reaction kinetics of Jhonson-Mehl-Avrami (JMA) equation.

Therefore, the theoretical model for the relaxation process in amorphous materials on the basis of a spectrum of available processes with a distribution of activation energy was proposed. In their model, the total change in the measured property, ΔP is given by

$$\Delta P = \int_0^E p(E)\mathrm{d}E \tag{1}$$

In the range of activation energy E to E+dE during the structural relaxation process, the total available property $p_0(E)$ changes such as

$$p(E)\mathrm{d}E = p_0(E)\left[1-\exp\left\{-\nu_0 t\exp\left(-\frac{E}{kT}\right)\right\}\right]\mathrm{d}E \tag{2}$$

where ν_0 is an order of the Debye frequency ($\nu_0 \cong 10^{12}$ Hz) . Primak [12] rewrites Eq. (2) as

$$p(E) = p_0(E)\,\theta\left(E,T,t\right), \tag{3}$$

where $\theta(E, T, t)$ is defined as the characteristic annealing function. Thus, the function of $\theta(E, T, t)$ is a measure of the proportion of available processes at the energy E. Proportion as $\theta(E, T, t)$ has contributed to the relaxation property after the time t at the annealing temperature T. The form of $\theta(E, T, t)$ is given in Fig. 2 (a) and (b) .

In the another paper [9], in a process for most simplifying assumption, the function $\theta(E, T, t)$ can be replaced by step function at an energy $E_0(T, t)$ as

$$E_0 = kT\ln\left(\nu_0 t\right) \tag{4}$$

These E_0 forms are given in Fig. 2 (c, d, e, f, g, h, i), then E_0 changes from 0 to 1 over one-step meanwhile θ (E) changes over a narrow range of E and T. If in the simplified calculations, the E_0 should have been applied. On the other hand, in the present work for calculation using specific 1st derivative-type relaxation ratio, as follows.

$$\frac{d\theta\left(E,T,t\right)}{dE} = \left\{\nu_0 t\exp\left(-\frac{E}{kT}\right)\right\}\left(-\frac{E}{kT}\right)\exp\left\{-\nu_0 t\exp\left(-\frac{E}{kT}\right)\right\} \tag{5}$$

The (5) function is shown in Fig 3 (a), on the contrary the derivative of $E_0(T)$ is shown in Fig 3 (b) . So it should be preferred replacement with a better approximation using the function (5). By using simple area summation method, furthermore the normalization for the linear function of S=0.234 T+0.244 in this case as shown in Fig 4 would be applied, we can estimate the actual activation energy spectra distributions using the method of Fig 5. It is so called normalized 1st derivative - type relaxation ratio function in our organized work.

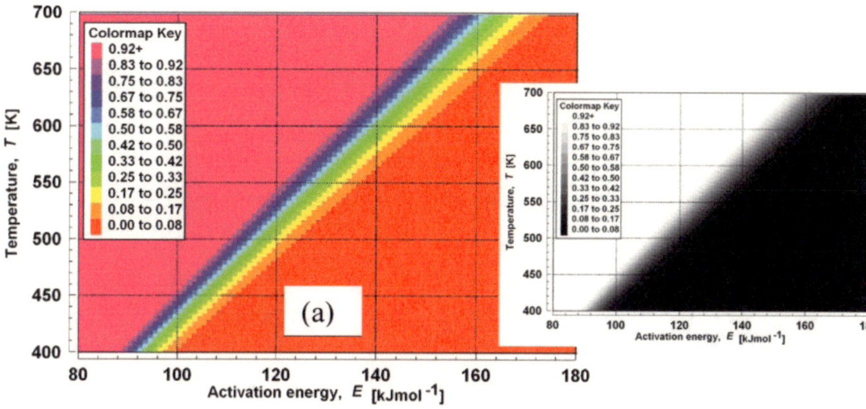
Colormap Key
0.92+
0.83 to 0.92
0.75 to 0.83
0.67 to 0.75
0.58 to 0.67
0.50 to 0.58
0.42 to 0.50
0.33 to 0.42
0.25 to 0.33
0.17 to 0.25
0.08 to 0.17
0.00 to 0.08
(a)
Temperature, T [K]
Activation energy, E [kJmol^-1]
700
650
600
550
500
450
400
80
100
120
140
160
180

(a)

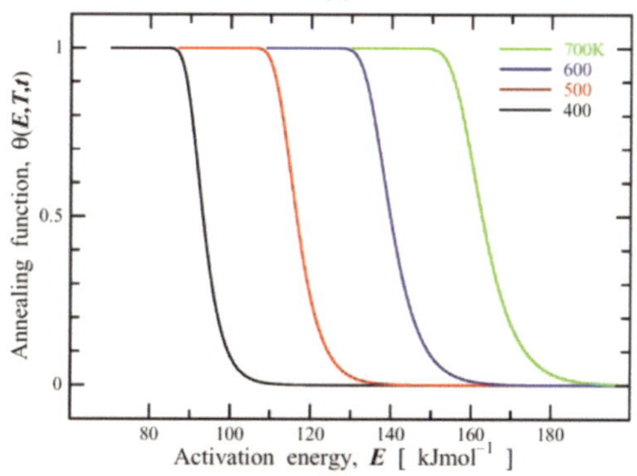
700K
600
500
400
Annealing function, θ(E,T,t)
1
0.5
0
80
100
120
140
160
180
Activation energy, E [kJmol^-1]

(b)

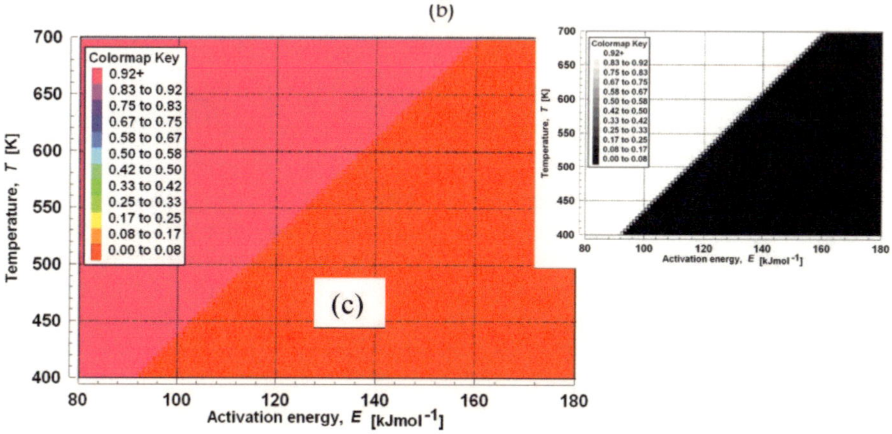
Colormap Key
0.92+
0.83 to 0.92
0.75 to 0.83
0.67 to 0.75
0.58 to 0.67
0.50 to 0.58
0.42 to 0.50
0.33 to 0.42
0.25 to 0.33
0.17 to 0.25
0.08 to 0.17
0.00 to 0.08
(c)
Temperature, T [K]
Activation energy, E [kJmol^-1]
700
650
600
550
500
450
400
80
100
120
140
160
180

(c)

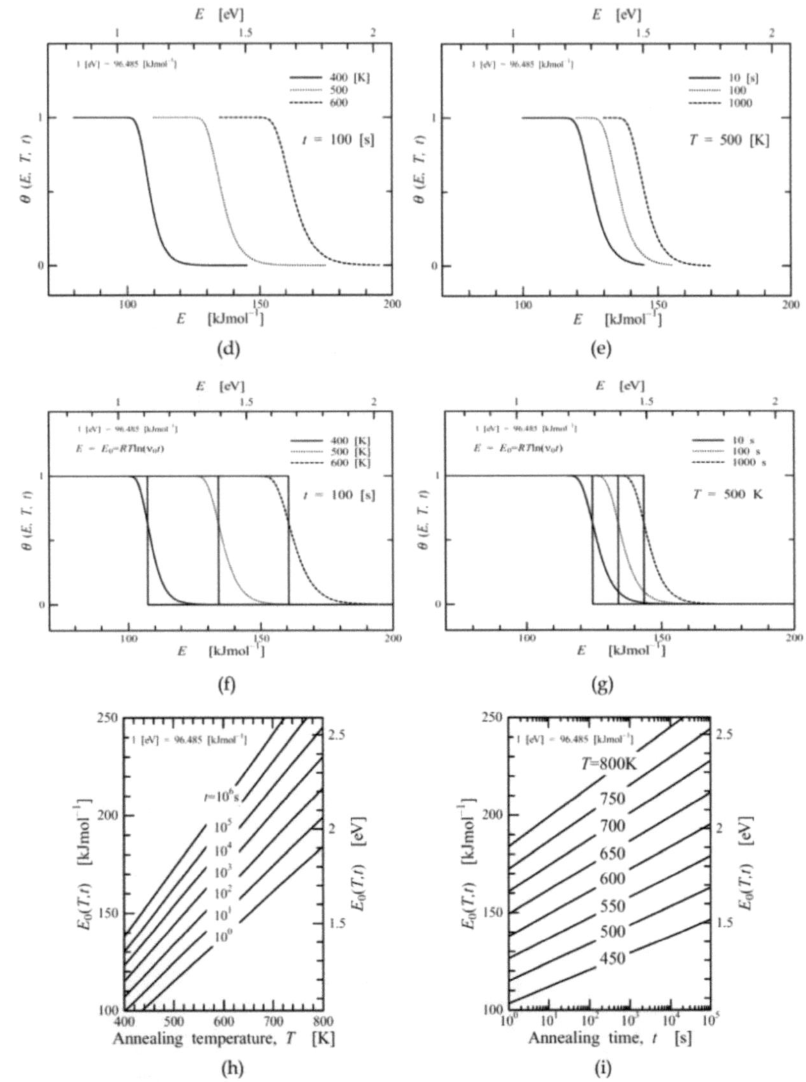

Figure 2. (a). Dependence of the characteristic annealing function $\theta(E, T, t\text{=}1\text{s})$ on the activation energy E and the temperature T [9,10]. Fig. 2(a) is used of $\theta(E, T, t\text{=}1\text{s})$ (b). Dependence of the characteristic annealing function $\Theta(E, T, t\text{=}1\text{s})$ on the activation energy E and the temperature T [9,10]. Fig. 2(b) is used of $\Theta(E, T, t\text{=}1\text{s})$ (c). Dependence of the characteristic annealing function $\theta(E, T, t\text{=}1\text{s})$ on the activation energy E and the temperature T [9,10]. Fig. 2(c) is used of $E_0(T, t\text{=}1\text{s})$. Right hand and left hand of charts are the same figure. Additionally right one is shown in grey scale for visible-contrasty (d,e,f,g). Dependence of the characteristic annealing function $\theta(E, T, t)$ and step function of threshold $E_0(T, t)$ on the horizontal axis of the activation energy E [9,10]. Fig. 2(d) at the upper left is used of $\theta(E, T, t\text{=}100\text{s})$. Fig. 2(e) at the upper right is used of $\theta(E, T\text{=}500\text{K}, t)$. Fig. 2(f) at the lower left is used with step functions superposed on $\theta(E, T, t\text{=}100\text{s})$. Fig. 2(g) at the lower right is used with step functions superposed on $\theta(E, T\text{=}500\text{K}, t)$ (h, i). Dependence of the simplifying assumption of characteristic annealing function, activation energy $E_0(T, t)$ vs the annealing time t and the isothermal annealing temperature T [9,10]. Fig. 2(h) at the left is used of $E_0(T, t\text{=}1, 10, 10^2, 10^3, 10^4, 10^5 \text{ and } 10^6 \text{ s})$. Fig. 2(i) at the right is used of $E_0(T\text{=}450, 500, 550, 600, 650, 700, 750 \text{ and } 800\text{K})$

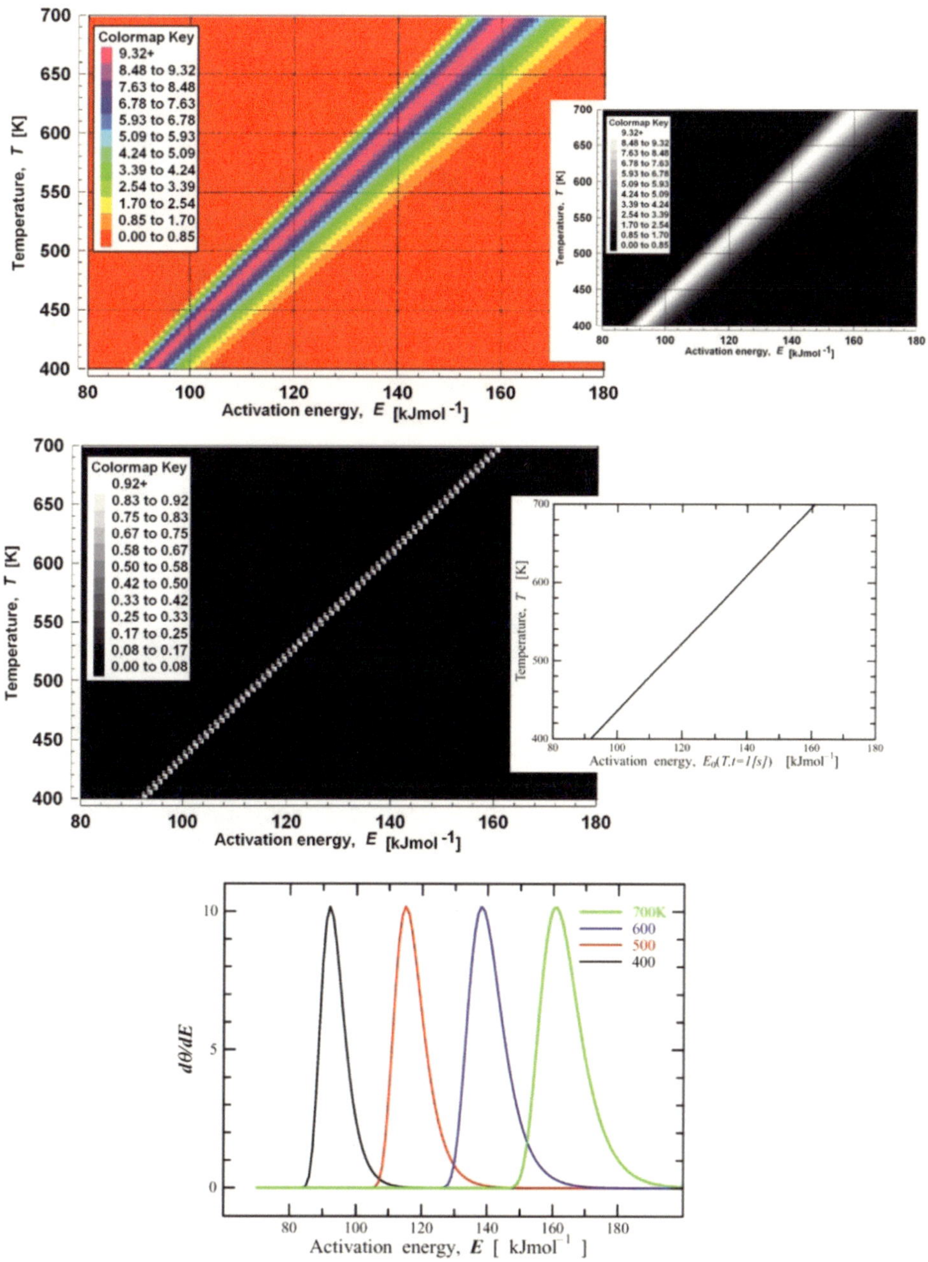

Figure 3. (a). Dependence of the 1st derivation of θ (E, T, t=1 s) on the activation energy E and the temperature T . Fig. 3 (a) Derivative type θ(E, T, t=1 s) . Right hand and left hand of charts are the same figure. Additionally right one is shown in grey scale for visible-contrasty (b). Dependence of the 1st derivation of θ (E, T, t=1 s) on the activation energy E and the temperature T . Fig. 3 (b) Approximation type $E_0(T, t=1\text{ s})$. Right hand and left hand of charts are the same figure. Additionally right one is for visible-contrasty (c). Dependence of the 1st derivation of θ (E, T=400, 500, 600 and 700K, t=1 s) on the activation energy E.

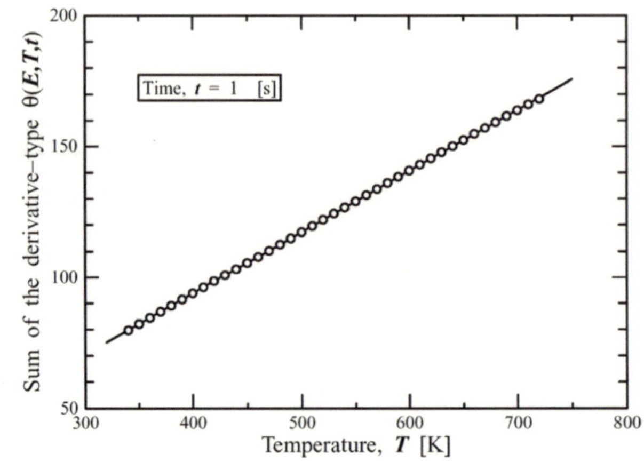

Figure 4. Dependence of the summation S of *1st* derivative type $\theta(E, T, t{=}1\text{s})$ on the temperature T. Plots can be replaced by linear function with a good approximation in $S = 0.234\ T + 0.244$

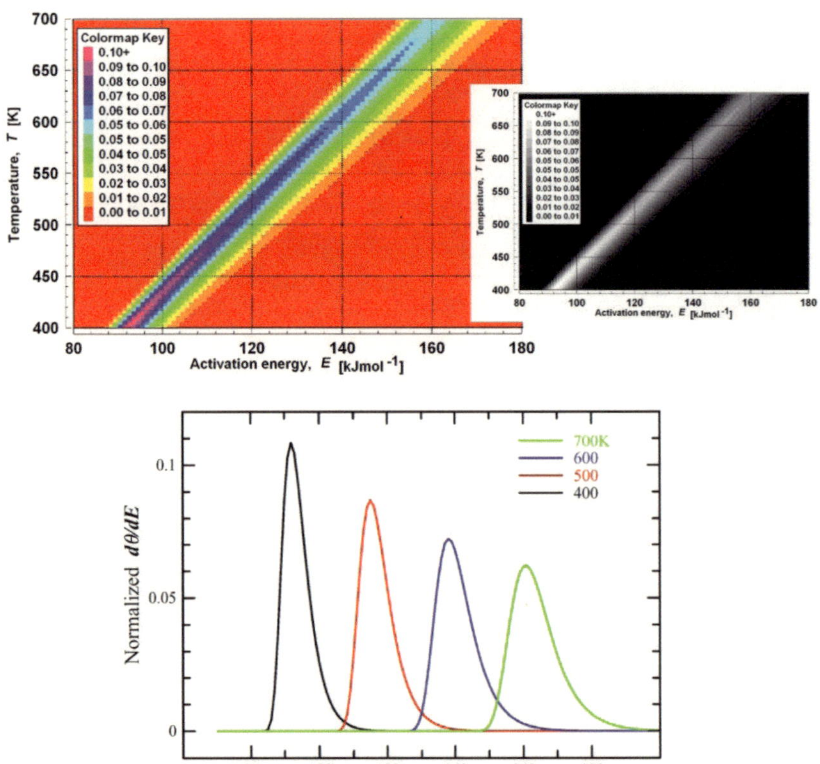

Figure 5. (a). Divided by the $S = 0.234\ T + 0.244$ of Fig 4, dependence of the normalized 1st derivative - type relaxation ratio function of $\theta(E, T, t{=}1\text{s})$ on the activation energy E and the temperature T. Right hand and left hand of charts are the same figure. Additionally right one is shown in grey scale for visible-contrasty (b). Divided by the $S = 0.234\ T + 0.244$ of Fig 4, dependence of the normalized 1st derivative - type relaxation ratio function of $\theta(E, T{=}400, 500, 600 \text{ and } 700\text{K}, t{=}1\text{ s})$ on the activation energy E

2.2. Experimental processes with DSC annealing

In the previous paper [9], we discussed at first the AES applied on the Cu-Hf-Ti system, because of the following reasons.

In Table 1 [2] the glass forming ability (GFA) related to the T_{rg} (=T_g / T_L) was observed that addition of Ti in Cu-Hf binary systems greatly increased. In this table, they were shown with the critical diameter dC for fully amorphous rods being at least 3 mm for $Cu_{60}Hf_{22.5}Ti_{17.5}$, $Cu_{60}Hf_{20}Ti_{20}$ and $Cu_{60}Hf_{17.5}Ti_{22.5}$.

Composition	d_C	T_g (K)	T_X (K)	T_L (K)	T_{rg} (=T_g / T_L)
$Cu_{60}Hf_{35}Ti_5$	Ribbon	775	828	1333	0.58
$Cu_{60}Hf_{30}Ti_{10}$	Ribbon	751	806	1263	0.59
$Cu_{60}Hf_{25}Ti_{15}$	Ribbon	748	788	1240	0.60
$Cu_{60}Hf_{22.5}Ti_{17.5}$	4mm	745	780	1234	0.60
$Cu_{60}Hf_{20}Ti_{20}$	4mm	740	767	1211	0.61
$Cu_{60}Hf_{17.5}Ti_{22.5}$	3mm	732	755	1229	0.59
$Cu_{60}Hf_{15}Ti_{25}$	Ribbon	722	745	1223	0.59
$Cu_{60}Hf_{10}Ti_{30}$	Ribbon	700	726	1233	0.56
$Cu_{60}Hf_5Ti_{35}$	Ribbon	690	712	1223	0.56

Table 1. Glass forming section thickness and thermal property for $Cu_{60}Hf_{40-X}Ti_X$ (X=from 5 to 35) alloy series [2]

After all, the aim of this research is also to clarify a quantitative evaluation in the structure relaxation processes focusing on the activation energy in $Cu_{60}Hf_{20}Ti_{20}$ based amorphous alloys with high GFA series.

Pre-annealing and main-annealing conditions were completely similar the way as Ref. 9. After that it will be noted that an atom to be in a higher energy position at stage 0 in Fig. 1, endothermic heat occurs at 1st run in the measurement scanning #1 even at 1st run in #2, that is to say the reversible relaxation processes, and the 1st minus 2nd run indicates the endothermic value (the 2nd minus 1st run indicates the exothermic value) that could be calculate the AES distributions that means an atom to be in a higher energy position at stage 0 in Fig. 1.

The liquidus temperature T_L has its minimum value for the 20at%Ti alloy in Table 1, probably corresponding to a ternary eutectic system, it is because only this composition has single melting peak in the DTA trace [2]. Thus this is a dominating factor in determining that T_{rg} has its maximum value at this composition clearly. The atomic diameter of Ti (0.289nm) is intermediate between those of Cu (0.256nm) and Hf (0.315nm) and, evidently, equal proportions of Hf and Ti result in maximum stabilization of the densely packed liquid structure [2] and normalized 1st derivative - type relaxation ratio function.

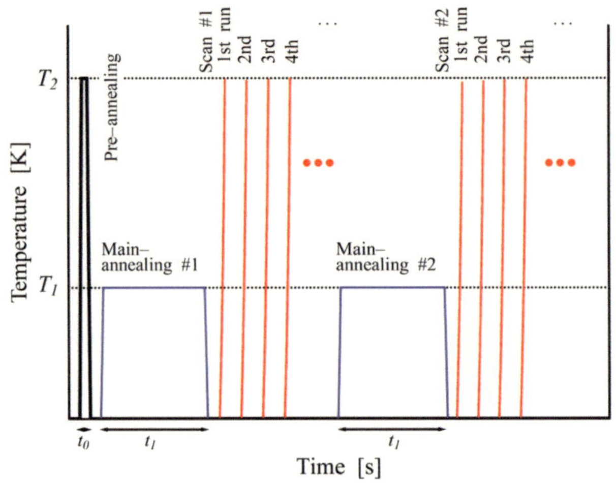

Figure 6. An example of annealing time and temperature history by use in fully electric furnace of the DSC. Specimens were prepared by annealing used in the DSC furnace, pre-annealing and following long-time main-annealing are at 700 K (=T_2) for 1800 s (=t_0) and recycled at a 580 K(=T_1) for 6000 s(=t_1), respectively. The maximum temperature of T_2 is enough to suppress the crystallization and to measure optimistically the structural relaxation following the continuous scan #1 1st run to n-th run. The "reversible" phenomena have been observed in the anneal process of scan #2 to scan #n as shown in Fig.7.

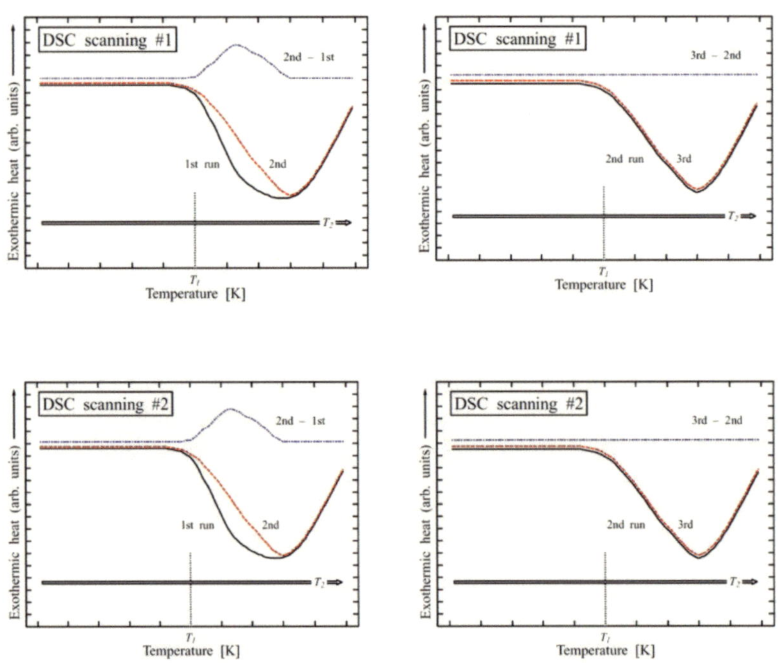

Figure 7. four-leafed schematic illustrations of reversible phenomena for DSC scanning #1 then #2, and included relaxation processes for scanning 1st run then 2nd run, on the other hand without relaxation processes for scanning 2nd run then 3rd run

3. Results through experimental procedure

In the presented work by use of the AES model, following above-mentioned, activation energies in structural relaxation processes have been determined of composition $Cu_{60}Hf_{20}Ti_{20}$ and related $(Cu_{60}Hf_{22}Ti_{18})_{0.99}B_1$ and $(Cu_{60}Hf_{22}Ti_{18})_{0.97}B_3$ amorphous alloys as shown in Fig. 8.

The maximum energies in AES have similar tendency among three kinds of alloy nearly at 160 $kJmol^{-1}$ (1.66 eV). Between the three kinds of B for 3, 1 and 0 % alloys, in an energy region less than 160 $kJmol^{-1}$, AES of only B 3 % alloy is higher than that of B 1% and 0 %. Meanwhile, in an energy region more than 160 $kJmol^{-1}$, AES of them are similar.

This suggests that the diffusion path size for the diffusant of Ti that atomic radius is smallest in the metallic compositions and the packing density of the covalent bonding matrix between the boron and metal are dominant in the relaxation processes. Consequently activation energy for the structural relaxation process has been determined in the $Cu_{60}Hf_{20}Ti_{20}$ with having the highest bulk glass-forming ability in $Cu_{60}Hf_{40-X}Ti_X$ (X are from 5 to 35 %) alloy series as almost 160 $kJmol^{-1}$ (1.66 eV) using the normalized derivative - type relaxed ratio function [9].

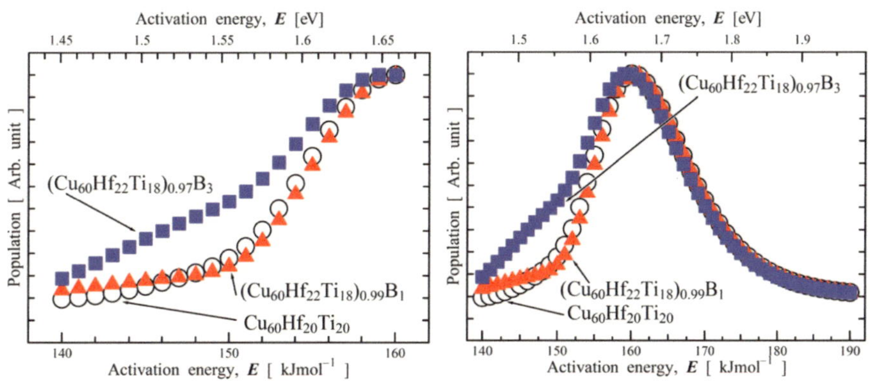

Figure 8. *In the present work for calculation using specific* normalized *1st derivative - type relaxation ratio function* of $\theta(E, T, t$=1s) , activation energy spectrum distributed in $Cu_{60}Hf_{20}Ti_{20}$, $(Cu_{60}Hf_{22}Ti_{18})_{0.99}B_1$ and $(Cu_{60}Hf_{22}Ti_{18})_{0.97}B_3$

4. Calculation technique

In the Fig. 7., four-leafed schematic illustrations show on the DSC scanning #1, #2, we could evaluate the value included relaxation processes for scanning 1st run then 2nd run, on the other hand we could estimate the value included not relaxation processes for scanning 2nd run then 3rd run that are almost without relaxation. So solving the value should be calculated by DSC exothermic heats of 2nd run minus 1st run. So it is very important to calculate the differences between 2nd and 1st run. But also it is difficult to calculated form the DSC exothermic heats of 2nd run minus 1st run because of the temperature scanning step problem. This section describes the way of calculation how to get the differences (distinction) data.

In the Fig. 9., if they were a typical numerical example for differential calculation with supplied as text-type file name, for example, TEST00.TXT of 1th DSC run and TEST01.TXT of 2nd one, it would be transformed from their differential calculation to result numerical data such as TEST04.TXT, to be free to use a program such as GP.EXE ver. 4.13 and DOSBox version 0.74. The 2-Dimension Graph Plotter GP.EXE version 4.13-PC/AT and Dos-emulator DOSBox version 0.74 for all kind of MS-windows OS (another DOSBox version exists for MAC OS probably) are both free software supported in English keyboard peripheral interface. Additionally information, the GP.EXE was built by Prof. Dr. K. Edamatsu(now at riec.tohoku.ac.jp) in 1980-99 year for design to plot scientific/engineering graphs using PCs. Prof. Edamatsu said in his GP's documentation "with GP.EXE, make smart graphs for your presentation and publication. Also, try GP's powerful data analysis capability such as general least-squares fitting, numerical differentiation and integration".

In around 2010 year, the GP.EXE with super high speed and powerful data analysis capability is born-again by use of high performance Dos-emulator DOSBox.

Presented process, to calculate the differential exo/endothermic heat supplied from DSC live scanning environmental with gas flow atmosphere. Overall, Relaxation processes for example has been tutorial as bellow description mainly using the freeware GP.EXE, further only using scanning data 1st run to 2nd run.

In the Fig. 10., for introduction to present calculation technique, typical complex 2-D tutorial graph samples are shown by using GP.EXE. A left chart is the typical Gaussian differentiation tutorial sample, 1st derivative and 2nd one and experimental data and calculation. A right chart is the typical Ahhrenius tutorial plot with inversed horizontal axis with logarithm vertical axis. If you were to use the GP.EXE, you should download from the site of `www.vector.co.jp/soft/dos/business/se004831.html`.

Then you could get the file of gpat431.lzh, you should make the directory for set the GP.EXE environments. It should be save and destination to (recommended): **`C:/prog/gp/gp.exe`**, **`C:/prog/gp/INIT.GPR`**, **`C:/prog/gp/DOC`**, **`C:/prog/gp/ DRIVERS`**, etc.

Note: GP.EXE system is so called legacy-DOS, overall generated user filename must be kept the name rule of 8 character letter filename and 3 character letter extensions around in the GP directory.

Addition you could get the file of gpsmp420.lzh of tutorial examples of GPR extension files, you could be easy to get the way the GP.EXE operating. As shown in Fig. 10., the tutorials exist in site of **`www.vector.co.jp/soft/dos/business/se010753.html`**.

In the Table 2. , Recommended Dos-emulator DOSBox version 0.74 configuration file descriptions are shown. You could edit (ie. MS-Win7) it in Program Menu, DOSBox options, editing the configuration, last lines for "autoexec" region. Addition, **`keyb`** command needs the user of Japanese JP106 keyboard peripheral interface only (addition the **`keyb`** program and keyboard-map should also be needed. The **`keyb`** system's useful information would be gathered in World Wide Web). Meanwhile for in English peripheral US101 user, the **`keyb`** command should be ignore. Furthermore the last **`gp`** command in table 2, it should not be need to user for non-automatic start of GP. For normal user, it should be ignore the last one. Otherwise all users command the type key of **gp** on DOSBox command line, GP.EXE starts anytime.

```
[autoexec]
# Lines in this section will be run at startup.
# You can put your MOUNT lines here.
@ECHO OFF
MOUNT c C:\
c:
```

Table 2. A sample of Dos-emulator DOSBox version 0.74 configuration file description (as shown in autoexec area only)

In the Fig. 11., GP.EXE column structure menu indicating live date column was shown. In the case, X, Y, YE of default column structure allow to use a delimiter also space and tabulator key. Live data should be minimum structure of X and Y with delimiter of space key. Meanwhile additional data column if include could be were specially galloped by use of "U" rule for GP column structure. The typical sample structure of live data is shown in Table 3a and 3b. Addition the 1st, 2nd and 3rd line were normally (default) galloped through a whole text-file for GP because of a purpose for a title and axis captions.

INIT.GPR and other GPR files would be able to modified by a text-type general-purpose editor, then directory file path, captions and so on in them could be also re-arranged and rapid setting for similar graph format preparations.

Note: GPR file always includes full-Path towards live data, but usually it is NOT often need to full-Path towards them but only Local-Path that means without non-Path description, then some of this full-Path should be deleted by use of a text-type general-purpose editor because of keeping the safety-connection between the live data and the GPR file.

In the Fig. 12., load file name menu indicates the live date formatted general-purpose text-style pursuant to table 3a,3b. As it was shown, 3 files (2 kinds of file) of TEST00.TXT, TEST01.TXT and TEST00.TXT are loaded in the live data tray in GP.EXE for 2 data differential calculations. The file #1 should be without differential calculations. The file #2 and #3 should be with differential calculation for #3 minus #2.

In the Fig. 13., load and save parameter's file (GPR of extensions) menu indicates graph structure list organized whole graphic design. Especially GPR file is also plain text-type, so we could arrange them before/afterward by use of a text-type general-purpose editor anytime.

Note: GP.EXE system is so called legacy-DOS, overall generated user filename must be kept the name rule of 8 character letter filename and 3 character letter extensions around in the GP directory.

In the Fig. 14., a Interfile Calculation Parameters and style-menu displayed, red-mark of TEST01 (Src. file # 2) on left-y-axis and white-mark of TEST00 (Src. file # 1) on left-y-axis, meanwhile blue-mark of TEST00 minus TEST01 (Src. file #3 minus #2) on right-y-axis. The aim of this computation is to process the file #3 minus #2 based rule on column X date for Temperature region. Finally a result of the processed data has been shown as blue-mark beside on right-y-axis.

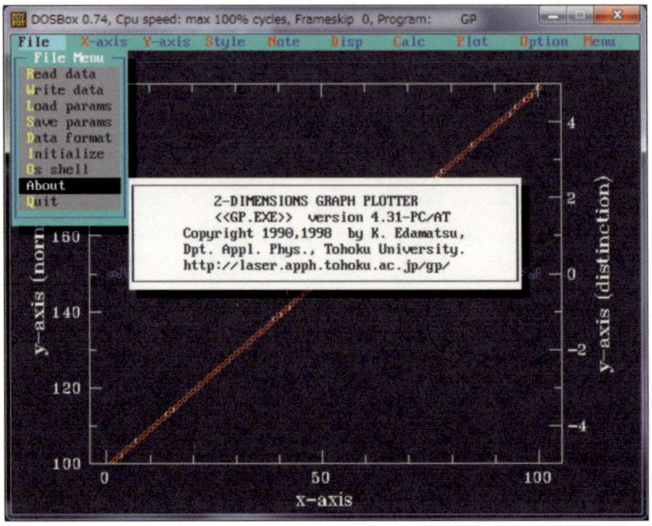

Figure 9. 2-Dimension Graph Plotter GP.EXE version 4.13-PC/AT and Dos-emulator DOSBox version 0.74 are both free software in English supported to calculate the differential exothermic heat data using DSC included relaxation processes for example, scanning 1st run to 2nd run.

GP.EXE: http://www.vector.co.jp/soft/dos/business/se004831.html
GP.EXE samples :http://www.vector.co.jp/soft/dos/business/se010753.html

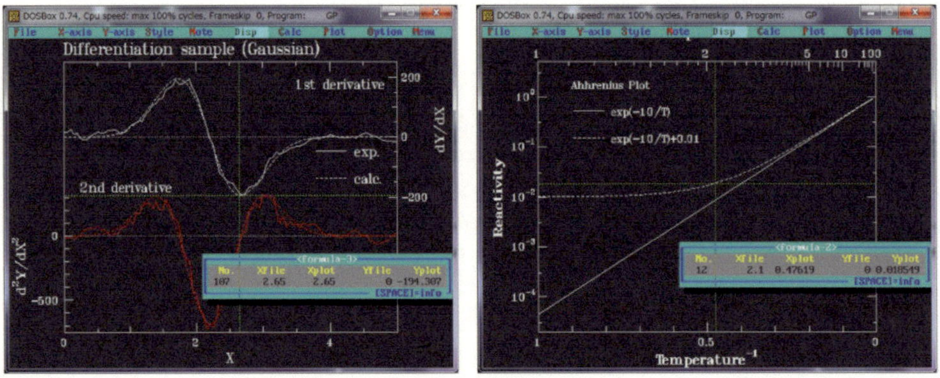

Figure 10. Typical samples of 2-Dimension Graph Plotter GP.EXE. A left chart is the typical Gaussian differentiation sample, 1st derivative and 2nd one and experimental data and calculation. A right chart is the typical Ahhrenius-type plot with inversed horizontal axis together with logarithm vertical axis. Green colour cross line indicator means the across point both live-data and translated-data.

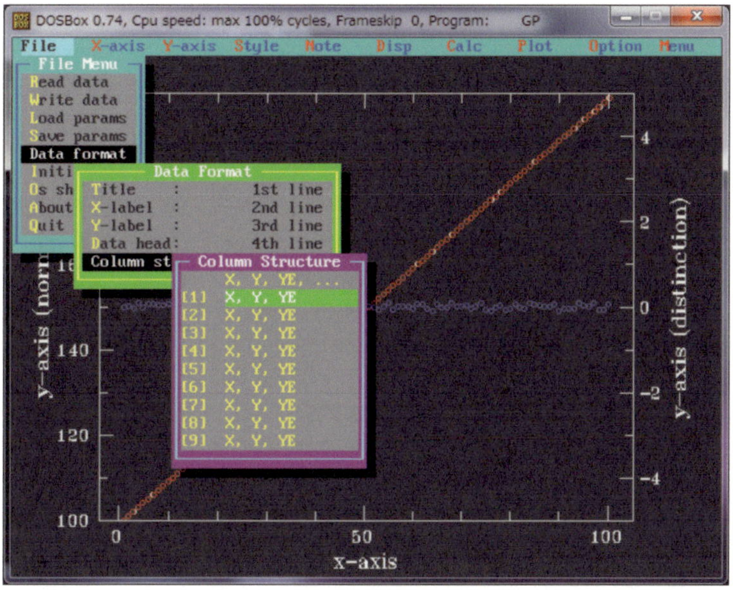

Figure 11. At first, column structure menu indicates the live date column structure

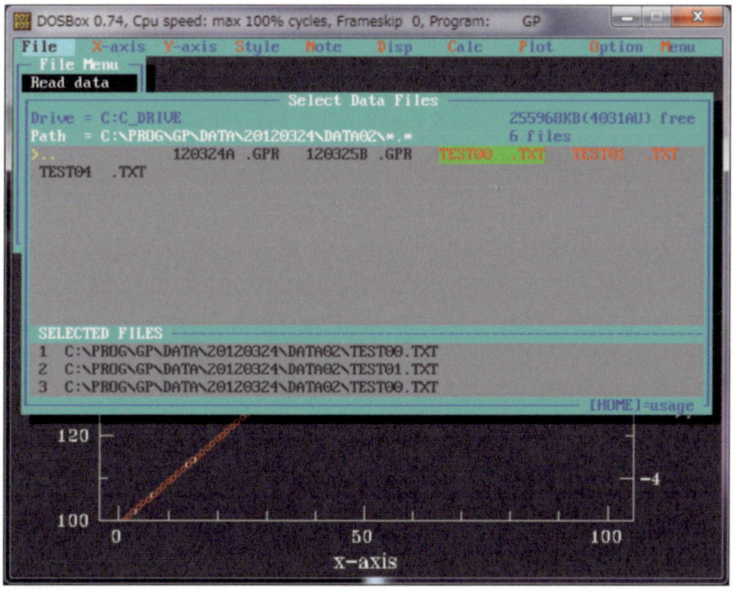

Figure 12. Second, load file name menu indicates the live date formatted general-purpose text-style pursuant to table 3a,3b

x-axis-data	y- axis-data
A main title of presented Graph	
A title of x-axis	
A title of y-axis	
1.022326732	100
2.010754735	101
2.943458995	102
4.094871558	103
4.993183668	104
6.073690076	105
7.054419829	106
•	•
•	•
•	•
97.94832211	197
98.91335378	198
100.0896175	199

a

x-axis-data	y- axis-data
A main title of presented Graph	
A title of x-axis	
A title of y-axis	
1.00	100
2.00	101
3.00	102
4.00	103
5.00	104
6.00	105
7.00	106
•	•
•	•
•	•
98.00	197
99.00	198
100.00	199

b

Table 3. a. Typical numerical example for differential calculation with random number generator only onside x-axis formatted for GP.exe as data filename TEST01.TXT b. Typical numerical example for differential calculation formatted for GP.exe as data filename TEST00.TXT

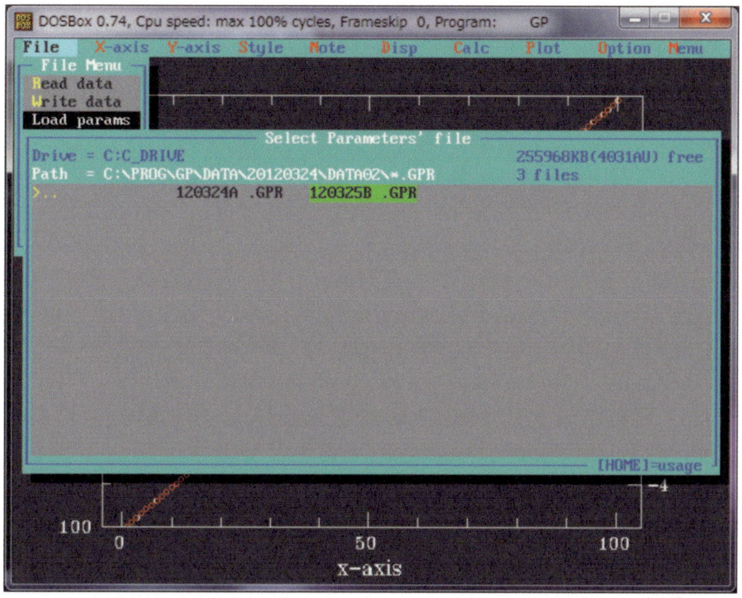

Figure 13. Third, load parameter's file menu indicates organized graph structure

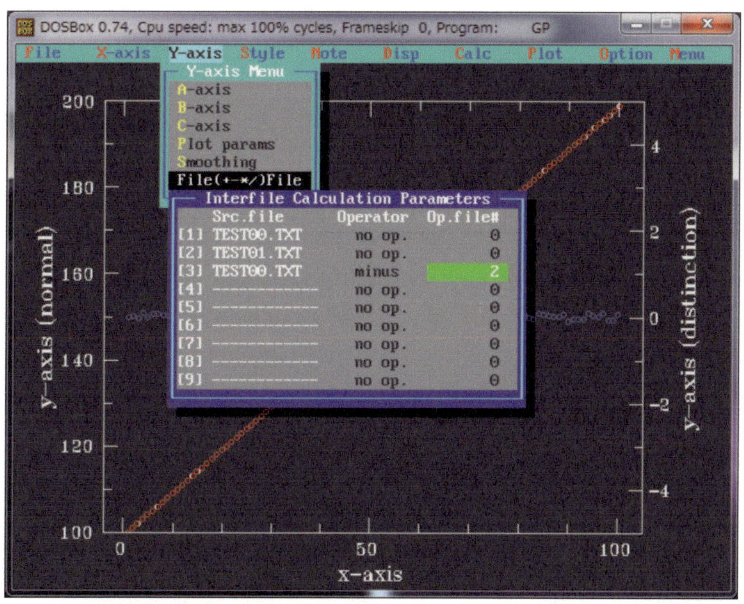

Figure 14. Forth, *an Interfile Calculation Parameters and* style-menu displayed, red-mark of TEST01 (Src. file # 2) on left-y-axis and white-mark of TEST00 (Src. file # 1) on left-y-axis, meanwhile blue-mark of TEST00 minus TEST01 (Src. file #3 minus #2) on right-y-axis. In the figure, a highlight area of green, the letter of the 2 means the #2 file that selected for Src. file #3 minus #2.

In the Fig. 15., Left and right axis, so called Y-axis Plotting Parameters are shown relation to Fig.14. For Src. file #1 and 2 are to belong to left-y-axis named A of Y-axis and further calculated Src. file #3 minus #2 is to belong to right-y-axis named B.

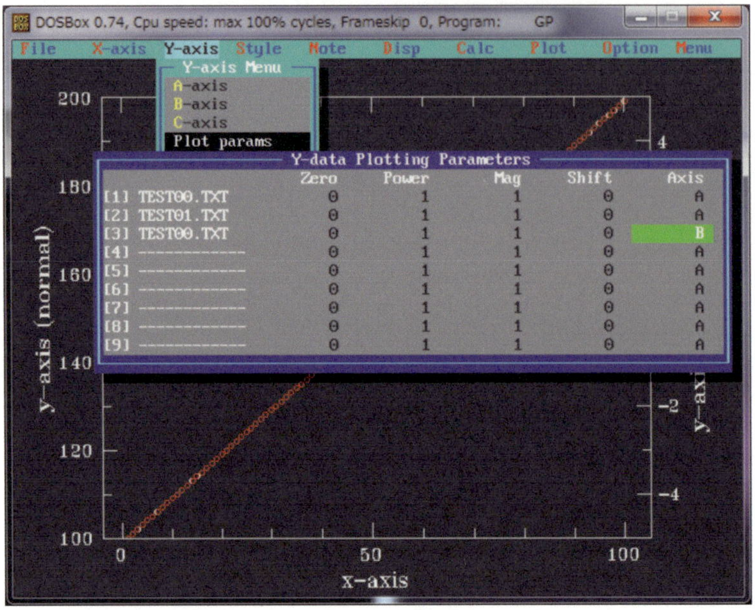

Figure 15. Fifth, Y-axis Plotting Parameters are shown. For Src. file #1 and 2 are to belong to left-y-axis named A and Src. file #3 minus #2 (calculated data) is to belong to right-y-axis named B

In the Fig. 16., Plotting green cross-line indicator means the calculated Src. file #3 minus #2 dots. Fig. 10s are also the similar for usage of cross-line indicator.

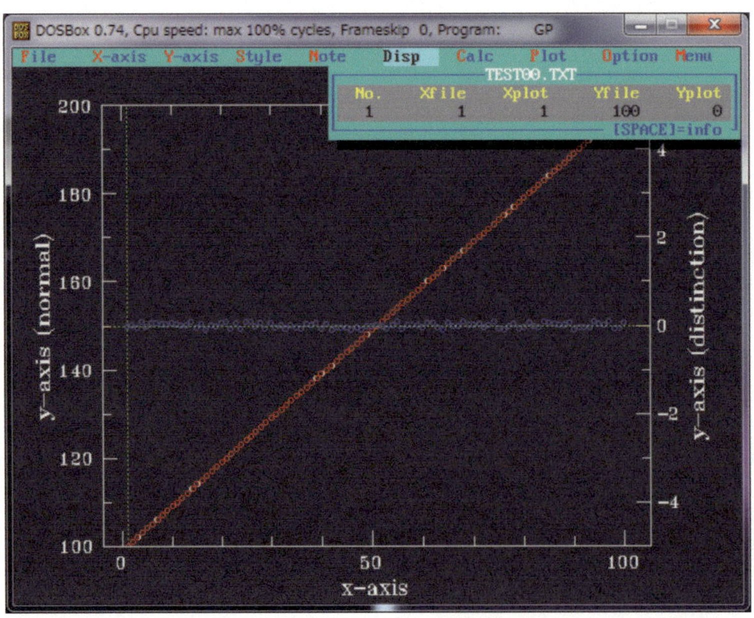

Figure 16. Sixth, Plotting green cross-line indicator means the calculated Src. file #3 minus #2 dots

In the Fig. 17., It is the most important method for calculating of relaxation process. Text-fire save-menu using blue-mark of TEST00 minus TEST01 (Src. file #3 minus #2) on right-y-axis should be describe for example as file name TEST04.TXT in write-data filename input region. Then the TEST04.TXT should be on further calculation process to equation (5) , t=1(s) , then should be normalized and multiplied by inverse reactor summation.

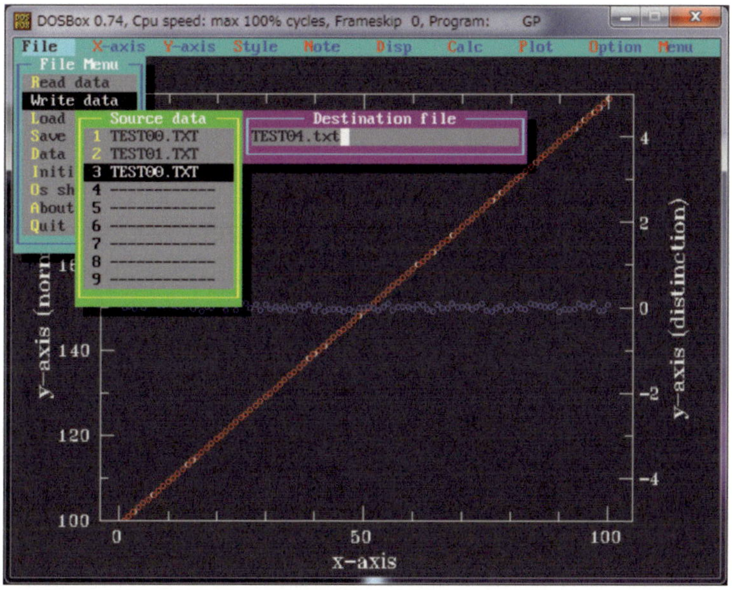

Figure 17. Seventh, text-fire save-menu using blue-mark of TEST00 minus TEST01 (Src. file #3 minus #2) on right-y-axis as file name TEST04.TXT displayed in write-data panel

In the Fig. 18., PostScript-file save-menu using a PostScript-file driver of PS.DLL, that include gpat431.lzh archive, displayed in Plot Parameters panel. Furthermore the useful information, if it assumed to be a 01.ps as saved file name for presented graph design, it would be transformed from PostScript-file to PDF-file, for example, from 01.ps to assumed 01gw.pdf, to be free to use a program ghostscript ver. 9.04. It should be typed on command-line supported by each OS in current directory of 01.ps (not use the command-line in DOSBox) as:

```
"C:\Program Files\gs\gs9.04\bin\gswin32c.exe" -dNOPAUSE -dBATCH -
sDEVICE = pdfwrite -r600 -sOutputFile = 01gw.pdf -c 300000
setvmthreshold save pop -f 01.ps
```

Assumed 01gw.pdf would be a graph with super-resolution quality attaching suitable for all kind of publications. For example, it could be transformed from their PDF to word-processor MS-Word, to be free to use a program such as "Acrobat Reader", and it should be typing keys of Control-a, then Zoom up to around 200%, then Control-c, after then in word-processor to be also typing keys Control-v for universal use.

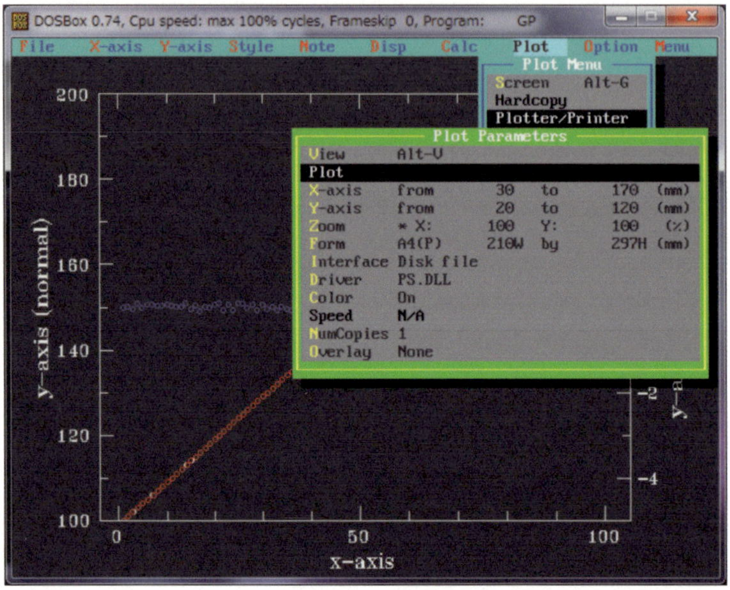

Figure 18. Final, PostScript-file save-menu using a PS.DLL PostScript-file driver displayed in Plot Parameters panel. If it assumed to be a 01.ps as saved file name, it would be transformed from PostScript-fire to PDF-file, for example, from 01.ps to assumed 01gw.pdf, to be free to use a program ghostscript ver. 9.04. It should be typed on command-line as:

```
"C:\Program Files\gs\gs9.04\bin\gswin32c.exe" -dNOPAUSE -dBATCH -sDEVICE=pdfwrite -r600
-sOutputFile=01gw.pdf -c 300000 setvmthreshold save pop -f 01.ps
```

5. Conclusion

In the present work for calculation using specific normalized 1st derivative - type relaxation ratio function of $\theta(E, T, t\text{=1s})$, activation energy spectrum distributed in $Cu_{60}Hf_{20}Ti_{20}$, $(Cu_{60}Hf_{22}Ti_{18})_{0.99}B_1$ and $(Cu_{60}Hf_{22}Ti_{18})_{0.97}B_3$ have been observed through the process of numerical-based discussion. It is so called rapid-type clarification between the temperature range T_2 and T_1 for almost around narrow 100 K region. In other words, even the above mentioned narrow temperature range induced the "reversible" phenomena, and it has been also observed in the anneal process of scan #1 to #n, repeatedly.

After it has been difficult in general to calculate numerical differences between any kinds of DSC live data. Because it has the time-domain problem for stepping accuracy and speed on temperature column region. So in second half of this paper, it was tutorial to short course calculation method for the differences using the freeware in Tohoku University Prof. K.

Edamatsu' GP.EXE that was designed until 1999 to make smart graphs for publication with powerful data analysis ability such as numerical complex differentiation. And now it is shown that the GP.EXE has been useful for genuine data processing even in the 2012's generation.

Author details

Kazu-masa Yamada
Hakodate National College of Technology,
Department of Electrical and Electronic Engineering,
Japan

Acknowledgement

The author was favoured to have the assistance of Dr. I. A. Figueroa in Universidad Nacional Autonoma de Mexico who contributed an experimental circumstance to the accomplishment of the amorphous sample preparations in the University of Sheffield UK. The author also would like to express the appreciation to Dr. Sergio Gonzalez Sanchez (Universitat Autonoma de Barcelona), Mr. P. J. J. Hawksworth and Dr. I. Todd in the University of Sheffield. The author is indebted to Professor H. A. Davies for drawing his attention to presented researches.

6. References

[1] Brochure for the Nanotechnology and Materials Technology Development Department, on March (2008), The New Energy and Industrial Technology Development Organization (NEDO) in Japan, processing technology for metallic glasses, p.61-62, http:// www.nedo.go.jp/ kankobutsu/ pamphlets/ nano/ nano_e2008.pdf

[2] I.A. Figueroa, R. Rawal, P. Stewart, P.A. Carroll, H.A. Davies, H. Jones and I. Todd, Journal of Non-Crystalline Solids, Vol. 353 (2007), pp. 839-841

[3] I.A. Figueroa, H.A. Davies, I. Todd and K. Yamada, Advanced Engineering Materials, Vol. 9 (2007), pp. 496-499

[4] K. Yamada, Y. Iijima and K. Fukamichi: Defect and Diffusion Forum, Vols. 143-147 (1997), pp.765-770

[5] K. Yamada, Y. Iijima and K. Fukamichi: J. Mater. Res., Vol. 8 (1993), pp.2231-2238

[6] Y. Takahara, A. Morita, T. Takeda and H. Matsuda: J. Japan Inst. Metals, Vol. 54 (1990) pp. 752-757 (in Japanese)

[7] K. Yamada, M. Ito, M. Tatsumiya, Y. Iijima and K. Fukamichi: Defect and Diffusion Forum, Vols. 194-199 (2001), pp.815-820

[8] K. Yamada, K. Fukamichi and Y. Iijima: J. Magn. Soc. Jpn., Vol.22 Suppl. S2 (1998), pp. 97-100

[9] K. Yamada et al, Defect and Diffusion Forum Vols. 283-286 (2009), pp 533-538

[10] J. A. Leake, E. Woldt, and J. E. Evetts, Mater. Sci. Eng. , Vol. 97 (1988), pp. 469-472
[11] M. R. J. Gibbs, J. E. Evetts and J. A. Leake, J. Mater. Sci., Vol. 18 (1983), pp. 278-288
[12] W. Primak: Phys. Rev. Vol. 100 (1955) pp. 1677-1689

Thermal Analysis of Sulfur and Selenium Compounds with Multiple Applications, Including Anticancer Drugs

Daniel Plano, Juan Antonio Palop and Carmen Sanmartín

Additional information is available at the end of the chapter

http://dx.doi.org/10.5772/53048

1. Introduction

Thermal methodologies are analytical and quantitative methods capable of providing reliable, fast and reproducible results. Thermogravimetry (TG), Differential Scanning Calorimetry (DSC) and Isothermal Titration Calorimetry (ITC) techniques are the chosen methods for several physicochemical determinations.

ITC is the most quantitative means available for measuring the thermodynamic properties of a protein-protein interaction. So, ITC is the calorimetric approach most used to investigate biomolecular interactions. ITC measures the binding equilibrium directly by determining the heat evolved on association of a ligand with its binding partner. In a single experiment, the values of the binding constant (K_a), the stoichiometry (n) and the enthalpy of binding (ΔH_b) are determined. The free energy and entropy of binding are determined from the association constant. The temperature dependence of the ΔH_b parameter, measured by performing the titration at varying temperatures, describes the ΔC_p term. Furthermore, binding of proteins and small molecules to nucleic acids is of course critical to all organisms, playing a role in replication, transcription, translation and DNA repair processes to name just a few. Protein association with nucleic acids has therefore been the subject of much study throughout the years, and ITC has been one of the most common tools used for such investigations. When used in conjunction with complementary techniques such as X-ray crystallography, ITC can provide an informative thermodynamic account of these systems [1]. Besides, ITC is a useful technique in the protein-lipid interactions studies, and two examples in 2008 were the study of the effect of cholesterol on an amphibian antimicrobial peptide interaction with membranes [2], and analysis of the interaction of mammalian bone-marrow derived peptides with model and natural membranes [3]. Finally, ITC is a powerful tool for the pursuit of higher affinity drugs with improved binding specificities [4]. In its simplest form, ITC is a rapid and convenient method for measuring affinities of new leads and optimized compounds. In addition, however, ITC is particularly useful in providing information about the mode of binding. The use of ITC as a general tool in drug design and characterization is exemplified in a study by McKew *et al.* [5] who demonstrated the efficacy of ITC for studying three classes of inhibitor to the cytosolic amphitropic enzyme phospholipase A2 alpha (cPLA2a).

TG is mainly employed to study thermal stability, kinetic parameters and degradation processes for a wide range of materials. DSC allows characterizing protein stability and folding, drug-protein interactions, as well as heat capacity, vapor pressures and polymorphism. Moreover, it has been pointed out the usefulness of DSC technique as a potential tool for the early diagnosis, monitoring and screening of cancer patients [6].

2. Application of thermal analysis to sulfur and selenium compounds with multiple applications

Sulfur (S) and selenium (Se) compounds present several applications in a great variety of fields. We consider the anticancer activity of these compounds as the most important application due to the burden, costs and mortality rates caused by cancer disease. Thus, we will treat in depth the application of thermal techniques to these anticancer compounds in the following section.

Due to the vast applications of S and Se compounds and to the great structural variability in each of these applications, the S and Se compounds are going to be classified according to their structural features.

2.1. Coordinated compounds

The study of the degradation process is one of the most common utility for thermal techniques in the study of metal complex derivatives. Coordination compounds with dithiocarbamates have attracted attention because of their potential biological activity [7-10]. In 2006, a novel dithiocarbamate ligand L (triammonium-*N*-dithiocarboxyiminodiacetate) was synthesized and the thermal decomposition of its cooper (II), niquel (II) and palladium (II) was studied by DSC and thermogravimetry [11]. The authors showed that thermal stability of $(NH_4)_3L$ is low and its decomposition starts with evaporation of an ammonia molecule. Of the three complexes, $Cu(H_2L)_2$ is the least thermally stable. Thermal decomposition of the complexes most likely begins with decarboxylation. It is endothermic up to 500 K, but exothermic oxidation processes are observed above this temperature. Thermal decomposition of the cooper (II) complex is accompanied by its melting and with an exothermic structural rearrangement [11]. During the last two years, these thermal techniques have been used to study the stability and to characterize the degradation process of several dithiocarbamate complexes [12-14]. On the other hand, a study of the degradation process for three novel selenocyanato complexes has been published recently [15]. The authors demonstrated that all compounds decompose in a single heating step without the formation of ligand-deficient intermediates.

Another application of thermal analysis is to study the chemical structure and the structure rearrangements in metallic complexes. The VA main group metal compounds including inorganic and metallorganic complex have showed interesting physical properties, medical and material functions. Some bismuth complexes can be used in medicine, microbiology and pharmacology [16-18]. Two studies with thiourea complexes of antimony and bismuth have evidenced several structure rearrangements or phase transformations for these complexes from 100 to 170 ºC [19, 20].

Several studies have been carried out in order to determine the specific and thermodynamic constants of methionine, which is one of the nine essential amino acids needed by human beings and contains a sulfur atom, and 2-mecaptonicotinic acid complexes [21, 22].

2.2. Glass materials

Chalcogenide semiconductors have been proposed for phase change nonvolatile random access memories, which is becoming the next generation for memory technology [23]. So, the proper description of thermal behavior of semiconducting chalcogenide glasses is crucial to understand their properties and functions. One of the crucial techniques to study the glass transition kinetics is the differential thermal analysis (DTA) and DSC.

Among the chalcogenidé systems, selenium and selenium based glassy alloys have been intensively studied due to their wide technical applications, especially in the field of electronics and optoelectronics. A recent study has used the DTA technique to study the glass transition kinetics of the two binary Se-In alloys in comparison with that of pure Se. The glass transition temperature was found to be shifting to a higher value with increasing of heating rates and indium content. It was observed an increase of the stability parameters accompanied with the introduction of In into the Se matrix [24]. Another interesting Se based glassy alloys are Se-Sb alloys owing to their electrical, optical dielectric and thermal properties. Mehta *et al.* have reported the thermal characterization with calorimetric measurements for some Se-Sb alloys [25]. They reported the Hruby number, which is the strong indicator of glass forming tendency, thermal stability parameter and the values of crystallization enthalpy and entropy.

Selenium-tellurium thin films have attractive semiconductors for device application. Se-Te form a continuous series of solid solution and the Se-Te system has an intermediate behavior between pure Se and pure Te. The addition of Te has a catalytic effect on the crystallization of Se. In 2009, the crystallization parameters of the bulk Se-Te chalcogenide glass have been studied using DSC [26]. The values of glass transition temperature, onset crystallization temperature, peak crystallization temperature and enthalpy released with and without laser irradiation for different exposure time have been studied. The films showed indirect allowed interband transition that is influenced by the laser irradiation.

Ternary systems of chalcogenide glasses containing metal elements possess unique optical, electrical and physicochemical properties [27]. The most popular metal is silver and its addition into chalcogenide glasses leads to a drastic change in the physical and chemical properties of the material, for instance, it increases the conductivity by several orders of magnitude and decreases the slope of frequency dependence of alternating current (AC) conductivity. A study conducted by Ogusu *et al.* carried out DSC, X-ray diffraction (XRD) and Raman scattering measurements for $Ag_x(As_{0.4}Se_{0.6})_{100-x}$ glasses with x = 0-35 at.% in order to investigate the crystallization kinetics and the local structure [28]. The DSC curves of the samples with Ag content x = 15-35 at.% were obtained at various heating rates for different Ag contents and two or three exothermic peaks for the crystallization were found depending on the Ag content. Furthermore, the dimension of crystal growth of sample particles and activation energy were determined using Matusita´s equation to analyze the DSC data. It was found that the surface and bulk crystallization take place depending on the Ag content and peak crystallization temperatures [28].

Glassy selenium has low sensitivity and thermal instability. These properties can be improved by alloying of some elements into selenium matrix, such as arsenic [29] and antimony [30, 31]. The proper description of thermal behavior of these glasses is important for understanding their properties and applications. Recently, the thermal properties and structure of As_xSe_{100-x} and Sb_xSe_{100-x} glass-forming systems (x = 0, 1, 2, 4, 8 and 16) were reported by conventional and StepScan DSC and Raman spectroscopy [32]. Among these thermal properties, the authors studied the glass transition temperature and the crystallization of undercooled melts. So, the glass transition temperature for As_xSe_{100-x} system increases almost linearly with increasing As content from 40 up to 93 ºC, because the glass structure becomes more stable due to cross-linking of Se chains by As. Nevertheless, the glass transition temperature of Sb_xSe_{100-x} changes only slightly from 40 to 48 ºC [32]. Concerning to the study of crystallization of undercooled melts, it was found that only selenium crystallizes from undercooled melts of As-Se system and its tendency to crystallize decreases markedly with increasing As content, for arsenic content higher than 4 at.% no crystallization was observed. In the case of Sb-Se system Sb_2Se_3 crystallizes in the first step followed by trigonal selenium crystallization from non-stoichiometric undercooled melt [32].

Another technologically important ternary system of chalcogen elements are the infrared transmitting glasses based on Ge-Sb-Se because they are good transmitters of radiation in the 2-16 μm wavelength region. The applications include fabrication of optical components like IR lenses, windows and filter used in thermal imaging systems. The Sb-Ge-Se films result sensitive for the UV exhibit mechanical, optical and structural changes [33, 34]. An understanding of the glass forming tendency and crystallization kinetics in these chalcogenide materials is very important to develop them for applications based on the amorphous to crystallization phase change and vice versa. So, one report evaluated the glass-forming ability of some alloys in $Sb_xGe_{25-x}Se_{75}$ ($0 \leq x \leq 10$) system by using various thermal stability criteria, based on characteristic temperatures [35]. It was observed that the thermal stability decrease with increasing Sb content in the glassy system.

2.3. Inorganic mixtures

Several reports have been published concerning to the solubility and thermal characterization of various metal-selenite systems. For example, some manganese(II) selenite are used for coloring glasses, enamel and glazes. On the basis of $MnSeO_3$, two manganese selenides (α-MnSe and $MnSe_2$) were obtained having very interesting semiconductor properties [36]. Vlaev *et al.* have studied the crystallization fields of manganese(II) selenites in the system $MnSeO_3$-SeO_2-H_2O in the temperature interval 25-300 ºC and characterized the observed phases [37]. Previously to this article, the same author reported the crystallization fields and the characterization of the observed phases for the system $NiSeO_3$-SeO_2-H_2O [38]. Another article studied the phase equilibrium in the system CdO-SeO_2-H_2O at 25 and 100ºC and the thermolysis mechanism of the compounds obtained.

The ytterbium selenites can serve as initial substances for obtaining selenides and oxyselenides having valuable photoconductive and superconductive properties. So, Gospodinov *et al.* have studied the solubility isotherm of the three-component system Yb_2O_3-SeO_2-H_20 at 100 ºC [39]. Furthermore, they have performed simultaneous TG and DTA curves of the compounds obtained in its fields of crystallization and the mechanism of the thermal decomposition [39].

Alkali metal sulfates, selenite and phosphate tellurate compounds having the formula $M_2XO_4Te(OH)_6$, where M is the metal and X is S, Se or P, form a broad families with interesting properties, such as superprotonic conduction and ferroelectricity [40-42]. So, synthesis, calorimetric and conductivity studies of new mixed solution of rubidium sulfate selenate tellurate [43] and thallium selenate tellurate [44] have been carried out.

2.4. Miscellaneous compounds

In the last two years, several thermal and structural investigations on crystal structures with thiourea have been carried out. The thermal decomposition of crystal structures with bisthiourea derivatives has been studied by TG-DSC [45]. Another study reported the growth and characterization of a new non-linear organometallic crystal (potassium thiourea thiocyanide or PTT) [46]. The TG curve showed the complete decomposition of PTT between 176 and 1000 ºC in three steps with corresponding three DTA peaks.

Some selenoesters present promising photophysical properties for optical device applications such as emissive liquid crystal displays (LCDs), polarized organic lasers and anisotropic Light-emitting diodes (LEDs). Rampon *et al.* have reported the synthesis and the study of the liquid crystalline and fluorescent properties of novel selenoesters [47]. So, these compounds were fluorescent in the blue region and exhibited their stability and liquid crystalline properties over a large range of temperatures. Moreover, these compounds showed a rich phase polymorphism.

Cooper chalcogenides are considered as promising in electronic technology due to their physicochemical properties [48, 49]. Chrissofis *et al.* [50] have reported the thermal behavior of samples with very slight divergence from stoichiometry ($Cu_{2-x}Se$). Also, they have studied the nature of the transformation with non-isothermal measurements at different heating and cooling rates.

Oligothiophenes and polythiophenes are another sulfur compounds that have attracted much attention due to their unusual electric and nonlinear optical properties as interesting materials for organic electronics and optoelectronics, LEDs, field-effect transistors, thin-film transistors… So, the relative stabilities of 2,2′- and 3,3′-bithiophenes (the main building blocks of these conducting organic materials) have been evaluated by experimental thermochemistry [51].

3. Application of thermal analysis to sulfur and selenium compounds with anticancer activity

In the last decade, among the wide range of compounds tested as potential anticancer agents, several structurally diverse derivatives that contain a sulfur or selenium template have been reported and have generated growing interest. For that reason, in this chapter we have focused on some relevant thermal studies in sulfur and selenium compounds with anticancer activity.

3.1. Sulfur amino acids and cysteine cathepsins.

The human family of cysteine cathepsins are a family of lysosomal proteases and has 11 members (cysteine cathepsin B, C, F, H, K, L, O, S, V, W and X), which share a conserved active site that is formed by cysteine, histidine and asparagine residues. Cysteine cathepsins are often upregulated in various human cancers, and have been implicated in distinct tumorigenic processes such as angiogenesis, proliferation, apoptosis and invasion. During cancer progression, cathepsins are often translocated to the cell surface of tumor cells or are secreted into the extracellular milieu, where they can promote tumor invasion through several possible mechanisms. Causal roles for cysteine cathepsins in cancer have been demonstrated by pharmacological and genetic techniques. This includes functional downregulation of cysteine cathepsin activity by increasing expression of endogenous inhibitors and administration of small-molecule cysteine protease inhibitors. Besides, causal roles for specific cysteine cathepsins in cancer have been demonstrated by downregulating their expression or crossing mouse models of cancer with mice in which the cysteine cathepsin has been genetically ablated. These studies have identified roles for cysteine cathepsins in both tumor cells and tumor-associated cells such as endothelial cells and macrophages.

Taking into account the causal roles for cysteine cathepsins, which present a cysteine residue in the active site, in cancer and the fact that the thiol-disulfide interchange reaction is important to a number of subjects in biochemistry, thermodynamic data regarding the relative energetics of the thiol and disulfide functional groups is essential for the understanding of the driving force and mechanism of biochemical processes. Temperature-induced changes in crystalline amino acids are of interest for their properties and because they reveal the intrinsic motions of these structural fragments and their contribution to the dynamic properties of proteins.

So, a thermophysical study of the sulfur containing amino acids *L*-cysteine and *L*-cystine by DSC has been reported [52]. Heat capacities of both compounds were measured in the temperature interval from T = 268 K to near their respective melting temperatures.

Furthermore, a solid-solid phase transition close to the melting point is only observed in the *L*-cysteine. Additionally, several polymorphic forms have been reported for both compounds. *L*-cysteine crystallizes in the monoclinic and orthorhombic forms and has been structurally characterized [53, 54]. Phase transitions have been detected when lowering the temperatures [55] and also when decreasing pressures up to 4.2 GPa and decreasing to 1-7 GPa [56]. *L*-cystine crystallizes in the tetragonal and hexagonal forms and has also been studied at ambient [57, 58] and at low temperature [59] and at high pressures although no solid-solid phase transition has been detected.

3.2. Metal complexes of sulfur compounds

During the past decade the study of mixed sulfur donor ligand complexes with main group metals has made a progressive development due to the development of new analytical and structural techniques [60-62]. These complexes present potential applications in areas such as fast ion conductivity, photocatalysis and electro-optics, among others as well as several biochemical applications [63-65].

One of these promising complexes is antimony(III) bis(pyrrolidinedithiocarbamato) alkyldithiocarbonates. The link of two active ligands was the rational design used for the design of these complexes. So, pyrrolidine dithiocarbamates which represent a class of antioxidants mediate a wide variety of effects in biological systems[66]. It is a multipotent synthetic compound well known for its metal chelation property and one of the most potent and specific NF-kB inhibitor [67]. Besides, antimony metal containing compounds are commonly used to treat parasitic infections and exhibit a broad spectrum of chemotherapeutic applications and cytotoxic activities. So, a study has reported the synthesis, spectroscopic, thermal and structural behavior of antimony(III) bis(pyrrolidinedithiocarbamato)alkyldithiocarbonates [12]. Thermogravimetric studies not only allows to determine purity and thermal stability of the complex but also composition of the complex as well which it is observed during different steps of weight losses as a fragment formed in different temperature ranges [12].

Another type of complexes with a potent anticancer activity which were designed using the link of two active ligands, are the palladium (II) and platinum (IV) complexes with active sulfur ligands. The use of antitumor drugs based on platinum(II) metal complexes, cisplatin and its analogues carboplatin and oxaliplatin is limited by two factors: installation of tumor drug resistance and severe adverse effects [68, 69]. Therapeutic strategies are oriented towards the development of new platinum- and non-platinum-based antitumor drugs with higher efficiency, reduced general toxicity and broader spectrum of activity [70]. Sulfur-containing molecules are studied as chemoprotectors in platinum-based chemotherapy. Dithiocarbamates have attracted particular attention for the use of chemical modulation of cisplatin nephrotoxicity [71-73].

The thermal behaviour of Pd(II) complexes with some dithiocarbamate derivatives was studied in order to establish the coordination mode of the ligand, to test the thermal stability [11] or to understand the effect of the alkyl chain attached to the nitrogen atom over the thermochemical parameters of the complexes [74, 75]. A series of Pd(II) complexes with *tert*-butylsarcosinedithiocarbamate [76], ethylsarcosinedithiocarbamate and 2-/3-picoline [77], dithiocarbamates and various amines [78] were developed as antitumor agents with low nephrotoxicity. Thermogravimetric analysis was used for the characterization of these compounds. Taking into accounts the kinetic inertness, high activity, low toxicity and suitability for oral administration of Pt(IV) complexes, some Pt(IV) complexes with dithiocarbamates have been synthesized. Morpholine dithiocarbamate, aniline dithiocarbamate and N-(methyl, cyclohexyl) dithiocarbamate alone [79] or with triphenyl phosphine as second ligand [80] were used in order to obtain Pt(IV) complexes with antitumour activity.

In 2012, Uivarosi *et al.* have reported the thermal and spectral studies of palladium(II) and platinum(IV) complexes with bis(dimethylthiocarbamoyl)sulphide and bis(diethylthiocarbamoyl)disulphide [81]. TG experiments revealed the nature of complex species as hydrated or anhydrous. Thermal decomposition of coordinated organic ligands occurs in one or two exothermic stages, the final residue being in all cases the free metal (Pd or Pt).

Other complexes with anticancer activity are de ruthenium (III) complexes with sulfur ligands. Ruthenium complexes with dimethyl sulfoxide (dmso) showed selective antitumor properties in preclinical testing [82]. Biological studies in *cis*- and *trans*-$RuX_2(dmso)_4$ complexes (X = Cl and Br) refer to different tumor toxicity and anti-metastasis properties of the isomers [83]. Dmso can be coordinated to ruthenium as a metal center either through the sulfur (dmso-S) or through the oxygen atom (dmso-O). Dmso provides a moderate acceptor site for π-electron donors and bound through sulfur stabilizes ruthenium in lower Ru(II) oxidation state, more reactive toward tumor cells [84]. The biological activity of complexes can be modified by addition or change of the ligands. Phenothiazines and their *N*-alkyl derivatives are themselves biological active compounds, suitable to take part in complex formation. Moreover, they exhibit a strong in vitro antitumor activity in numerous and various tumor cell lines [85].

Recently, thermal decomposition of chlorpromazine hydrochloride (CP·HCl), trifluoperazine dihydrochloride (TF·2HCl) and thioridazine hydrochloride (TR·HCl), and the ruthenium complexes with dimethyl sulfoxide (dmso) of composition $[RuCl_2(dmso)_4]$ and $L[RuCl_3(dmso)_3]\cdot x EtOH$, L = CP·HCl, TF·2HCl or TR·HCl is described [86]. The phenothiazines are stable to temperature range of 200–280 ºC with an increasing stability order of TF·2HCl < CP·HCl < TR·HCl. The decomposition of all the compounds takes place in superposing steps.

3.3. Alkylimidothio- and alkylimidoselenocarbamates

During the last five years, our research group reported the promising and potent anticancer effects for several alkylimidothio- and alkylimidoselenocarbamate derivatives [87-89]. These compounds showed a remarkable cytotoxic activity *in vitro* against prostate cancer cells and other several cancer cell lines. One of these derivatives, the quinoline imidoselenocarbamate

EI201, inhibits the PI3K/AKT/mTOR pathway, which is persistently activated and contributes to malignant progression in various cancers, and contributes to the loss of maintenance of the selfrenewal and tumorigenic capacity of cancer stem cells. This compound (EI201) suppressed almost 80% prostate tumor growth *in vivo* ($p < 0.01$) compared to controls at a relatively low dose (10 mg/kg) in a mouse xenograft model [90].

Degradation and fusion temperatures for 20 of these anticancer derivatives were determined using TG and DSC [91]. Analysis of the thermal data indicated that: (a) in general, sulfur compounds are more stable than selenium compounds; (b) the pyridine ring diminished stability of sulfur and selenium compounds much more than the carbocyclic aromatic rings did; (c) selenomethyl derivatives are more stable than selenoethyl and selenoisopropyl compounds; (d) a chlorine atom on selenocompounds has surprising effects. So, the presence of intermolecular bonds was pointed out between chlorine atom and selenium atom [91]. With regard some substituents present on aromating ring and the ramification and length of chain, it can be concluded that the presence of electron-withdrawing groups in selenocompound structures improves their stability. Besides, selenomethyl derivatives are more stable than selenoethyl and selenoisopropyl compounds [91].

The determination of the polymorphism of a substance is of great importance due to the strong influence of the crystalline form on the physicochemical properties, bioavailability and stability of drugs [92], and, in some compounds with biological activity, can even become metastable forms, being twice as active as the stable form [93]. So, our research group has carried out the study of the physicochemical properties of polymorphic forms of a serie of alkylimidothio- and alkylimidoselenocarbamate derivatives with a combination of DSC, thermomicroscopy and X-ray diffractometry [94]. In this study we observed that polymorphs could be formed when the compounds are heated above their melting points. The results showed that there are four types of thermal behavior for alkylimidothio- and alkylmidoselenocarbamate derivatives: (a) compounds which do not evidence any polymorphic forms (behavior I); (b) compounds which solidify into an amorphous solid form (behavior II); (c) compounds which present a new polymorphic form at a T_{onset} lower than the original one (behavior III); (d) finally, compounds which have three polymorphic forms with three different T_{onset} values (behavior IV). Calorimetric studies demonstrated that sulfur and selenium analogs have the same thermal behavior. So, the different thermal behaviors observed for these alkylimidothio- and alkylimidoselenocarbamates are caused by the substituent groups in the aromatic ring, although there is no relationship between electron-withdrawing and electron-donating groups and the thermal behavior.

3.4. Case study: Thermal analysis of selenyl acetic derivatives.

In 2009, our research group reported the cytotoxic and antiproliferative activities *in vitro* of selenyl acetic derivatives against several cancer cell lines [95]. Considering the structure of these derivatives and their inefficacy to induce apoptosis and to affect to cell cycle, we decided to perform a thermal analysis for these derivatives. So, we carried out a thermal stability and calorimetric studies for some of these anticancer selenyl acetic acids (**Figure 1**).

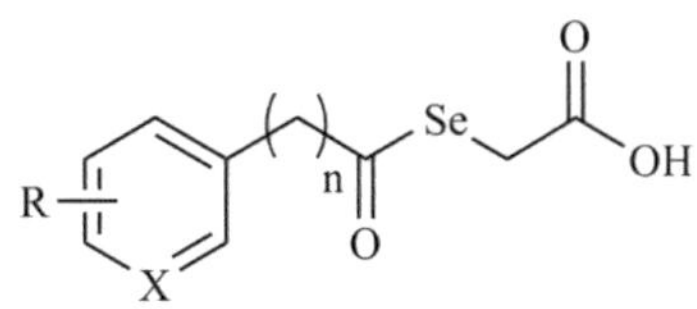

Compound 1:	R = 2-Cl;	X = N;	n = 0
Compound 2:	R = H;	X = C;	n = 0
Compound 3:	R = 3,5-diOCH_3;	X = C;	n = 0
Compound 4:	R = 4-CF_3;	X = C;	n = 0
Compound 5:	R = 4-CN	X = C;	n = 0
Compound 6:	R = 4-CH_3;	X = C;	n = 0
Compound 7:	R = H;	X = C;	n = 1
Compound 8:	R = Phenyl;	X = C;	n = 0

Figure 1. General structure for studied selenyl acetic acids.

3.4.1. Thermal stability studies

The thermogravimetric studies were carried out with a Perkin-Elmer TGA-7. The thermobalance was calibrated with alumel and nickel at 10 ºC min^{-1}. The calibration of the oven temperatures was carried out automatically. Mass calibration was carried out with a certified mass of 10 mg (ASTM E617).

The calorimeter was calibrated with indium and zinc (provided by Perkin-Elmer and fabricated according to guideline ISO35) at 10 ºC min^{-1} and a nitrogen flow of 20 mL min^{-1}. The gases connected to the equipment were nitrogen and air with a purity of 99.999%.

Thermogravimetric analyses were carried out under nitrogen atmosphere with a gas flow of 40 mL min^{-1} at 10 ºC min^{-1}, using a sample of approximately 3 mg.

All the compounds sublimated before the degradation process start. So, it is not possible to study the thermal stability of these compounds using thermogravimetric techniques.

3.4.2. Calorimetric studies

The calorimetric studies were carried out with a Perkin-Elmer DSC Diamond. Calorimetric analyses were carried out in aluminium capsules for volatiles of 10 μL, at a heating rate of 10 ºC min^{-1}, using a sample of approximately 3 mg, in order to establish the T_{onset} and the enthalpy of fusion ΔH_f. All of the experiments were performed at least three times and the values were expressed as mean ± standard deviation.

The obtained data (**Table 1**) allow us to point out the following calorimetric behaviors:

1. Regarding to the T_{onset} values:
   a. The substitution on the aromatic ring causes an increase in the fusion temperatures of these compounds.
   b. The inclusion of a methylene group between aromatic ring and carbonyl group seem to lead to a diminution in the T_{onset} values.
   c. The presence of groups such as cyano and chloro on the ring, which can form hydrogen bonds, significantly increase the fusion temperatures of these derivatives.
2. Regarding to the enthalpy of fusion values:

a. Compounds with phenyl ring (2 and 7) possess values for enthalpy of fusion higher than T_{onset} ones. It could be caused by the presence of strong π-π stacking interactions between two molecules in the crystal packaging.
b. Compounds with groups that can establish hydrogen interactions, such as CN and Cl, substituted over the ring present the highest enthalpy values, owing to these hydrogen interactions.
c. The substitution on the *para* position of the phenyl ring with groups that cannot establish hydrogen interactions significantly diminished the enthalpy values. It seems that these substituents alter the electronic distribution over the ring, affecting to the π-π stacking interactions strength.

Reference	T_{onset} (ºC ± SD)	ΔH_f (Jg^{-1})
Compound 1	157.2 ± 0.2	126.8 ± 4.8
Compound 2	83.3 ± 0.6	92.3 ± 2.2
Compound 3	117.4 ± 0.3	108.3 ± 5.5
Compound 4	107.7 ± 1.1	73.7 ± 1.0
Compound 5	146.5 ± 0.1	112.3 ± 1.7
Compound 6	92.2 ± 0.2	78.7 ± 0.8
Compound 7	73.8 ± 0.3	74.9 ± 7.2
Compound 8	130.1 ± 0.4	105.9 ± 4.6

Table 1. T_{onset} and enthalpy of fusion values for selenyl acetic acid derivatives studied.

The calorimetric data (**Table 1**) demonstrated that compounds 1 and 5 present a very significant higher T_{onset} value for the fusion process than the other selenyl acetic derivatives. Both compounds present two groups substituted in the *para* position of the aromatic ring that can act as hydrogen bonding donors. So, these groups could form a different hydrogen bond interaction with the proton of carboxylic acid group (**Figure 2**) and this interaction should be stronger than the interactions established in the rest of the compounds.

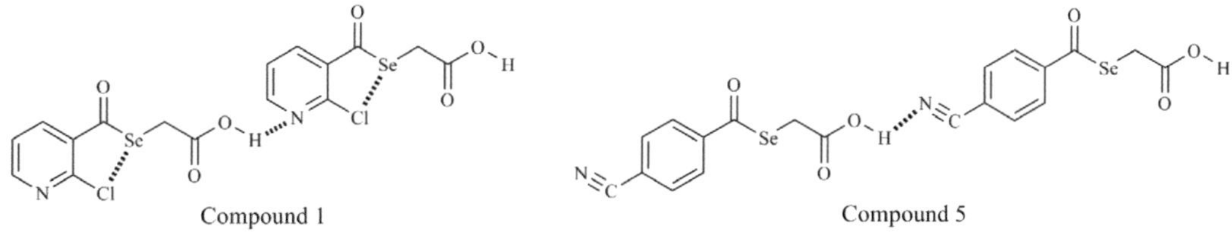

Figure 2. Possible interactions in the crystal packaging for compounds 1 and 5.

If we compare the calorimetric data for selenyl acetic derivatives with other organoselenium compounds (alkylimidoselenocarbamates) synthesized and published by our research group [91], we observed that the T_{onset} values for the first ones were significantly lower that their imidoselenocarbamate analogs (**Table 2**). The selenyl acetic derivatives are smaller molecules than imidoselenocarbamates and hence they possess lower T_{onset} values. Nevertheless, the enthalpy of fusion values for selenyl acetic acids are significantly higher compared with the imidoselenocarbamates in all cases (**Table 2**). These data seem to point to the existence of intramolecular hydrogen bonds for selenyl acetic derivatives (**Figure 3**).

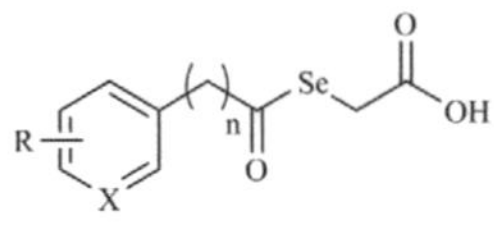

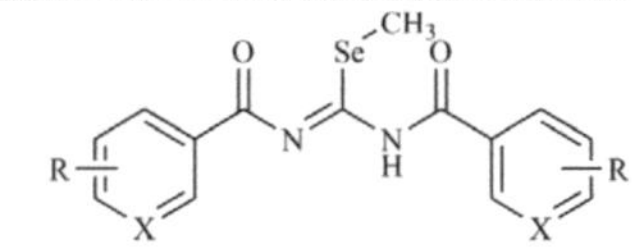

				Selenyl acetic acids		**Imidoselenocarbamates**	
Ref.	**R**	**X**	**n**	**T_{onset} (ºC ± SD)**	**ΔH_f (Jg^{-1})**	**T_{onset} (ºC ± SD)**	**ΔH_f (Jg^{-1})**
1	2-Cl	N	0	157.2 ± 0.2	126.8 ± 4.8	186.8 ± 0.2	42.4 ± 1.3
2	H	C	0	83.3 ± 0.6	92.3 ± 2.2	139.5 ± 0.2	30.6 ± 0.6
3	3,5-diOCH$_3$	C	0	117.4 ± 0.3	108.3 ± 5.5	163.6 ± 0.3	39.6 ± 0.3
4	4-CF$_3$	C	0	107.7 ± 1.1	73.7 ± 1.0	172.9 ± 0.3	30.7 ± 0.8
5	4-CN	C	0	146.5 ± 0.1	112.3 ± 1.7	219.5 ± 2.8	44.4 ± 14.3
6	4-CH$_3$	C	0	92.2 ± 0.2	78.7 ± 0.8	148.5 ± 0.3	32.7 ± 4.9
7	H	C	1	73.8 ± 0.3	74.9 ± 7.2	---	---
8	Phenyl	C	0	130.1 ± 0.4	105.9 ± 4.6	---	---

Table 2. T_{onset} and enthalpy of fusion values for selenyl acetic acids and imidoselenocarbamates.

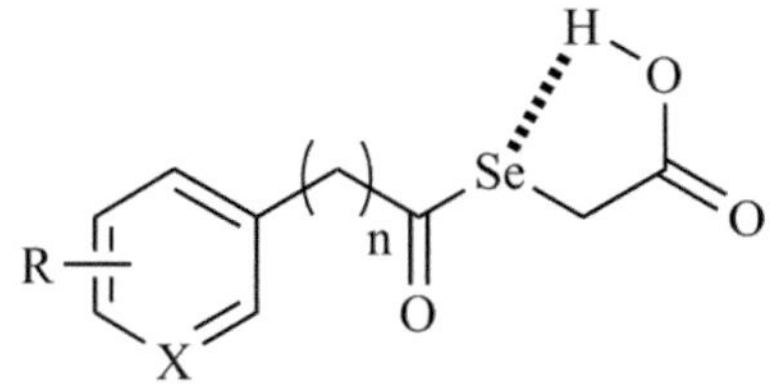

Figure 3. Intramolecular hydrogen bonds for selenyl acetic derivatives.

4. Conclusions

Sulfur (S) and selenium (Se) compounds present several applications in a great variety of fields and the study of their thermal behavior is important for their usefulness in these applications. So, the thermal data are necessary to understand the properties and functions for most of these derivatives. The application of the thermal techniques to these S and Se compounds allows, among others: (a) the study of degradation process, as well as the quantification of the purity and composition for coordination compounds; (b) the characterization of thermal behavior for semiconducting glasses; (c) the determination of solubility isotherms and the field of crystallization of inorganic metal-selenite mixtures.

In the last decade, among the wide range of compounds tested as potential anticancer agents, several structurally diverse derivatives that contain a sulfur or selenium template have been reported and have generated growing interest. For that reason, in this chapter we have focused on some relevant thermal studies in sulfur and selenium compounds with anticancer activity. The thermal techniques, and particularly the DSC, are especially useful in these compounds since they can point out future issues in its pharmaceutical development. The determination of the polymorphism of a substance is of great importance due to the strong influence of the crystalline form on the physicochemical properties, bioavailability and stability of drug, and, in some compounds with biological activity, can even become metastable forms, being twice as active as the stable form.

Finally, we report the unpublished thermal analysis data for eight selenyl acetic acid derivatives, which possess cytotoxic activity *in vitro* against several cancer cell lines. All the compounds sublimated before the degradation process start. So, it is not possible to study the thermal stability of these compounds using thermogravimetric techniques. Nevertheless, the obtained results for calorimetric studies allow to point out some calorimetric behaviors concerning to their stability in the fusion process: (a) the substitution on the aromatic ring causes an increase in the fusion temperatures of these compounds; (b) the inclusion of a methylene group between aromatic ring and carbonyl group seem to lead to a diminution in the T_{onset} values; (c) the presence of groups such as cyano and chloro on the ring, which can form hydrogen bonds, significantly increase the fusion temperatures of these derivatives.

Author details

Daniel Plano*
Synthesis Section, Department of Organic and Pharmaceutical Chemistry, University of Navarra, Pamplona, Spain
Department of Pharmacology, Penn State Hershey College of Medicine, Hershey, PA, USA

Juan Antonio Palop and Carmen Sanmartín
Synthesis Section, Department of Organic and Pharmaceutical Chemistry, University of Navarra, Pamplona, Spain

5. References

[1] Zhou Y Z, Larson J D, Bottoms C A, Arturo E C, Henzl M T, Jenkins J L, Nix J C, Becker D F, Tanner J J (2008) Structural Basis of the Transcriptional Regulation of the Proline Utilization Regulon by Multifunctional PutA. Journal of Molecular Biology. 381: 174-188.

[2] Verly R M, Rodrigues M A, Daghastanli K R P, Denadai A M L, Cuccovia I M, Bloch C, Frezard F, Santoro M M, Pilo-Veloso D, Bemquerer M P (2008) Effect of Cholesterol on the Interaction of the Amphibian Antimicrobial Peptide DD K with Liposomes. Peptides. 29: 15-24.

[3] Andrushchenko V V, Aarabi M H, Nguyen L T, Prenner E J, Vogel H J (2008) Thermodynamics of the Interactions of Tryptophan-Rich Cathelicidin Antimicrobial Peptides with Model and Natural Membranes. Biochimica Et Biophysica Acta-Biomembranes. 1778: 1004-1014.

* Corresponding Author

[4] Chaires J B. In *Annual Review of Biophysics*, 2008; Vol. *37*, pp 135-151.

[5] McKew J C, Lee K L, Shen M W H, Thakker P, Foley M A, Behnke M L, Hu B, Sum F W, Tam S, Hu Y, Chen L, Kirincich S J, Michalak R, Thomason J, Ipek M, Wu K, Wooder L, Ramarao M K, Murphy E A, Goodwin D G, Albert L, Xu X, Donahue F, Ku M S, Keith J, Nickerson-Nutter C L, Abraham W M, Williams C, Hegen M, Clark J D (2008) Indole Cytosolic Phospholipase a(2) Alpha Inhibitors: Discovery and in Vitro and in Vivo Characterization of 4-{3-5-Chloro-2-(2-{(3,4-dichlorobenzyl)sulfonyl amino}ethyl)-1-(diphenylmethyl)-1H-indol-3-yl propyl}benzoic Acid, Efipladib. Journal of Medicinal Chemistry. 51: 3388-3413.

[6] Garbett N C, Mekmaysy C S, Helm C W, Jenson A B, Chaires J B (2009) Differential Scanning Calorimetry of Blood Plasma for Clinical Diagnosis and Monitoring. Experimental and Molecular Pathology. 86: 186-191.

[7] Manoussakis G, Bolos C, Ecateriniadou L, Sarris C (1987) Synthesis, Characterization and Anti-Bacterial Studies of Mixed-Ligand Complexes of Dithiocarbamato—Thiocyanato and Iron(III), Nickel(II), Copper(II) and Zinc(II). European Journal of Medicinal Chemistry. 22: 421-425.

[8] Montagner D, Marzano C, Gandin V (2011) Synthesis, Characterization and Cytotoxic Activity of Palladium (II) dithiocarbamate Complexes with Alpha,Omega-Diamines. Inorganica Chimica Acta. 376: 574-580.

[9] Chen Y-W, Chen K-L, Chen C-H, Wu H-C, Su C-C, Wu C-C, Way T-D, Hung D-Z, Yen C-C, Yang Y-T, Lu T-H (2010) Pyrrolidine Dithiocarbamate (PDTC)/Cu Complex Induces Lung Epithelial Cell Apoptosis through Mitochondria and ER-Stress Pathways. Toxicology Letters. 199: 333-340.

[10] Mohamed G G, Ibrahim N A, Attia H A E (2009) Synthesis and Anti-Fungicidal Activity of Some Transition Metal Complexes with Benzimidazole Dithiocarbamate Ligand. Spectrochimica Acta Part A-Molecular and Biomolecular Spectroscopy. 72: 610-615.

[11] Leka Z B, Leovac V M, Lukic S, Sabo T J, Trifunovic S R, Szecsenyi K M (2006) Synthesis and Physico-Chemical Characterization of New Dithiocarabamato Ligand and Its Complexes with Copper(II), Nickel(II) and Palladium(II). Journal of Thermal Analysis and Calorimetry. 83: 687-691.

[12] Chauhan H, Bakshi A (2011) Synthetic, Spectroscopic, Thermal, and Structural Studies of Antimony(III) bis(pyrrolidinedithiocarbamato)alkyldithiocarbonates. Journal of Thermal Analysis and Calorimetry. 105: 937-946.

[13] Onwudiwe D C, Ajibade P A (2011) Synthesis, Characterization and Thermal Studies of Zn(II), Cd(II) and Hg(II) Complexes of N-Methyl-N-Phenyldithiocarbamate: The Single Crystal Structure of $[(C_6H_5)(CH_3)NCS_2]_4Hg_2$. International Journal of Molecular Sciences. 12: 1964-1978.

[14] Onwudiwe D C, Ajibade P A, Omondi B (2011) Synthesis, Spectral and Thermal Studies of 2,2′-Bipyridyl Adducts of Bis(N-Alkyl-N-Phenyldithiocarbamato)Zinc(II). Journal of Molecular Structure. 987: 58-66.

[15] Wriedt M, Naether C (2011) Synthesis, Crystal Structures, Thermal and Magnetic Properties of New Selenocyanato Coordination Polymers with Pyrazine as Co-Ligand. Zeitschrift Fur Anorganische Und Allgemeine Chemie. 637: 666-671.

[16] Köpf-Maier P, Klapötke T (1988) Antitumor Activity of Some Organomettalic Bismuth(III)Thiolates. Inorganica Chimica Acta. 152: 49-52.
[17] Cantos G, Barbieri C L, Iacomini M, Gorin P A J, Travassos L R (1993) Synthesis of Antimony Complexes of Yeast Mannan and Mannan Derivatives and Their Effect on Leishmania-Infected Macrophages. Biochemical Journal. 289: 155-160.
[18] Briand G G, Burford N (1999) Bismuth Compounds and Preparations with Biological or Medicinal Relevance. Chemical Reviews. 99: 2601-2658.
[19] Luan S R, Zhu Y H, Jia Y Q, Cao Q (2010) Characterization and Thermal Analysis of Thiourea and Bismuth Trichloride Complex. Journal of Thermal Analysis and Calorimetry. 99: 523-530.
[20] Zhong G Q, Luan S R, Wang P, Guo Y C, Chen Y R, Jia Y Q (2006) Synthesis, Characterization and Thermal Decomposition of Thiourea Complexes of Antimony and Bismuth Triiodide. Journal of Thermal Analysis and Calorimetry. 86: 775-781.
[21] Swain T (2012) Synthesis and Thermal Characterization of Sulfur Containing Methionine Bridged Cobalt(III) and Copper(II) Complex. Journal of Thermal Analysis and Calorimetry. 109: 365-372.
[22] Yang Q, Chen S, Gao S (2007) Syntheses and Thermal Properties of Some Complexes with 2-Mercaptonicotinic Acid. Journal of Thermal Analysis and Calorimetry. 90: 881-885.
[23] Lee M L, Shi L P, Tian Y T, Gan C L, Miao X S (2008) Crystallization Behavior of $Sb_{70}Te_{30}$ and $Ag_3In_5Sb_{60}Te_{32}$ Chalcogenide Materials for Optical Media Applications. Physica Status Solidi (a). 205: 340-344.
[24] Kotkata M F, Mansour S A (2011) Study of Glass Transition Kinetics of Selenium Matrix Alloyed with up to 10% Indium. Journal of Thermal Analysis and Calorimetry. 103: 555-561.
[25] Mehta N, Tiwari R S, Kumar A (2006) Glass Forming Ability and Thermal Stability of Some Se–Sb Glassy Alloys. Materials Research Bulletin. 41: 1664-1672.
[26] Bahishti A A, Majeed Khan M A, Patel B S, Al-Hazmi F S, Zulfequar M (2009) Effect of Laser Irradiation on Thermal and Optical Properties of Selenium–Tellurium Alloy. Journal of Non-Crystalline Solids. 355: 2314-2317.
[27] Ohto M, Itoh M, Tanaka K (1995) Optical and Electrical Properties of Ag-As-S Glasses. Journal of Applied Physics. 77: 1034-1039.
[28] Ogusu K, Kumagai T, Fujimori Y, Kitao M (2003) Thermal Analysis and Raman Scattering Study on Crystallization and Structure of $Ag_x(As_{0.4}Se_{0.6})_{100-x}$ Glasses. Journal of Non-Crystalline Solids. 324: 118-126.
[29] Saiter J M, Ledru J, Hamou A, Saffarini G (1998) Crystallization of As_xSe_{1-x} from the Glassy State ($0.005<x<0.03$). Physica B: Condensed Matter. 245: 256-262.
[30] Tonchev D, Fogal B, Belev G, Johanson R E, Kasap S O (2002) Properties of a-Sb_xSe_{1-x} Photoconductors. Journal of Non-Crystalline Solids. 299–302, Part 2: 998-1001.
[31] Mikla V I, Mikhalko I P, Mikla V V (2001) Laser-Induced Amorphous-to-Crystalline Phase Transition in Sb_xSe_{1-x} Alloys. Materials Science and Engineering: B. 83: 74-78.

[32] Holubová J, Černošek Z, Černošková E (2009) The Selenium Based Chalcogenide Glasses with Low Content of As and Sb: DSC, Stepscan DSC and Raman Spectroscopy Study. Journal of Non-Crystalline Solids. 355: 2050-2053.

[33] Savage J A, Webber P J, Pitt A M (1978) An Assessment of Ge-Sb-Se Glasses as 8 to 12μm Infra-Red Optical Materials. Journal of Materials Science. 13: 859-864.

[34] Giridhar A, Narasimham P S L, Mahadevan S (1980) Electrical Properties of Ge-Sb-Se Glasses. Journal of Non-Crystalline Solids. 37: 165-179.

[35] Shaaban E, Tomsah I (2011) The Effect of Sb Content on Glass-Forming Ability, the Thermal Stability, and Crystallization of Ge–Se Chalcogenide Glass. Journal of Thermal Analysis and Calorimetry. 105: 191-198.

[36] Peng Q, Dong Y, Deng Z, Kou H, Gao S, Li Y (2002) Selective Synthesis and Magnetic Properties of α-MnSe and $MnSe_2$ Uniform Microcrystals. The Journal of Physical Chemistry B. 106: 9261-9265.

[37] Vlaev L T, Tavlieva M P (2007) Structural and Thermal Studies on the Solid Products in the System $MnSeO_3$-SeO_2-H_2O. Journal of Thermal Analysis and Calorimetry. 90: 385-392.

[38] Vlaev L T, Genieva S D, Georgieva V G (2006) Study of the Crystallization Fields of Nickel(II) Selenites in the System $NiSeO_3$-SeO_2-H_2O. Journal of Thermal Analysis and Calorimetry. 86: 449-456.

[39] Gospodinov G G, Stancheva A G (2004) Physicochemical Study on Selenites of the Three-Component System Yb_2O_3-SeO_2-H_2O. Journal of Thermal Analysis and Calorimetry. 76: 537-542.

[40] Dammak M, Khemakhem H, Mhiri T (2001) Superprotonic Conduction and Ferroelectricity in Addition Cesium Sulfate Tellurate $Cs_2SO_4.Te(OH)_6$. Journal of Physics and Chemistry of Solids. 62: 2069-2074.

[41] Dammak M, Khemakhem H, Mhiri T, Kolsi A W, Daoud A (1999) Structural and Vibrational Study of $K_2SeO_4{\cdot}Te(OH)_6$ Material. Journal of Solid State Chemistry. 145: 612-618.

[42] Dammak M, Khemakhem H, Mhiri T, Kolsi A W, Daoud A (1998) Structure and Characterization of a Mixed Crystal $Rb_2SO_4.Te(OH)_6$. Journal of Alloys and Compounds. 280: 107-113.

[43] Abdelhedi M, Dammak M, Cousson A, Kolsi A W (2005) Structural, Calorimetric and Conductivity Study of the New Mixed Solution $Rb_2(SO_4)_{0.5}(SeO_4)_{0.5}Te(OH)_6$. Journal of Alloys and Compounds. 398: 55-61.

[44] Ktari L, Abdelhedi M, Bouhlel N, Dammak M, Cousson A (2009) Synthesis, Calorimetric, Structural and Conductivity Studies in a New Thallium Selenate Tellurate Adduct Compound. Materials Research Bulletin. 44: 1792-1796.

[45] Pansuriya P, Parekh H, Friedrich H, Maguire G (In Press) Bisthiourea: Thermal and Structural Investigation. Journal of Thermal Analysis and Calorimetry. DOI: 10.1007/s10973-012-2309-3.

[46] Ramamurthi K, Madhurambal G, Ravindran B, Mariappan M, Mojumdar S (2011) The Growth and Characterization of a Metal Organic Crystal, Potassium Thiourea Thiocyanide. Journal of Thermal Analysis and Calorimetry. 104: 943-947.

[47] Rampon D S, Rodembusch F S, Schneider J M F M, Bechtold I H, Goncalves P F B, Merlo A A, Schneider P H (2010) Novel Selenoesters Fluorescent Liquid Crystalline Exhibiting a Rich Phase Polymorphism. Journal of Materials Chemistry. 20: 715-722.
[48] Xie Y, Zheng X, Jiang X, Lu J, Zhu L (2001) Sonochemical Synthesis and Mechanistic Study of Copper Selenides $Cu_{2-x}Se$, β-CuSe, and Cu_3Se_2. Inorganic Chemistry. 41: 387-392.
[49] Jiang Y, Xie B, Wu J, Yuan S, Wu Y, Huang H, Qian Y (2002) Room-Temperature Synthesis of Copper and Silver, Nanocrystalline Chalcogenides in Mixed Solvents. Journal of Solid State Chemistry. 167: 28-33.
[50] Chrissafis K, Paraskevopoulos K, Manolikas C (2006) Studying $Cu_{2-x}Se$ Phase Transformation through DSC Examination. Journal of Thermal Analysis and Calorimetry. 84: 195-199.
[51] Ribeiro da Silva M A V, Santos A F L O M, Gomes J R B, Roux M a V, Temprado M, Jiménez P, Notario R (2009) Thermochemistry of Bithiophenes and Thienyl Radicals. A Calorimetric and Computational Study. The Journal of Physical Chemistry A. 113: 11042-11050.
[52] Foces-Foces C, Roux M, Notario R, Segura M (2011) Thermal Behavior and Polymorphism in Medium–High Temperature Range of the Sulfur Containing Amino Acids *L*-Cysteine and *L*-Cystine. Journal of Thermal Analysis and Calorimetry. 105: 747-756.
[53] Kerr K A, Ashmore J P (1973) Structure and Conformation of Orthorhombic *L*-Cysteine. Acta Crystallographica Section B. 29: 2124-2127.
[54] Gorbitz C H, Dalhus B (1996) *L*-Cysteine, Monoclinic Form, Redetermination at 120K. Acta Crystallographica Section C. 52: 1756-1759.
[55] Kolesov B A, Minkov V S, Boldyreva E V, Drebushchak T N (2008) Phase Transitions in the Crystals of L- and DL-Cysteine on Cooling: Intermolecular Hydrogen Bonds Distortions and the Side-Chain Motions of Thiol-Groups. 1. L-Cysteine. The Journal of Physical Chemistry B. 112: 12827-12839.
[56] Moggach S A, Allan D R, Clark S J, Gutmann M J, Parsons S, Pulham C R, Sawyer L (2006) High-Pressure Polymorphism in L-Cysteine: The Crystal Structures of L-Cysteine-III and L-Cysteine-IV. Acta Crystallographica Section B. 62: 296-309.
[57] Oughton B M, Harrison P M (1959) The Crystal Structure of Hexagonal L-Cystine. Acta Crystallographica. 12: 396-404.
[58] Chaney M O, Steinrauf L K (1974) The Crystal and Molecular Structure of Tetragonal L-Cystine. Acta Crystallographica Section B. 30: 711-716.
[59] Dahaoui S, Pichon-Pesme V, Howard J A K, Lecomte C (1999) Ccd Charge Density Study on Crystals with Large Unit Cell Parameters: The Case of Hexagonal L-Cystine. The Journal of Physical Chemistry A. 103: 6240-6250.
[60] Regis Botelho J, Duarte Gondim A, Garcia dos Santos I, Dunstan P, Souza A, Fernandes V, Araújo A (2004) Thermochemical Parameters of Dimethyl and Di-Iso-Propyl Dithiocarbamate Complexes of Palladium(II). Journal of Thermal Analysis and Calorimetry. 75: 607-613.

[61] Airoldi C, Chagas A P (1992) Some Features of the Thermochemistry of Coordination-Compounds. Coordination Chemistry Reviews. 119: 29-65.

[62] de Souza A, Neto F, de Souza J, Macedo R, de Oliveira J, Pinheiro C (1997) Thermochemical Parameters of Complexes of Di-Isobutyldithiocarbamate with Phosphorus-Group Elements. Journal of Thermal Analysis and Calorimetry. 49: 679-684.

[63] Feng M-L, Xie Z-L, Huang X-Y (2009) Two Gallium Antimony Sulfides Built on a Novel Heterometallic Cluster. Inorganic Chemistry. 48: 3904-3906.

[64] Yao H-G, Ji M, Ji S-H, Zhang R-C, An Y-L, Ning G-l (2009) Solvothermal Syntheses of Two Novel Layered Quaternary Silver-Antimony(III) Sulfides with Different Strategies. Crystal Growth & Design. 9: 3821-3824.

[65] Ribeiro da Silva M A V, Santos A F L O M (2010) Thermochemical Properties of Two Nitrothiophene Derivatives 2-Acetyl-5-Nitrothiophene and 5-Nitro-2-Thiophenecarboxaldehyde. Journal of Thermal Analysis and Calorimetry. 100: 403-411.

[66] Morais C, Pat B, Gobe G, Johnson D W, Healy H (2006) Pyrrolidine Dithiocarbamate Exerts Anti-Proliferative and Pro-Apoptotic Effects in Renal Cell Carcinoma Cell Lines. Nephrology Dialysis Transplantation. 21: 3377-3388.

[67] Morais C, Gobe G, Johnson D W, Healy H (2009) Anti-Angiogenic Actions of Pyrrolidine Dithiocarbamate, a Nuclear Factor Kappa B Inhibitor. Angiogenesis. 12: 365-379.

[68] Kostova I (2006) Platinum Complexes as Anticancer Agents. Recent Patents on Anti-Cancer Drug Discovery. 1: 1-22.

[69] Yao X, Panichpisal K, Kurtzman N, Nugent K (2007) Cisplatin Nephrotoxicity: A Review. American Journal of the Medical Sciences. 334: 115-124.

[70] Galanski M, Arion V B, Jakupec M A, Keppler B K (2003) Recent Developments in the Field of Tumor-Inhibiting Metal Complexes. Current Pharmaceutical Design. 9: 2078-2089.

[71] Jones M M, Basinger M A, Mitchell W M, Bradley C A (1986) Inhibition of Cis-Diamminedichloroplatinum(II)-Induced Renal Toxicity in the Rat. Cancer Chemotherapy and Pharmacology. 17: 38-42.

[72] Gandara D R, Wiebe V J, Perez E A, Makuch R W, Degregorio M W (1990) Cisplatin Rescue Therapy - Experience with Sodium Thiosulfate, WR2721, and Diethyldithiocarbamate. Critical Reviews in Oncology/Hematology. 10: 353-365.

[73] Hidaka S, Tsuruoka M, Funakoshi T, Shimada H, Kiyozumi M, Kojima S (1994) Protective Effects of Dithiocarbamates against Renal Toxicity of Cis-Diamminedichloroplatinum in Rats. Renal Failure. 16: 337-349.

[74] Botelho J R, Gondim A D, Santos I M G, Dunstan P O, Souza A G, Fernandes V J, Araujo A S (2004) Thermochemical Parameters of Dimethyl and Di-Iso-Propyl Dithiocarbamate Complexes of Palladium(II). Journal of Thermal Analysis and Calorimetry. 75: 607-613.

[75] Botelho J, Souza A, Nunes L, Chagas A, Garcia dos Santos I, da Conceição M, Dunstan P (2002) Thermochemical Properties of Palladium(II) Chelates Involving Dialkyldithiocarbamates. Journal of Thermal Analysis and Calorimetry. 67: 413-417.

[76] Fregona D, Giovagnini L, Ronconi L, Marzano C, Trevisan A, Sitran S, Biondi B, Bordin F (2003) Pt(II) and Pd(II) Derivatives of Ter-Butylsarcosinedithiocarbamate. Synthesis, Chemical and Biological Characterization and in Vitro Nephrotoxicity. Journal of Inorganic Biochemistry. 93: 181-189.

[77] Giovagnini L, Marzano C, Bettio F, Fregona D (2005) Mixed Complexes of Pt(II) and Pd(II) with Ethylsarcosinedithiocarbamate and 2-/3-Picoline as Antitumor Agents. Journal of Inorganic Biochemistry. 99: 2139-2150.

[78] Faraglia G, Fregona D, Sitran S, Giovagnini L, Marzano C, Baccichetti F, Casellato U, Graziani R (2001) Platinum(II) and Palladium(II) Complexes with Dithiocarbamates and Amines: Synthesis, Characterization and Cell Assay. Journal of Inorganic Biochemistry. 83: 31-40.

[79] Manav N, Mishra A K, Kaushik N K (2006) In Vitro Antitumour and Antibacterial Studies of Some Pt(IV) Dithiocarbamate Complexes. Spectrochimica Acta Part A-Molecular and Biomolecular Spectroscopy. 65: 32-35.

[80] Manav N, Mishra A K, Kaushik N K (2004) Triphenyl Phosphine Adducts of Platinum(IV) and Palladium(II) Dithiocarbamates Complexes: A Spectral and in Vitro Study. Spectrochimica Acta Part A-Molecular and Biomolecular Spectroscopy. 60: 3087-3092.

[81] Uivarosi V, Badea M, Aldea V, Chirigiu L, Olar R (In Press) Thermal and Spectral Studies of Palladium(II) and Platinum(IV) Complexes with Dithiocarbamate Derivatives. Journal of Thermal Analysis and Calorimetry.

[82] Rademaker-Lakhai J M, van den Bongard D, Pluim D, Beijnen J H, Schellens J H M (2004) A Phase I and Pharmacological Study with Imidazolium-Trans-Dmso-Imidazole-Tetrachlororuthenate, a Novel Ruthenium Anticancer Agent. Clinical Cancer Research. 10: 3717-3727.

[83] Alessio E, Mestroni G, Nardin G, Attia W M, Calligaris M, Sava G, Zorzet S (1988) Cis-Dihalotetrakis(Dimethyl Sulfoxide)Ruthenium(II) and Trans-Dihalotetrakis(Dimethyl Sulfoxide)Ruthenium(II) Complexes ($RuX_2(Dmso)_4$, X = Cl, Br) - Synthesis, Structure, and Antitumor-Activity. Inorganic Chemistry. 27: 4099-4106.

[84] Alessio E, Iengo E, Geremia S, Calligaris M (2003) New Geometrical and Linkage Isomers of the Ru(II) Precursor *Cis,Cis,Trans*-$RuCl_2(Dmso\text{-}S)_2(Dmso\text{-}O)(CO)$: A Spectroscopic and Structural Investigation. Inorganica Chimica Acta. 344: 183-189.

[85] Zhelev Z, Ohba H, Bakalova R, Hadjimitova V, Ishikawa M, Shinohara Y, Baba Y (2004) Phenothiazines Suppress Proliferation and Induce Apoptosis in Cultured Leukemic Cells without Any Influence on the Viability of Normal Lymphocytes - Phenothiazines and Leukemia. Cancer Chemotherapy and Pharmacology. 53: 267-275.

[86] Hollo B, Krstic M, Sovilj S P, Pokol G, Szecsenyi K M (2011) Thermal Decomposition of New Ruthenium(II) Complexes Containing N-Alkylphenothiazines. Journal of Thermal Analysis and Calorimetry. 105: 27-32.

[87] Plano D, Sanmartin C, Moreno E, Prior C, Calvo A, Palop J A (2007) Novel Potent Organoselenium Compounds as Cytotoxic Agents in Prostate Cancer Cells. Bioorganic & Medicinal Chemistry Letters. 17: 6853-6859.

[88] Plano D, Baquedano Y, Ibáñez E, Jiménez I, Palop J A, Spallholz J E, Sanmartín C (2010) Antioxidant-Prooxidant Properties of a New Organoselenium Compound Library. Molecules. 15: 7292-7312.

[89] Ibáñez E, Plano D, Font M, Calvo A, Prior C, Palop J A, Sanmartín C (2011) Synthesis and Antiproliferative Activity of Novel Symmetrical Alkylthio- and Alkylseleno-Imidocarbamates. European Journal of Medicinal Chemistry. 46: 265-274.

[90] Ibáñez E, Agliano A, Prior C, Nguewa P, Redrado M, González-Zubeldia I, Plano D, Palop J A, Sanmartin C, Calvo A (2012) The Quinoline Imidoselenocarbamate EI201 Blocks the Akt/Mtor Pathway and Targets Cancer Stem Cells Leading to a Strong Antitumor Activity. Current Medicinal Chemistry. 19: 3031-3043.

[91] Plano D, Lizarraga E, Font M, Palop J A, Sanmartin C (2009) Thermal Stability and Decomposition of Sulphur and Selenium Compounds. Journal of Thermal Analysis and Calorimetry. 98: 559-566.

[92] Kitamura S, Miyamae A, Koda S, Morimoto Y (1989) Effect of Grinding on the Solid-State Stability of Cefixime Trihydrate. International Journal of Pharmaceutics. 56: 125-134.

[93] Perrenot B, Widmann G (1994) Polymorphism by Differential Scanning Calorimetry. Thermochimica Acta. 234: 31-39.

[94] Plano D, Lizarraga E, Palop J, Sanmartín C (2011) Study of Polymorphism of Organosulfur and Organoselenium Compounds. Journal of Thermal Analysis and Calorimetry. 105: 1007-1013.

[95] Sanmartin C, Plano D, Dominguez E, Font M, Calvo A, Prior C, Encio I, Palop J A (2009) Synthesis and Pharmacological Screening of Several Aroyl and Heteroaroyl Selenylacetic Acid Derivatives as Cytotoxic and Antiproliferative Agents. Molecules. 14: 3313-3338.

Chapter 17

Calorimetric Determination of Heat Capacity, Entropy and Enthalpy of Mixed Oxides in the System $CaO–SrO–Bi_2O_3–Nb_2O_5–Ta_2O_5$

Jindřich Leitner, David Sedmidubský, Květoslav Růžička and Pavel Svoboda

Additional information is available at the end of the chapter

http://dx.doi.org/10.5772/54064

1. Introduction

Mixed oxides in the system $CaO–SrO–Bi_2O_3–Nb_2O_5–Ta_2O_5$ possess many extraordinary electric, magnetic and optical properties for which they are used in fabrication of various electronic components. For example $Sr_2(Nb,Ta)_2O_7$ and $(Sr,Ca)Bi_2(Nb,Ta)_2O_9$ are used for ferroelectric memory devices, $CaNb_2O_6$, $Sr_5(Nb_{1-x}Ta_x)_4O_{15}$ and $Bi(Nb,Ta)O_4$ for microwave dielectric resonators and $Ca_2Nb_2O_7$ as non-linear optical materials and hosts for rare-earth ions in solid-state lasers. Ternary strontium bismuth oxides $SrBi_2O_4$, $Sr_2Bi_2O_5$, and $Sr_6Bi_2O_9$ are of considerable interest due to a visible light driven fotocatalytic activity.

To assess the thermodynamic stability and reactivity of these oxides under various conditions during their preparation, processing and operation, a complete set of consistent thermodynamic data, including heat capacity, entropy and enthalpy of formation, is necessary. Some of these data are available in literature. Akishige et al. [1] have been measured the heat capacities of $Sr_2Nb_2O_7$ and $Sr_2Ta_2O_7$ single crystals in the temperature range 2-600 K. The results have been only plotted and the values of $S°_m(298)$ have not been calculated. A commensurate transformation of $Sr_2Nb_2O_7$ at T_{INC} = 495 K has been observed accompanied by changes in enthalpy and entropy of ΔH = 291 J mol^{-1} and ΔS = 0.587 J K^{-1} mol^{-1}. The heat capacity of $Sr_2Nb_2O_7$ has been also measured by Shabbir at al. [2] in the temperature range 375-575 K. They have observed a phase transition at T_{INC} = 487 ± 2 K connected with ΔH = 147 ± 14 J mol^{-1} and ΔS = 0.71 ± 0.10 J K^{-1} mol^{-1}. The heat capacities of polycrystalline and monocrystalline $SrBi_2Ta_2O_9$ and $Sr_{0,85}Bi_{2,1}Ta_2O_9$ have been measured by Onodera at al. [3–5] at 80-800 K. Morimoto at al. [6] have reported the results of the heat capacity measurements of $SrBi_2(Nb_xTa_{1-x})_2O_9$ (x = 0, 1/3, 2/3 a 1). The temperature dependences of heat capacities show lambda-transitions with maxima at the Currie temperature T_C = 570 ± 1 K, 585 ± 2 K, 625 ± 3 K a 690 ± 2 K for x = 0, 1/3, 2/3 and 1, respectively. Using EMF (electromotive force) measurements, Raghavan has obtained the values of the Gibbs energy of formation from binary oxides, $\Delta_{ox}G$, for some niobates [7,8] and tantalates [9,10] of calcium. His results are summarized in Table 1. The same technique has been employed by Dneprova et al. [11] for $\Delta_{ox}G$ measurement for $CaNb_2O_6$ and $Ca_2Nb_2O_7$. Their results presented in Table 1 are not significantly different from the results of Raghavan. Using the CALPHAD approach [13], Yang et al. [14] have assessed thermodynamic data for various mixed oxides in the $SrO–Nb_2O_5$ system. The same approach has been used by Hallstedt et al. for the assessment of thermodynamic data of mixed oxides in the systems $CaO–Bi_2O_3$ [14] and $SrO–Bi_2O_3$ [15]. Besides equilibrium data, values of the enthalpy of formation [16] of mixed oxide have been considered. Later on, these systems have been studied by EMF method by Jacob and Jayadevan [17,18] and temperature dependences of $\Delta_{ox}G$ for various mixed oxides have been derived. These data have been included into the thermodynamic re-assessment of the $CaO-SrO-Bi_2O_3$ system [19].

Oxide	$\Delta_{ox}G$ (kJ mol^{-1})	T (K)	$\Delta_{ox}H$ (kJ mol^{-1})	$\Delta_{ox}S$ (J K^{-1} mol^{-1})	Ref.
$CaNb_2O_6$	–75.82 – 0.03345T	1245-1300	–75.82	33.45	[7]
$Ca_2Nb_2O_7$	–178.44	1256			[8]
$Ca_3Nb_2O_8$	–209.94	1256			[8]
$CaTa_4O_{11}$	–36.982 – 0.029T	1250-1300	–36.98	29.0	[9]
$CaTa_2O_6$	–65.14	1250			[10]
$Ca_2Ta_2O_7$	–102.82	1250			[10]
$Ca_4Ta_2O_9$	–165.05	1250			[10]
$CaNb_2O_6$	–175.73 + 0.02259T	1100-1276	–175.73	–22.59	[11]
$Ca_2Nb_2O_7$	–212.54 – 0.02218T	1100-1350	–212.54	22.18	[11]
$Sr_2Nb_{10}O_{27}$	–1125.69 + 0.35069T	298-5000	–1125.69	–350.69	[12]
$SrNb_2O_6$	–325.04 + 0.05865T	298-5000	–325.04	–58.65	[12]
$Sr_2Nb_2O_7$	–367.43 + 0.03993T	298-5000	–367.43	–39.93	[12]
$Sr_5Nb_4O_{14}$	–746.72 + 0.05101T	298-5000	–746.72	–51.01	[12]
$Ca_5Bi_{14}O_{26}$	–125.90 – 0.055T	298-1300	–125.9	55.0	[19]
$CaBi_2O_4$	–27.60 – 0.003T	298-1300	–27.6	3.0	[19]
$Ca_4Bi_6O_{13}$	–97.60 – 0.008T	298-1300	–97.6	8.0	[19]
$Ca_2Bi_2O_5$	–42.20 – 0.003T	298-1300	–42.2	3.0	[19]
$SrBi_2O_4$	–63.86 – 0.0018T	298-1300	–63.86	1.8	[19]
$Sr_2Bi_2O_5$	–118.75 + 0.024T	298-1300	–118.75	–24.0	[19]
$Sr_3Bi_2O_6$	–109.60 + 0.0024T	298-1300	–109.60	–2.4	[19]

Table 1. Published values of $\Delta_{ox}G$, $\Delta_{ox}H$ a $\Delta_{ox}S$ for some mixed oxides in the system $CaO-SrO-Bi_2O_3-Nb_2O_5-Ta_2O_5$

This review brings a summary of our results [20–30] focused on calorimetric determination of heat capacity, entropy end enthalpy of mixed oxides in the system $CaO–SrO–Bi_2O_3–Nb_2O_5–Ta_2O_5$. Temperature dependences of molar heat capacity in a broad temperature range were evaluated from the experimental heat capacity and relative enthalpy data. Molar entropies at T = 298.15 K were calculated from low temperature heat capacity measurements. Furthermore, the results of calorimetric measurements of the enthalpies of drop-solution in a sodium oxide-molybdenum oxide melt for several stoichiometric mixed oxides in the above mentioned system are reported from which the values of enthalpy of formation from constituent binary oxides were derived. Finally, some empirical estimation and correlation methods (the Neumann-Kopp's rule, entropy-volume correlation and electronegativity-differences method) for evaluation of thermodynamic data of mixed oxides are tested and assessed.

2. Experimental

Nineteen mixed oxides in the system $CaO–SrO–Bi_2O_3–Nb_2O_5–Ta_2O_5$ with stoichiometry $CaBi_2O_4$, $Ca_4Bi_6O_{13}$, $Ca_2Bi_2O_5$, $SrBi_2O_4$, $Sr_2Bi_2O_5$, $CaNb_2O_6$, $Ca_2Nb_2O_7$, $SrNb_2O_6$, $Sr_2Nb_2O_7$, $Sr_2Nb_{10}O_{27}$, $Sr_5Nb_4O_{15}$, $BiNbO_4$, $BiNb_5O_{14}$, $BiTaO_4$, $Bi_4Ta_2O_{11}$, $Bi_7Ta_3O_{18}$, Bi_3TaO_7, $SrBi_2Nb_2O_9$, and $SrBi_2Ta_2O_9$ were prepared, characterized and examined. The samples were prepared by conventional solid state reactions from high purity precursors ($CaCO_3$, $SrCO_3$ Bi_2O_3, Nb_2O_5 and Ta_2O_5). A three step procedure was used consisting of an initial calcination run of mixed powder precursors and subsequent double firing of prereacted mixtures pressed into pellets. The phase composition of the prepared samples was checked by X-ray powder diffraction (XRD). XRD data were collected at room temperature with an X'Pert PRO (PANalytical, the Netherlands) θ-θ powder diffractometer with parafocusing Bragg-Brentano geometry using CuK$_\alpha$ radiation (λ = 1.5418 nm). Data were scanned over the angular range 5–60° (2θ) with an increment of 0.02° (2θ) and a counting time of 0.3 s step^{-1}. Data evaluation was performed by means of the HighScore Plus software.

The PPMS equipment 14 T-type (Quantum Design, USA) was used for the heat capacity measurements in the low temperature region [31-35]. The measurements were performed by the relaxation method [36] with fully automatic procedure under high vacuum (pressure ~10^{-2} Pa) to avoid heat loss through the exchange gas. The samples were compressed powder pellets. The densities of the samples were about 65 % of the theoretical ones.

The samples were mounted to the calorimeter platform with cryogenic grease Apiezon N (supplied by Quantum Design). The procedure was as follows: First, a blank sample holder with the Apiezon only was measured in the temperature range approx. 2–280 K to obtain background data, then the sample plate was attached to the calorimeter platform and the measurement was repeated in the same temperature range with the same temperature steps. The sample heat capacity was then obtained as a difference between the two data sets. This procedure was applied, because the heat capacity of Apiezon is not negligible in comparison with the sample heat capacity (~8 % at room temperature) and exhibits a peak-shaped transition below room temperature [37]. The manufacturer claims the precision of this measurement better then 2 % [38]; the control measurement of the copper sample (99.999 % purity) confirmed this precision in the temperature range 50–250 K. However, the precision of the measurement strongly depends on the thermal coupling between the sample and the calorimeter platform. Due to unavoidable porosity of the sample plate this coupling is rapidly getting worse as the temperature raises above 270 K and Apiezon diffuses into the porous sample. Consequently, the uncertainty of the obtained data tends to be larger.

A Micro DSC III calorimeter (Setaram, France) was used for the heat capacity determination in the temperature range of 253–352 K. First, the samples were preheated in a continuous mode from room temperature up to 352 K (heating rate 0.5 K min^{-1}). Then the heat capacity was measured in the incremental temperature scanning mode consisting of a number of 5–10 K steps (heating rate 0.2 K min^{-1}) followed by isothermal delays of 9000 s. Two subsequent step-by-step heating were recorded for each sample. Synthetic sapphire, NIST Standard reference material No. 720, was used as the reference. The uncertainty of heat capacity measurements is estimated to be better than ±1 %.

Enthalpy increment determinations were carried out by drop method using high-temperature calorimeter, Multi HTC 96 (Setaram, France). All measurements were performed in air by alternating dropping of the reference material (small pieces of synthetic sapphire, NIST Standard reference material No. 720) and of the sample (pressed pellets 5 mm in diameter) being initially held at room temperature, through a lock into the working cell of the preheated calorimeter. Endothermic effects are detected and the relevant peak area is proportional to the heat content of the dropped specimen. The delays between two subsequent drops were 25–30 min. To check the accuracy of measurement, the enthalpy increments of platinum in the temperature range 770–1370 K were measured first and compared with published reference values [39]. The standard deviation of 22 runs was 0.47 kJ mol^{-1}, the average relative error was 2.0 %. Estimated overall accuracy of the drop measurements is ±3 %.

The heats of drop-solution were determined using a Multi HTC 96 high-temperature calorimeter (Setaram, France). A sodium oxide-molybdenum oxide melt of the stoichiometry $3Na_2O + 4MoO_3$ was used as the solvent. The ratio of solute/solvent varied from 1/250 up to 1/500. The measurements were performed at temperatures of 973 and 1073 K in argon or air atmosphere. The method consists in alternating dropping of the reference material (small spherules of pure platinum) and of the sample (small pieces of pressed tablets 10–40 mg), being initially held near room temperature (T_0), through a lock into the working cell (a platinum crucible with the solvent) of the preheated calorimeter at temperature T. Two or three samples were examined during one experimental run. The delays between two subsequent drops were 30–60 min. The total heat effect ($\Delta_{ds}H$) includes the heat of solution ($\Delta_{sol}H$), the heat content of the sample ($\Delta_T H$), and, for the carbonates, the heat of decomposition ($\Delta_{decomp}H$) to form solid CaO or SrO and gaseous CO_2. Using appropriate thermochemical cycles, the values of the enthalpy of formation of mixed oxides from the binary oxides and from the elements at 298 K were evaluated. The temperature dependence of the heat capacity of platinum [39] was used for the calculation of the sensitivity of the calorimeters.

2.1. Characterization of prepared samples

The XRD analysis revealed that the prepared samples were without any observable diffraction lines from unreacted precursors or other phases. The lattice parameters of the oxides were evaluated by Rietveld refinement [40] and are summarized in Table 2 together with the values of theoretical density calculated from the lattice parameters.

2.2. Evaluation of temperature dependence of heat capacity at low temperatures

The fit of the low-temperature heat capacity data (LT fit) consists of two steps. Assuming the validity of the phenomenological formula $C_{pm} = \beta T^3 + \gamma_{el}T$, at $T \rightarrow 0$ where β is proportional to the inverse cube root of the Debye temperature Θ_D and $\gamma_{el}T$ is the Sommerfeld term, we plotted the C_{pm}/T *vs.* T^2 dependence for $T < 8$ K to estimate the Θ_D and γ_{el} values. Since all compounds under study are semiconductors with a sufficiently large band gap, the non-zero γ_{el} values are supposed to be either due to some metallic impurities or to a series of Schottky-like transitions resulting from structure defects. Nevertheless, they are negligible in most cases (typically < 0.5 mJ K^{-2} mol^{-1}) and can be ignored in further analysis. As an example, the results of heat capacity measurements on $CaNb_2O_6$ and LT fit for $T < 10$ K is shown in Fig. 1.

Oxide	a (nm)	b (nm)	c (nm)	α (°)	β (°)	γ (°)	d (g cm^{-3})	Ref.
$CaBi_2O_4$	1.66143	1.15781	1.39915	90	134.03	90	6.631	[20]
$Ca_4Bi_6O_{13}$	0.59308	1.73512	0.72192	90	90	90	6.540	[20]
$Ca_2Bi_2O_5$	1.01074	1.01249	1.04618	116.88	107.16	92.98	6.468	[20]
$SrBi_2O_4$	1.92635	0.43437	0.61444	90	95.50	90	7.392	[29]
$Sr_2Bi_2O_5$	1.42935	0.61715	0.76478	90	90	90	6.628	[29]
$CaNb_2O_6$	1.49698	0.57472	0.52202	90	90	90	4.760	[26]
$Ca_2Nb_2O_7$	0.76853	1.33587	0.54959	90	90	98.29	4.496	[26]
$SrNb_2O_6$	0.77209	0.55930	1.09821	90	90.37	90	5.174	[24]
$Sr_2Nb_2O_7$	0.39544	2.67735	0.57004	90	90	90	5.206	[26]
$Sr_2Nb_{10}O_{27}$	3.715	3.697	0.3943	90	90	90	5.653	a)
$Sr_5Nb_4O_{15}$	0.56576	0.56576	1.14536	90	90	120	5.490	[27]
$BiNbO_4$	0.56893	1.1728	0.49915	90	90	90	7.297	[21]
$BiNb_5O_{14}$	1.76762	1.72072	0.39610	90	90	90	4.948	b)
$BiTaO_4$	0.56394	1.1776	0.49626	90	90	90	9.149	[21]
$Bi_4Ta_2O_{11}$	0.66159	0.76528	0.98781	101.39	90.10	89.99	9.306	[28]
$Bi_7Ta_3O_{18}$	3.40162	0.76054	0.66354	90	109.16	90	9.395	[28]
Bi_3TaO_7	0.54711	0.54711	0.54711	90	90	90	9.327	[28]
$SrBi_2Nb_2O_9$	0.55160	0.55087	2.51020	90	90	90	7.275	[22]
$SrBi_2Ta_2O_9$	0.55224	0.55266	2.50124	90	90	90	8.801	[22]

a) Quoted according to JCPDS 035-1220.
b) Quoted according to JCPDS 048-0986

Table 2. Structural characterization of prepared samples

In the second step of the LT fit, both sets of the C_{pm} data (relaxation time + DSC) were considered. Analysis of the phonon heat capacity was performed as an additive combination of Debye and Einstein models. Both models include corrections for anharmonicity, which is responsible for a small, but not negligible, additive term at higher temperatures and which accounts for the difference between isobaric and isochoric heat capacity. According to literature [41], the term $1/(1 - \alpha T)$ is considered as a correction factor.

The acoustic part of the phonon heat capacity is described using the Debye model

$$C_{\mathrm{phD}} = \frac{9R}{1-\alpha_{\mathrm{D}}T}\left(\frac{T}{\Theta_{\mathrm{D}}}\right)^{3}\int_{0}^{x_{\mathrm{D}}}\frac{x^{4}\exp(x)}{\left[\exp(x)-1\right]^{2}}\mathrm{d}x \tag{1}$$

where R is the gas constant, Θ_D is the Debye characteristic temperature, α_D is the coefficient of anharmonicity of acoustic branches and $x_D = \Theta_D/T$. Here the three acoustic branches are taken as one triply degenerate branch. Similarly, the individual optical branches are described by the Einstein model

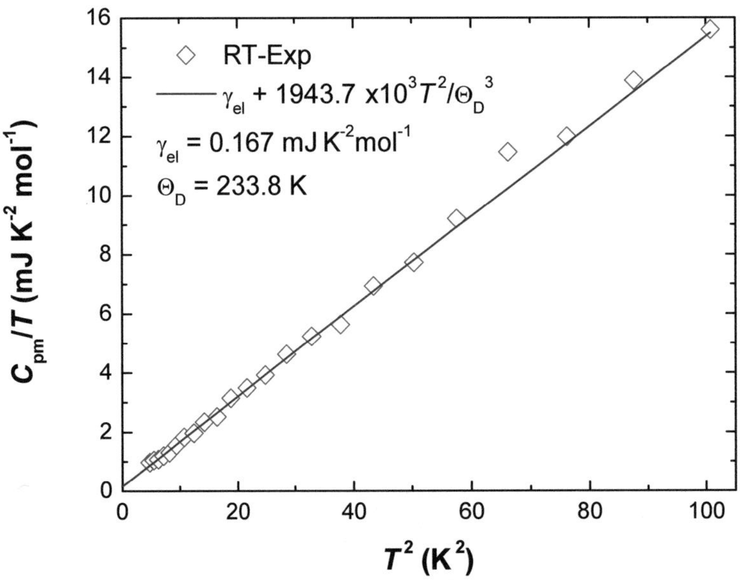

Figure 1. Temperature dependence of C_{pm}/T function for $CaNb_2O_7$ at low temperatures

$$C_{\mathrm{phE}i} = \frac{R}{1-\alpha_{\mathrm{E}i}T}\frac{x_{\mathrm{E}i}^{2}\exp(x_{\mathrm{E}i})}{\left[\exp(x_{\mathrm{E}i})-1\right]^{2}} \tag{2}$$

where α_{Ei} and $x_{Ei} = \Theta_{Ei}/T$ have analogous meanings as in the previous case. Several optical branches are again grouped into one degenerate multiple branch with the same Einstein characteristic temperature and anharmonicity coefficient. The phonon heat capacity then reads

$$C_{ph} = C_{phD} + \sum_{i=1}^{3n-3} C_{phEi} \quad (3)$$

All the estimated values were further treated by a simplex routine and a full non-linear fit was performed on all adjustable parameters.

The values of relative enthalpies at 298.15 K, $H_m(298.15) - H_m(0)$, were evaluated from the low-temperature C_{pm} data (LT fit) by numerical integration of the $C_{pm}(T)$ dependences from zero to 298.15 K. Standard deviations (2σ) were calculated using the error propagation law. The values of standard molar entropies at 298.15 K, $S_m(298.15)$, were derived from the low-temperature C_{pm} data (LT fit) by numerical integration of the $C_{pm}(T)/T$ dependences from zero to 298.15 K. A numerical integration was used with the boundary conditions $S_m = 0$ and $C_{pm} = 0$ at $T = 0$ K. Standard deviations (2σ) were calculated using the error propagation law. All calculated values are summarized in Table 3.

Oxide	$C_{pm}(298)$ (J K^{-1} mol^{-1})	$H_m(298)–H_m(0)$ (J mol^{-1})	$S_m(298)$ (J K^{-1} mol^{-1})	$\Delta_{ox}S$ (J K^{-1} mol^{-1})	Ref.
$CaBi_2O_4$	151.3	26470 ± 158	188.5 ± 3.3	1.9	[20]
$Ca_4Bi_6O_{13}$	504.1	85079 ± 507	574.1 ± 8.8	–23.8	[20]
$Ca_2Bi_2O_5$	197.4	33735 ± 201	231.3 ± 2.9	6.6	[20]
$SrBi_2O_4$	155.6	29601 ± 169	206.1 ± 1.1	4.0	[29]
$Sr_2Bi_2O_5$	201.9	38199 ± 219	261.2 ± 1.4	5.5	[29]
$CaNb_2O_6$	171.8	28159 ± 170	167.3 ± 0.9	–8.1	[26]
$Ca_2Nb_2O_7$	218.1	35631 ± 215	212.4 ± 1.2	–1.1	[26]
$SrNb_2O_6$	170.2	28722 ± 174	173.9 ± 0.9	–17.0	[24]
$Sr_2Nb_2O_7$	216.6	37977 ± 266	238.5 ± 1.3	–5.9	[26]
$Sr_2Nb_{10}O_{27}$	746.8	124150 ± 740	759.7 ± 4.1	–33.9	[27]
$Sr_5Nb_4O_{15}$	477.2	83340 ± 490	524.5 ± 2.8	–18.4	[27]
$BiNbO_4$	121.3	22120 ± 134	147.9 ± 0.8	5.0	[21]
$BiNb_5O_{14}$	386.8	62639 ± 362	397.2 ± 2.1	–25.8	[23]
$BiTaO_4$	119.3	22021 ± 132	149.1 ± 0.8	3.3	[21]
$Bi_4Ta_2O_{11}$	363.2	66566 ± 384	449.6 ± 2.3	9.5	[28]
$Bi_7Ta_3O_{18}$	602.7	109760 ± 634	743.0 ± 3.8	8.6	[28]
Bi_3TaO_7	235.2	44265 ± 254	304.3 ± 1.6	10.0	[28]
$SrBi_2Nb_2O_9$	286.4	49230 ± 292	327.2 ± 1.7	–12.2	[22]
$SrBi_2Ta_2O_9$	286.6	49060 ± 289	339.2 ± 1.8	–5.9	[22]

Table 3. Heat capacity, relative enthalpy, entropy and entropy of formation from binary oxides at temperature 298.15 K of various mixed oxides in the system CaO–SrO–Bi_2O_3–Nb_2O_5–Ta_2O_5

A comparison is given in Table 4 of the values of entropy of formation from binary oxides $\Delta_{ox}S$ at 298 K calculated from our results and those from literature. The values of $\Delta_{ox}S$ are calculated using the relation

$$\Delta_{ox}S = S_m(\mathrm{MO}) - \sum_i b_i S_m(\mathrm{BO},i) \tag{4}$$

where $S_m(\mathrm{MO})$ and $S_m(\mathrm{BO},i)$ stand for the molar entropies of a mixed oxide and a binary oxide i, respectively, and b_i is a constitution coefficient representing the number of formula units of a binary oxide i per formula unit of the mixed oxide. The following values were used for calculation: $S_m(\mathrm{CaO}, 298.15\ \mathrm{K}) = 38.1$ J K^{-1} mol^{-1} [42], $S_m(\mathrm{SrO}, 298.15\ \mathrm{K}) = 53.58$ J K^{-1} mol^{-1} [43], $S_m(\mathrm{Bi_2O_3}, 298.15\ \mathrm{K}) = 148.5$ J K^{-1} mol^{-1} [44], $S_m(\mathrm{Nb_2O_5}, 298.15) = 137.30$ J K^{-1} mol^{-1} [45] $S_m(\mathrm{Ta_2O_5}, 298.15) = 143.09$ J K^{-1} mol^{-1} [45]. Furthermore, $S_m(\mathrm{Sr_2Nb_2O_7}, 298.15) = 238.5$ J K^{-1} mol^{-1} from this work can be directly compared with the value 232.37 J K^{-1} mol^{-1} obtained by numeric integration of the $C_{pm}(T)/T$ dependences from zero to 298.15 K given in Ref. [1]. It should be noted that the values of entropy assessed by thermodynamic optimization of phase equilibrium data are generally considered as less reliable as the values derived from low temperature heat capacity measurements. It is due to possible strong correlation between the enthalpy and entropy contributions to the Gibbs energy. So the obvious discrepancies between our values and data from assessments [12,19] could be explain in this way.

Oxide	$\Delta_{ox}S$ [a] (J K^{-1} mol^{-1})	$\Delta_{ox}S$ (J K^{-1} mol^{-1})	Ref.
$\mathrm{CaNb_2O_6}$	–8.1	33.45	[7]
		–22.59	[11]
$\mathrm{Ca_2Nb_2O_7}$	–1.1	22.18	[11]
$\mathrm{Sr_2Nb_{10}O_{27}}$	–34.0	–350.69	[12]
$\mathrm{SrNb_2O_6}$	–17.0	–58.65	[12]
$\mathrm{Sr_2Nb_2O_7}$	–6.0	–39.93	[12]
$\mathrm{Sr_5Nb_4O_{14}}$	–18.0	–51.01	[12]
$\mathrm{CaBi_2O_4}$	1.9	3.0	[19]
$\mathrm{Ca_4Bi_6O_{13}}$	–23.8	8.0	[19]
$\mathrm{Ca_2Bi_2O_5}$	6.6	3.0	[19]
$\mathrm{SrBi_2O_4}$	4.0	1.8	[19]
$\mathrm{Sr_2Bi_2O_5}$	5.5	–24.0	[19]

[a] This work

Table 4. The values of entropy of formation from binary oxides at 298.15 K: a comparison of our results and data from literature

It should be noted that the thorough analysis of the Debye and Einstein contributions to the heat capacities reveals that the different vibrational modes contribute to the total values of $\Delta_{ox}S$ to a different extent and partial compensation is possible in some cases.

2.3. Evaluation of heat capacity at temperatures above 298 K

For the assessment of temperature dependences of C_{pm} above room temperature, the heat capacity data from DSC and the enthalpy increment data from drop calorimetry were treated simultaneously by the linear least-squares method (HT fit). The temperature dependence of C_{pm} was considered in the form

$$C_{pm} = A + BT + C/T^2 \tag{5}$$

thus the related temperature dependence of $\Delta H_m(T) = H_m(T) - H_m(T_0)$ is given by equation

$$\Delta H_{\mathrm{m}}(T) = H_{\mathrm{m}}(T) - H_{\mathrm{m}}(T_0) = \int_{T_0}^{T} C_{pm}\mathrm{d}T = A(T - T_0) + B(T^2 - T_0^2)/2 - C(1/T - 1/T_0) \tag{6}$$

The sum of squares which is minimized has the following form

$$\begin{gathered} F = \sum_{i=1}^{N(C_p)} w_i^2 \left[C_{pm,i} - A - BT_i - C/T_i^2 \right]^2 + \\ + \sum_{j=1}^{N(\Delta H)} w_j^2 \left[\Delta H_{m,j} - A(T_j - T_{0,j}) - B(T_j^2 - T_{0,j}^2)/2 + C(1/T_j - 1/T_{0,j}) \right]^2 \to \min \end{gathered} \tag{7}$$

where the first sum runs over the C_{pm} experimental points while the second sum runs over the ΔH_m experimental points. Different weights w_i (w_j) were assigned to individual points calculated as $w_i = 1/\delta_i$ ($w_j = 1/\delta_j$) where δ_i (δ_j) is the absolute deviation of the measurement estimated from overall accuracies of measurements (1 % for DSC and 3 % for drop calorimetry). Both types of experimental data thus gain comparable significance during the regression analysis. To smoothly connect the LT fit and HT fit data the values of $C_{pm}(298.15)$ from LT fit were used as constraints and so Eq. (7) is modified

$$F_{\mathrm{constr}} = F - \lambda\left[C_{pm}(298.15) - A - 298.15B - C/298.15^2 \right] \to \min \tag{8}$$

The numerical values of parameters A, B and C are now obtained by solving a set of equations deduced as derivatives of F_{constr} with respect of these parameters and a multiplier λ which are equal to zero at the minimum of F_{constr}. Assessed values of parameters A, B and C of Eq. (4) for mixed oxides are presented in Table 5.

As an example, the results of heat capacity measurements and relative enthalpy measurements on $Bi_7Ta_3O_{18}$ [28] are shown in Fig. 2. Empirical estimation according to the Neumann-Kopp's rule (NKR) is also plotted for comparison.

The empirical Neumann-Kopp's rule (NKR) is frequently used for estimation of unknown values of the heat capacity of mixed oxides [46–48]. According to NKR, heat capacity of a mixed oxide is calculated as a sum of heat capacities of the constituent binary ones

$$C_{pm}(\mathrm{MO}) = \sum_i b_i C_{pm}(\mathrm{BO},i) \tag{9}$$

It was concluded [47,48] that NKR predicts the heat capacities of mixed oxides remarkably well around room temperature but the deviations (mostly positive) from NKR become substantial at higher temperatures. Mean relative error of the estimated values of $C_{pm}(298.15\ K)$ is 1.4 %. Calculated temperature dependences of $\Delta_{ox}C_{pm} = C_{pm}(MO) - \sum b_i C_{pm}(BO,i)$ for various mixed oxides in the systems $CaO–Nb_2O_5$, $SrO–Nb_2O_5$ and $Bi_2O_3–Ta_2O_5$ are shown in Fig. 3.

Oxide	$C_{pm} = A + B \cdot T + C/T^2$ (J K^{-1} mol^{-1})			Temperature range (K)	Ref.
	A	10^3 B	10^{-6} C		
$CaBi_2O_4$	157.161	38.750	–1.546	298-1000	[20]
$Ca_4Bi_6O_{13}$	550.808	114.890	–7.201	298-1200	[20]
$Ca_2Bi_2O_5$	226.096	33.374	–3.432	298-1100	[20]
$SrBi_2O_4$	161.97	45.936	–1.7832	298–1100	[29]
$Sr_2Bi_2O_5$	197.48	87.463	–1.9282	298–1200	[29]
$CaNb_2O_6$	200.40	34.32	–3.45	298-1500	[26]
$Ca_2Nb_2O_7$	257.20	36.21	–4.435	298-1400	[26]
$SrNb_2O_6$	200.47	29.37	–3.473	298-1500	[24]
$Sr_2Nb_2O_7$	248.00	43.50	–3.948	298-1400	[26]
$Sr_2Nb_{10}O_{27}$	835.351	227.648	–13.904	298-1400	[27]
$Sr_5Nb_4O_{15}$	504.796	147.981	–6.376	298-1400	[27]
$BiNbO_4$ a)	128.628	33.400	–1.991	150-1200	[21]
$BiNb_5O_{14}$	455.840	60.160	–7.734	298-1400	[23]
$BiTaO_4$ b)	133.594	25.390	–2.734	150-1200	[21]
$Bi_4Ta_2O_{11}$	445.8	5.451	–7.489	298–1400	[28]
$Bi_7Ta_3O_{18}$	699.0	52.762	–9.956	298–1400	[28]
Bi_3TaO_7	251.6	67.05	–3.237	298–1400	[28]
$SrBi_2Nb_2O_9$	324.470	63.710	–5.076	298-1400	[22]
$SrBi_2Ta_2O_9$	320.220	64.510	–4.700	298-1400	[22]

a) An extra term $1.363 \times 10^8/T^3$ was added.
b) An extra term $2.360 \times 10^8/T^3$ was added.

Table 5. Parameters of temperature dependence of molar heat capacities of various mixed oxides in the system $CaO–SrO–Bi_2O_3–Nb_2O_5–Ta_2O_5$

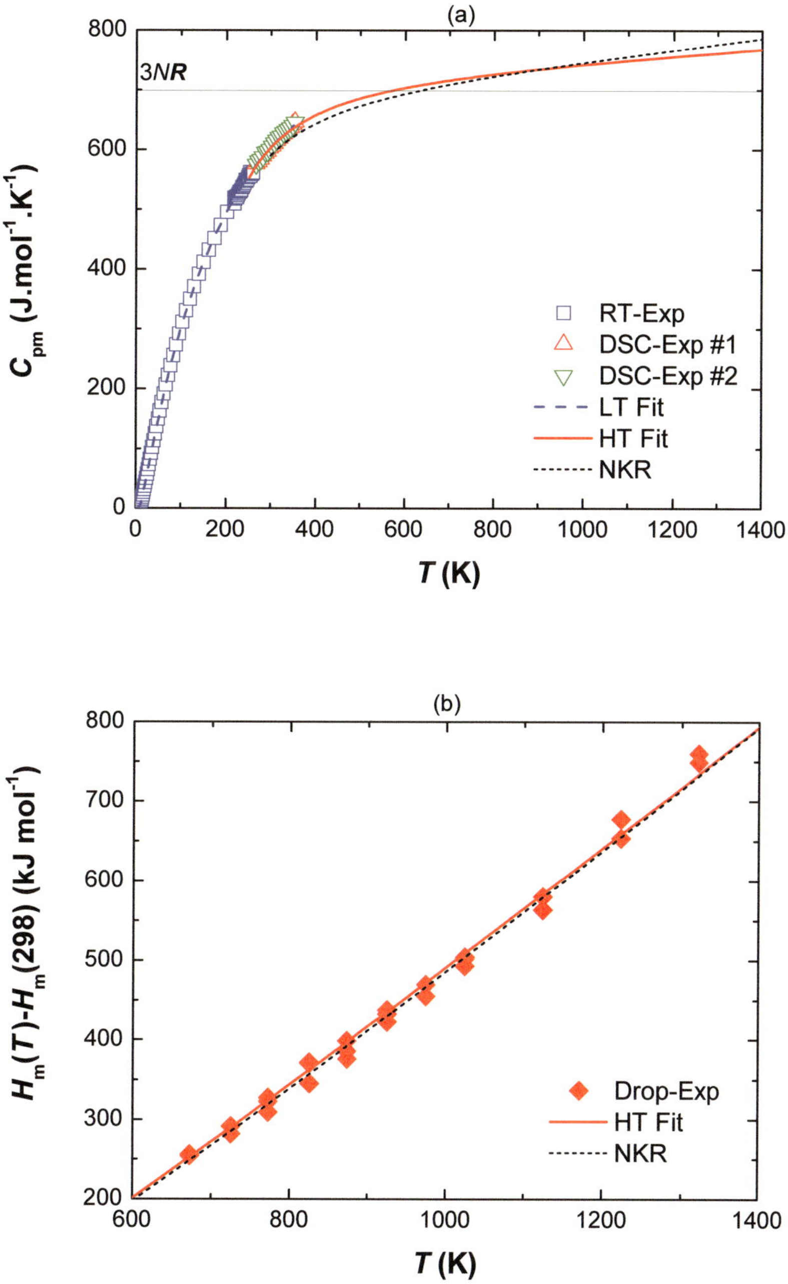

Figure 2. Temperature dependence of heat capacity (a) and relative enthalpy (b) of $Bi_7Ta_3O_{18}$ (3*NR* means the Dulong-Petit limit).

2.4. Evaluation of enthalpy of formation

The heats of drop-solution for the calcium and strontium carbonates and for the bismuth and niobium oxides were measured first. These data are necessary for the evaluation of the $\Delta_{ox}H$ values for the mixed oxides, and furthermore, these data could be compared with the literature data [49–52]. For the $AECO_3$ carbonates, the measured heat effect consists of three contributions:

$$\Delta_{ds}H(AECO_3,T) = \Delta_T H(AECO_3,T_0 \to T) + \Delta_{decomp}H(AECO_3,T) + \Delta_{sol}H(AEO,T) \quad (10)$$

The measurements were performed at 973 K. The values of $\Delta_{ds}H(AECO_3$, 973 K) are given in Table 6 along with the values of $\Delta_{ds}H$(AEO, 973 K), which were derived based on the following thermochemical cycle ($T_0 \approx 298$ K):

$$AECO_3(s,T_0) \rightarrow AEO(melt,T) + CO_2(g,T), \qquad \Delta_{ds}H(AECO_3) \quad (11)$$

$$AECO_3(s,T_0) \rightarrow AEO(s,T_0) + CO_2(g,T_0), \qquad \Delta_{decomp}H(AECO_3) \quad (12)$$

$$CO_2(g,T_0) \rightarrow CO_2(g,T), \qquad \Delta_T H(CO_2) \quad (13)$$

$$AEO(s,T_0) \rightarrow AEO(melt,T), \qquad \Delta_{ds}H(AEO) \quad (14)$$

$$\Delta_{ds}H(AEO) = \Delta_{ds}H(AECO_3) - \Delta_{decomp}H(AECO_3) - \Delta_T H(CO_2) \quad (15)$$

The values $\Delta_{decomp}H(CaCO_3$, 298 K) = 178.8 kJ mol^{-1}, $\Delta_{decomp}H(SrCO_3$, 298 K) = 233.9 kJ mol^{-1} and $\Delta_T H(CO_2$, 298 $\rightarrow$ 973 K) = 32.0 kJ mol^{-1} [53] were used for the calculations.

Next, the $\Delta_{ds}H$ values of the binary oxides Bi_2O_3 and Nb_2O_5 were measured. Because the dissolution of Nb_2O_5 and of the mixed oxides at 973 K proceeds rather slowly, the higher temperature of 1073 K was used. The measured values $\Delta_{ds}H$ are also given in Table 6.

The experimental values of $\Delta_{ds}H$ for $SrCO_3$ and $CaCO_3$ are in quite good agreement with the literature data [49–51]. On the other hand, our results and the published [52] values of $\Delta_{ds}H(Nb_2O_5)$ are quite different. It should be noted that a more endothermic value $\Delta_{decomp}H(SrCO_3$, 298 K) = 249.4 kJ mol^{-1} is presented in the literature [45], which results in more exothermic value for $\Delta_{ds}H$(SrO) by 15.5 kJ mol^{-1}.

$\Delta_{ds}H$ for the mixed oxides was measured at 1073 K. The following thermochemical cycle was used for the calculation of $\Delta_{ox}H$ for calcium and strontium niobates ($T_0 \approx 298$ K):

$$AE_xNb_2O_{5+x}(s,T_0) \rightarrow xAEO(melt,T) + Nb_2O_5(melt,T), \qquad \Delta_{ds}H(AEO) \quad (16)$$

$$AEO(s,T_0) \rightarrow AEO(melt,T), \qquad \Delta_{ds}H(AEO) \quad (17)$$

$$Nb_2O_5(s,T_0) \rightarrow Nb_2O_5(melt,T), \qquad \Delta_{ds}H(Nb_2O_5) \quad (18)$$

$$xAEO(s,T_0) + Nb_2O_5(s,T_0) \rightarrow AE_xNb_2O_{5+x}(s,T_0), \qquad \Delta_{ox}H(AE_xNb_2O_{5+x}) \quad (19)$$

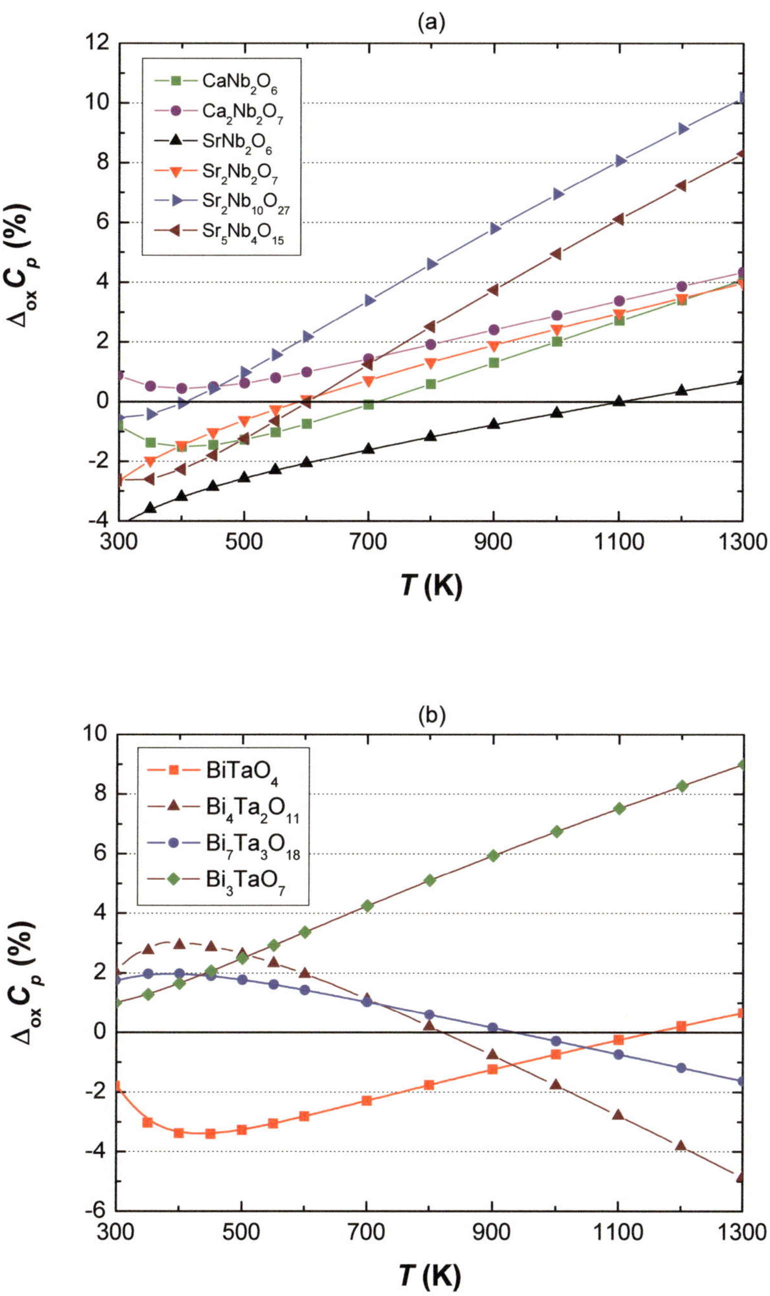

Figure 3. Temperature dependences of $\Delta_{ox}C_{pm}$ for various mixed oxides in the systems CaO–Nb_2O_5, and SrO–Nb_2O_5 (a) and Bi_2O_3–Ta_2O_5 (b)

$$\Delta_{ox}H(AE_xNb_2O_{5+x}) = x\Delta_{ds}H(AEO) + \Delta_{ds}H(Nb_2O_5) - \Delta_{ds}H(AE_xNb_2O_{5+x}) \quad (20)$$

An analogous scheme was applied to calculate $\Delta_{ox}H(BiNbO_4)$. All of the experimental and calculated values are summarized in Table 7. The $\Delta_{ox}H$(298 K) values derived from high-temperature EMN measurements [7,8,11] for the CaO-Nb_2O_5 oxides and the assessed values from the phase diagram for the SrO-Nb_2O_5 oxides [12] are also presented in Table 7.

Substance	T (K)	$\Delta_{ds}H$ (kJ mol^{-1}) a)	$\Delta_{ds}H$ (kJ mol^{-1})
$CaCO_3$	973	128.4 ± 10.1 (10)	119.70 ± 1.02 b)
CaO	973	–82.39	–90.70 ± 1.69 b)
CaO	1073	–77.04 c)	
$SrCO_3$	973	131.4 ± 9.1 (7)	130.16 ± 1.66 d) 134.48 ± 1.89 e)
SrO	973	–134.47	–135.82 ± 2.48 d) –131.42 ± 1.89 e)
SrO	1073	–129.25 f)	
Bi_2O_3	973	26.0 ± 2.9 (12)	
Bi_2O_3	1073	39.6 g)	
Nb_2O_5	1073	141.8 ± 6.0 (11)	91.97 ± 0.78 h)

a) Data from the present work. The uncertainty is two standard deviations of the mean (95% confidence level), the number in parentheses is the number of experiments performed, b) From ref. [49], T = 976 K, c) The value $\Delta_T H$(CaO, 973 → 1073 K) = 5.35 kJ mol^{-1} [26] was used for the calculation, d) From ref. [50], T = 975 K, e) From ref. [51], T = 974 K, f) The value $\Delta_T H$(SrO, 973 → 1073 K) = 5.35 kJ mol^{-1} [27] was used for the calculation, g) The value $\Delta_T H$(Bi_2O_3, 973 → 1073 K) = 13.61 kJ mol^{-1} [28] was used for the calculation, h) From ref. [52], T = 973 K.

Table 6. Enthalpy of drop-solution in $3Na_2O + 4MoO_3$ melts [30]

Substance	T (K)	$\Delta_{ds}H$ (kJ mol^{-1}) a)	$\Delta_{ox}H$(298 K) (kJ mol^{-1}) b)	$\Delta_{ox}H$ (298 K) (kJ mol^{-1})
$CaNb_2O_6$	1073	196.8 ± 20.7 (8)	–132.0 ± 23.8	–159.8 c) –130.1 d)
$Ca_2Nb_2O_7$	1073	195.7 ± 27.8 (8)	–208.0 ± 31.9	–147.3 c) –177.5 e)
$SrNb_2O_6$	1073	180.50 ± 15.7 (4)	–167.9 ± 19.1	–325.0 f)
$Sr_2Nb_2O_7$	1073	167.54 ± 34.7 (4)	–289.2 ± 37.5	–367.4 f)
$BiNbO_4$	1073	132.61 ± 8.9 (7)	–41.9 ± 11.1	

a) Data from the present work. The uncertainty is two standard deviations of the mean (95% confidence level), the number in parentheses is the number of experiments performed, b) The experimental data from the present work. The uncertainty was calculated according to the error propagation law, c) From ref. [11], d) From ref. [7], e) From ref. [8], f) From ref. [12].

Table 7. Enthalpies of drop-solution in $3Na_2O + 4MoO_3$ melt ($\Delta_{ds}H$) and enthalpy of formation from constituent binary oxides ($\Delta_{ox}H$) [30]

Our values for the calcium niobates are in good agreement with Raghavan's data [7,8], while the data from Dneprova et al. [11] are quite different. Moreover, a relation, $\Delta_{ox}H(CaNb_2O_6) > \Delta_{ox}H(Ca_2Nb_2O_7)$, that holds for the values from the work of Dneprova et al. is rather unexpected. The $\Delta_{ox}H$ values for strontium niobates obtained based on the binary SrO-Nb_2O_5 phase diagram evaluation [12] are substantially more exothermic than our calorimetric data. These large differences in the $\Delta_{ox}H$ values are not surprising in view of simultaneous differences in the $\Delta_{ox}S$ values from the assessment [12] and those derived from low temperature dependences of the molar heat capacity of $SrNb_2O_6$ and $Sr_2Nb_2O_7$ [24,26].

3. Empirical correlation *S–V*

A linear correlation between the standard molar entropy at 298.15 K and the formula unit volume $V_{f.u}$ has been proposed by Jenkins and Glaser [54–56]. This approach was used in this work for mixed oxides in the CaO–SrO–Bi_2O_3–Nb_2O_5–Ta_2O_5 system. The linear relation is obvious (see Fig. 4) and the straight line almost naturally passes through the origin:

$$S_m(\mathrm{J\,K^{-1}mol^{-1}}) = 1680.5\, V_{f.u.}(\mathrm{nm^3\,f.u.^{-1}}) \tag{21}$$

The average relative error in entropy is 8.2 %, the binary oxides CaO and Nb_2O_5 show the deviations around 20 %. It should be noted that, in this set of values, the simple analogy of NKR (Eq.(9)) provides a better prediction with an average relative error in entropy of 4.2 %.

Eq. (21) can be used for estimation of missing data. So, the estimated value $S_m(Sr_2Ta_2O_7)$ = 256.06 J K^{-1} mol^{-1} can be compared with the value 245.41 J K^{-1} mol^{-1} obtained by numeric integration of the $C_{pm}(T)/T$ dependences from zero to 298.15 K given in Ref. [1] (relative deviation of –4.3 %). Simple calculation $S_m(Sr_2Ta_2O_5) = 2S_m(SrO) + S_m(Ta_2O_5)$ = 250.25 J K^{-1} mol^{-1} gives more reliable value (relative deviation 2.0 %).

4. Empirical estimation of enthalpy of formation

There are other mixed oxides in the system CaO–SrO–Bi_2O_3–Nb_2O_5–Ta_2O_5 for which the values of enthalpy of formation Δ_fH or enthalpy of formation from binary oxides $\Delta_{ox}H$ have not yet been determined. As a rough estimate, the values of $\Delta_{ox}H$ calculated according to an empirical method proposed by the authors [56] can be used. In the case of Ca, Sr and Bi niobates the following relation holds for $\Delta_{ox}H$:

$$\frac{\Delta_{ox}H}{n_{Nb}+n_{Me}} = -2\cdot 96.5\alpha y\, x_{Nb}\, x_{Me}^{\delta}\left(X_{Nb}-X_{Me}\right)^2 \tag{22}$$

where X_{Nb} and X_{Me} (Me = Ca, Sr or Bi) are Pauling's electronegativities of the relevant elements, x_{Nb} and x_{Me} are the molar fractions of the oxide-forming elements ($x_{Nb} = n_{Nb}/(n_{Nb} + n_{Me})$ etc.), y is the number of oxygen atoms per one atom of oxide-forming elements and α

and δ are the model parameters. Using Pauling's electronegativities, $X_{Nb} = 1.60$, $X_{Ca} = 1.00$, $X_{Sr} = 0.95$, and $X_{Bi} = 2.02$, and the calorimetric values of $\Delta_{ox}H$ obtained in this work, the values of $\alpha = 2.576$ and $\delta = 1.50$ were derived from the least-squares fit. The estimated $\Delta_{ox}H$ values for calcium and strontium niobates are shown in Fig. 5. The values of $\Delta_{ox}H$ that were calculated according to an empirical method proposed by Zhuang et al. [57] are displayed for comparison.

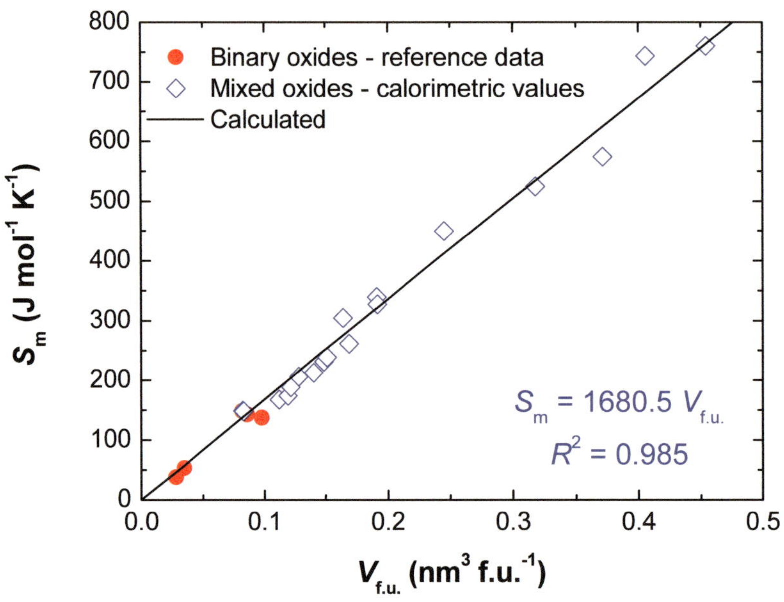

Figure 4. Correlation between the standard molar entropy at 298.15 K and the formula unit volume $V_{f.u.}$ for various mixed oxides in the system CaO–SrO– Bi_2O_3–Nb_2O_5–Ta_2O_5 (data from table Table 3)

5. Conclusion

The above presented data derived from calorimetric measurements became the basis for thermodynamic database FS-FEROX [58] compatible with the FactSage software [59,60]. Missing data for other stoichiometric mixed oxides were estimated by the empirical methods described before: the Neumann-Kopp's rule for heat capacities, the entropy-volume correlation for molar entropies and electronegativity-differences method for enthalpies of formation. At the same time, thermodynamic description of a multicomponent oxide melt was obtained analyzing relevant binary phase diagrams published in literature. The database and the FactSage software were subsequently used for various equilibrium calculations including binary *T-x* phase diagrams and ternary phase diagrams in subsolidus region. Thermodynamic modeling of $SrBi_2Ta_2O_9$ and $SrBi_2Nb_2O_9$ thin layers deposition from the gaseous phase were also performed to optimize the deposition conditions.

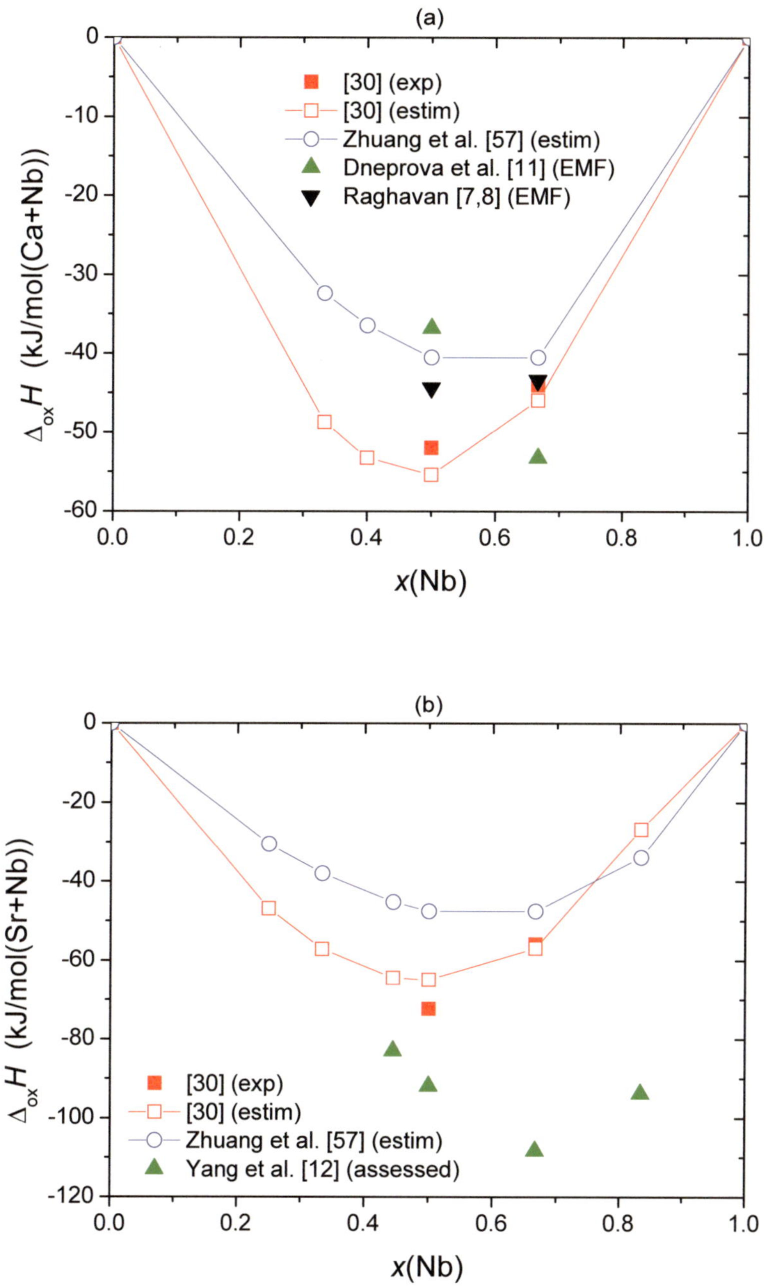

Figure 5. Values of enthalpy of formation of the mixed oxides from constituent binary oxides in the $CaO–Nb_2O_5$ (a) and $SrO–Nb_2O_5$ (b) systems (lines serve only as a guide for the eyes)

Author details

Jindřich Leitner, David Sedmidubský and Květoslav Růžička
Institute of Chemical Technology, Prague, Czech Republic

Pavel Svoboda
Charles University in Prague, Faculty of Mathematics and Physics, Prague, Czech Republic

Acknowledgment

This work was supported by the Ministry of Education of the Czech Republic (research projects N° MSM6046137302 and N° MSM6046137307). Part of this work was also supported from the Grant Agency of the Czech Republic, grant No P108/10/1006. Low temperature experiments were performed in MLTL (http://mltl.eu/), which is supported within the program of Czech Research Infrastructures (project no. LM2011025).

6. References

[1] Akishige Y, Shigematsu H, Tojo T, Kawaji H, Atake T (2005) Specific heat of $Sr_2Nb_2O_7$ and $Sr_2Ta_2O_7$. J. Therm. Anal. Calorim. 81: 537–540.

[2] Shabbir G, Kojima S (2003) Acoustic and thermal properties of strontium pyroniobate single crystals. J. Phys. D: Appl. Phys. 36: 1036–1039.

[3] Onodera A, Yoshio K, Myint CC, Kojima S, Yamashita H, Takama T (1999) Thermal and structural studies of phase transitions in layered perovskite $SrBi_2Ta_2O_9$. Jpn. J. Appl. Phys. 38: 5683–5685.

[4] Onodera A, Yoshio K, Myint CC, Tanaka M, Hironaka K, Kojima S (2000) Thermal behavior in feroelectric $SrBi_2Ta_2O_9$ thin films. Ferroelectrics 241: 159–166.

[5] Yoshio K, Onodera A, Yamashita H (2003) Ferroelectric phase transition and new intermediate phase in bi-layered perovskite $SrBi_2Ta_2O_9$. Ferroelectrics 284: 65–74.

[6] Morimoto K, Sawai S, Hisano K, Yamamoto T (1999) Simultaneous measurements of specific heat capacity and dielectric constant of ferroelectric $SrBi_2(Nb_xTa_{1-x})_2O_9$ ceramics. Ferroelectrics 227: 133–140.

[7] Raghavan S (1991) Thermodynamic stability of monocalcium niobate using calcium fluoride solid electrolyte galvanic cell. Trans. Indian Inst. Met. 44: 285–286.

[8] Raghavan S (1992) Thermodynamics of formation of high calcium niobates from emf measurements. J. Alloys Compd. 179: L25–L27.

[9] Raghavan S (1991) Electrochemical determination of the stability of calcium ditantalate. Indian J. Technol. 29: 313–314.

[10] Raghavan S (1992) Thermodynamics of the formation of high calcium tantalates from emf measurements. J. Alloys Compd. 189: L39–L40.

[11] Dneprova VG, Rezukhina TN, Gerasimov YI (1968) Thermodynamic properties of some calcium niobates. Dokl. Akad. Nauk SSSR 178: 135–137.

[12] Yang Y, Yu H, Jin Z (1999) Thermodynamic calculation of the SrO-Nb_2O_5 system.J. Mater. Sci. Technol. 15: 203– 207.

[13] Saunders N, Miodownik AP (1998) Calphad (Calculation of phase diagrams): A comprehensive guide. Pergamon Materials Series, Vol. 1., Pergamon.

[14] Hallstedt B, Risold D, Gauckler LJ (1997): Thermodynamic assessment of the bismuth–calcium–oxygen system. J. Am. Ceram. Soc. 80: 2629–2636.

[15] Hallstedt B, Risold D, Gauckler LJ (1997) Thermodynamic assessment of the bismuth–strontium–oxygen system. J. Am. Ceram. Soc. 80: 1085–1094.

[16] Idemoto Y, Shizuka K, Yasuda Y, Fueki K (1993) Standard enthalpies of formation of member oxides in the Bi–Sr–Ca–Cu–O system. Physica C 211: 36–44.

[17] Jacob KT, Jayadevan KP (1997) Combined use of oxide and fluoride solid electrolytes for the measurement of Gibbs energy of formation of ternary oxides: System Bi–Ca–O. Mater. Trans. JIM 38: 427–436.

[18] Jacob KT, Jayadevan KP (198) System Bi–Sr–O: Synergistic measurements of thermodynamic properties using oxide and fluoride solid electrolytes. J. Mater. Res. 13: 1905–1908.

[19] Hallstedt B, Gauckler L (2003) Revision of the thermodynamic descriptions of the Cu–O, Ag–O, Ag–Cu–O, Bi–Sr–O, Bi–Ca–O, Bi–Cu–O, Sr–Cu–O, Ca–Cu–O and Sr–Ca–Cu–O systems CALPHAD 27: 177–191.

[20] Abrman P, Sedmidubský D, Strejc A, Voňka P, Leitner J (2002) Heat capacity of mixed oxides in the Bi_2O_3–CaO system. Thermochim. Acta 381: 1–7.

[21] Hampl M, Strejc A, Sedmidubský D, Růžička K, Hejtmánek J, Leitner J (2006) Heat capacity, enthalpy and entropy of bismuth niobate and bismuth tantalate. J. Solid State Chem. 179: 77–80.

[22] Leitner J, ,Hampl M, Růžička K, Sedmidubský D, Svoboda P, Vejpravová J (2006) Heat capacity, enthalpy and entropy of strontium bismuth niobate and strontium bismuth tantalate. Thermochim. Acta 450: 105–109.

[23] Hampl M, Leitner J, Růžička K, Straka M, Svoboda P (2007) Heat capacity and heat content of $BiNb_5O_{14}$. J. Thermal Anal. Calorimetry 87: 553–556.

[24] Leitner J, Hampl M, Růžička K, Straka M, Sedmidubský D, Svoboda P. (2008) Thermodynamic properties of strontium metaniobate $SrNb_2O_6$. J. Thermal. Anal. Calorimetry 91: 985–990.

[25] Leitner J, Hampl M, Růžička K, Straka M, Sedmidubský D, Svoboda P (2008) Heat capacity, enthalpy and entropy of strontium niobate $Sr_2Nb_2O_7$ and calcium niobate $Ca_2Nb_2O_7$. Thermochim. Acta 475: 33–38.

[26] Leitner J, Růžička K, Sedmidubský D, Svoboda P (2009) Heat capacity, enthalpy and entropy of calcium niobates. J. Thermal. Anal. Calorimetry 95: 397–402.

[27] Leitner J, Šipula I, Růžička K, Sedmidubský D, Svoboda P (2009) Heat capacity, enthalpy and entropy of strontium niobates $Sr_2Nb_{10}O_{27}$ and $Sr_5Nb_4O_{15}$. J. Alloys Compd. 481: 35–39.

[28] Leitner J, Jakeš V, Sofer Z, Sedmidubský D, Růžička K, Svoboda P (2011) Heat capacity, enthalpy and entropy of ternary bismuth tantalum oxides. J. Solid State Chem. 184: 241–245.

[29] Leitner J, Sedmidubský D, Růžička K, Svoboda P (2012) Heat capacity, enthalpy and entropy of $SrBi_2O_4$ and $Sr_2Bi_2O_5$. Thermochim. Acta. 531: 60–65.

[30] Leitner J, Nevřiva M, Sedmidubský D, Voňka P (2011) Enthalpy of formation of selected mixed oxides in a $CaO–SrO–Bi_2O_3–Nb_2O_5$ system. J. Alloys Compd. 509: 4940–4943.

[31] Lashley J.C., Hundley MF, Migliori A, Sarrao JL, Pagliuso PG, Darling TW, Jaime M, Cooley JC, Hults WL, Morales L, Thoma DJ, Smith JL, Boerio-Goates J, Woodward BF, Stewart GR, Fisher RA, Phillips NE (2003) Critical examination of heat capacity measurements made on a Quantum Design physical property measurement system. Cryogenics 43: 369–378.

[32] Dachs E, Bertoldi C (2005) Precision and accuracy of the heat-pulse calorimetric technique: low-temperature heat capacities of milligram-sized synthetic mineral samples. Eur. J. Mineral. 17: 251–259.

[33] Marriott RA, Stancescu M, Kennedy CA, White MA (2006) Technique for determination of accurate heat capacities of volatile, powdered, or air-sensitive samples using relaxation calorimetry. Rev. Sci. Instrum. 77: 096108 (3 pp).

[34] Kennedy CA, Stancescu M, Marriott RA, White MA (2007) Recommendations for accurate heat capacity measurements using a Quantum Design physical property measurement system. Cryogenics 47: 107–112.

[35] Shi Q, Snow CL, Boerio-Goates J, Woodfield BF (2010) Accurate heat capacity measurements on powdered samples using a Quantum Design physical property measurement system. J. Chem. Thermodyn. 42: 1107–1115.

[36] Hwang JS, Lin KJ, Tien C (1997) Measurement of heat capacity by fitting the whole temperature response of a heat-pulse calorimeter. Rev. Sci. Instrum. 68: 94–101.

[37] Schnelle W, Engelhardt J, Gmelin E (1999) Specific heat capacity of Apiezon N high vacuum grease and of Duran borosilicate glass. Cryogenics 39: 271–275.

[38] Quantum Design, Physical Property Measurement System – Application Note, http://www.qdusa.com/sitedocs/productBrochures/heatcapacity-he3.pdf (accessed 21-12-2011).

[39] Arblaster JW (1994) The thermodynamic properties of platinum on ITS-90, Platinum Metals Rev. 38: 119–125.

[40] Rodriguez-Carvajal J (1993) Recent advances in magnetic structure determination by neutron powder diffraction. Physica B 192: 55–69.

[41] Martin CA (1991) Simple treatment of anharmonic effects on the specific heat. J. Phys.: Condens. Matter 3: 5967–5974.

[42] Taylor JR, Dinsdale AT (1990) Thermodynamic and phase diagram data for the $CaO-SiO_2$ system. CALPHAD 14: 71–88.

[43] Risold D, Hallstedt B, Gauckler LJ (1996) The strontium-oxygen system. CALPHAD 20: 353–361.

[44] Risold D, Hallstedt B, Gauckler LJ, Lukas HL, Fries SG (1995) The bismuth-oxygen system. J. Phase Equilib. 16: 223–234.

[45] Knacke O, Kubaschewski O, Hesselmann K (1991) Thermochemical Properties of Inorganic Substances, 2nd Ed. Berlin: Springer.

[46] Qiu L, White MA (2001) The constituent additivity method to estimate heat capacities of complex inorganic solids. J. Chem. Educ. 78: 1076–1079.

[47] Leitner J, Chuchvalec P, Sedmidubský D, Strejc A, Abrman P (2003) Estimation of heat capacities of solid mixed oxides. Thermochim. Acta 395: 27–46.

[48] Leitner J, Voňka P, Sedmidubský D, Svoboda P (2010) Application of Neumann-Kopp rule for the estimation of heat capacity of mixed oxides. Thermochim. Acta 497: 7–13.

[49] Helean KB, Navrotsky A, Vance ER, Carter ML, Ebbinghaus B, Krikorian O, Lian J, Wang LM, Catalano JG (2002) Enthalpies of formation of Ce-pyrochlore, $Ca_{0.93}Ce_{1.00}Ti_{2.035}O_{7.00}$, U-pyrochlore, $Ca_{1.46}U^{4+}{}_{0.23}U^{6+}{}_{0.46}Ti_{1.85}O_{7.00}$ and Gd-pyrochlore, $Gd_2Ti_2O_7$: three materials relevant to the proposed waste form for excess weapons plutonium. J. Nuclear Mater. 303: 226–239.

[50] Cheng J, Navrotsky A (2004) Energetics of magnesium, strontium, and barium doped lanthanum gallate perovskites. J. Solid State Chem. 177: 126–133.

[51] Xu H, Navrotsky A, Su Y, Balmer ML (2005) Perovskite Solid Solutions along the NaNbO3-SrTiO3 Join: Phase Transitions, Formation Enthalpies, and Implications for General Perovskite Energetics. Chem. Mater. 17: 1880–1886.

[52] Pozdnyakova I, Navrotsky A, Shilkina L, Reznitchenko L (2002) Thermodynamic and structural properties of sodium lithium niobate solid solutions. J. Am. Ceram. Soc. 85: 379–384.

[53] Robbie RA, Hemingway BS (1995) Thermodynamic properties of minerals and related substances at 298.15 K and 1 bar pressure and at higher temperatures, U.S. Geological Survey Bulletin, Vol. 2131,Washington.

[54] Jenkins HDB, Glasser L (2003) Standard absolute entropy, $S°_{298}$, values from volume or density. 1. Inorganic materials. Inorg. Chem. 42: 8702–8708.

[55] Jenkins HDB, Glasser L (2006) Volume-based thermodynamics: Estimations for 2:2 salts, Inorg. Chem. 45: 1754–1756.

[56] Voňka P, Leitner J (2009) A method for the estimation of the enthalpy of formation of mixed oxides in $Al_2O_3–Ln_2O_3$ systems. J. Solid State Chem. 182: 744–748.

[57] Zhuang W, Liang J, Qiao Z, Shen J, Shi Y, Rao G (1998) Estimation of the standard enthalpy of formation of double oxide. , J. Alloys Compd. 267: 6-10.

[58] Leitner J, Sedmidubský D, Voňka P. (2009) Thermodynamic database for the oxide system $CaO–SrO–Bi_2O_3–Nb_2O_5–Ta_2O_5$. CALPHAD XXXVIII, 17.-22.5.2009, Praha, ČR.

[59] Bale CW, Chartrand P, Degterov SA, Eriksson G, Hack K, Ben Mahfoud R, Melançon J, Pelton AD, Petersen S (2002) FactSage thermochemical software and databases. Calphad 26: 189-228.

[60] Bale CW, Bélisle E, Chartrand P, Decterov SA, Eriksson G, Hack K, Jung IH, Kang YB, Melançon J, Pelton AD, Robelin C, Petersen S (2009) FactSage thermochemical software and databases — recent developments. Calphad 33: 295-311.

Chapter 18

Differential Scanning Calorimetry Studies of Phospholipid Membranes: The Interdigitated Gel Phase

Eric A. Smith and Phoebe K. Dea

Additional information is available at the end of the chapter

http://dx.doi.org/10.5772/51882

1. Introduction

DSC is a versatile technique and has been used for decades to study hydrated phospholipid membranes [1-4]. It can even be used to analyze whole cell samples [5]. For pure lipids, DSC can accurately determine the phase transition temperatures and the associated enthalpies. As a consequence, how the chemical structure of lipids translates into thermodynamic properties can be systematically studied. In addition to determining the physical properties of pure lipids, the miscibility and phase behavior of lipid mixtures can be determined.

The detailed review of the interdigitated phase written by Slater and Huang in 1988 provides an excellent outline of the properties of the interdigitated phase and the relevant analytical techniques [6]. Furthermore, the meticulous studies of Koynova and Caffrey describe how systematic changes in lipid chemistry can affect their phase behavior [7-9]. Lipids with asymmetrical acyl chains that form either mixed- or partially-interdigitated phases have also been thoroughly investigated [7,10-12]. This review focuses on the interdigitated phase of fully hydrated phospholipids with hydrocarbon chain lengths of equal size. We pay special attention to recently discovered interdigitated systems and the chemicals that can induce or inhibit lipid interdigitation.

For simplicity, we have centered our review around the extensively studied lipid, 1,2-dipalmitoyl-*sn*-glycero-3-phosphocholine (DPPC). DPPC is naturally occurring and has thermodynamic phase behavior that is typical for saturated phosphatidylcholines (PCs) [7]. Although DPPC does not spontaneously interdigitate when hydrated, it can be reliably transformed into the fully interdigitated gel phase (Tables 1 and 2). Alterations in the lipid hydrocarbon chains (Figure 1) and the lipid head group (Figure 2) substantially affect spontaneous interdigitation (Figure 3). The predisposition for interdigitation is a finely tuned balance of interactions among hydrocarbon chains, between polar head groups, and at the interfacial area with the aqueous phase.

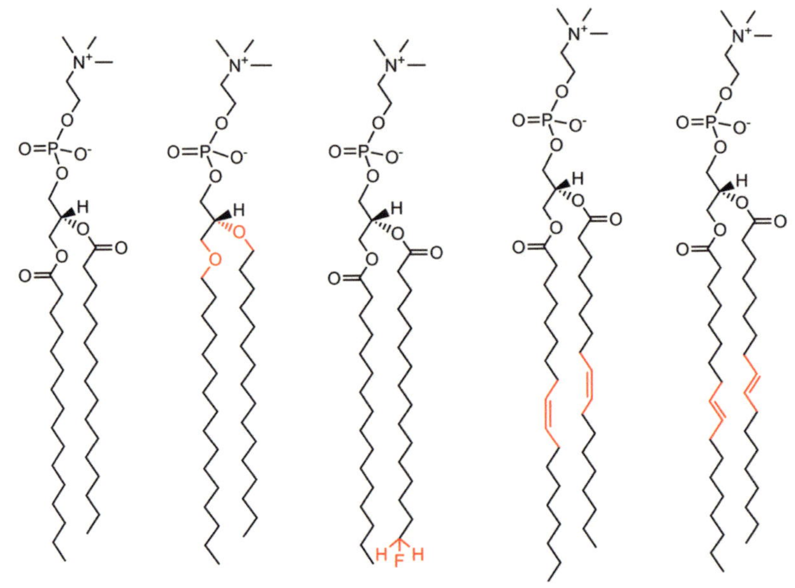

DPPC DHPC F-DPPC DOPC DEPC

Figure 1. Representative examples of fatty acid modifications of DPPC. The segments in red identify modifications to the structure of DPPC.

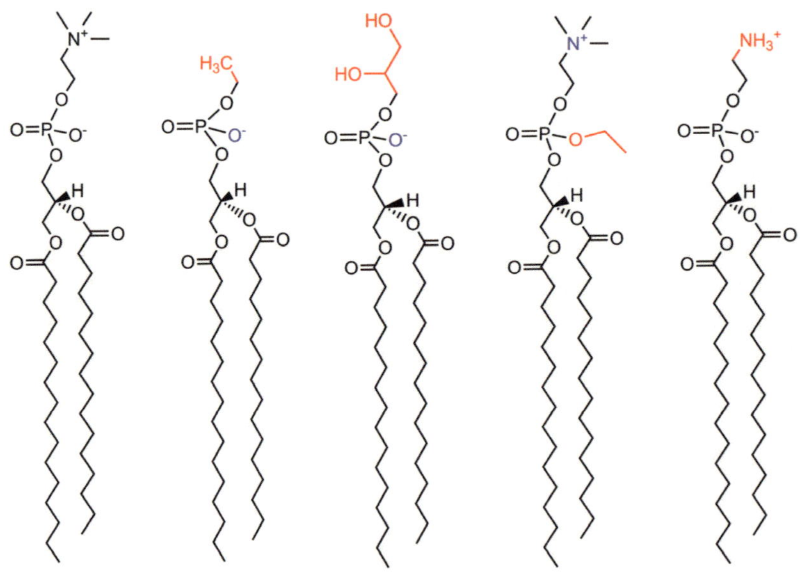

DPPC DPPeth DPPG EDPPC DPPE

Figure 2. Representative examples of head group modifications of DPPC. The segments in red identify modifications to the structure of DPPC. The segments in blue demonstrate the resulting change in charge.

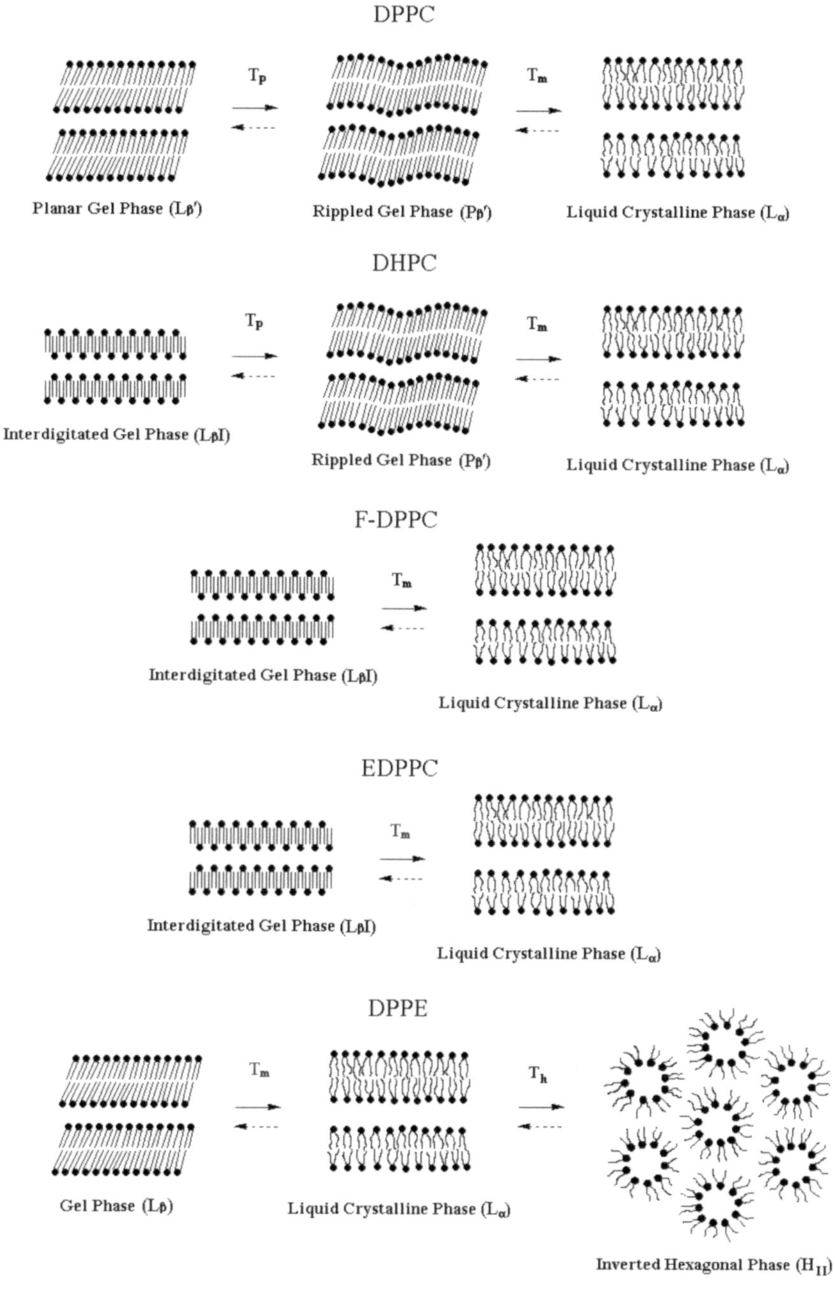

Figure 3. Phase transitions of representative lipids. Solid arrows indicate heating transitions and dotted arrows indicate cooling transitions. Note: subgel phase transitions are not shown.

DPPC/alcohol	Threshold concentration (M)	References
Methanol	2.75 ± 0.35	[13]
Ethanol	1.10 ± 0.10	[13]
1-propanol	0.39 ± 0.03	[14]
2-propanol	0.52 ± 0.03	[14]
1-butanol	0.16 ± 0.02	[15]
Isobutanol	0.17 ± 0.02	[15]
sec-butanol	0.22 ± 0.02	[15]
tert-butanol	0.27 ± 0.02	[15]
1-pentanol	0.07 ± 0.01	[16]
2-pentanol	0.10 ± 0.01	[16]
3-pentanol	0.11 ± 0.01	[16]
3-methyl-2-butanol	0.10 ± 0.01	[16]
2-methyl-1-butanol	0.08 ± 0.01	[16]
3-methyl-1-butanol	0.08 ± 0.01	[16]
2-methyl-2-butanol	0.13 ± 0.01	[16]
neopentanol	0.08 ± 0.01	[16]

Table 1. Threshold concentrations for some alcohol-induced interdigitation.

	Chemicals	References
Anesthetics	bupivacaine	[17]
	dibucaine	[17]
	lidocaine	[17]
	procaine	[17]
	tetracaine	[17-19]
Drugs	labdanes	[20]
	chlorpromazine	[19]
	valsartan	[21]
Organic solvents	glycerol	[22-24]
	ethylene glycol	[25]
	acetone	[26,27]
	acetonitrile	[27,28]
	propionaldehyde	[27]
	petrahydrofuran	[27]
Salts	KSCN	[29,30]
Pressure	N/A	[31-34]

Table 2. Induced interdigitation of PC membranes.

There are multiple types of interdigitation (Figure 4). The type of interdigitation that forms is heavily dependent on the structure and symmetry of the hydrocarbon chains [6]. Interdigitated lipid systems can further be separated into two classes: spontaneous and induced. We use "spontaneous" to describe lipids that self-assemble into the interdigitated gel phase when fully hydrated under typical preparation procedures and at ambient pressure. Some notable recent examples are highlighted, such as cationic lipids and lipids with monofluorinated acyl chains. Whether or not a particular lipid spontaneously interdigitates is determined by the balance of properties that favor and disfavor interdigitation. Lipids often have conflicting characteristics regarding the ability to form the interdigitated phase. Consequently, there is no simple formula for determining which lipids will spontaneously interdigitate without relying on experimental data.

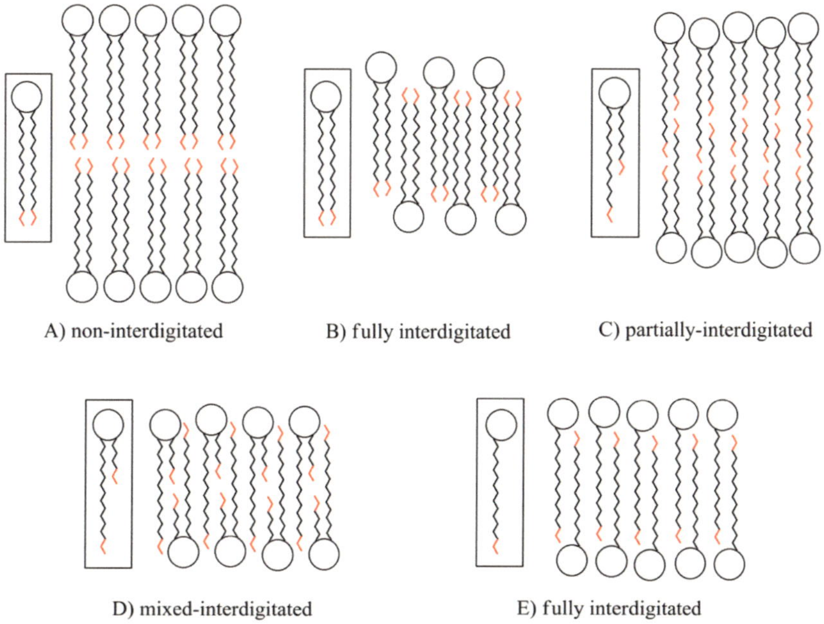

Figure 4. Schematic representation of the different types of membrane interdigitation: A) symmetrical lipid, non-interdigitated; B) symmetrical lipid, fully interdigitated; C) asymmetrical lipid, partially interdigitated; D) highly asymmetrical lipid, mixed-interdigitated; E) lysolipid, fully interdigitated. For clarity, the terminal ends of the hydrocarbon chains are labeled in red.

As can be seen in Figure 4, the structural difference between the interdigitated and non-interdigitated phases can be substantial. In the non-interdigitated membrane, both ends of the hydrocarbon chains meet in the membrane midplane (Figure 4A). Two well-defined leaflets are formed and there is a thick hydrophobic core. In the fully-interdigitated membrane, the thickness of the interdigitated phase is greatly reduced and there is the loss of the membrane midplane. There is an increase in the spacing between the polar lipid head groups and the ends of the lipid hydrocarbon chains become more exposed to the aqueous interface [6]. The difference is most dramatic in the fully interdigitated phase compared to the non-interdigitated membrane (Figure 4A and 4B). In the partially-interdigitated system, the longer chain extends to the other side of the membrane and aligns with the apposing shorter chain (Figure 4C). In the mixed-interdigitated membrane, the short hydrocarbon chains line up with each other and the full-length chain extends to the other side of the membrane (Figure 4D). Lyso lipids also form a fully interdigitated structure (Figure 4E) [35].

2.

2.1. Thermodynamics of phosphatidylcholine membranes

PCs are common in mammalian membranes and have well known phase transitions [7]. The most ubiquitous of these is the gel-to-liquid crystalline transition, often referred to as the melting or main transition. This transition is relatively rapid and is highly reversible [36]. It is characterized by the co-operative melting of the hydrocarbon chains and a high enthalpy DSC peak [2]. The liquid crystalline (L_α) phase has an increased number of gauche conformers and a large increase in membrane fluidity and disorder [2,37]. The pre-transition from the planar gel (L_β') to the rippled gel phase (P_β') has a low enthalpy and is sensitive to sample preparation and the presence of impurities [7,36]. It is also more sensitive to the scan rate, with lower scan rates resulting in lower T_P temperatures [3]. Some PCs also have subgel phases. The subgel transition is slow and is dependent on sample preparation, especially incubation temperature and time [36]. All of the above phases are strongly affected by changes in lipid structure. This review will show that such alterations also have a profound effect on the interdigitated phase.

2.2. Chemically-induced interdigitation of phosphatidylcholines

The most widely studied chemical inducer of interdigitation is ethanol. In non-interdigitated phospholipid membranes, ethanol tends to adsorb to the head groups, especially the region near the hydrocarbon chains [38,39]. In particular, the carbonyl groups of the glycerol backbone of phospholipids are thought to be the favored hydrogen bonding sites for ethanol [40]. Ethanol displaces water when it adsorbs to the head group, which increases the head group volume and decreases the order of the hydrocarbon chains [13,41,42]. The increase in head group volume leads to increased chain tilting and creates energetically unfavorable voids in the hydrocarbon region of non-interdigitated membranes, encouraging the creation of the $L_\beta I$ phase at high concentrations [13,39,42-44]. Once the $L_\beta I$ phase is formed, ethanol can bind to the exposed hydrocarbon chains, replacing the unfavorable interaction of the acyl chains with water [45]. Also, it is typical for alcohols to increase the main transition enthalpy above the threshold concentration for interdigitation [15,16].

There are three main characteristics of the chemically-induced interdigitated phase in the DSC thermograms of saturated PCs: the presence of biphasic phase behavior, an increase in T_m hysteresis, and the suppression of the pre-transition. The combination of these can be used to determine the threshold concentration for interdigitation.

The "biphasic effect" indicates two independent interactions within different concentration ranges [46,47]. The biphasic effect is most strongly characterized by an initial decrease in the T_m, but an increase in or stabilization of the T_m once the $L_\beta I$ phase is formed. The first interaction lies below the threshold concentration. Here, ethanol preferentially partitions into the liquid crystalline phase, lowering the phase transition temperature [47]. The secondary interaction above the threshold concentration stabilizes the interdigitated gel phase. The main transition co-operativity (sharpness of transition peak) can also be enhanced above the threshold concentration [47].

The shape and magnitude of the T_m biphasic effect is also dependent on the alcohol isomer used as shown in Figure 5 [15,16]. For example, *tert*-butanol is an effective inducer of interdigitation and has a pronounced biphasic effect [15]. Other alcohols, such as the pentanol isomers have more "stunted" biphasic behavior [16]. While a difference in trend can still be observed above and below the threshold concentration, the distinction is less pronounced.

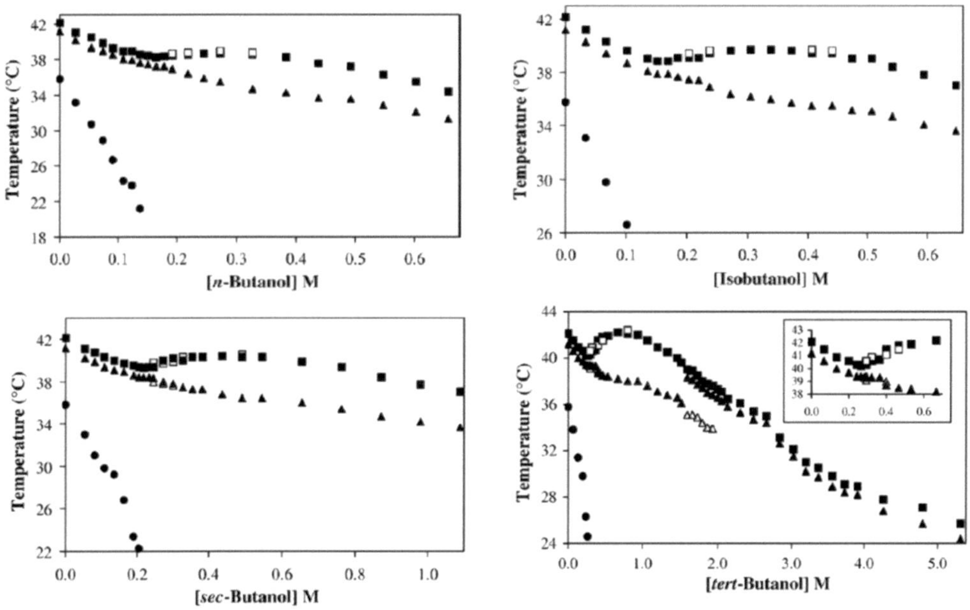

Figure 5. Effects of *n*-butanol, isobutanol, *sec*-butanol, and *tert*-butanol on DPPC phase transition temperatures. (■, heating scan main peak; □, heating scan shoulder peak; ▲, cooling scan main peak; Δ, cooling scan shoulder peak; •, pre-transition peak). Reprinted from Biophys. Chem., 128, Reeves MD, Schawel AK, Wang W, Dea PK, Effects of butanol isomers on dipalmitoylphosphatidylcholine bilayer membranes, Pages No. 13-18, Copyright (2007), with permission from Elsevier [15].

Increasing the alcohol content well above the threshold concentration lowers the T_m [15,16]. At these concentrations, additional alcohol destabilizes the $L_{\beta}I$ phase relative to the L_{α} phase. The membrane bilayer structure can also break down for alcohols that are highly soluble in water [15]. For example, above 2.00 M *tert*-butanol in DPPC, the main transition hysteresis is absent (Figure 5). Additionally, the heating main transition DSC peaks above 2.00 M *tert*-butanol become increasingly broad. Changes in the ^{31}P-NMR spectra at high concentrations confirm the loss of lamellar structure [15].

The biphasic behavior is also reflected in the increase in the main transition enthalpy as the alcohol concentration increases [15,16]. Often, the rate of change in the transition enthalpy above and below the threshold concentration is different (Figure 6). This effect also depends on the alcohol chain length and isomer used. For instance, this difference is clear with *n*-butanol but not *tert*-butanol [15].

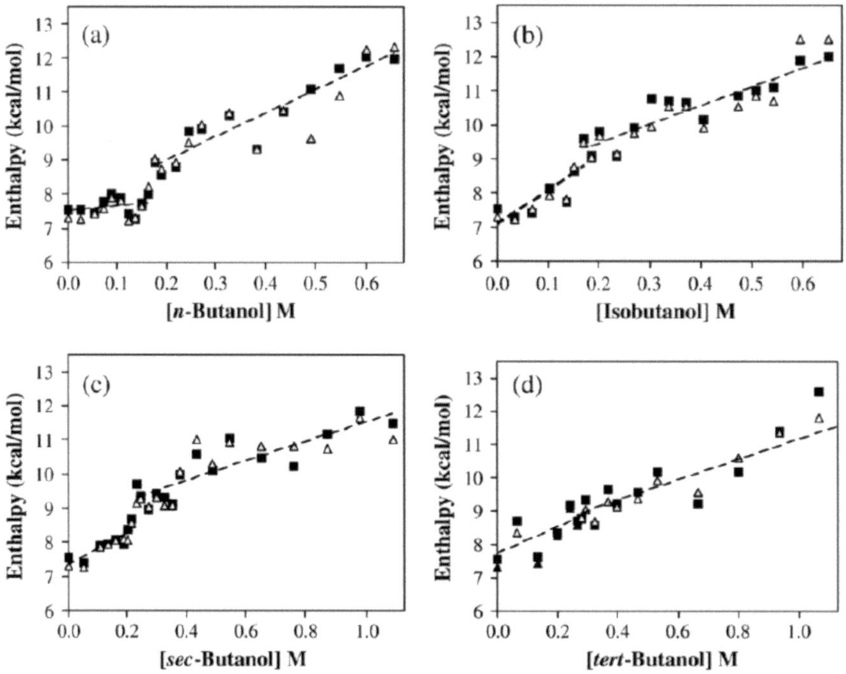

Figure 6. Effects of (a) *n*-butanol, (b) isobutanol, (c) *sec*-butanol and (d) *tert*-butanol on DPPC main transition enthalpies (■, heating scan main transition enthalpy; Δ, cooling scan main transition enthalpy). Reprinted from Biophys. Chem., 128, Reeves MD, Schawel AK, Wang W, Dea PK, Effects of butanol isomers on dipalmitoylphosphatidylcholine bilayer membranes, Pages No. 13-18, Copyright (2007), with permission from Elsevier [15].

A property that accompanies the biphasic effect is the emergence of hysteresis in the main transition [15,16,48,49]. The hysteresis as it relates to DSC is defined as the difference in the transition temperature between heating and cooling scans. This corresponds to the reversibility and kinetics of the transition. The return to the interdigitated phase with a decrease in temperature is a slow process and is therefore less reversible [48,51]. For systems that do not interdigitate, such as phosphatidylethanolamine (PE) lipids, the addition of alcohol does not affect the transition hysteresis [48].

The disappearance of the pre-transition is another consistent property of alcohol-induced interdigitation of saturated PCs. The decrease in the T_P follows a well defined trend below the threshold concentration until it is finally abolished (Figure 5). The rate at which the T_P is depressed depends on the efficacy of the chemical inducer.

By comparing the threshold concentrations of different chemicals, they can be ranked on their effectiveness at inducing the interdigitated phase. For instance, the threshold concentrations for alcohol-induced interdigitation systematically decreases as the lipid hydrocarbon chain length increases (Table 1). The isomers with the most solubility in water are the least effective at inducing interdigitation, as shown by the increase in threshold concentrations [15]. Additionally, the more soluble an isomer is in water, the less effectively it depresses both the temperature of the pre-transition and the main transition prior to the threshold concentration [15].

The ether-linked analogue of DPPC, 1,2-di-*O*-hexadecyl-*sn*-glycero-3-phosphocholine (DHPC), is also a useful model membrane for studying interdigitation. DHPC goes through a low temperature pre-transition from the $L_{\beta}I$ phase to the non-interdigitated rippled gel phase P_{β}' (Figure 3) [52,53]. Therefore, chemicals that stabilize the $L_{\beta}I$ phase increase the T_p until it merges with the main transition into the L_{α} phase [49]. This process occurs at lower concentrations for more effective inducers of interdigitation.

2.3. Pressure-induced interdigitation

It is well established that the application of hydrostatic pressure favors interdigitation in a multitude of lipid systems (Tables 2 and 3). As hydrostatic pressure is applied, the intermolecular distance between adjacent lipids is reduced and molecular packing becomes denser [34]. By changing the packing structure of the membrane, interdigitation can relieve the stress caused by the increased steric hindrance.

Pressure-induced interdigitation is dependent on lipid hydrocarbon chain length and the chemical structure, much like chemically-induced interdigitation. Ether- and ester-linked lipids with longer chains require less pressure to interdigitate [51,54]. Under high temperature and pressure conditions, ester-linked lipids behave similarly to the equivalent ether-linked lipids at normal pressure [51]. Pressure-induced interdigitation is not universal, however. As with chemically-induced interdigitation, certain lipids do not interdigitate even under high pressure [34,55,56].

2.4. Spontaneous interdigitation in ether-linked lipids and 1,3-DPPC

The type of bond that connects the hydrocarbon chain to the lipid head group also affects the thermodynamic properties. Switching either or both of the ester bonds of DPPC with ether linkages results in a small increase in the T_m (<4 °C) and enthalpy (<1 kcal/mol) [57]. A single ether linkage can be sufficient to allow the formation of the interdigitated gel phase [57,58]. Furthermore, in ether lipids that spontaneously interdigitate, the interdigitated phase is stable up to higher temperatures as the chain length increases (Figure 7) [51]. There is an increased amount of head group repulsion in DHPC, which favors the interdigitated phase [51,59,60]. Conversely, the stronger interactions in the head groups of ester lipids hinder interdigitation [34,51]. DHPC is an especially useful lipid for studying the interdigitated phase because its transition from the interdigitated gel to non-interdigitated ripple gel phase is highly sensitive to its environment.

The similarity of DHPC to DPPC also allows for the comparison between ether- and ester-linked lipids. It is consistently easier to interdigitate ether-linked lipids whether through chemical means [17,48,49] or by the application of pressure [51,61]. Furthermore, the ether-linked 1,2-di-*O*-hexadecyl-*sn*-glycero-3-phosphoethanolamine (DHPE) demonstrates the result of competing influences on the interdigitated gel phase. In DHPE, the ether linkages favor interdigitation, but the PE head group is more strongly opposed to interdigitation [62]. Therefore, while DHPC spontaneously interdigitates, the PE head group of DHPE prevents interdigitation.

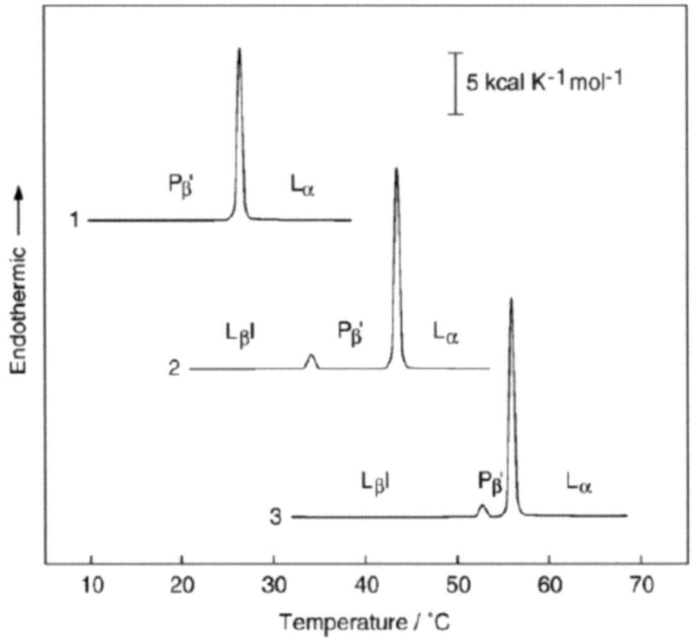

Figure 7. DSC heating thermograms of ether-linked PC bilayer membranes: (1) *O*-14:0-PC, (2) *O*-16:0-PC, (3) *O*-18:0-PC Reprinted from Biochim. Biophys. Acta, 1768, Matsuki H, Miyazaki E, Sakano F, Tamai N, Kaneshina S, Thermotropic and barotropic phase transitions in bilayer membranes of ether-linked phospholipids with varying alkyl chain lengths, Pages No. 479-489, Copyright (2007), with permission from Elsevier [51].

While the majority of PC lipid studies use lipids with the hydrocarbon chains on the *sn*-1 and *sn*-2 positions, there are some examples of experiments using synthetic lipids with the chains located at *sn*-1 and *sn*-3. One intriguing example is the positional isomer of DPPC, 1,3-dipalmitoyl-*sn*-glycero-2-phosphocholine (1,3-DPPC or β-DPPC), which has unique properties including the ability to spontaneously interdigitate [63-65]. However, the phase diagram is different from the ether-linked lipids that spontaneously interdigitate [51,66]. At lower temperatures, 1,3-DPPC exists in a non-interdigitated "crystalline" bilayer phase termed (L_c). At higher temperatures, but below the T_m, 1,3-DPPC can form a fully interdigitated structure [63]. The ability to interdigitate may be due to greater head group repulsion resulting from a different phosphocholine tilt or conformation relative to the glycerol backbone [63-65]. As with most interdigitated systems, 1,3-DPPC converts into a non-interdigitated structure during the heating transition into the L_α phase. The cooling transition from the L_α phase into the interdigitated phase has considerable hysteresis [63].

2.5. The monofluorinated analogue of DPPC: F-DPPC

The monofluorinated analogue of DPPC, 1-palmitoyl-2-(16-fluoropalmitoyl)*sn*-glycero-3-phosphocholine (F-DPPC), spontaneously forms the $L_{\beta}I$ phase below the main transition temperature (T_m) [67-69]. The main transition temperature of F-DPPC also occurs at a higher temperature (~50 °C) and with a higher transition enthalpy (9.8 kcal/mol) compared to DPPC [67]. The endothermic peak of F-DPPC is also broader than DPPC. The transition can be split into two overlapping peaks, with the peak centered at 50.6 °C accounting for 36% of the area and the peak at 52.0 °C accounting for 64% of the area [67]. The lower transition peak component does not correspond to a change in the hydrocarbon chains as detected by

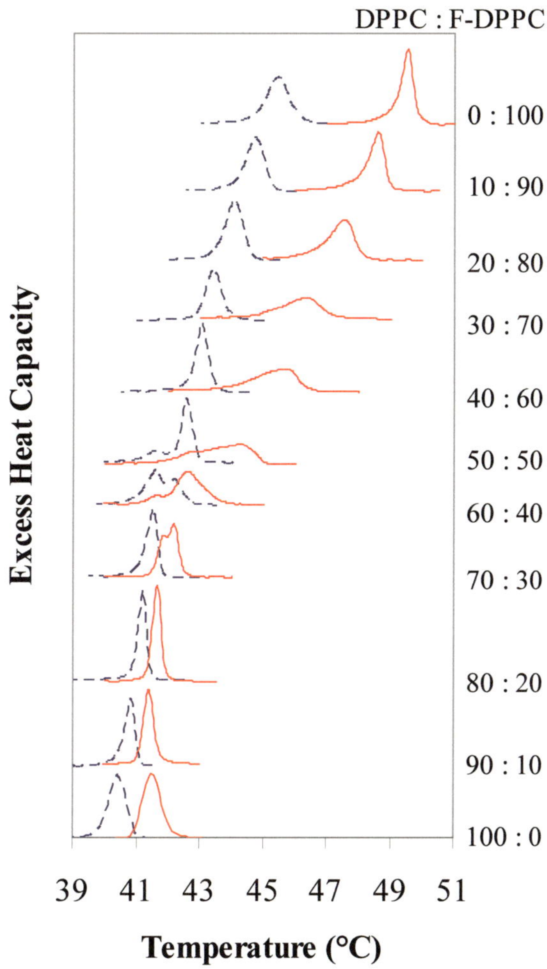

Figure 8. Heating (red lines) and cooling (dashed blue lines) DSC thermograms of the T_m of the DPPC/F-DPPC system are shown. The cooling scans have been inverted to allow comparison with the heating thermograms. For clarity, the thermograms are also offset vertically. Reprinted from Biophys. Chem., 147, Smith E.A., van Gorkum C.M., Dea P.K., Properties of phosphatidylcholine in the presence of its monofluorinated analogue, Pages No. 20-27, Copyright (2010), with permission from Elsevier [69].

FTIR spectroscopy. It is possible that this relates to a conversion from interdigitated to non-interdigitated gel right before the transition into the liquid crystalline phase [67]. The main transition is also characterized by a large main transition hysteresis (Figures 8 and 9) [67,69]. Additionally, the $L_{\beta}I$ phase has high conformational order and tight lipid packing [68].

It appears that the fluorine must be located on the terminal hydrocarbon chain to have a dramatic effect on interdigitation. When the fluorine substitution is not located on the terminal carbon, DSC data reveal that the physical properties are only modestly changed and they are largely miscible with the non-fluorinated parent lipid [70]. Lipids with more fluorine, such as when the 13-16 carbons are perfluorinated, do not spontaneously interdigitate either [71,72]. Therefore, it is the interaction of the polar terminal C-F bond with the aqueous interface that encourages interdigitation [67]. The large dipole moment is the most likely culprit for stabilizing the interdigitated phase by reducing the unfavorable exposure of the hydrophobic acyl chains to water. However, the slightly larger van der Waals radius and the possibility of weak hydrogen bonding may also play a role [73-77].

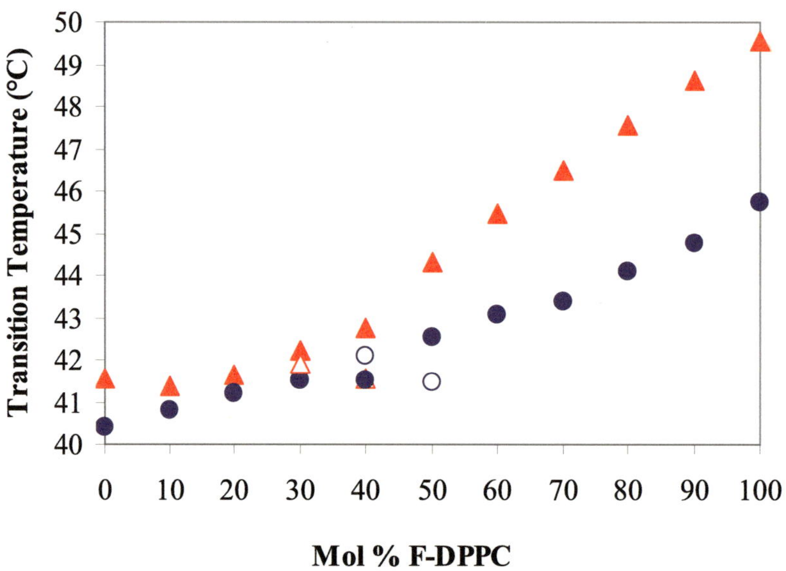

Figure 9. The main transition temperature (T_m) of DPPC/F-DPPC mixtures determined by DSC. Heating scans shown by filled red triangles (▲). Cooling scans shown by filled blue circles (•). Shoulder peaks indicated by unfilled triangles for heating scans (△) and unfilled circles for cooling scans (○). Reprinted from Biophys. Chem., 147, Smith EA, van Gorkum CM, Dea PK, Properties of phosphatidylcholine in the presence of its monofluorinated analogue, Pages No. 20-27, Copyright (2010), with permission from Elsevier [69].

2.6. The interdigitated gel phase in anionic lipids

As with PCs, di-saturated long chain phosphatidylglycerols (PGs) have a strong propensity towards interdigitation (Table 3) [78]. The negatively charged PGs are commonly found in microbial membranes [90]. The interaction of some peptides with lipids is heavily dependent on the composition of the membrane [82]. This contributes to the ability of antimicrobial peptides to selectively target microbial membranes [91]. Recently, it was found that DPPG has the ability to form a quasi-interdigitated gel phase with the addition of the human multifunctional peptide LL-37 [81,82]. The antimicrobial peptide peptidyl-glycylleucine-carboxyamide (PGLa) has a similar effect below the main transition temperature of saturated PGs [83]. In these instances, the peptide shields the acyl chains of the interdigitated lipid from the aqueous layer by orienting in the interfacial region below the T_m (Figure 10).

Furthermore, other chemicals such as Tris-HCl induce interdigitation in DPPG by binding between lipids, resulting in the increased area per head group that favors interdigitation [85]. As in zwitterionic lipids, interdigitation relieves head group repulsion in charged lipids by allowing for a larger area per head group [84]. Charge repulsion in DPPG leads to tilted acyl chains in the non-interdigitated bilayer [85]. This is similar to the ethanol-induced interdigitation of DPPC, where the increased head group size increases the tilt in the gel phase and which ultimately results in the interdigitated gel phase [43]. Ethanol further enhances interdigitation in DPPG, most likely by partitioning into the interfacial region and reducing the exposure of the terminal methyl groups to water [84].

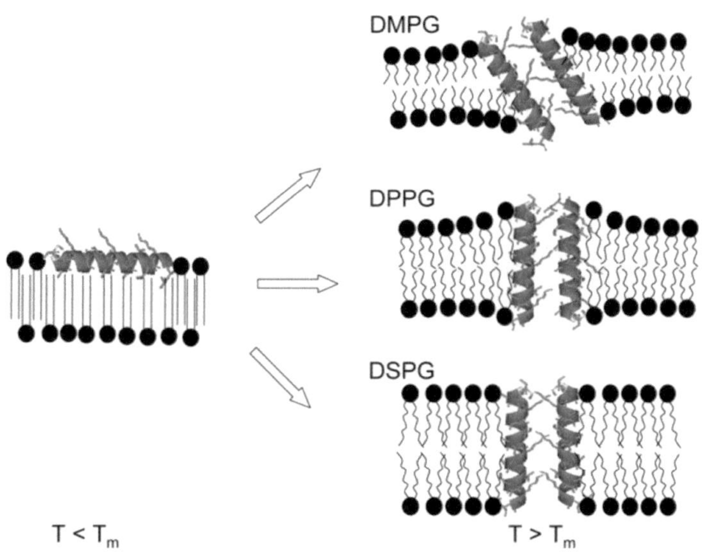

Figure 10. Schematic representation of the peptide PGLa-associated structural changes in PG bilayers. Below the main phase transition ($T < T_m$), the lipids of different hydrocarbon chain lengths exhibit a quasi-interdigitated phase in the presence of PGLa. Reprinted from Biophys. J. 95, Pabst G, Grage SL, Danner-Pongratz S, Jing W, Ulrich AS, Watts A, Lohner K, Hickel A, Membrane thickening by the antimicrobial peptide PGLa, Pages No. 5779-5788, Copyright (2008), with permission from Elsevier [83].

Inducer	References
polymyxin B	[24,79,80]
peptide LL-37	[81,82]
peptide PGLa	[83]
myelin basic protein	[80]
Tris HCl	[79,84,85]
choline and acetylcholine	[86]
atropine	[87,88]
anisodamine	[89]
Pressure	[50]

Table 3. Induced interdigitation of PG membranes

When ethanol substitutes for water in the transphosphatidylation reaction catalyzed by phospholipase D, phosphatidylethanols (Peth) are formed [84,92]. Peth lipids are unique because they have a small anionic lipid headgroup (Figure 2). These lipids are biologically relevant since Peths accumulate in membranes of animal models of alcoholism [93]. Like DPPG, DPPeth can be chemically induced to interdigitate with Tris-HCl and the interdigitated phase is stabilized with the addition of ethanol [84].

2.7. The interdigitated gel phase in cationic lipids

Cationic lipids with modified head groups can spontaneously form interdigitated gel phases below the main transition. One recent example is the positively charged lysyl-DPPG, which is DPPG with a lysine moiety attached. Lysyl-DPPC forms an interdigitated phase primarily due to the large repulsion between head groups [94].

Another modification is the esterification of the phosphate head group, which increases the steric bulk and changes the molecule from zwitterionic to positively charged, allowing interdigitation [95,96]. For example, the P-O-ethyl ester analogue of DPPC, 1,2-dipalmitoyl-*sn*-glycero-3-ethylphosphocholine (EDPPC or Et-DPPC), is fully interdigitated in the gel phase and has a main transition at 42.5 °C with an enthalpy of 9.6 kcal/mol [95-97]. These values are slightly higher than those for DPPC, which has a T_m around 41.3 °C and a corresponding enthalpy of 8.2 kcal/mol ([7] and references therein). The thermodynamic behavior of these cationic triesters of phosphatidylcholine can be attributed to the net positive charge and the absence of intermolecular hydrogen bonding [98]. Furthermore, the overall polarity of the lipid is decreased, which may decrease the interfacial polarity. This would reduce the energetic cost when the ends of the hydrocarbon chains are exposed to the polar head group and the aqueous phase in the interdigitated phase [96]. The additional ethyl group in the head group may also mitigate the unfavorable exposure of the acyl chains [96]. Infrared spectroscopic data lend some support to this conclusion, since the polar/apolar interfaces of cationic PCs are less polar than the parent PC lipids [96]. These lipids have complex phase diagrams that are dependent on temperature, mechanical agitation, and kinetics (Figure 11) [97].

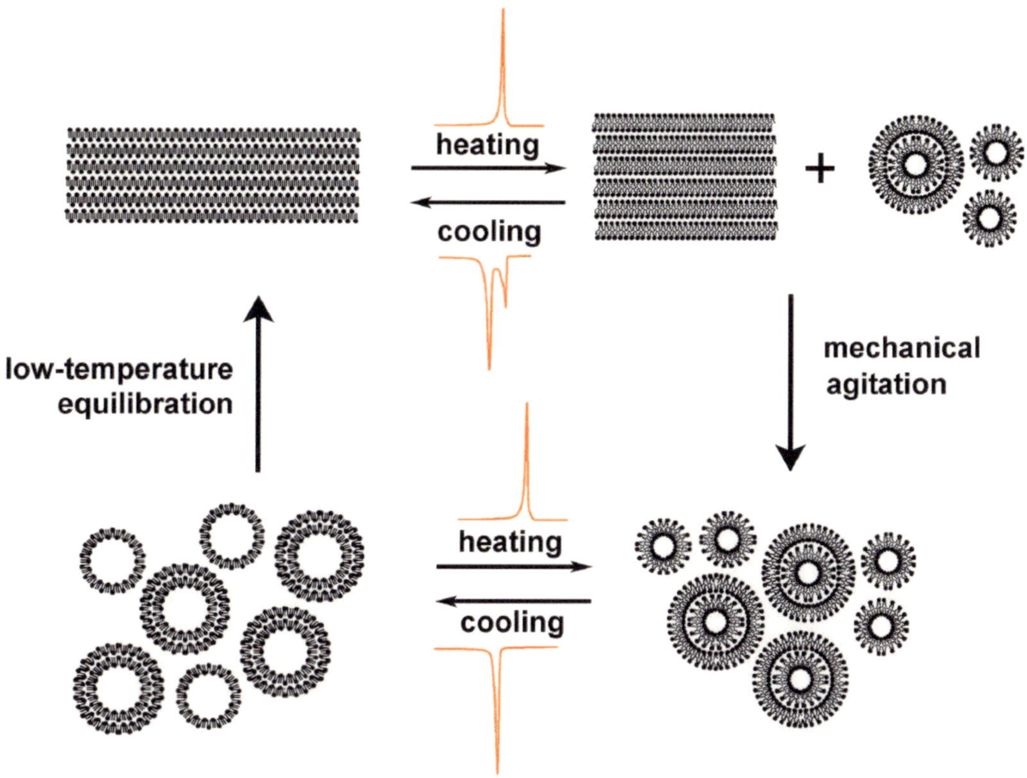

Figure 11. Diagram of the morphological changes in EDPPC dispersions. The equilibrium low temperature arrangement appears to be lamellar sheets, with chain interdigitation. Upon heating, liposomes and lamellar sheets (both non-interdigitated) coexist, whose mixture fully converts into liposomes (apparently the equilibrium liquid crystalline phase arrangement) only after mechanical treatment. Cooling back to the gel phase produces gel-phase liposomes which convert back into lamellar sheets only after prolonged low-temperature exposure. Reprinted from Biochim. Biophys. Acta., 1613, Koynova R, MacDonald RC, Cationic *O*-ethylphosphatidylcholines and their lipoplexes: phase behavior aspects, structural organization and morphology, Pages No. 39-48, Copyright (2003), with permission from Elsevier [97].

Vesicles formed from cationic triester lipids readily fuse with anionic lipids [98]. This may help explain why lipoplexes made from cationic *o*-ethylphosphatidylcholines with di-saturated hydrocarbon chains are effective transfection agents [98,99]. The structure and transfection capability of cationic phospholipid-DNA complexes are dependent on preparation conditions and ionic strength [98,100]. These lipids confer multiple advantages: they are non-viral, metabolized by cells, have low toxicity, and closely resemble naturally occurring phospholipids [101].

In some instances, DNA can be sandwiched between interdigitated gel phase lipid sheets into a rectangular columnar two-dimensional superlattice [97,99]. The gel-to-liquid crystalline phase transition results in the contraction of the DNA strand arrays so that the mean charge density is balanced with the increased positive charge of the non-interdigitated lipid. Additionally, in the non-interdigitated liquid crystalline phase, the interlamellar correlation in DNA ordering is no longer observed [99].

2.8. Phosphatidylethanolamines

Phosphatidylethanolamines (PEs) are distinct due to their strong reluctance to interdigitate. PEs are not susceptible to alcohol-induced interdigitation [48] or pressure-induced interdigitation [55]. Even the ether-linked DHPE does not interdigitate with pressure [34,56]. A major reason for this is that the PE headgroup can form hydrogen bonds [34]. PC headgroups interact through a weaker electrostatic attraction between the positively charged quaternary nitrogen and the negatively charged oxygen of a neighboring lipid headgroup [34]. Additionally, the smaller size of the PE headgroup also allows for closer interaction (less repulsion) [2,8,102].

2.9. Unsaturated phospholipids

Unsaturated lipids with common head groups and acyl chains of nearly equal length are strongly disfavored to interdigitate spontaneously. In general, unsaturated lipids are also resistant to both pressure- and chemically-induced interdigitation [44,103,104]. Even under high pressure, unsaturated lipids typically retain the transition from the non-interdigitated lamellar gel phase (L_β) into the liquid crystalline phase (L_α) [104,105]. Pressure does stabilize the L_β phase to the detriment of the L_α phase, but it is not sufficient to induce interdigitation [105].

Furthermore, unsaturated lipids have substantially lower main transition temperatures [7,10]. As interdigitation is highly unfavorable in the liquid crystalline phase, the relevant temperature range of the gel phase where interdigitation is likely to occur is much smaller. The main transition tends to be lowered the most when the double bond is located near the middle of the fatty acid chain [2,10,36]. Although double bonds that are *trans* usually have less influence than *cis* bonds [7,36], no ethanol-induced interdigitation was found in the *trans* lipid 1,2-dielaidoyl-*sn*-glycero-3-phosphocholine (DEPC) [106]. It was postulated that the increased cross-sectional area due to the double bond and the restriction in sliding motions contributes to the lack of interdigitation [106].

The inhibition of interdigitation also applies to lipid mixtures involving unsaturated lipids. In a model membrane of DPPC/DOPC/ergosterol, increasing the unsaturated lipid or sterol component co-operatively hinders the formation of the interdigitated phase [107]. DOPC is known to result in a more disordered and less tilted gel phase and can lead to phase separation at higher concentrations [108]. As a consequence, it was hypothesized that changes in the plasma membrane composition may play a role in the ethanol tolerance of yeast cells during fermentation ([107] and references therein).

There are some exceptions, however. While lipids with double bonds on both chains are particularly unlikely to interdigitate, there are a few examples of interdigitation where only one chain has a double bond. For example, McIntosh et al. tested the ethanol-induced interdigitation of five positional isomers of 1-eicosanoyl-2-eicosenoyl-*sn*-glycero-3-phosphocholine (C(20):C(20:1Δ^n)PC) with a single *cis* bond on the *sn*-2 chain at position $n = 5$, 8, 11, 13 and 17 [109]. Ethanol-induced interdigitation can be induced when the position of the *cis* bond is at $n= 5$ or 8, but not at $n= 11$, 13, or 17 [109]. In contrast, the fully saturated lipid with the same chain length can easily be interdigitated with a small amount of ethanol [47].

Additionally the *cis* mono-unsaturated 1-stearoyl,2-oleoyl-phosphatidylcholine (SOPC) can be interdigitated with glycerol [24]. The PG lipid, 1-palmitoyl,2-oleoyl-phosphatidylglycerol (POPG), can also be interdigitated with the addition of polymyxin B [24]. Certain mixtures of unsaturated zwitterionic and charged lipids, such as 1-palmitoyl-2-oleoyl-phosphatidylcholine (POPC) and POPG, can form the interdigitated phase at high concentrations of ethanol and at low hydration [110].

Therefore, it can be concluded that while it is possible to induce the interdigitated gel phase in unsaturated lipids, they are more resistant to interdigitation compared to the equivalent saturated lipid. Additionally, when the interdigitated phase does occur in unsaturated lipids, the phase appears to be less stable and less ordered than in saturated lipids [24].

2.10. Membrane curvature and interdigitation

The curvature of the membrane due to the macromolecular size and shape affects the thermodynamic properties [36]. For instance, on DSC scans, small unilamellar vesicles (SUVs) have lower enthalpic peaks and greater widths compared to multilamellar vesicles (MLVs) [2,3]. SUVs also have more mobility and less order in the hydrocarbon chains [2].

The degree of membrane curvature also affects the ability to interdigitate. Bending in the membrane causes increased steric interference in opposing lipid monolayers [44]. As a consequence, ethanol-induced interdigitation is dependent on curvature, with the more highly curved vesicles requiring more ethanol to interdigitate [111,112]. Sonicated DPPC SUVs are not stable in the presence of ethanol above the threshold concentration for interdigitation [112]. Furthermore, SUVs have a tendency to fuse into large unilamellar vesicles (LUVs), which have properties more similar to MLVs [36,113]. The more planar MLVs allow interdigitated lipids to slide by each other with low steric interference and therefore have the lowest threshold concentrations [44,112].

2.11. Inhibition of the interdigitated gel phase by cholesterol

Just as there are chemicals that induce interdigitation, there are chemicals that inhibit the formation of the $L_{\beta}I$ phase. Cholesterol is a chemical inhibitor of interdigitation in a wide variety of lipid systems (Table 4).

Interdigitated Lipid System	References
DPPC/EtOH	[114]
DPPC/Pressure	[31]
DPPeth/Tris-HCl	[115]
DPPG/Tris-HCl	[115]
DHPC	[66,116]
EDPPC	[95]
F-DPPC	[117]
3:7 ratio of 16:0 LPC:DPPC	[118]

Table 4. The inhibition of interdigitation by cholesterol

In non-interdigitated membranes, cholesterol increases the fluidity of the gel phase, broadens the main transition, and decreases the main transition enthalpy [119]. Figure 12 demonstrates that these effects are also seen in membranes where cholesterol eliminates the interdigitated phase [115-117]. The amount of cholesterol required to prevent interdigitation is related to the stability of the $L_{\beta}I$ phase. For DHPC, only ~5 mol% cholesterol is required to eliminate interdigitation [66,116]. However, the amount of cholesterol required to prevent interdigitation is approximately quadrupled for F-DPPC, which exists in the $L_{\beta}I$ phase around 15 °C higher than DHPC [117]. At high cholesterol concentrations the $L_{\beta}I$ phase of F-DPPC is replaced by a non-interdigitated liquid-ordered (l_o) phase with properties similar to DPPC/cholesterol [117]. On DSC scans, this effect can be observed by the broadening of the main transition peak and a reduction in the T_m hysteresis (Figure 12). The interdigitated phase of cationic EDPPC is especially resilient in the presence of cholesterol, with interdigitated domains still present at 30 mol% cholesterol [95].

There are multiple reasons why cholesterol-rich membranes disfavor interdigitation. For example, lipid head group crowding is mitigated by cholesterol serving as a spacer between lipids [115]. If cholesterol is placed within an interdigitated membrane, the increased spacing also increases the likelihood that the terminal lipid methyl groups will be exposed at the aqueous interface. Since the interdigitated phase lacks the thick membrane midplane region of non-interdigitated membranes, hydrophobic cholesterol located within the interdigitated phase is more likely to come in contact with water [115]. Furthermore, cholesterol significantly disrupts the lattice structure of gel phase lipids [108,117,120]. Lastly, in the interdigitated phase of highly asymmetrical lyso-lipids, cholesterol can take the place of the missing acyl chain thereby compensating for the size mismatch between the head group and the hydrocarbon chains [118,121].

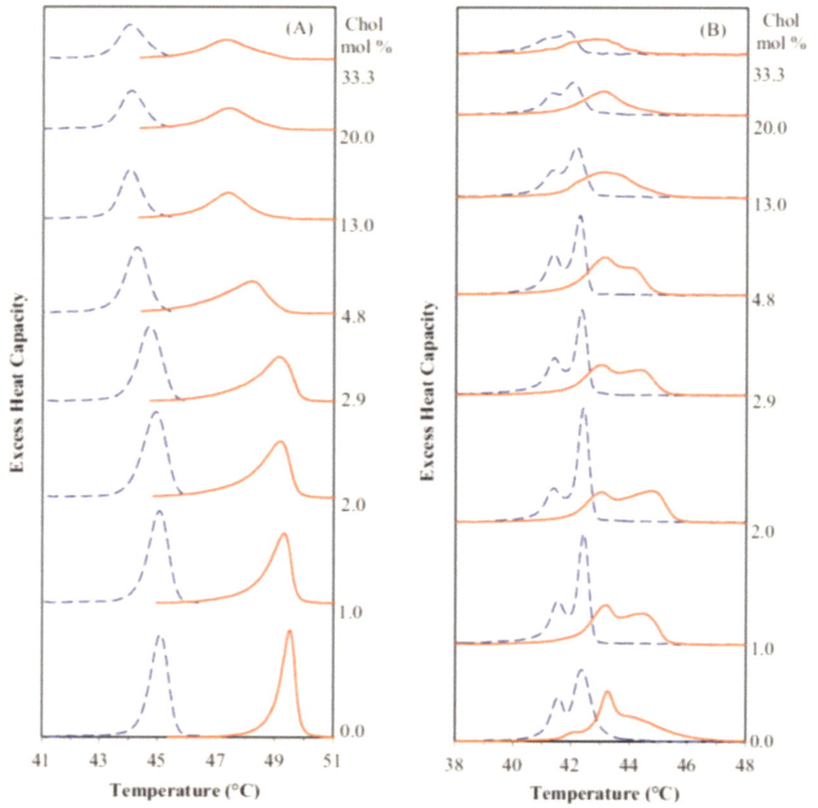

Figure 12. Heating (solid red lines) and cooling (dashed blue lines) DSC thermograms of: (A) F-DPPC and (B) 1:1 F-DPPC/DPPC with various concentrations of cholesterol. The cooling scans have been inverted to allow comparison with the heating peaks. For clarity, the thermograms are also offset vertically. Reprinted from Chem. Phys. Lipids, 165, Smith EA, Wang W, Dea PK, Effects of cholesterol on phospholipid membranes: Inhibition of the interdigitated gel phase of F-DPPC and F-DPPC/DPPC, Pages No. 151-159, Copyright (2012), with permission from Elsevier [117].

2.12. Chemical inhibition of the interdigitated gel phase

In a general sense, solvent inhibitors of interdigitation work in the opposite fashion as chemical inducers. Some researchers have focused on the difference of how kosmotropic and chaotropic solutes interact with lipid membranes [122-124]. Kosmotropes deplete the solution at the interface and increase the interfacial tension whereas chaotropes accumulate in the interface and decrease surface tension [125]. The changes in the structure of water due to these types of chemicals can be attributed to alterations in the hydrogen bonding network of water [126,127]. Kosmotropic substances are classified as water-structure makers, meaning that they stabilize the structure of bulk water. When kosmotropes interact with hydrated lipids, they tend to reduce the interfacial area and inhibit interdigitation [122]. Chaotropic chemicals are classified as water-structure breakers and increase the surface area per lipid, favoring interdigitation [122,124].

The differences between chemicals that induce or inhibit interdigitation have also been illustrated according to the interaction free energy of the lipid membrane interface with solvents ([128] and references therein). The solvent free energy relationship can be further split into interactions with the polar head groups and interactions with the hydrophobic lipid chains. In this model, when "good" solvents are added, the interfacial area swells to increase the total contract with the solvent. For organic solvents that are water-miscible and have a high solubility for alkanes, such as acetone and ethanol, the interaction increases the interfacial area by reducing the interaction free energy between the solvent and the interfacial alkyl chains [128]. On the other hand, the interaction of "poor" solvents with lipids is unfavorable and has larger free energy penalty. As a result, the interfacial segments shrink in size to prevent contact with the solvent. Consequently, "good" solvents will favor interdigitation while "poor" solvents will destabilize the $L_{\beta}I$ phase.

The inhibition of interdigitation has also been described in terms of osmotic stress. Chemicals that apply osmotic stress, such as poly(ethylene glycol) tend to inhibit interdigitation [60]. As in the other models described above, this has been proposed to occur because of a decrease in the repulsive interaction between the lipid head groups.

Dimethyl sulfoxide (DMSO) is an example of a solvent inhibitor of interdigitation. The interaction of DMSO with membranes is of great interest because it can be used as a cryoprotectant for biological material, such as stem cells [129]. DMSO can also enhance the permeability of membranes [130]. The mechanism by which DMSO inhibits interdigitation is by decreasing the repulsion between head groups [131]. The ability of DMSO to form unusually strong hydrogen bonds may explain this effect [132]. This phenomenon can be clearly seen in the phase behavior of DHPC. Just as chemicals that favor interdigitation shift the pre-transition of DHPC to a higher temperature; factors that disfavor interdigitation shift the pre-transition to a lower temperature. The suppression of the pre-transition clearly demonstrates that DMSO destabilizes the $L_{\beta}I$ phase (Figure 13) [131].

A major caveat with these solvent models is that the interactions with lipids are often concentration-dependent. For instance, in the DPPC/DMSO/water system, three distinct effects are found within different DMSO concentration ranges [133]. Perhaps the most remarkable is at above mol fractions of ~0.9 DMSO, the T_m temperature is greatly elevated and an interdigitated gel phase is formed [133].

The disaccharide trehalose is another example of a chemical inhibitor of interdigitation [124]. Like DMSO, the interactions of trehalose with membranes show promise in the cryopreservation of biological material [129]. In the yeast *Saccharomyces cerevisiae*, trehalose appears to increase viability during ethanol fermentation and provide protection against oxidative stress [134-137]. Similar to DMSO, trehalose disfavors interdigitation by increasing the packing density of the lipid head groups [124,138,139]. However, there is disagreement over the exact molecular interaction with lipids. The main dispute is over whether or not sugars are directly bound to or excluded from the membrane surface [140,141]. Recently, Andersen et al. have tried to explain this discrepancy by proposing that there are two concentration-dependent interactions. In this explanation, trehalose binds strongly to the bilayer at low concentrations, but is gradually expelled above ~0.2 M [141].

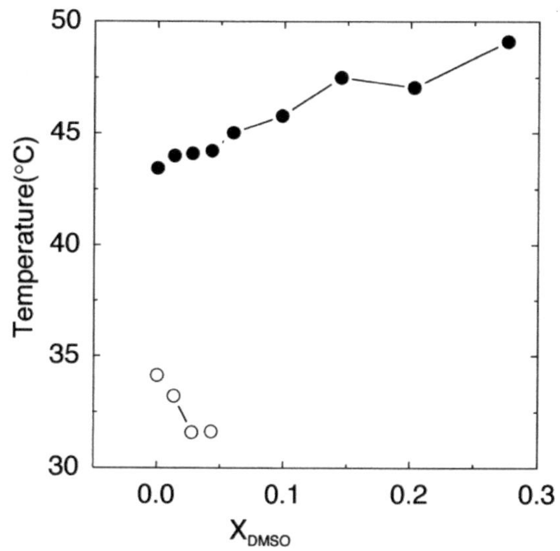

Figure 13. Phase transition temperatures of DHPC-MLV at various concentrations of DMSO (mole fraction) determined by DSC. (•) shows gel to liquid-crystalline phase transition temperatures and (O) shows $L_{\beta}I$ to P_{β}' phase transition temperatures. Reprinted from Biochim. Biophys. Acta., 1467, Yamashita Y, Kinoshita K, Yamazaki M, Low concentration of DMSO stabilizes the bilayer gel phase rather than the interdigitated gel phase in dihexadecylphosphatidylcholine membrane, Pages No. 395-405, Copyright (2000), with permission from Elsevier [131].

2.13. The interdigitated gel phase versus the inverted hexagonal phase

A clear inverse relationship exists between the interdigitated phase gel phase and the inverted hexagonal phase (H_{II}) [56,128]. The major structural factor is the relative size of the lipid headgroup and the attraction/repulsion between headgroups. A lipid that forms the inverted hexagonal phase is unlikely to interdigitate and vice versa. The temperature dependence of these phases is also opposite. For example, with DHPC, the interdigitated phase is present only below the pre-transition. The interdigitated phase requires predominately *trans* confirmations in the hydrocarbon chains, so it is unlikely to form in the liquid crystalline phase where there are abundant *gauche* confirmations and a high degree of disorder [2,37]. In contrast, the inverted hexagonal phase typically forms well above the main transition into the liquid crystalline phase [8].

This relationship also extends to environmental factors that encourage or discourage interdigitation (Table 5). Chemicals that favor the interdigitated phase such as ethanol tend to destabilize the H_{II} phase [128,124 and references therein]. Interdigitation is favored because the surface area per lipid head group in the $L_{\beta}I$ phase is substantially larger versus non-interdigitated membranes [124]. The H_{II} is the opposite because it requires a small head group area. Solvents that stabilize the H_{II} phase like dimethyl sulfoxide therefore also inhibit interdigitation [56,122,128]. This relationship appears to apply to hydrostatic pressure as well. While increased pressure favors interdigitation (Tables 2 and 3), pressure destabilizes the inverted hexagonal phase in PE lipids [56].

$L_\beta I$	H_{II}
Large head group repulsion	Small head group repulsion
Chaotropic chemicals	Kosmotropic chemicals
High hydrostatic pressure	Low hydrostatic pressure
Low Temperatures	High Temperatures

Table 5. Factors that stabilize the interdigitated gel ($L_\beta I$) phase and the inverted hexagonal (H_{II}) phase.

2.14. Influence of hydration and pH on the $L_\beta I$ phase

While most interdigitated systems are studied in excess water, interdigitation can be affected at less than full hydration. For instance, interdigitation of DHPC is reliant on hydration, as coexisting interdigitated and non-interdigitated phases are found at low hydration [142,143]. However, the cationic EDPPC may be interdigitated in the dry state [97].

Furthermore, substituting deuterium oxide (D_2O) for water slightly disfavors the spontaneous interdigitated phase of DHPC [144]. Using D_2O also increases the threshold concentration for the chemically-induced interdigitation of DPPC [27] and increases threshold pressure for interdigitation [145]. These phenomena are explained by the different hydrophobic interactions and interfacial energies in H_2O versus D_2O [27,144,145].

Changing the pH of the aqueous solution can also affect interdigitation. In DHPC membranes a low pH will inhibit interdigitation [59]. As the pH is lowered the phosphate groups are protonated and ultimately the total repulsive force between head groups is decreased, disfavoring interdigitation [59]. The pH is also highly relevant to the interdigitated phase in charged lipids, such as PGs. At a high pH, the electrostatic repulsion between head groups that encourages interdigitation in PGs is increased [78].

2.15. Lipid mixtures and interdigitated/non-interdigitated gel phase coexistence

Under certain circumstances, interdigitated and non-interdigitated phases can coexist within a membrane even though the boundaries between these domains are considered to be energetically unfavorable [115,116]. The uneven structure between these domains can significantly increase the membrane permeability [146,147]. With the variety of lipids now known to interdigitate, there are many possible lipid systems that will have complex phase diagrams involving the $L_\beta I$ phase.

Gel phase coexistence can often be found in binary mixtures of a lipid that can spontaneously interdigitate (e.g. F-DPPC or EDPPC) and one that cannot (e.g. DPPC or PE lipids). For example, at equimolar amounts of F-DPPC and DPPC, interdigitated F-DPPC-rich domains create a phase-segregated system [69,117]. On DSC scans this manifests itself as multiple peaks (Figure 8). The peaks with the greatest transition hysteresis likely correspond to interdigitated domains rich in F-DPPC. When the F-DPPC molar fraction is large, the hysteresis is also increased [69]. Additionally, gel phase coexistence occurs in the mixture of 1,2-dielaidoyl-sn-glycero-3-phosphoethanolamine (DEPE) and EDPPC [95].

However, this is not true of all such binary mixtures. Outside of the phase transition regions in DPPC/EDPPC, for instance, there is no gel phase segregation [95].

Mixing an interdigitated lipid with cholesterol can also produce gel phase coexistence. Cholesterol-poor interdigitated domains and cholesterol-rich non-interdigitated domains have been found in DHPC/cholesterol [116], F-DPPC/cholesterol [117], and EDPPC/cholesterol [95]. For these mixtures, the lipids with the most stable interdigitated phase tend to have a larger region of phase coexistence within the phase diagram.

Alternatively, a lipid such as DPPC that can be chemically induced to interdigitate can be mixed with lipids that cannot, such as PE lipids [147]. The DPPC-rich domains will interdigitate with ethanol, but domains composed of mostly PE lipid will not. A similar result can be achieved in mixtures of DPPC/cholesterol/ethanol, where the cholesterol-rich domains remain non-interdigitated in the presence of ethanol [39,146].

It is also possible to have coexistence in membranes with only one lipid. For instance, coexisting interdigitated and non-interdigitated phases form in supported F-DPPC membranes where the lateral expansion of the lipid film is restricted [68]. This results in a "frustrated" state, where the energetically favorable interdigitated phase cannot fully form due to constraints in topology and the available surface area [68]. Additionally, while the 16-carbon chain length DPPG does not spontaneously interdigitate, the 18-carbon chain DSPG spontaneously forms an interdigitated gel phase that coexists with a non-interdigitated gel phase [78]. This two-phase coexistence was attributed to a kinetically trapped system that is not at thermal equilibrium [78].

2.16. Applications of the interdigitated gel phase

One of the most promising applications for the interdigitated gel phase is the creation of large unilamellar vesicles termed interdigitation-fusion (IF) vesicles [44,148]. Figure 14 demonstrates the process for the creation of IF liposomes using ethanol [148]. Below the main transition, the ethanol causes the formation rigid and flat interdigitated sheets [149]. These sheets are surprisingly stable under the T_m, even when ethanol is removed [149]. When the temperature is raised above the main transition, the sheets fuse into large vesicles. This fusion encapsulates particles from the surrounding solution [149,150]. These materials include other small vesicles, biological macromolecules, colloids, and nanoparticles [149,150]. The amphiphilic nature of lipids allows for the capture of hydrophobic materials [151]. Transmembrane insertion of protein into IF vesicles has also been achieved using electropulsation [152].

The IF procedure can also be used to create multicompartment vesicle-in-vesicle structures called "vesosomes" [150]. These multicompartment vesicles should be closer replicas of eukaryotic cells than regular vesicles [150,153]. Therefore, vesosomes have the potential to more closely mimic biological conditions and reactions in artificial cells [150,154,155]. Furthermore, the retention of encapsulated material can be substantially increased in vesosomes [151,156,157]. These vesicles are highly customizable because the composition of the inner and outer components can be varied [149,150,154]. As a result, it is theoretically possible to use vesosomes as controlled nanoreactors [153,155]. For complex and expensive chemistry such as enzyme reactions, vesosomes should be able to optimize reaction conditions and drastically reduce the amount of reagents needed [155].

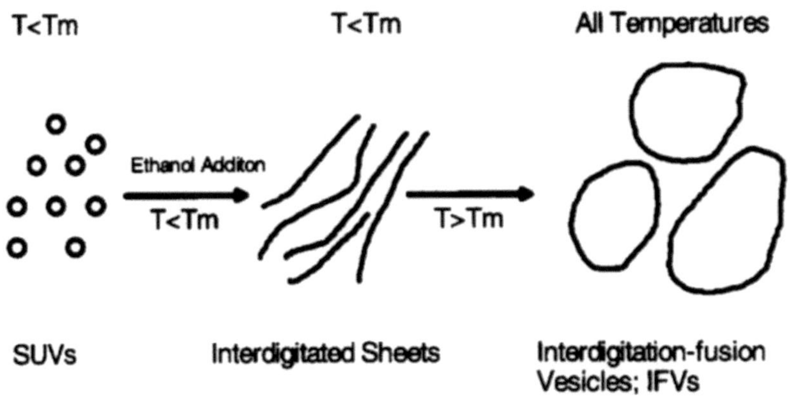

Figure 14. Liposome formation by interdigitation fusion (IF) using ethanol. Reprinted from Biochim. Biophys. Acta., 1195, Ahl PL, Chen L, Perkins WR, Minchey SR, Boni LT, Taraschi TF, Janoff AS, Interdigitation-fusion: a new method for producing lipid vesicles of high internal volume, Pages No. 237-244, Copyright (1994), with permission from Elsevier [148].

As described by Ahl et al. [44], there are four general guidelines for IF liposomes: (1) the lipids must be able to form the interdigitated phase; (2) the precursor liposomes should be small, preferably sonicated SUVs; (3) the temperature of the precursor SUV suspension after the addition of the alcohol must be below the T_m of the phospholipids; and (4) the temperature should be raised above the T_m of the phospholipids after the formation of the interdigitated sheets. Therefore, the creation of these liposomes is dependent on the lipid composition. Adding cholesterol and lipids containing *cis* double bonds can compromise the formation of IF liposomes [103,148]. PE lipids are also unsuitable because of their reluctance to interdigitate [44].

A similar result can be achieved using pressure to create pressure-induced fusion (PIF) liposomes [103]. An advantage of this technique is that no organic solvent is required and it is an effective sterilization method [103]. The captured volume of the IF or PIF vesicles is larger than other techniques for liposome preparation ([44] and references therein).

3. Conclusions

As an analytical instrument, DSC offers many advantages. One advantage is the simplicity of the sample preparation procedure. Samples do not have to be supported or spatially oriented and do not require the insertion of a membrane probe. For sensitive low enthalpy phase transitions, it is a great benefit not to need a probe so that the purity of the sample can be maintained. The importance of this can be seen in alcohol-induced interdigitation, where the low enthalpy pre-transition is an important aspect of the analysis (Figure 5) [15]. Moreover, the effects of pressure can be measured concomitantly with calorimetry data with the appropriate equipment. This greatly expands the range of the phase diagram that can be experimented with.

We have shown that DSC can accurately measure changes in the thermodynamic properties of phospholipid membranes with the addition of chemicals that either encourage or discourage interdigitation. DSC is particularly well-suited for the study of chemically-induced interdigitation because it is sensitive enough to detect small, incremental changes in phase transition temperatures (Figure 5). With the capability to perform heating and cooling scans at a constant rate, the transition hysteresis can also be easily determined. In addition, the transition enthalpy can highlight the "biphasic" behavior above and below the threshold concentration for interdigitation (Figure 6).

Moreover, DSC can reveal how changes in either the hydrocarbon chains (Figure 1) or in the polar head group (Figure 2) will affect the thermodynamics. Modifications that either encourage or discourage interdigitation are summarized in Figure 15. Understanding the importance of structural differences reveals the importance of lipid diversity in biological membranes. Lipid composition can help explain why, for example, a peptide might interact differently with human versus microbial membranes [81]. With the increasing popularity of liposomes for pharmaceutical applications and research, it also is essential to find suitable lipid candidates. For instance, calorimetry can be applied to screen potential IF vesicles by determining whether interdigitation is present and by determining the T_m temperature.

In addition, more information can be inferred from DSC data than the phase transition temperature. With careful analysis, the nature of the lipid/solvent interaction and the properties of the chemicals themselves can be derived. For example, the characteristics of kosmotropic and chaotropic chemicals are clearly reflected in their effects on lipid membranes (see section 3.12.). This analysis can also increase the understanding of how chemicals interact with biological membranes, such as why chemicals like DMSO and trehalose can protect cells during cryopreservation [129].

However, DSC also has limitations when analyzing phospholipid samples. Perhaps the greatest weakness is the lack of direct structural information. As a consequence, relying solely on DSC data can be misleading. For instance, the pre-transition peaks of DPPC and DHPC look similar on DSC thermograms. However, the actual nature of the transition is substantially different (Figure 3). While the structure can often be reasonably inferred from thermodynamic properties, it is not as robust as other experimental techniques [6]. Additionally, while alterations in the macromolecular structure can be reflected in DSC data (see section 3.10.), the changes are not specific enough to be able to infer the true structure.

Overlapping or multiple transitions can also present a problem. In F-DPPC/DPPC, the multiple peaks reflect the presence of phase segregation (Figure 8), but this is not always the case. Multiple DSC peaks can also indicate separate phase transitions that involve the entire membrane. In the case of EDPPC, different morphologies result in separate DSC peaks (Figure 11) [97]. Overlapping peaks can also obscure individual transitions, especially when there are multiple components in the membrane and the transition peaks are broad.

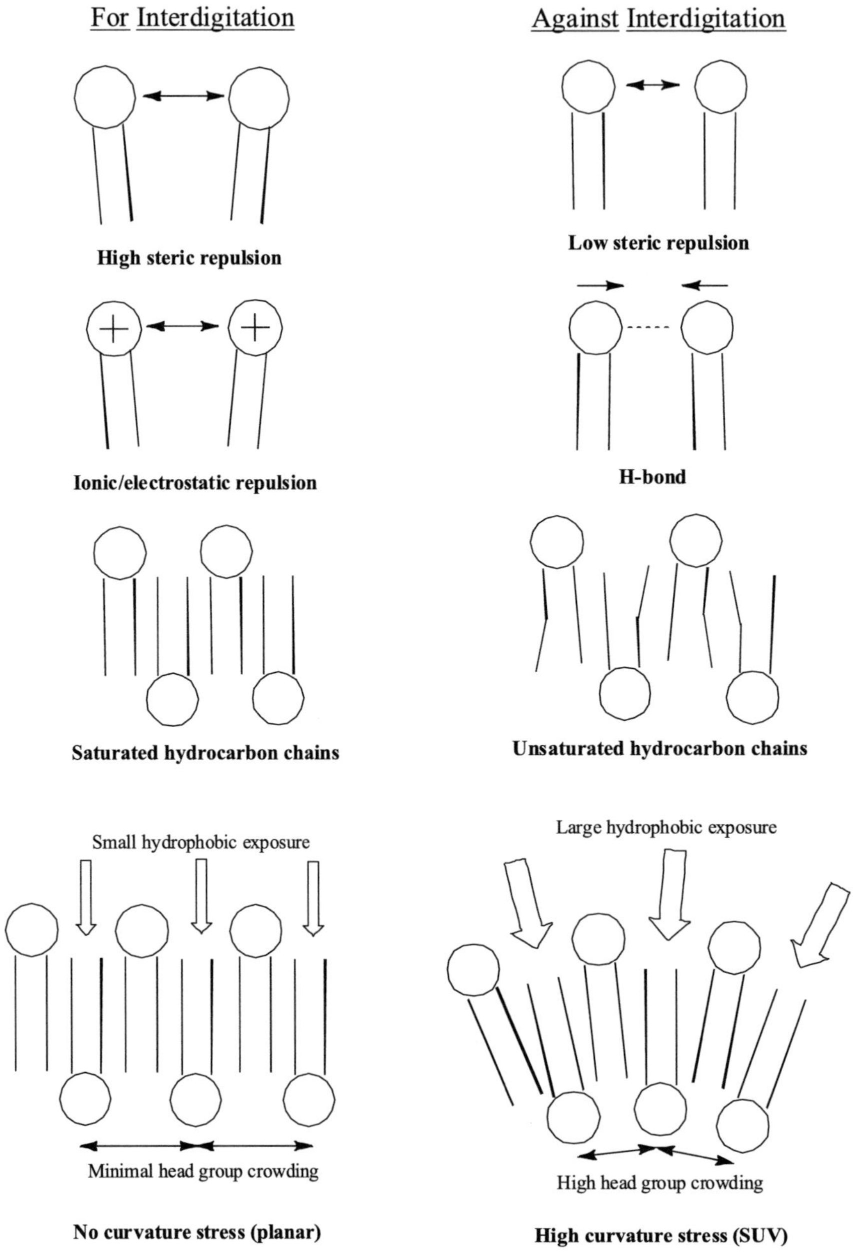

Figure 15. Schematic representation of factors that favor or disfavor interdigitation.

Fortunately, one of the greatest strengths of DSC data is that it is highly compatible with other analytical techniques. In the case of the $L_{\beta}I$ phase, methods such as x-ray diffraction, nuclear magnetic resonance, and fluorescence techniques can fill in the gaps ([6] and references therein). Additionally, DSC is highly valuable in determining the relevant temperature range to use for the other experimental techniques.

The stability of the interdigitated phase plainly demonstrates the balance of forces within the membrane. Factors as varied as electrostatic and steric interactions, van der Waals forces, solvent binding at the interface, and the presence of double bonds all contribute to the properties of hydrated phospholipid membranes. DSC provides a way to judge the resulting balance of these forces by measuring the stability of different thermodynamic phases. Consequently, the wealth of information calorimetric analysis provides ensures that DSC will remain an invaluable tool for the study of membrane biophysics.

Author details

Eric A. Smith and Phoebe K. Dea*
Department of Chemistry, Occidental College, Los Angeles, USA

Abbreviations

differential scanning calorimetry (DSC)
main transition temperature (T_m)
pre-transition temperature (T_P)
small unilamellar vesicle (SUV)
large unilamellar vesicle (LUV)
multilamellar vesicle (MLV)
interdigitation-fusion vesicle (IFV)
interdigitated gel phase ($L_\beta I$)
planar gel phase (L_β')
ripple gel phase (P_β')
liquid crystalline phase (L_α)
inverted hexagonal phase (H_{II})
crystalline bilayer phase (L_c)
liquid-ordered (l_o)
phosphatidylcholine (PC)
phosphatidylglycerol (PG)
phosphatidylethanolamine (PE)
phosphatidylethanol (Peth)
1,2-dipalmitoyl-*sn*-glycero-3-phosphocholine (DPPC)
1,3-dipalmitoyl-*sn*-glycero-2-phosphocholine (1,3-DPPC or β-DPPC)
1,2-di-*O*-hexadecyl-*sn*-glycero-3-phosphocholine (DHPC)
1-palmitoyl-2-(16-fluoropalmitoyl)*sn*-glycero-3-phosphocholine (F-DPPC)
1,2-dipalmitoyl-*sn*-glycero-3-phosphoethanol (DPPeth)

* Corresponding Author

1,2-dipalmitoyl-*sn*-glycero-3-phospho-(1'-*rac*-glycerol) (DPPG)
1-palmitoyl-2-hydroxy-*sn*-glycero-3-phosphocholine (16:0 LPC)
1,2-dipalmitoyl-*sn*-glycero-3-ethylphosphocholine (EDPPC or Et-DPPC)
1,2-dipalmitoyl-*sn*-glycero-3-phosphoethanolamine (DPPE)
1,2-di-*O*-hexadecyl-*sn*-glycero-3-phosphoethanolamine (DHPE)
1,2-dielaidoyl-*sn*-glycero-3-phosphoethanolamine (DEPE)
1,2-dioleoyl-*sn*-glycero-3-phosphocholine (DOPC)
1,2-dielaidoyl-*sn*-glycero-3-phosphocholine (DEPC)
1-stearoyl-2-oleoyl-phosphatidylcholine (SOPC)
1-palmitoyl-2-oleoyl-phosphatidylcholine (POPC)
1-palmitoyl-2-oleoyl-phosphatidylglycerol (POPG)

4. References

[1] Mabrey S, Sturtevant JM (1976) Investigation of Phase Transitions of Lipids and Lipid Mixtures by High Sensitivity Differential Scanning Calorimetry. Proc. natl. acad. sci. USA. 73: 3862-3866.

[2] McElhaney RN (1982) The Use of Differential Scanning Calorimetry and Differential Thermal Analysis in Studies of Model and Biological Membranes. Chem. phys. lipids. 30: 229-259.

[3] Chiu MH, Prenner EJ (2011) Differential Scanning Calorimetry: An Invaluable Tool for a Detailed Thermodynamic Characterization of Macromolecules and their Interactions. J. pharm. bioallied sci. 3: 39-59.

[4] Demetzos C (2008) Differential Scanning Calorimetry (DSC): A Tool to Study the Thermal Behavior of Lipid Bilayers and Liposomal Stability. J. liposomal res. 18: 159-173.

[5] Jackson MB, Sturtevant JM (1977) Studies of the Lipid Phase Transitions of *Escherichia coli* by High Sensitivity Differential Scanning Calorimetry. J. biol. chem. 252: 4749-4751.

[6] Slater JL, Huang CH (1988) Interdigitated Bilayer Membranes. Prog. lipid res. 27: 325–359.

[7] Koynova R, Caffrey M (1998) Phases and Phase Transitions of the Phosphatidylcholines. Biochim. biophys. acta. 1376: 91-145.

[8] Koynova R, Caffrey M (1994) Phases and Phase-Transitions of the Hydrated Phosphatidylethanolamines. Chem. phys. lipids 69: 1-34.

[9] Koynova R, Caffrey M (2002) An Index of Lipid Phase Diagrams. Chem. phys. lipids. 115: 107-219.

[10] Marsh D (2010) Structural and Thermodynamic Determinants of Chain-Melting Transition Temperatures for Phospholipid and Glycolipids Membranes. Biochim. biophys. acta. 1789: 40-51.

[11] Xu H, Huang CH (1987) Scanning Calorimetric Study of Fully Hydrated Asymmetric Phosphatidylcholines with One Acyl Chain Twice as Long as the Other. Biochem. 26: 1036-1043.

[12] Huang C, McIntosh TJ (1997) Probing the Ethanol-Induced Chain Interdigitation in Gel-State Bilayers of Mixed-Chain Phosphatidylcholines. Biophys. j. 70: 2702-2709.
[13] Löbbecke L, Cevc G (1995) Effects of Short-Chain Alcohols on the Phase Behavior and Interdigitation of Phosphatidylcholine Bilayer Membranes. Biochim. biophys. acta. 1237: 59-69.
[14] Wang Y, Dea P (2009) Interaction of 1-propanol and 2-propanol with Dipalmitoylphosphatidylcholine Bilayer: A Calorimetric Study. J. chem. eng. data. 54: 1447-1451.
[15] Reeves MD, Schawel AK, Wang W, Dea P (2007) Effect of Butanol Isomers on Dipalmitoylphosphatidylcholine Bilayer Membranes. Biophys. chem. 128: 13-18.
[16] Griffin KL, Cheng C-Y, Smith EA, Dea PK (2010) Effects of Pentanol Isomers on the Phase Behavior of Phospholipid Bilayer Membranes. Biophys. chem. 152: 178-183.
[17] Hata T, Matsuki H, Kaneshina S (2000) Effect of Local Anesthetics on the Phase Transition Temperatures of Ether- and Ester-Linked Phospholipid Bilayer Membranes. Colloid surf. b. biointer. 18: 41-50.
[18] Maruyama S, Hata T, Matsuki H, Kaneshina S (1997) Effects of Pressure and the Local Anesthetic Tetracaine on Dihexadecylphosphatidylcholine Bilayer Membrane. Coll. surf. b. 8: 261-266.
[19] McIntosh TJ, McDaniel RV, Simon SA (1983) Induction of an Interdigitated Gel Phase in Fully Hydrated Phosphatidylcholine Bilayers. Biochim. biophys. acta. 731: 109-114.
[20] Matsingou C, Demetzos C (2007) Calorimetric Study on the Induction of Interdigitated Phase in Hydrated DPPC Bilayers by Bioactive Labdanes and Correlation to their Liposomal Stability: The Role of Chemical Structure. Chem. phys. lipids. 145: 45-62.
[21] Potamitis C, Chatzigeorgiou P, Siapi E, Viras K, Mavromoustakos T, Hodzic A, Pabst G, Cacho-Nerin F, Laggner P, Rappolt M (2011) Interactions of the AT_1 Antagonist Valsartan with Dipalmitoyl-phosohatidylcholine Bilayers. Biochim. biophys. acta. 1808: 1753-1763.
[22] Swamy MJ, Marsh D (1995) Thermodynamics of Interdigitated Phases of Phosphatidylcholine in Glycerol. Biophys. j. 69: 1402-1408.
[23] Boggs JM, Rangaraj G, Watts A (1989) Behavior of Spin Labels in a Variety of Interdigitated Lipid Bilayers. Biochim. biophys. acta biomembr. 981: 243-253.
[24] Boggs JM, Tümmler B (1993) Interdigitated Gel Phase Bilayers Formed by Unsaturated Synthetic and Bacterial Glycerolipids in the Presence of Polymyxin B and Glycerol. Biochim. biophys. acta. 1145: 42-50.
[25] Yamazaki M, Ohshika M, Kashiwagi N, Asano T (1992) Phase Transitions of Phospholipid Vesicles under Osmotic Stress and in the Presence of Ethylene Glycol. Biophys. chem. 43: 29-37.
[26] Kinoshita K., Asano T, Yamazaki M (1997) Interaction of the Surface of Biomembrane with Solvents: Structure of Multilamellar Vesicles of Dipalmitoylphosphatidylcholine in Acetone-Water Mixtures. Chem. phys. lipids. 85: 53-65.
[27] Kinoshita K, Yamazaki M. (1996) Organic Solvents Induce Interdigitated Gel Structures in Multilamellar Vesicles of Dipalmitoylphosphatidylcholine. Biochim. biophys. acta biomembr. 1284: 233-239.

[28] Wu F-G, Wang N-N, Tao L-F, Yu Z-W (2010) Acetonitrile Induces Nonsynchronous Interdigitation and Dehydration of Dipalmitoylphosphatidylcholine Bilayers. J. phys. chem. b. 114: 12685-12691.

[29] Cunningham BA, Tamura-Lis W, Lis LJ, Collins JM (1989) Thermodynamic Properties of Acyl Chain and Mesophase Transition for Phospholipids in KSCN. Biochim. biophs. acta biomembr. 984: 109-112.

[30] Cunningham BA, Quinn PJ, Wolfe DH, Tamura-Lis A, Lis LJ, Kucuk O, Westerman MP (1995) Real-Time X-ray Diffraction Study at Different Scan Rates of Phase Transitions for Dipalmitoylphosphatidylcholine in KSCN. Biochim. biophs. acta biomembr. 1233: 68-74.

[31] Tamai N, Matsui T, Moribayashi N, Goto M, Matsuki H, Kaneshina S (2008) Cholesterol Suppresses Pressure-Induced Interdigitation of Dipalmitoylphosphatidylcholine Bilayer Membrane. Chem. lett. 37: 604-605.

[32] Zeng J, Chong PLG (1991) Interactions between Pressure and Ethanol on the Formation of Interdigitated DPPC Liposomes: A Study with Prodan Fluorescence. Biochem. 30: 9485-9491.

[33] Braganza LF, Worcester DL (1986) Hydrostatic Pressure Induces Hydrocarbon Chain Interdigitation in Single-Component Phospholipid Bilayers. Biochem. 25: 2591-2596.

[34] Tamai N, Goto M, Matsuki H, Kaneshina S (2010) A Mechanism of Pressure-Induced Interdigitation of Lipid Bilayers. J. phys. conf. ser. 215: 012161 1-7.

[35] Hui SW, Huang C-H (1986) X-ray Diffraction Evidence for Fully Interdigitated Bilayers of 1-stearoyllysophosphatidylcholine. Biochem. 25: 1330-1335.

[36] Biltonen RL, Lichtenberg D (1993) The Use of Differential Scanning Calorimetry as a Tool to Characterize Liposome Preparations. Chem. phys. lipids. 64: 129-142.

[37] Heerklotz H (2004) The Microcalorimetry of Lipid Membranes. J. phys. condens. matter 16: R441-R467.

[38] Zeng J, Smith K, Chong PL (1993) Effects of Alcohol-Induced Lipid Interdigitation on Proton Permeability in L-α-Dipalmitoylphosphatidylcholine Vesicles. Biophys. j. 65: 1404-1414.

[39] Tierney KJ, Block DE, Longo ML (2005) Elasticity and Phase Behavior of DPPC Membrane Modulated by Cholesterol, Ergosterol, and Ethanol. Biophys. j. 89: 2481-2493.

[40] Barry JA, Gawrisch K (1995) Effects of Ethanol on Lipid Bilayers Containing Cholesterol, Gangliosides, and Sphingomyelin. Biochem. 34: 8852-8860.

[41] Kõiv A, Kinnunen PKJ (1992) Influence of Ca^{2+} and Ethanol on the Aggregation and Thermal Phase Behavior of l-dihecadecylphosphatidylcholine Liposomes. Chem. phys. lipids. 62: 253-261.

[42] Vierl U, Löbbecke L, Nagel N, Cevc G (1994) Solute Effects on the Colloidal and Phase Behavior of Lipid Bilayer Membranes: Ethanol-dipalmitoylphosphatidylcholine Mixtures. Biophys. j. 67: 1067-1079.

[43] Nagel NE, Cevc G, Kirchner S (1992) The Mechanism of the Solute-Induced Chain Interdigitation in Phosphatidylcholine Vesicles and Characterization of the Isothermal

Phase Transitions by Means of Dynamic Light Scattering. Biochim. biophys. acta biomembr. 1111: 263-269.

[44] Ahl PL, Perkins WR (2003) Interdigitation-Fusion Liposomes. Methods enzymol. 367: 80-98.

[45] Adachi T, Takahashi H., Ohki K, Hatta I (1995) Interdigitated Structure of Phospholipid-Alcohol Systems Studied by X-ray Diffraction. Biophys. j. 68: 1850-1855.

[46] Simon SA, McIntosh TJ (1984) Interdigitated Hydrocarbon Chain Packing Causes the Biphasic Transition Behavior in Lipid/alcohol Suspensions. Biochim. biophys. acta. 773: 169-172.

[47] Rowe ES (1983) Lipid Chain Length and Temperature Dependence of Ethanol-Phosphatidylcholine Interaction. Biochem. 22: 3299-3305.

[48] Rowe ES (1985) Thermodynamic Reversibility of Phase Transitions: Specific Effects of Alcohols on Phosphatidylcholines. Biochim. biophys. acta. 813: 321-330.

[49] Veiro JA, Nambi P, Rowe ES (1988) Effect of Alcohols on the Phase Transitions of Dihexadecylphosphatidylcholine. Biochim. biophys. acta biomembr. 943: 108-111.

[50] Singh H, Emberley J, Morrow MR (2008) Pressure Induces Interdigitation Differently in DPPC and DPPG. Eur. biophys. j. 37: 783-792.

[51] Matsuki H, Miyazaki E, Sakano F, Tamai N, Kaneshina S (2007) Thermotropic and Barotropic Phase Transitions in Bilayer Membranes of Ether-linked Phospholipids with Varying Alkyl Chain Lengths. Biochim. biophys. acta. 1768: 479-489.

[52] Laggner P, Lohner K, Degovics G, Müller, Schuster KA (1987) Structure and Thermodynamics of the Dihexadecylphosphatidylcholine–Water System. Chem. phys. lipids. 44: 31-60.

[53] Kim JT, Mattai J, Shipley GG (1987) Bilayer Interactions of Ether- and Ester-Linked Phospholipids: Dihexadecyl- and Dipalmitoylphosphatidylcholines. Biochem. 26: 6599-6603.

[54] Ichimori H, Hata T, Matsuki H, Kaneshina S (1998) Barotropic Phase Transitions and Pressure-induced Interdigitation on Bilayer Nembranes of Phospholipids with Varying Acyl Chain Lengths, Biochim. biophys. acta biomembr. 1414: 165-174.

[55] Kusube M, Matsuki H, Kaneshina S (2005) Thermotropic and Barotropic Phase Transitions of *N*-methylated Dipalmitoylphosphatidylethanolamine Bilayers. Biochim. biophys. acta biomembr. 1668: 25-32.

[56] Cheng A, Mencke A, Caffrey M (1996) Manipulating Mesophase Behavior of Hydrated DHPE: An X-ray Diffraction Study of Temperature and Pressure Effects. J. phys. chem. 100: 299-306.

[57] Lewis RNAH, Pohle W, McElhaney RN (1996) The Interfacial Structure of Phospholipid Bilayers: Differential Scanning Calorimetry and Fourier Transform Infrared Spectroscopic Studies of 1,2-dipalmitoyl-*sn*-glycero-3-phosphorylcholine and its Dialkyl and Acyl-alkyl Analogs. Biophys. j. 70: 2736-2746.

[58] Haas NS, Sripada PK, Shipley GG (1990) Effect of Chain-linkage on the Structure of Phosphatidylcholine Bilayers. Biophys. j. 57: 117-124.

[59] Furuike S, Levadny VG, Li SJ, Yamazaki M (1999) Low pH Induces an Interdigitated Gel to Bilayer Gel Phase Transition in Dihexadecylphosphatidylcholine Membrane, Biophys. j. 77: 2015-2023.

[60] Hatanaka Y, Kinoshita K, Yamazaki M (1997) Osmotic Stress Induces a Phase Transition from Interdigitated Gel Phase to Bilayer Gel Phase in Multilamellar Vesicles of Dihexadecylphosphatidylcholine, Biophys. chem. 65: 229-233.

[61] Kaneshina S, Maruyama S, Matsuki H (1996) Effect of Pressure on the Phase Behavior of Ester- and Ether-linked Phospholipid Bilayer Membranes. Prog. biotechnol. 13: 175-180.

[62] Hing FS, Maulik PR, Shipley GG (1991) Structure and Interactions of Ether- and Ester-linked Phosphatidylethanolamines. Biochem. 30: 9007-9015.

[63] Serrallach EN, Dijkman R, de Haas GH, Shipley GG (1983) Structure and Thermotropic Properties of 1,3-dipalmitoyl-glycero-2-phosphocholine. J. mol. biol. 170: 155-174.

[64] Dluhy RA, Chowdhry BZ, Cameron DG (1985) Infrared Characterization of Conformational Differences in the Lamellar Phases of *1,3*-dipalmitoyl-*sn*-glycero-*2*-phosphocholine. Biochim. biophys. acta biomembr. 821: 437-444.

[65] Seelig J, Dijkman R, de Haas GH (1980) Thermodynmaic and Conformational Studies on *sn*-2-phosphatidylcholines in Monolayers and Bilayers. Biochem. 19: 2215-2219.

[66] Cunningham BA, Midmore L, Kucuk O, Lis LJ, Westerman MP, Bras W, Wolfe DH, Quinn PJ, Qadri SB (1995) Sterols Stabilize the Ripple Phase Structure in Dihexadecylphosphatidylcholine. Biochim. biophys. acta. 1233: 75-83.

[67] Hirsh DJ, Lazaro N, Wright LR, Boggs JM, McIntosh TJ, Schaefer J, Blazyk J (1998) A New Monofluorinated Phosphatidylcholine Forms Interdigitated Bilayers. Biophys. j. 75: 1858-1868.

[68] Sanii B, Szmodis AW, Bricarello DA, Oliver AE, Parikh AN (2010) Frustrated Phase Transformations in Supported, Interdigitating Lipid Bilayers. J. phys. chem. b. 114: 215-219.

[69] Smith EA, van Gorkum CM, Dea PK (2010) Properties of Phosphatidylcholine in the Presence of its Monofluorinated Analogue. Biophys. chem. 147: 20-27.

[70] McDonough B, Macdonald PM, Sykes BD, McElhaney RN (1983) Fluorine-19 Nuclear Magnetic Resonance Studies of Lipid Fatty Acyl Chain Order and Dynamics in *Acholeplasma laidlawii B* Membranes. A Physical, Biochemical, and Biological Evaluation of Monofluoropalmitic Acids as Membrane Probes. Biochem. 22: 5097-5103.

[71] Santaella C, Vierling P, Riess JG, Gulik-Krzywicki T, Gulik A, Monasse B (1994) Polymorphic Phase Behavior of Perfluoroalkylated Phosphatidylcholines. Biochim. biophys. acta. 1190: 25–39.

[72] McIntosh TJ, Simon SA, Vierling P, Santaella C, Ravily V (1996) Structure and Interactive Properties of Highly Fluorinated Phospholipid Bilayers. Biophys. j. 71: 1853-1868.

[73] Barbarich TJ, Rithner CD, Miller SM, Anderson OP, Strauss SH (1999) Significant Inter- and Intramolecular O—H···FC Hydrogen Bonding. J. am. chem. soc. 121: 4280-4281.

[74] Caminati W, Melandri S, Maris A, Ottaviani P (2006) Relative Strengths of the O—H···Cl and O—H···F Hydrogen Bonds. Angew. chem. int. ed. 45: 2438-2442.

[75] Hyla-Kryspin I, Haufe G, Grimme S (2004) Weak Hydrogen Bridges: A Systematic Theoretical Study on the Nature and Strength of C—H···F—C Interactions. Chem. eur. j. 10: 3411-3422.
[76] O'Hagan D (2008) Understanding Organofluorine Chemistry: An Introduction to the C—F bond. Chem. soc. rev. 37: 308-319.
[77] Toimil P, Prieto G, Miñones Jr. J, Sarmiento F (2010) A Comparative Study of F-DPPC/DPPC Mixed Monolayers: Influence of Subphase Temperature on F-DPPC and DPPC Monolayers. Phys. chem. chem. phys. 12: 13323-13332.
[78] Pabst G, Danner S, Karmakar S, Deutsch G, Raghunathan VA (2007) On the Propensity of Phosphatidylglycerols to Form Interdigitated Phases. Biophys. j. 93: 513-525.
[79] Wang P-Y, Lu J-Z, Chen J-W, Hwang F (1994) Interaction of the Interdigitated DPPG or DPPG/DMPC Bilayer with Human Erythrocyte Band 3: Differential Scanning Calorimetry and Fluorescence Studies. Chem. phys. lipids. 69: 241-249.
[80] Boggs JM, Rangaraj G (1997) Greater Partitioning of Small Spin Labels into Interdigitated than into Non-interdigitated Gel Phase Bilayers. Chem. phys. lipids. 87: 1-15.
[81] Sevcsik E, Pabst G, Jilek A, Lohner K (2007) How Lipids Influence the Mode of Action of Membrane-active Peptides. Biochim. biophys. acta. 1768: 2586-2595.
[82] Sevcsik E, Pabst G, Richter W, Danner S, Amenitsch H, Lohner K (2008) Interaction of LL-37 with Model Membrane Systems of Different Complexity: Influence of the Lipid Matrix. Biophys. j. 94: 4688-4699.
[83] Pabst G, Grage SL, Danner-Pongratz S, Jing W, Ulrich AS, Watts A, Lohner K, Hickel A (2008) Membrane Thickening by the Antimicrobial Peptide PGLa. Biophys. j. 95: 5779-5788.
[84] Bondar OP, Rowe ES (1996) Thermotropic Properties of Phosphatidylethanols. Biophys. j. 71: 1440-1449.
[85] Wilkinson DA, Tirrell DA, Turek AB, McIntosh TJ (1987) Tris Buffer Causes Acyl Chain Interdigitation in Phosphatidylglycerol. Biochim. biophys. acta. 905: 447-453.
[86] Ranck JL, Tocanne JF (1982) Choline and Acetylcholine Induce Interdigitation of Hydrocarbon Chains in Dipalmitoylphosphatidylglycerol Lamellar Phase with Stiff Chains. FEBS lett. 143: 171-174.
[87] Hao Y-H, Xu Y-M, Chen J-W, Huang F (1998) A Drug-Interaction Model: Atropine Induces Interdigitated Bilayer Structure. Biochem. biophys. res. commun. 245: 439-442.
[88] Boon JM, McClain RL, Breen JJ, Smith BD (2001) Inhibited Phospholipid Translocation Across Interdigitated Phosphatidylglycerol Vesicle Membranes. J. supramol. chem. 1: 17-21.
[89] Wang P-Y, Chen J-W, Hwang F (1993) Anisodamine Causes Acyl Chain Interdigitation in Phosphatidylglycerol. FEBS lett. 332: 193-196.
[90] Fang J, Barcelona MJ, Alvarez PJJ (2000) A Direct Comparison Between Fatty Acid Analysis and Intact Phospholipid Profiling for Microbial Identification. Org. geochem. 31: 881-887.

[91] Lohner K, Blondelle SE (2005) Molecular Mechanisms of Membrane Perturbation by Antimicrobial Peptides and the Use of Biophysical Studies in the Design of Novel Peptide Antibiotics. Comb. chem. high throughput screen. 8: 241-246.
[92] Gustavsson L, Alling C (1987) Formation of Phosphatidylethanol in Rat Brain by Phospholipase D. Biochem. biophys. res. commun. 142: 958-963.
[93] Gustavsson E (1995) Phosphatidylethanol Formation: Specific Effects of Ethanol Mediated via Phospholipase D. Alcohol alcoholism. 30: 391-406.
[94] Danner S, Pabst G, Lohner K, Hickel A (2008) Structure and Thermodynamic Behavior of *Staphylococcus aureus* Lipid Lysyl-dipalmitoylphosphatidylglycerol, Biophys. j. 94: 2150-2159.
[95] Koynova R, MacDonald RC (2003) Mixtures of Cationic Lipid *O*-ethylphosphatidylcholine with Membrane Lipids and DNA: Phase Diagrams, Biophys. j. 85: 2449-2465.
[96] Lewis RNAH, Winter I, Kriechbaum M, Lohner K, McElhaney RN (2001) Studies of the Structure and Organization of Cationic Lipid Bilayer Membranes: Calorimetric, Spectroscopic, and X-ray Diffraction Studies of Linear Saturated P-O-ethyl Phosphatidylcholines, Biophys. j. 80: 1329-1342.
[97] Koynova R, RC MacDonald (2003) Cationic O-ethylphosphatidylcholines and their Lipoplexes: Phase Behavior Aspects, Structural Organization and Morphology. Biochim. biophys. acta. 1613: 39-48.
[98] MacDonald RC, Ashley GW, Shida MM, Rakhmanova VA, Tarahovshy YS, Pantazatos DP, Kennedy MT, Pozharski EV, Baker KA, Jones RD, Rosenzweig HS, Choi KL, Qiu R, McIntosh TJ (1999) Physical and Biological Properties of Cationic Triesters of Phosphatidylcholine. Biophys. j. 77: 2612-2629.
[99] Koynova R, MacDonald RC (2004) Columnar DNA Superlattices in Lamellar *O*-Ethylphosphatidylcholine Lipoplexes: Mechanism of the Gel-liquid Crystalline Lipid Phase Transition. Nano lett. 4: 1475-1479.
[100] Kennedy MT, Pozharski EV, Rakmanova VA, MacDonald RC (2000) Factors Governing the Assembly of Cationic Phospholipid-DNA Complexes. Biophys. j. 78: 1620-1633.
[101] MacDonald RC, Rakhmanova VA, Choi KL, Rosenzweig HS, Lahiri MK (1999) *O*-ethylphosphatidylcholine: A Metabolizable Cationic Phospholipid which is a Serum-Compatible DNA Transfection Agent. J. pharm. sci. 88: 896-904.
[102] McIntosh TJ (1996) Hydration Properties of Lamellar and Non-lamellar Phases of Phosphatidylcholine and Phosphatidylethanolamine. Chem. phys. lipids. 81: 117-131.
[103] Perkins WR, Dause R, Li X, Davis TS, Ahl PL, Minchey SR, Taraschi TF, Erramilli S, Gruner SM, Janoff AS (1995) Pressure Induced Fusion (Pif) Liposomes: A Solventless Sterilizing Method for Producing Large Phospholipid Vesicles. J. liposome res. 5: 605-626.
[104] Tada K, Goto M, Tamai N, Matsuki H, Kaneshina S (2010) Pressure Effect on the Bilayer Phase Transition of Asymmetric Lipids with an Unsaturated Acyl Chain. Ann. N.Y. acad. sci. 1180: 77-85.

[105] Ichimori H, Hata T, Matsuki H, Kaneshina S (1999) Effect of Unsaturated Acyl Chains on the Thermotropic and Barotropic Phase Transitions of Phospholipid Bilayer Membranes. Chem. phys. lipids. 100: 151-164.

[106] Dalton LA, Miller KW (1993) *Trans*-unsaturated Lipid Dynamics: Modulation of Dielaidoylphosphatidylcholine Acyl Chain Motion by Ethanol. Biophys. j. 65: 1620-1631.

[107] Vanegas JM, Contreras MF, Faller R, Longo ML (2012) Role of Unsaturated Lipid and Ergosterol in Ethanol Tolerance of Model Yeast Biomembranes. Biophys. j. 102: 507-516.

[108] Mills TT, Huang J, Feigenson GW, Nagle JF (2009) Effects of Cholesterol and Unsaturated DOPC Lipid on Chain Packing of Saturated Gel-phase DPPC Bilayers. Gen. physiol. biophys. 28: 126-139.

[109] McIntosh TJ, Lin H, Li S, Huang C-H (2001) The Effect of Ethanol on the Phase Transition Temperature and the Phase Structure of Monounsaturated Phosphatidylcholines. Biochim. biophys. acta 1510: 219-230.

[110] Polozova A, Li X, Shangguan T, Meers P, Schuette DR, Ando N, Gruner SM, Perkins WR (2005) Formation of Homogeneous Unilamellar Liposomes from an Interdigitated Matrix. Biochim. biophys. acta. 1668: 117-125.

[111] Boni LT, Minchey SR, Perkins WR, Ahl PL, Slater JL, Tate MW, Gruner SM, Janoff AS (1993) Curvature Dependent Induction of the Interdigitated Gel Phase in DPPC Vesicles. Biochim. biophys. acta. 1146: 247-257.

[112] Komatsu H, Guy PT, Rowe ES (1993) Effect of Unilamellar Vesicle Size on Ethanol-Induced Interdigitation in Dipalmitoylphosphatidylcholine. Chem. phys. lipids. 65: 11-21.

[113] Mason JT, Huang CH, Biltonen RL (1983) Effect of Liposomal Size on the Calorimetric Behavior of Mixed-chain Phosphatidylcholine Bilayer Dispersions. Biochem. 22: 2013-2018.

[114] Komatsu H, Rowe ES (1991) Effect of Cholesterol on the Ethanol-Induced Interdigitated Gel Phase in Phosphatidylcholine: Use of Fluorophore Pyrene-Labeled Phosphatidylcholine. Biochem. 30: 2463-2470.

[115] Bondar OP, Rowe ES (1998) Role of Cholesterol in the Modulation of Interdigitation in Phosphatidylethanols. Biochim. biophys. acta. 1370: 207-217.

[116] Laggner P, Lohner K, Koynova R, Tenchov B (1991) The Influence of Low Amounts of Cholesterol on the Interdigitated Gel Phase of Hydrated Dihexadecylphosphatidylcholine. Chem. phys. lipids. 60: 153-161.

[117] Smith EA, Wang W, Dea PK (2012) Effects of Cholesterol on Phospholipid Membranes: Inhibition of the Interdigitated Gel Phase of F-DPPC and F-DPPC/DPPC. Chem. phys. lipids. 165: 151-159.

[118] Lu JZ, Hao YH, Chen JW (2001) Effect of Cholesterol on the Formation of an Interdigitated Gel Phase in Lysophosphatidylcholine and Phosphatidylcholine Binary Mixtures. J. biochem. 129: 891-898.

[119] McMullen TPW, Lewis RNAH, McElhaney RN (1994) Comparative Differential Scanning Calorimetric and FTIR and ^{31}P-NMR Spectroscopic Studies of the Effects of

Cholesterol and Androstenol on the Thermotropic Phase Behavior and Organization of Phosphatidylcholine Bilayers. Biophys. j. 66: 741-752.

[120] Clarke JA, Heron AH, Seddon JM, Law RV (2006) The Diversity of the Liquid Ordered (L_o) Phase of Phosphatidylcholine/Cholesterol Membranes: A Variable Temperature Multinuclear Solid-state NMR and X-ray Diffraction Study. Biophys. j. 90: 2383-2393.

[121] Rand RP, Pangborn WA, Purdon AD, Tinker DO (1975) Lysolecithin and Cholesterol Interact Stoichiometrically Forming Bimolecular Lamellar Structures in the Presence of Excess Water. Can. j. biochem. 53: 189-195.

[122] [122] Koynova R, Brankov J, Tenchov B (1997) Modulation of Lipid Phase Behavior by Kosmotropic and Chaotropic Solutes. Eur. biophys. j. 25: 261-274.

[123] Yu Z-W, Quinn PJ (1995) Phase Stability of Phosphatidylcholines in Dimethylsulfoxide Solutions. Biophys. j. 69: 1456-1463.

[124] Takahashi H, Ohmae H, Hatta I (1997) Trehalose-induced Destabilization of Interdigitated Gel Phase in Dihexadecylphosphatidylcholine. Biophys. j. 73: 3030-3038.

[125] Söderlund T, Alakoskela JM, Pakkanen AL, Kinnunen PKJ (2003) Comparison of the Effects of Surface Tension and Osmotic Pressure in the Interfacial Hydration of a Fluid Phospholipid Bilayer. Biophys. j. 85: 2333-2341.

[126] Luu DV, Cambon L, Mathlouthi M (1990) Perturbation of Liquid-Water Structure by Ionic Substances. J. mol. struct. 237: 411-419.

[127] Collins KD (1997) Charge Density-dependent Strength of Hydration and Biological Structure. Biophys. j. 72: 65-76.

[128] Kinoshita K, Li SJ, Yamazaki M (2001) The Mechanism of the Stabilization of the Hexagonal II (H_{II}) Phase in Phosphatidylethanolamine Membranes in the Presence of Low Concentrations of Dimethyl Sulfoxide. Eur. biophys. j. 30: 207-220.

[129] Scheinkönig C, Kappicht S, Kolb H-J, Schleuning M (2004) Adoption of Long-term Cultures to Evaluate the Cryoprotective Potential of Trehalose for Freezing Hematopoietic Stem Cells, Bone marrow transplant. 34: 531-536.

[130] Notman R, Noro M, O'Malley B, Anwar J (2006) Molecular Basis for Dimethylsulfoxide (DSMO) Action on Lipid Membranes. J. am. chem. soc. 128: 13982-13983.

[131] Yamashita Y, Kinoshita K, Yamazaki M (2000) Low Concentration of DMSO Stabilizes the Bilayer Gel Phase Rather than the Interdigitated Gel Phase in Dihexadecylphosphatidylcholine Membrane. Biochim. biophys. acta. 1467: 395-405.

[132] Yu Z-W, Chen L, Sun S-Q, Noda I (2002) Determination of Selective Molecular Interactions Using Two-dimensional Correlation FT-IR Spectroscopy. J. phys. chem. a. 106: 6683-6687.

[133] Gordeliy VI, Kiselev MA, Lesieur P, Pole AV, Teixeira J (1998) Lipid Membrane Structure and Interactions in Dimethyl Sulfoxide/Water Mixtures. Biophys. j. 75: 2343-2351.

[134] Mansure JJ, Souza RC, Panek AD (1997) Trehalose Metabolism in *Saccharomyces cerevisiae* During Alcoholic Fermentation. Biotechnol. lett. 19: 1201-1203.

[135] Lucero P, Peñalver E, Moreno E, Lagunas R (2000) Internal Trehalose Protects Endocytosis from Inhibition by Ethanol in *saccharomyces cerevisiae*. Appl. environ. microbiol. 66: 4456-4461.

[136] Gibson BR, Lawrence SJ, Leclaire JPR, Powell CD, Smart KA (2007) Yeast Responses to Stresses Associated with Industrial Brewery Handling. FEMS microbiol. rev. 31: 535-569.

[137] Trevisol ETV, Panek AD, Mannarino SC, Eleutherio ECA (2011) The Effect of Trehalose on the Fermentation Performance of Aged Cells of *Saccharomyces cerevisiae*. Appl. microbiol. biotechnol. 90: 697-704.

[138] Nishiwaki T, Sakurai M, Inoue Y, Chŭjŏ R, Koybayashi S (1990) Increasing Packing Density of Hydrated Dipalmitoylphosphatidylcholine Unilamellar Vesicles Induced by Trehalose. Chem. lett. 19: 1841-1844.

[139] di Gregorio GM, Mariani P (2005) Rigidity and Spontaneous Curvature of Lipidic Monolayers in the Presence of Trehalose: A Measurement in the DOPE Inverted Hexagonal Phase. Eur Biophys. j. 34: 67-81.

[140] Villarreal MA, Díaz SB, Disalvo EA, Montich GG (2004) Molecular Dynamics Simulation Study of the Interaction of Trehalose with Lipid Membranes. Langmuir. 20: 7844-7851.

[141] Andersen HD, Wang C, Arleth L, Peters GH, Westh P (2011) Reconciliation of Opposing Views on Membrane-Sugar Interactions. Proc. natl. acad. sci. 108: 1874-1878.

[142] Kim JT, Mattai J, Shipley GG (1987) Gel Phase Polymorphism in Ether-linked Dihexadecylphosphatidylcholine Bilayers. Biochem. 26: 6592-6598.

[143] Laggner P, Lohner K, Degovics G, Müller K, Schuster A (1987) Structure and Thermodynamics of the Dihexadecylphosphatidylcholine-Water System. Chem. phys. lipids. 44: 31-60.

[144] Ohki K (1991) Effect of Substitution of Hydrogen Oxide by Deuterium Oxide on Thermotropic Transition Between the Interdigitated Gel phase and the Ripple Gel Phase of Dihexadecylphosphatidylcholine. Biochem. biophys. res. commun. 174: 102-106.

[145] Ichimori H, Sakano F, Matsuki H, Kaneshina S (2002) Effect of Deuterium Oxide on the Phase Transitions of Phospholipid Bilayer Membranes Under High Pressure. Prog. biotechnol. 19: 147-152.

[146] Komatsu H, Okada S (1997) Effects of Ethanol on Permeability of Phosphatidylcholine/Cholesterol Mixed Liposomal Membranes. Chem. phys. lipids. 85: 67-74.

[147] Komatsu H, Okada S (1996) Ethanol-Enhanced Permeation of Phosphatidylcholine/phosphatidylethanolamine Mixed Liposomal Membranes Due to Ethanol-induced Lateral Phase Separation. Biochim. biophys. acta. 1283: 73-79.

[148] Ahl PL, Chen L, Perkins WR, Minchey SR, Boni LT, Taraschi TF, Janoff AS (1994) Interdigitation-fusion: A New Method for Producing Lipid Vesicles of High Internal Volume. Biochim. biophys. acta biomembr. 1195: 237-244.

[149] Kisak ET, Coldren B, Zasadzinski JA (2002) Nanocompartments Enclosing Vesicles, Colloids and Macromolecules via Interdigitated Lipid Bilayers. Langmuir. 18: 284-288.

[150] Kisak ET, Coldren B, Evans CA, Boyer C, Zasadzinski JA (2004) The Vesosome- A Multicompartment Drug Delivery Vehicle. Curr. med. chem. 11: 199-219.

[151] Zasadzinski JA, Wong B, Forbes N, Braun G, Wu G (2011) Novel Methods of Enhanced Retention in and Rapid, Targeted Release from Liposomes, Curr. opin. colloid interface sci. 16: 203-214.

[152] Raffy S, Teissié J (1997) Electroinsertion of Glycophorin A in Interdigitation-fusion Giant Unilamellar Lipid Vesicles. J. biol. chem. 272: 25524-25530.

[153] Paleos CM, Tsiourvas D, Sideratou Z (2012) Preparation of Multicompartment Lipid-based Systems Based on Vesicle Interactions. Langmuir. 28: 2337-2346.

[154] Chandrawati R, van Koeverden MP, Lomas H, Caruso F (2011) Multicompartment Particle Assemblies for Bioinspired Encapsulated Reactions. J. phys. chem. lett. 2: 2639-2649.

[155] Bolinger P-Y, Stamou D, Vogel H (2008) An Integrated Self-assembled Nanofluidic System for Controlled Biological Chemistries. Angew. chem. int. ed. 47: 5544-5549.

[156] Boyer C, Zasadzinski JA (2007) Multiple Lipid Compartments Slow Content Release in Lipases and Serum. ACS nano. 1: 176-182.

[157] Wong B, Boyer C, Steinbeck C, Peters D, Schmidt J, van Zanten R, Chmelka B, Zasadzinski JA (2011) Design and In Situ Characterization of Lipid Containers with Enhanced Drug Retention. Adv. Mater. 23: 2320-2325.

Chapter 19

Oxidative Stability of Fats and Oils Measured by Differential Scanning Calorimetry for Food and Industrial Applications

M.D.A. Saldaña and S.I. Martínez-Monteagudo

Additional information is available at the end of the chapter

http://dx.doi.org/10.5772/54486

1. Introduction

Fats and oils are important ingredients in the human diet for nutritional and sensory contributions. The terms fats and oils commonly refer to their phase being solid and liquid, respectively. In addition, lipids are the main ingredients to manufacture various products, such as soups, butter, ready to eat food, among others for the food industry and other products, such as lipstick, creams, etc, for the cosmetic and pharmaceutical industries. Most of these products use vegetable oils from seeds, beans, and nuts, which are important due to their high content in polyunsaturated fatty acids compared to animal fats. However, oxidation of unsaturated fatty acids is the main reaction responsible for the lipid degradation, which is related to the final quality of the product. Furthermore, lipids undergo oxidation, developing unpleasant taste, off flavour and undesirable changes in quality, decreasing the nutritional value of the product and compromising safety of the product that might even affect health and well-being.

In general, oxygen reacts with the double bonds present in lipids, following a free radical mechanism, known as autooxidation. This reaction is quite complex and depends on the lipid type used and the processing conditions. The use of thermal processes, such as frying, sterilization, hydrolysis, etc, accelerate the oxidation of lipids. Various different reactions during lipid oxidation occur simultaneously at different rates. These reactions release heat that can be measured using differential scanning calorimetry (DSC).

Oxidation temperatures and kinetic parameters obtained from DSC can be used to rank and classify lipids in terms of their oxidative stability. Therefore, the reproducibility of oxidation experiments is crucial to evaluate the oxidative stability of lipids using DSC since variables, such as pre-treatment and amount of sample, the heating protocol, among others, strongly influence the results. Other methods that assess the extent of oxidative deterioration are peroxide value (PV) that measures volumetrically the concentration of hydroperoxides, anisidine value (AV), spectrophotometric measurements in the UV region and gas chromatography (GC) analysis for volatile compounds [1-3]. Over the years, thousands of studies have focused on monitoring and evaluating the oxidation of lipids using the Rancimat method, PV, AV, spectrophotometric and GC analysis of fats and oils from various sources. However, it is beyond the scope of this chapter to provide a comprehensive listing of all research using those methods. Comprehensive reviews on the oxidative lipid deterioration using those methods are well discussed somewhere else [4-6].

Among all these methods that measure the extent of lipid oxidation, DSC is widely used as an analytical, diagnostic and research tool from which relevant information, such as onset temperature of oxidation (T_{on}), height, shape and position of peaks are obtained and used for subsequent kinetic calculations. Kinetic information of lipid oxidation has been reported for a number of lipid systems, such as soybean/anhydrous milk fat blends, unsaturated fatty acids (oleic, linoleic, and linolenic acids), saturated fatty acids (lauric, myristic, palmitic, and stearic acids), "natural" vegetable oils (canola, corn, cottonseed, and soybean oils) and genetically modified vegetable oils.

This chapter focuses on the principles of lipid oxidation, the use of DSC technique to evaluate lipid oxidation, and recent studies on oxidative stability of fats and oils for food and industrial applications, addressing the generation and analysis of DSC thermograms for kinetic studies, where a method to analyse DSC data is described in detail, as well as the interpretation of kinetic parameters obtained at isothermal and non-isothermal conditions. In addition, some results on oxidation kinetics of milk fat after the use of traditional and emerging technologies, such as enzymatic hydrolysis and pressure assisted thermal processing are discussed in detail. Finally, conclusions on lipid oxidation analysis by DSC are provided.

2. Fundamentals of lipid oxidation

Lipid oxidation is a free radical chain reaction that leads to the development of unpleasant flavour and taste, loss of nutrients and formation of toxic compounds [7, 8]. Consequently, the shelf life and the final use of any lipid depend on its resistance to oxidation or oxidative stability [9]. The term lipid oxidation usually refers to a three consecutive reactions or stages, known as initiation, propagation and termination (Figure 1). In the initiation stage, free radicals are formed through thermolysis, where the break of covalent bonds is induced by heat. In addition, free radicals can also be formed due to the presence of enzymes, light, metal ions (Ca^{2+} and Fe^{3+}) and reactive oxygen species. A list of initiators of lipid oxidation and their standard reduction potentials are provided somewhere else [10]. Compounds that homolyze at relative low temperature (<100°C) are important initiators of radical-based chain reactions. Unsaturated fatty acids are compounds that homolyze at lower temperatures compare to saturated fatty acids. The homolytic products of unsaturated fatty acids are hydroxyl radical ($HO^{\bullet}$), alkyl radical ($RO^{\bullet}$) and hydroperoxyl radical ($HOO^{\bullet}$), which further reacts with triplet oxygen to form peroxyl radicals. Among the starters, hydroxyl radicals are mainly responsible for the initiation of lipid oxidation due to its strong tendency to acquire electrons [10]. These radical products that have high energy, bond to a hydrogen molecule from the lipid structure, forming hydroperoxides (primary oxidation products). The formation of hydroperoxides can be repeated several times, propagating the oxidation reactions. Conjugated fatty acids have more than one type of primary oxidation products and more than one oxidation pathway, as previously reported [11, 12]. A kinetic analysis on autoxidation of methyl-conjugated linoleate showed that monomeric and cyclic peroxides are the major primary oxidation products rather than hydroperoxides [12]. Consequently, addition by Diels Alder-type reaction was earlier suggested as a reaction mechanism.

The next oxidation stage is the propagation, which consists in the further degradation of hydroperoxides or any other primary oxidation product [1]. There are mainly two types of degradation products from hydroperoxides. First, hydroperoxides interact with double bonds to form monomeric degradation products, such as ketones. The reaction occurs through the reduction of the hydroperoxyl group to hydroxyl derivative. Second, low molecular weight products, that results from the cleavage of the hydroperoxide chain, form aldehydes, ketones, alcohols and hydrocarbons. These low molecular weight compounds are responsible for the rancid and off-flavour produced by oxidized fats [1-4]. Finally, hydroperoxides and primary oxidation products homolyze to form peroxyl or alkoxy radicals that further reacts to form stable dimer-like products. In addition, alcohols, and unsaturated fatty acids (secondary oxidation products) also lead to termination products. The resulting compounds form viscous materials through polymerization as the oxidation proceeds. These polymers are oil insoluble and represent the termination stage of oxidation [2, 15].

Radical formation

$$\text{A-B} \xrightarrow{\text{Homolysis}} \text{I}^\bullet$$

Initiation

$$\text{R-H} + \text{I}^\bullet \longrightarrow \text{R}^\bullet + \text{I-H}$$

Propagation

$$\text{R}^\bullet + \text{O}_2 \longrightarrow \text{ROO}^\bullet$$

$$\text{ROO}^\bullet + \text{R-H} \longrightarrow \text{ROOH} + \text{R}^\bullet$$

Termination

$$\text{ROO}^\bullet + \text{ROO}^\bullet \longrightarrow \text{ROOR} + \text{O}_2$$

$$\text{RO}^\bullet + \text{R}^\bullet \longrightarrow \text{R-O-R}$$

$$\text{ROO}^\bullet + \text{R}^\bullet \longrightarrow \text{Stable product}$$

$$\text{R}^\bullet + \text{R}^\bullet \longrightarrow \text{R-R (dimers)}$$

Figure 1. Main lipid oxidation reactions.

3. Fundamentals on the use of DSC to study lipid oxidation

The oxidation mechanisms presented in Figure 1 is an oversimplification because lipids consist of a non-homogeneous mixture of fatty acids. For example, anhydrous milk fat (AMF) is composed of more than 400 fatty acids with extremely diverse chain lengths, position and number of unsaturations of their fatty acids [13, 14]. Consequently, several reactions occur simultaneously at different rates as the oxidation proceeds. Although several methods have been used to analyze and monitor lipid oxidation [1], the oxidation reactions cannot be measured by a single method due to their complexity. Some of the available methods allow quantifying one or more reaction products of the different oxidation stages. Methods, such as oxidative stability index (OSI) and peroxide value (PV) are officially accepted by the American of Analytical Communities (AOAC) [1-6], while other methods are routinely used, such as the Racimat, chemilominescent, and volumetric methods [16].

An important overlooked characteristic of the oxidation reactions of lipids is the exothermal effect as the oxidation occurs. The released heat from a particular reaction can be measured using DSC in either isothermal or non-isothermal mode. For DSC oxidation measurements, the heat released from the oxidized oil is compared to the heat flowing from an inert reference (empty pan) both heated at the same rate. When the oxidation of the sample occurs, the recorded heat shows a peak which area is proportional to the amount of heat released by the sample. Figure 2 shows an ideal thermogram for the non-isothermal oil oxidation with the three consecutive reaction stages of initiation, propagation and termination.

The heat released by the oxidized oil is recorded as the heat flow signal (y-axis) as a function of temperature (x-axis). The period of time where no change in the heat flow signal occurs is known as the induction time (Figure 2) that is exemplified at the beginning of the thermogram. An excellent review on the theory and application of induction time is provided elsewhere [17]. The length of the induction time is often considered as a measurement of oil stability. During this period, no chemical reaction occurs. At the point in

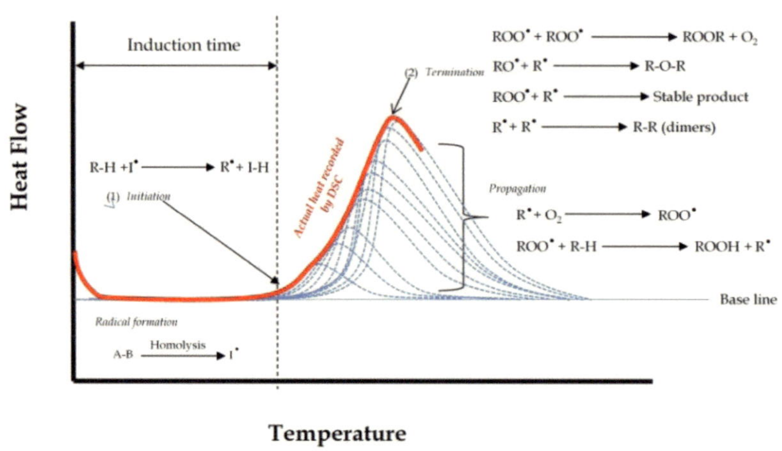

Figure 2. Ideal thermogram of non-isothermal oil oxidation.

which the heat flow signal separates from the baseline (straight line) is considered to be the end of the induction time (arrow (1)). Arrow (1) also indicates the start of oxidation or initiation stage. This stage is short and can be theoretically interpreted as the reaction between the radical, formed during the induction time, and the unsaturated fatty acid. The products of this reaction are unstable hydroperoxides that further react propagating the oxidation. A sudden increase in the heat flow signal is related to the propagation stage. The blue dashed lines illustrate oxidation reactions that occur and cannot be detected by the DSC because they are less exothermal. Finally, arrow (2) illustrates the termination stage, where stable products are formed. The red line is the actual heat flow recorded by the DSC.

In the next section, a set of recommended guidelines and laboratory practices are reviewed for the good use and analysis of data using DSC.

4. Important considerations for measurement of lipid oxidation using DSC

Although DSC is a simple, convenient and fast technique to measure lipid oxidation, some recommended guidelines should be considered to obtain reliable and reproducible results. Among those guidelines are the pre-treatment and sample preparation, amount of sample used, heating protocol, gas flow rate, and interpretation of DSC thermograms.

Sample pre-treatment – a representative amount of lipid sample should be used for DSC oxidation measurements. The lipid should be melted at a temperature that ensures that its thermal memory is erased. The melting temperature for lipids is quite diverse, ranging from -25 to 80°C. Melting temperatures of some vegetable oils are provided elsewhere [18]. In some lipid-based products, the pre-treatment of sample involves the fat extraction from the food matrix. For example, fat from commercial baby formulas was first extracted with chloroform [19], and then the fat was further dried under vacuum. Therefore, the behavior of these extracted fats might be different from the fat in the original matrix. In addition, a proper chemical description of any pre-treatment must be provided together with the thermal history.

Sample preparation – a liquid sample should be loaded into the DSC pan using a syringe or a Pasteur pipette. The lipid oxidation measured by DSC can be conducted in an open aluminum pan or in a hermetic sealed pan with a pinhole (Figure 3). The main difference between an open pan and a sealed pan with a pinhole is the diffusion of oxygen and the amount of oxygen that is in contact with the sample. This is because the thermal conductivity of the air is smaller than that of the metal of the pan. Indeed, numerical simulations showed that the energy transmitted to the sample comes from the plate, which transmits the heat to the pan [20].

A comparison of glycerol oxidation obtained in an open pan and a hermetic sealed pan with a pinhole was earlier reported [21]. For experiments conducted in an open pan, the maximum heat flow temperature of oxidation of glycerol was around 40°C lower than those obtained with sealed pans. In an open pan, the vapor produced during lipid oxidation

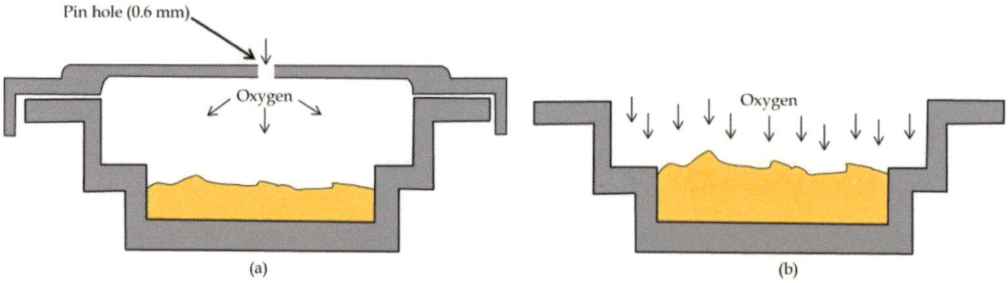

Figure 3. Illustrations of common DSC pans: (a) hermetically sealed pan with a pinhole, and (b) an open aluminium pan.

leaves the pan as it is formed. This is because the purge of oxygen acts as a carrier of the vapor. Consequently, part of the mass is lost before reaching the temperature at which the oxidation starts. On the other hand, in a sealed pan with a pinhole, the vapor produced cannot escape from the pan, remaining in the oil. This vapor increases the pressure inside the pan and elevates the oxidation temperature. Contradictory results were reported for the melting temperature of benzoic acid and vanillin obtained in an open pan and a sealed pan with a pinhole [22]. No significant differences were observed in the onset temperature (T_{on}) and the peak maximum temperature (T_P). Unfortunately, studies with a direct comparison between the types of pans (open and sealed with a pinhole) for oxidation of are scarce in the literature. Indeed, the international organization for standardization does not specify the types of pans for the determination of oxidation induction time [23]. Although hermetic sealed pans with a pinhole have an additional cost to each experimental run, their use avoids contamination of the DSC chamber.

Sample size – the amount of sample has significant effect on the shape of the thermogram and reproducibility of the DSC oxidation experiments as it is related to heat transfer within the pan. Figure 4 illustrates the effect of the amount of sample in hermetic sealed pans with a pinhole. For the sample with an optimum sample thickness (Figure 4a), the oil is in contact with excess of oxygen, facilitating oxygen diffusion within the oil sample. An earlier study [7] recommended 1 mm of sample thickness (approximately 1.5 mg of oil) to yield consistent results in non-isothermal oil oxidation. Similarly, no changes on the DSC thermograms using samples between 1 to 4 mg were reported [24]. A ratio of 1:3 (oil:oxygen) not only avoids diffusional limitations [6, 7, 25] but also allows the vapor molecules formed during the oxidation reaction to rapidly escape from the pan [26]. This enhances the baseline and the resolution of the oxidation thermogram [27].

On the other hand, an excess of sample (Figure 4b) creates a temperature gradient within the sample, especially at high heating rates [27]. Also, the diffusion of oxygen is limited, which broader the DSC oxidation curves. This is illustrated in Figure 5 where samples of 3 and 13 mg of anhydrous milk fat (AMF) were oxidized at 12°C/min from 100 to 250°C. Interestingly, the T_{on} of oxidation is quite similar between samples (181.12 and 179.33°C, respectively). But, the use of 13 mg of AMF leads to a broader and less resolved curve. The deviation of the heat flow signal from the vertical edge (broadening) is attributed to a longer time needed to oxidize a larger sample and the development of a temperature gradient within the sample. Another reason is that the probability of vapor molecules to escape from the pan through the pinhole is considerable reduced. Consequently, the peak maximum temperature shifts to a lower temperature because some of the vapor molecules react, accelerating the termination stage of oxidation.

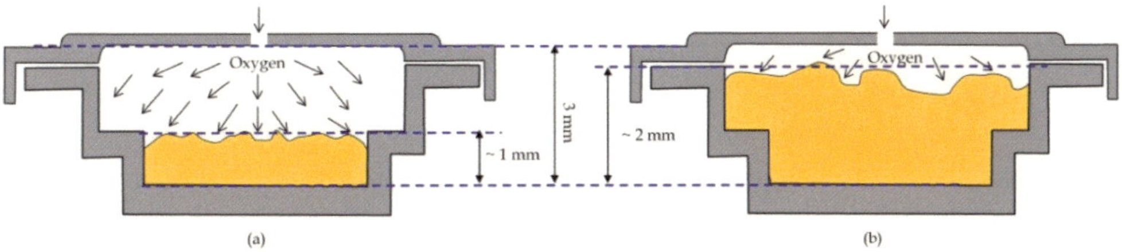

Figure 4. Illustration of (a) optimum sample size, and (b) excess sample size, in hermetic sealed pans with a pinhole.

Heating rate – this is one of the most important parameters to determine the oxidative stability and oxidation kinetics of lipids. Before the start of the heating rate, an equilibrium period between 3 and 5 min is recommended to enhance the baseline. At slow heating rates, primary oxidation products, such as hydroperoxides generated during the initial oxidation stage, react with excess of oxygen to form low molecular weight compounds (intermediate oxidation products), accelerating the degradation process. At fast heating rates, these intermediate products are lost through evaporation before they react with the lipid, shifting to a high value the threshold DSC signal [6, 9]. This phenomenon is the basis to calculate the kinetic parameters from a DSC thermogram [28]. However, it is important to highlight that the heating rate should not exceed 25°C/min since the temperature of the sample is different from the furnace temperature, creating a temperature gradient which affects the oxidation kinetics [29]. An important overlooked consideration is the temperature range at which the oxidation study should be conducted. In general, the temperature range should start and end as far as possible from the T_{on} and T_p (approximately a difference of at least 50°C).

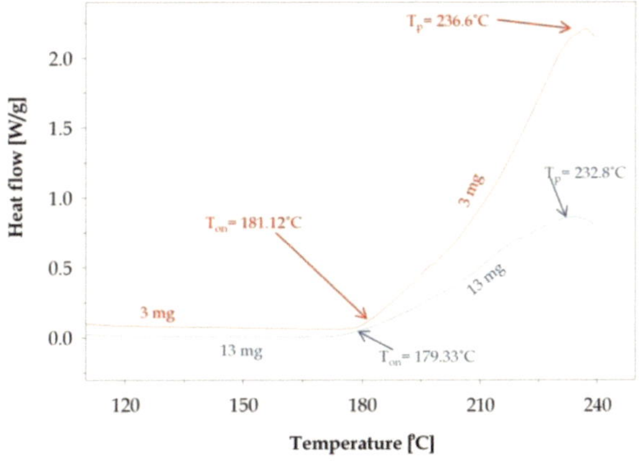

Figure 5. Effect of sample size on non-isothermal oxidation of anhydrous milk fat at 12°C/min. T_{on} – onset temperature of oxidation; T_p – maximum heat flow temperature.

DSC mode – oxidation experiments can be conducted in either isothermal or non-isothermal mode. Both methods provide analytical information, such as the oxidation induction time in the case of isothermal measurements and the oxidation onset temperature in the case of non-isothermal measurements [28, 29]. Further comparisons on the kinetic studies conducted in either isothermal or non-isothermal mode are discussed in the following sections.

5. Analysis of DSC thermograms

5.1. Location of key parameters

The analysis and interpretation of the generated DSC spectra consist in identifying key parameters, such as the induction time, onset and peak maximum temperatures. These key parameters are manually obtained from the DSC spectra. The T_{on} is obtained extrapolating the tangent drawn on the steepest slope of T_p. This procedure is usually performed

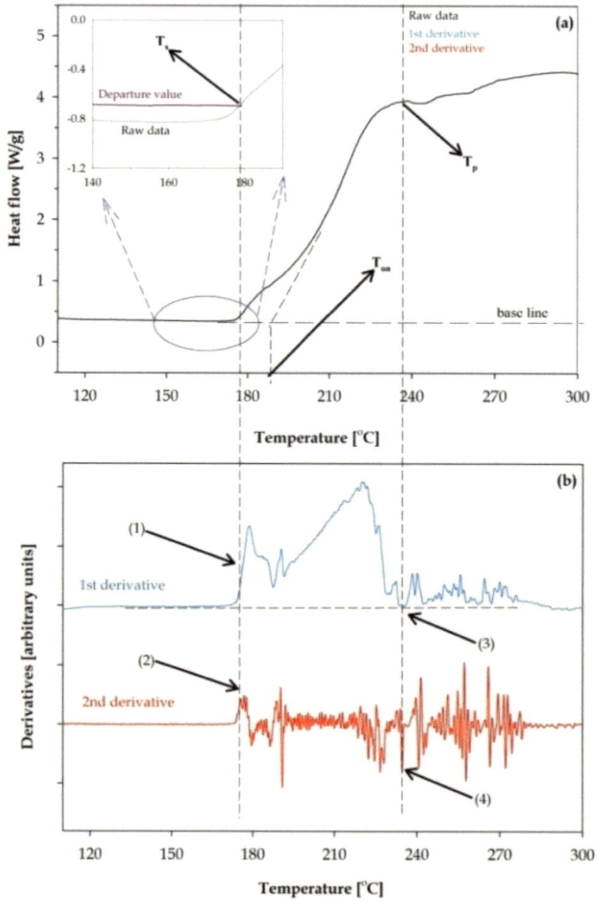

Figure 6. Anhydrous milk fat rich in conjugated linoleic acid oxidized at 15°C min^{-1}. (a) Determination of the start temperature (T_s) (inlet), onset temperature (T_{on}) and maximum heat flow temperature (T_p), and (b) zoom on the first and second derivatives that precisely locates the T_s, T_{on}, and T_p

manually by the DSC operator, relying in the equipment software. Inherently, there is certain degree of uncertainty associated with this procedure. A method that accurately and unambiguously determines those key parameters from the DSC spectra was early proposed for lipid crystallization [30]. This method was first developed to calculate onset, offset, and peak maximum temperatures in crystallization of binary mixtures of different triacylglycerols. More recently, the same methodology was adapted to calculate the start, onset and peak maximum temperatures in non-isothermal oxidation of anhydrous milk fat [9]. Figure 6 exemplifies the location of the start, onset and peak maximum temperatures for anhydrous milk fat rich in conjugated linoleic acid oxidized at 15°C/min in a hermetic sealed pan with a pinhole.

Once the DSC curves are generated, the error associated with the raw data is calculated through standard deviation. Then, the first and second derivatives are calculated. The error is obtained from the baseline, which in Figure 6 corresponds to the segment of 140 to 178°C. This is essential since the signal variability can be misinterpreted as a thermal event. In this method, a true thermal event was considered when the heat flow signal is twice greater than the standard deviation of the baseline. This criterion is known as the departure value (inlet Figure 6a). To locate the start temperature of oxidation, three criteria were considered. Firstly, the first derivative of the signal shows an inflexion point between a maximum and a minimum point of the signal (arrow (1)). Secondly, the second derivative reaches a maximum point on the heat flow signal (arrow (2)). Finally, the heat flow signal should be greater than the departure value (inlet). T_P was obtained when the first derivative of the signal intersects with the x-axis (arrow (3)) and the second derivative reached a maximum point on the signal (arrow (4)). T_{on} was obtained extrapolating the tangent drawn on the steepest slope of T_P.

5.2. Iso-conversional method

In chemical reactions, the degree of conversion ($0 \leq \alpha \leq 1$) or extent of reaction of a particular compound is defined by moles at a given time divided by the initial moles [29]. Similarly, α in thermal analysis is defined by the heat flow at a given time or temperature divided by the heat flow at time or temperature at which the maximum heat flow signal is reached. The heat flow from the DSC spectra is converted to α based on the initial ($signal_o$) and final ($signal_f$) heat flow signals, as shown in equation (1).

$$\alpha = \frac{signal_o - signal}{signal_o - signal_f} \quad (1)$$

At a given degree of conversion, the reaction kinetics is described by a single-step reaction that follows an Arrhenius type equation within a narrow range of temperature. Then, the overall kinetics [31-34] is the result of multiple single-step reactions in the form:

$$\frac{d\alpha}{dt} = A\, exp^{\left(\frac{-E_a}{RT}\right)} f(\alpha) \quad (2)$$

where t is the time, A is the pre-exponential factor, E_a is the effective activation energy, T is the temperature, and $f(\alpha)$ is the reaction model. This procedure, known as iso-conversional or model-free method, is used to calculate kinetic triplet parameters (effective activation energy, E_a, pre-exponential factor, A and constant rate, k) in thermally stimulated reactions [29-34]. As known, reactions are the sequence of physical changes that can be measured by thermal techniques.

In non-isothermal oxidation of lipids, the consumption of oxygen can be neglected due to the large excess of oxygen generated by a constant flow rate (>25 mL/min). Such condition allows the formation of peroxides, being independently of the oxygen concentration, which also means that the autoxidation is a first order reaction [7, 9]. This is an essential assumption for the calculation of the kinetic triplet parameters (E_a, A, and k). A commonly used iso-conversional method is the Ozawa-Flynn-Wall method. Using this method, a set of data (T_s, T_{on}, and T_p) was obtained for constant heating rates ($\beta = dT/dt$) from which the kinetic parameters were calculated using the following equations:

$$\log \beta = a\frac{1}{T} + b \tag{3}$$

where β is the heating rate (K/min) and T is the temperature T_s, T_{on}, or T_p (K). By plotting log β against $1/T$, the effective activation energy (E_a) and the pre-exponential factor (A) were determined directly from the slope and intercept according to:

$$a = -0.4567\frac{E_a}{R} \tag{4}$$

$$b = -2.315 + log\left(A\frac{E_a}{T}\right) \tag{5}$$

where a and b are the slope and intercept from equation (3), respectively, and R is the universal gas constant (8.31 J/mol K). Therefore, the effective activation energy (E_a) and the constant rate (k) are calculated from:

$$E_a = -2.19R\frac{d\log \beta}{dT^{-1}} \tag{6}$$

$$k = A\,exp^{\left(\frac{E_a}{RT}\right)} \tag{7}$$

5.3. Induction time

As mentioned earlier, the period of time where no change in the heat flow signal occurs is known as the induction time and its length is considered as a measurement of lipid stability [17]. The determination of the induction period is routinely conducted to evaluate stability of oils, lubricants, biodiesel, and pharmaceutical products [34-38]. Unfortunately, no standard protocol for the determination of the oxidation induction time has been developed for lipids.

Figure 7a illustrates the standard protocol for determination of oxidation induction time (OIT). The polymer sample (15 mg) is rapidly heated under nitrogen atmosphere (≤ 20°C/min) until it reaches the temperature that corresponds to time, t_1, which is the starting point for OIT determination. At t_1, the atmosphere is switched to oxygen and the sample is held at the same temperature until t_2 is reached. The difference between t_2 and t_1 is the OIT. A disadvantage of this protocol is to find an adequate temperature for the isothermal stage. For example, the use of a low temperature might considerably increase the OIT while the use of a high temperature might oxidize the sample immediately, making difficult to obtain a reliable baseline. In tests conducted in 16 different laboratories, it was demonstrated that OIT is associated with a high degree of uncertainty. On the other hand, Figure 7b illustrates the experimental protocol for the determination of oxidation induction temperature (OIT*). In this case, the sample is continuously heated at a constant heat rate (for example, 12°C/min) under oxygen atmosphere.

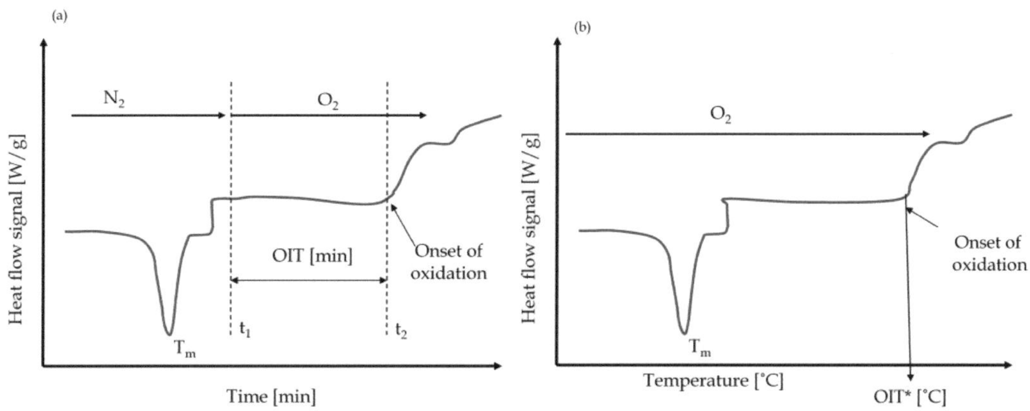

Figure 7. Determination of oxidation induction time, OIT (a), and oxidation induction temperature, OIT* (b). T_m – melting temperature, t_1 – start of the oxidation induction time, t_2 – end of the oxidation induction time. Adapted from reference [35].

6. Kinetic studies of oxidation of fats and oils

Tables 1 and 2 summarize oxidation studies of fats and oils using DSC. Most of the isothermal studies were conducted between 80 and 180°C with different flow rates varying from 10 to 100 mL/min. The amount of sample used also varied from 3 to 30 mg and most of these studies used open pans. The oxidation onset times were quite diverse (23-108 min) and therefore the E_a values ranged from 50 to 130 kJ/mol, depending mainly on the oil composition, temperature, and amount of sample. For the non-isothermal studies, the temperature ranged from 50 to 350°C and the heating rate used ranged from 1 to 25°C/min. The onset oxidation temperatures depended on the heating rate and oil composition.

6.1. Isothermal studies

For isothermal oxidation, the heat flow signal generated at constant temperature is plotted against time (Fig 7a). From these curves, the start of the oxidation (t_1) and the end of the oxidation induction time (t_2) are first located and then used for analysis [39].

Table 1 summarizes the isothermal oxidation studies of fats and oils, such as peanut, safflower seed, blackcurrant seed, rice bran, cotton seed, Buriti seed, passion fruit seed, sunflower seed, soybean, linseed, canola seed, coconut, grape seed, palm seed, and sesame seed. Earlier studies on isothermal oxidation were imprecise because the baselines obtained were highly unstable, making it difficult to obtain oxidation onset times [40, 41]. However, stable baselines were currently obtained in the isothermal oxidation of linoleic, linolenic, oleic, stearic, peanut, safflower and blackcurrant seed oils [42]. These oils were heated under argon flow. After thermal equilibrium was reached, the gas flow was switched to oxygen, allowing the oxidation to start. The temperature used to conduct the oxidation test should be far below the self-ignition temperature for fats and oils (~ 350°C). Thus, the recorded thermal events are due to lipid oxidation rather than combustion. Using this approach, the obtained induction times were reproducible and highly influenced by the test temperature and sample composition. Attempting to validate the isothermal DSC oxidation, an earlier study [43] isothermally oxidized purpurea seed, rice bran and cotton seed oils using the DSC and the Rancimat methods. At each tested temperature, the DSC oxidation times were shorter than those obtained using the Rancimat method for the same oil. Despite these differences, the oxidation times were satisfactorily correlated with the Rancimat induction times. Similarly, longer induction times were obtained with the use of the Rancimat method for Buriti pulp seed oil, rubber seed oil and passion fruit seed oil [44]. In these studies, DSC rapidly reaches the threshold of the heat flow signal. This difference is attributed to the small sample used in DSC experiments, allowing a higher oil-oxygen ratio compared to the Rancimat method. Moreover, to oxidize the oil, DSC employs pure oxygen (99% purity) while the Rancimat method uses air (~21% of oxygen).

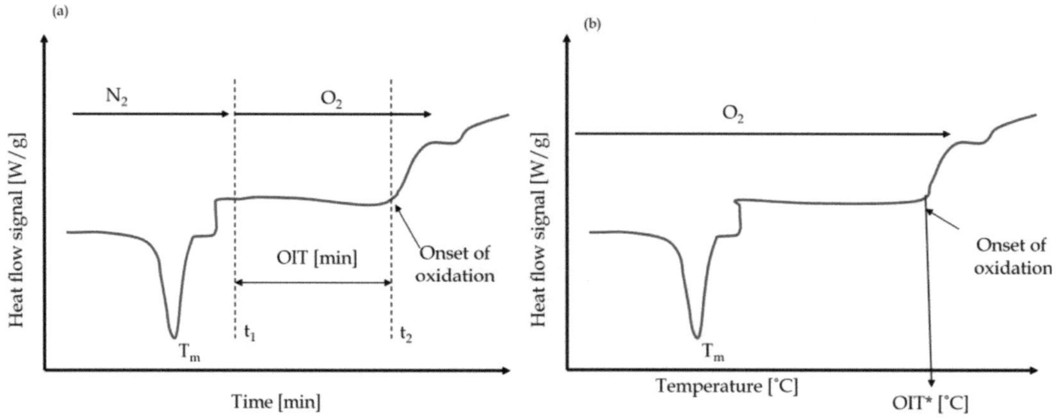

Lipid	Experimental protocol	Oxidation induction times (OIT, min)	Ref.
Linoleic, linolenic, peanut, oleic, stearic, safflower and blackcurrant seed	- 80-160°C - 5 mg in an open pan - Flow rate = 8.3 mL/min -Argon flow to equilibrate	At 120°C: - Olive = 108; - Blackcurrant = 95; - Corn = 78; - Peanut = 68; - Sunflower = 69; - Linseed = 50; - Safflower = 43 At 130°C: - Olive = 105; - Blackcurrant = 50; - Corn = 50; - Peanut = 54; - Sunflower = 22; - Linseed = 21; - Safflower = 18	[42]
Rice, cotton seed and *B. purpurea*	- 110, 120, 130 and 140°C - 5 mg in an open pan - Flow rate = 50 mL/min	At 110°C: - *B. purpurea* = 483; - Rice = 132; - Cotton = 172 At 120°C: - *B. purpurea* = 269; - Rice = 72; - Cotton = 92 At 130°C: - *B. purpurea* = 99; - Rice = 36; - Cotton = 42 At 140°C: - *B. purpurea* = 48; - Rice = 18; - Cotton = 20	[43]
Buriti, rubber seed and passion fruit oil	- 100, 110, 120, 130 and 140°C - 5 mg in an open pan - Flow rate = 50 mL/min	100°C: - Buriti = 116; - Rubber = 106; - Passion = 778 110°C: - Buriti = 42; - Rubber = 51; - Passion = 369 120°C: - Buriti = 23; - Rubber = 25; - Passion = 163 130°C: - Buriti = 7.8; - Rubber = 7.5; - Passion = 66.5 140°C: - Buriti = 3.3; - Rubber = 3.3; - Passion = 30.3	[44]
Rapeseed and sunflower	- 120°C - 30 mg in a sealed pan - Flow rate = 600 mL/min	- Rapeseed: $OIT_{120^{\circ}C}$ = 1.5-1.9 min - Sunflower: $OIT_{120^{\circ}C}$ = 0.6-1.2 min	[45]
Linseed	- 130°C - 5 mg in a sealed pan - Flow rate = 60 mL/min	- Linseed: $OIT_{130^{\circ}C}$ = 23.1 - Linseed + 0.05% of antioxidant blend = 25.1 - Linseed + 0.2% of antioxidant blend = 33.1 - Linseed + 0.01% Butylated hydroxyl anisole (BHA) = 24 - Linssed + 0.02% BHA = 30.1	[49]
Lipid	**Experimental protocol**	**Kinetic parameters (E_a, kJ/mol; A min^{-1}; k, min)**	**Ref.**
Rapeseed, soybean, corn and sunflower	- 121 to 149°C - 3.5-6.5 mg in an open pan - Flow rate = 100 mL/min	- Soybean: $OIT_{121-150^{\circ}C}$ = 11-2.9; E_a = 67.2; A= $2.7x10^7$ - Rapeseed: $OIT_{121-130^{\circ}C}$ = 16.2-10.4; E_a = 71.9; A = $1.2x10^8$ - Corn: $OIT_{114-130^{\circ}C}$ = 23-3; E_a = 98.6; A= $5.1x10^{11}$ - Sunflower: $OIT_{106-130^{\circ}C}$ = 11-2.8; E_a = 84.6; A= $1.4x10^{10}$	[47, 48]
Soybean, rapeseed, and sunflower oil	- 110-150°C. - 3-4 mg in an open pan - Flow rate = 100 mL/min	- Soybean: $OIT_{110-150^{\circ}C}$ = 139-11; E_a = 72.4; A= $6.3x10^7$; $k_{150^{\circ}C}$ = 0.073 - Rapeseed: $OIT_{110-150^{\circ}C}$ = 122-14; E_a = 83.4; A= $1.6x10^9$; $k_{150^{\circ}C}$ = 0.084 - Sunflower: $OIT_{110-150^{\circ}C}$ = 64-4.4; E_a = 86.2; A= $8.7x10^9$; $k_{150^{\circ}C}$ = 0.201	[50]
Canola, coconut, corn, grapeseed, peanut, palm kernel, palm olein, safflower, sesame and soybean	- 110-140°C. - 5 mg in an open pan - Flow rate = 50 mL/min	- Canola: $OIT_{110-140^{\circ}C}$ = 259-37; E_a = 86; $k_{128^{\circ}C}$ = 0.016 - Coconut: $OIT_{110-140^{\circ}C}$ = 325-44; E_a = 86.9; $k_{128^{\circ}C}$ = 0.011 - Corn: $OIT_{110-140^{\circ}C}$ = 166-21.4; E_a = 88.1; $k_{128^{\circ}C}$ = 0.021 - Grape seed: $OIT_{110-140^{\circ}C}$ = 74-7.5; E_a = 99.9; $k_{128^{\circ}C}$ = 0.056 - Peanut: $OIT_{110-140^{\circ}C}$ = 127-12.3; E_a = 99.1; $k_{128^{\circ}C}$ = 0.025 - Palm kernel: $OIT_{110-140^{\circ}C}$ = 539-70; E_a = 89.4; $k_{128^{\circ}C}$ = 0.007 - Palm olein: $OIT_{110-140^{\circ}C}$ = 515-82; E_a = 79.9; $k_{128^{\circ}C}$ = 0.006 - Safflower: $OIT_{110-140^{\circ}C}$ = 88-8; E_a = 104.3; $k_{128^{\circ}C}$ = 0.055 - Sesame: $OIT_{110-140^{\circ}C}$ = 542-20; E_a = 88.8; $k_{128^{\circ}C}$ = 0.007 - Soybean: $OIT_{110-140^{\circ}C}$ = 124-20; E_a = 80.8; $k_{128^{\circ}C}$ = 0.029	[51, 52]
Soybean, rapeseed, corn and peanut oil	- 108-162°C. - 2-5 mg in an open pan. - Flow rate = 100 mL/min.	- Rapeseed: $OIT_{130-150^{\circ}C}$ = 33-11; E_a = 64.4 - Soybean: $OIT_{138-162^{\circ}C}$ = 8.9-2.3; E_a = 51.3 - Corn: $OIT_{147-160^{\circ}C}$ = 13.5-5.7; E_a = 77.5 - Palm: $OIT_{108-121^{\circ}C}$ = 24-6.5; E_a = 82.3	[53]
Lauric, myristic, palmitic, stearic acids and their ester	- 155-180°C. - 5-8 mg in an open pan. - Flow rate = 100 mL/min.	- Palmitic: E_a = 125.1 ± 11.2 - Ethyl palmitate: E_a = 126.6 ± 5.0 - Glycerol tripalmitate: E_a = 105.3 ± 7.7 - Stearic acid: E_a = 134.3 ± 12.0 - Ethyl stearate: E_a = 128.5 ± 2.6 - Glycerol tristearate: E_a = 102.5 ± 10.9 - Lauric acid: E_a = 97.3 ± 7.3 - Ethyl laurate: E_a = 127.3 ± 6.3 - Ethyl myristate: E_a = 117.1 ± 5.4	[54]
Blends of cocoa butter/cocoa butter fat like	- 130, 140, 150, and 160°C. - 5 mg in an open pan. - Flow rate = 100 mL/min.	Cocoa butter/cocoa butter fat like - 100/0: E_a = 106.2; A = $5.6x10^{10}$; $k_{160^{\circ}C}$ = 0.008 - 90/10: E_a = 105.1; A = $4.4x10^{10}$; $k_{160^{\circ}C}$ = 0.009 - 80/20: E_a = 105.9; A = $6.7x10^{10}$; $k_{160^{\circ}C}$ = 0.011 - 70/30: E_a = 95.1; A = $3.9x10^9$; $k_{160^{\circ}C}$ = 0.013 - 60/40: E_a = 99.5; A = $1.5x10^{10}$; $k_{160^{\circ}C}$ = 0.015 - 50/50: E_a = 104.4; A = $6.7x10^{10}$; $k_{160^{\circ}C}$ = 0.017 - 0/100: E_a = 120.5; A = $1.9x10^{13}$; $k_{160^{\circ}C}$ = 0.055	[55]

Table 1. Summary of isothermal oxidation studies of lipids using differential scanning calorimetry

A comparative study of sunflower seed oil and rapeseed oil oxidation using DSC and volatile analysis showed that the ratio of hexanal/2-trans-nonenal linearly correlates with the onset heat flow signal of the DSC spectra [44]. In addition, a correlation between the peroxide values with the oxidation onset time was developed to monitor rapeseed oil oxidation [44]. Unfortunately, the correlation was valid only for oils with peroxide values lower than 30 mmol O_2/kg oil. In other study, the experimental oxidation data obtained from electron spin spectroscopy was compared with the data obtained from DSC oxidation [46]. The onset oxidation times were highly correlated for different fat and oil blends. However, the obtained correlations were valid only at moderate temperatures of 60°C, which considerably limits the applicability of electron spin spectroscopy for oxidation analysis.

An investigation of isothermal oxidation of oils (e.g. rapeseed, soybean, corn and sunflower oils) used onset time (t_{on}) to rank the oxidized oils in terms of their oxidative stability [47]. The high maximum heat flow time (t_p) value indicates that the oil is more stable. Although these relationships were statistically validated, a single parameter to evaluate the oxidative stability can lead to overestimation of the oxidative stability. In the same study, the authors considered t_p to be proportional to the rate of oxidation, which might not be valid. Indeed, peroxide value determinations showed that t_p represents the oxidation termination stage while t_{on} is associated with the rate of initiation [48]. Furthermore, the addition of antioxidants prolongs only t_{on} while t_p values were minimally affected [48]. The same behavior was also observed in the isothermal oxidation of linseed oil with the use of antioxidants [49]. The addition of BHA (butylated hydroxyl anisole) and a mixture of antioxidants (tocopherol, ascorbyl palmitate, citric acid and ascorbic acid) prolonged the onset time at 130°C. All these approaches provided information of great value for validation of the DSC isothermal oxidation. However, these equations are limited for a set of temperatures and specific oils (Table 1). Additional factors, such as degree of saturation, amount of free fatty acids, chain length and the presence of natural antioxidants were not considered.

Using the isothermal method, the sample is heated at a constant temperature and the released heat is recorded as a function of temperature. Such situation allows the identification of the maximum heat flow time (t_m), which linearly correlates with the temperature [39,48, 49].

$$\log t_m = A \cdot T^{-1} + B \tag{8}$$

where A and B are regression parameters. Due to the excess of oxygen generated by a constant flow rate, the formation of peroxides is considered to be independent of the oxygen concentration, which also means that the autoxidation is a first order reaction [48, 49]. This is an essential assumption for the calculation of kinetic parameters, such as effective activation energy (E_a), pre-exponential factor (A), and reaction rate (k).

$$E_a = 2.19 \cdot R \cdot \frac{d\log(t_m)}{dT^{-1}} \tag{9}$$

Equations 8 and 9 have been applied not only to obtain the maximum heat flow signal [47-49] but also to obtain the oxidation onset time. An attempt to correlate induction times and kinetic parameters of isothermally heated rapeseed, soybean and sunflower seed oils at different temperatures was proposed [50]. The authors correlated the oxidation induction times with the tested temperature using Arrhenius-like equations. The obtained kinetic parameters were comparable to those obtained by the Rancimat method. The onset time values of 12 different oils obtained by DSC were reduced by half of their previous values for every increase of 10°C in the oxidation temperature [51]. These relationships were further used to obtain kinetic parameters of DSC oxidation [52]. The E_a values of rapeseed, soybean, corn and peanut oils were strongly influenced by the amount of saturated fatty acids [53]. The oxidation of saturated fatty acids (C_{12}-C_{18}) and their esters revealed that the E_a values (100-125 kJ/mol) were within the same range for all the tested oils [54]. This suggests that the isothermal oxidation is not influenced by the carbon chain length. Another important conclusion from this investigation [54] is that the start of oxidation is similar for fatty acids, their esters and triglycerides. The kinetic oxidation parameters of cocoa butter blends were obtained using the oxidation onset time [55]. The blends were oxidized from 130 to 160°C in an open pan. Interestingly, the E_a and A values were slightly affected by the addition of saturated fatty acids. But, k values considerably changed with the amount of saturated fatty acids. Thus, k values can be used to rank the oxidative stability of cocoa butter blends. However, the use of k values to evaluate oxidative stability might be valid only at the temperature tested since changes in the reaction mechanisms can occur as a function of temperature.

6.2. Non-isothermal studies

For non-isothermal oxidation studies, two maximum heat flow peaks are commonly observed in the DSC spectra [56-59]. But, only the first peak is related to lipid oxidation. This was demonstrated in non-isothermal oxidation studies of corn and linseed oils with different peroxide values [57]. A decrease in the onset temperature was observed as the peroxide value increased. Contrary, the first and second peak temperatures were not affected by increasing the peroxide value. Consequently, the first peak can be related to hydroperoxides formation while the second peak can be due to further oxidation of peroxides. In an earlier study [58], the weight loss of lecithin during non-isothermal heating under nitrogen flow rate was analyzed. The thermogravimetric analysis showed that in the temperature range of the first peak only 4% of weight was lost but above that temperature, the weight loss considerably increased. Therefore, changes in the DSC signal within the range of the first peak temperature were attributed to oxidation and those changes above that temperature corresponded to thermal degradation rather than oxidation (Figure 8).

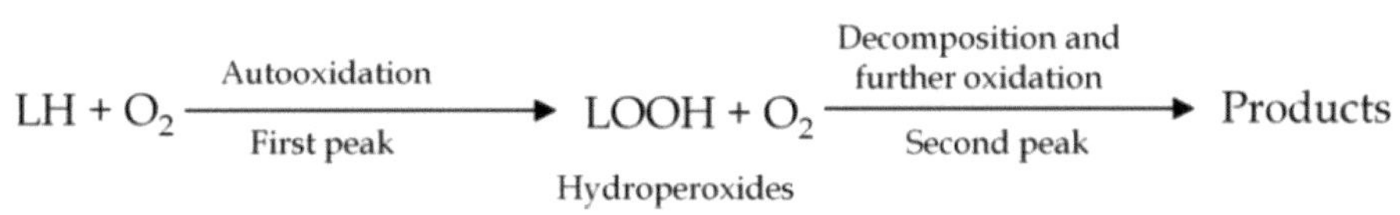

Figure 8. Reaction mechanism proposed for lipid oxidation under non-isothermal conditions.

The proposed reactions in Figure 8 resemble to an autocatalytic reaction scheme. Indeed, computer simulated DSC oxidation curves using autocatalytic scheme fitted well the experimental data [56]. Similarly, the non-isothermal oxidation of mustard oil was best described by an autocatalytic reaction scheme [58]. Therefore, it was proposed that T_{on} is the most representative reference point of lipid oxidation under non-isothermal conditions. In addition, kinetic parameters calculated from T_{on} were comparable to those obtained at isothermal DSC conditions (Tables 1 and 2) [56-59]. Similarly, non-isothermal kinetics was used to evaluate the oxidative stability of commercial olive oil samples [59-62]. The obtained kinetic parameters were comparable to those obtained with the Rancimat method.

Lipid	**Experimental protocol**	**Kinetics parameters (E_a, kJ/mol; A min^{-1}; k, min)**	**Ref.**
Corn and linseed oil	- 50-300˚C - 5 mg in an open pan - β = 5-20˚C/min - Flow rate = 166 mL/min	- Corn: $E_a = 69.5\pm9.7$; $A = 1.25\times10^{8}$ - Linseed: $E_a = 132.7\pm8.8$; $A = 1.15\times10^{16}$ *Onset temperature* - Corn + 47 mmol O_2/kg = 146˚C - Corn + 70 mmol O_2/kg = 142˚C - Corn + 93 mmol O_2/kg = 137˚C - Corn + 136 mmol O_2/kg = 137˚C - Corn + 145 mmol O_2/kg = 128˚C - Corn + 158 mmol O_2/kg = 127˚C - Linseed + 31 mmol O_2/kg = 144˚C - Linseed + 119 mmol O_2/kg = 131˚C - Linseed + 180 mmol O_2/kg = 129˚C - Linseed + 252 mmol O_2/kg = 125˚C - Linseed + 349 mmol O_2/kg = 117˚C - Linseed + 383 mmol O_2/kg = 121˚C	[56]
Linolenic acid and soy lecithin	- 50-300˚C - 5 mg in an open pan - β = 2-20˚C/min - Flow rate = 6600 mL/min	- Linolenic: E_a from T_{on}: 65±4; $A = 2.3\times10^{7}$; $k_{100˚C} = 0.016$ - Linolenic: E_a from T_p: 78.9±6.9; $A = 9.9\times10^{7}$ - Lecithin: E_a from T_{on}: 97±8l; $A = 2.4\times10^{11}$; $k_{100˚C} = 0.001$ - Linolenic: E_a from T_p: 141.4±4; $A = 9.4\times10^{15}$	[57]
Mustartd oil	- 140-350˚C - 3-5 mg in an open pan - β = 2, 5, 7.5, 10, and 15˚C/min - Flow rate = 100 mL/min	- From onset: $E_a = 90.6$; $A = 3.4\times10^{9}$; $k_{225˚C} = 0.97$ - From 1st peak: $E_a = 88.5$; $A = 1.4\times10^{9}$; $k_{225˚C} = 0.90$ - From 2nd peak: $E_a = 84.6$; $A = 4.4\times10^{7}$; $k_{225˚C} = 1.12$	[58]
Olive oil	- 2-3 mg in an open pan - β = 4, 5, 7.5, 10, 12.5, 15˚C/min - Flow rate = 100 mL/min	- E_a calculated from T_{on}: 72-104; $A = 3.31\times10^{8}$ -1.1×10^{12} ; $k_{120˚C} = 0.09$-0.015	[59]
Blends of soybean/AMF	- 100 to 350˚C - 5-15 mg in an open pan - β = 2.5-12.5˚C/min - Flow rate = 100 mL/min	*Soybean/AMF* - 100/0: $E_a = 93.5$; $A = 1.25.6\times10^{10}$; $k_{200˚C} = 0.57$ - 90/10: $E_a = 59.5$; $A = 2.5\times10^{6}$; $k_{200˚C} = 0.68$ - 80/20: $E_a = 58.4$; $A = 2.1\times10^{6}$; $k_{200˚C} = 0.71$ - 70/30: $E_a = 64.5$; $A = 1.1\times10^{7}$; $k_{200˚C} = 0.80$ - 60/40: $E_a = 70.2$; $A = 5.2\times10^{7}$; $k_{200˚C} = 0.88$ - 50/50: $E_a = 73.2$; $A = 1.1\times10^{8}$; $k_{200˚C} = 0.95$ - 40/60: $E_a = 102.7$; $A = 2.6\times10^{11}$; $k_{200˚C} = 1.18$ - 30/70: $E_a = 102.8$; $A = 2.7\times10^{11}$; $k_{200˚C} = 1.20$ - 20/80: $E_a = 105.8$; $A = 5.3\times10^{11}$; $k_{200˚C} = 1.26$ - 10/90: $E_a = 117.4$; $A = 1.4\times10^{13}$; $k_{200˚C} = 1.55$ - 0/100: $E_a = 89.5$; $A = 8.4\times10^{9}$; $k_{200˚C} = 1.09$	[62]
Oleic, erucic, linoleic, linolenic and their ethyl esters and glycerol trioleate and trilinoleate	- Temperature = 50-300˚C - 5 mg in an open pan - β = 2-20˚C/min - Flow rate = 600 mL/min	- Erucic: $E_a = 89.6\pm4.4$; $A = 4.9\times10^{10}$; $k_{90˚C} = 0.006$ - Oleic: $E_a = 88.4\pm4.7$; $A = 1.0\times10^{11}$; $k_{90˚C} = 0.021$ - Oleate: $E_a = 95\pm4.7$; $A = 9.4\times10^{10}$; $k_{90˚C} = 0.002$ - Trioleate: $E_a = 91.8\pm13.3$; $A = 4.4\times10^{10}$; $k_{90˚C} = 0.002$ - Linoleic: $E_a = 72\pm2.9$; $A = 1.6\times10^{9}$; $k_{90˚C} = 0.071$ - Linoleate: $E_a = 76.4\pm5$; $A = 2.4\times10^{9}$; $k_{90˚C} = 0.024$ - Trilinoleate: $E_a = 74.3\pm3$; $A = 9.6\times10^{8}$; $k_{90˚C} = 0.020$ - Linolenic: $E_a = 62.4\pm3.7$; $A = 2.6\times10^{8}$; $k_{90˚C} = 0.027$ - Linolenate: $E_a = 74.5\pm8.2$; $A = 4.2\times10^{9}$; $k_{90˚C} = 0.082$	[63, 64]
AMF with low, medium and high CLA content	- 100 to 350˚C - 2 mg in an sealed pan - β = 3, 6, 9, 12 and 15˚C/min - Flow rate = 50 mL/min	- Low CLA: $E_a = 146.5$; $A = 4.1\times10^{14}$; $k_{200˚C} = 0.026$ - Med CLA: $E_a = 112.4$; $A = 3.6\times10^{10}$; $k_{200˚C} = 0.014$ - High CLA: $E_a = 87.6$; $A = 6.3\times10^{7}$; $k_{200˚C} = 0.013$	[9]
Cotton, corn, canola, safflower, high oleic safflower, high	- 1-1.5 mg in a sealed pan - β = 1, 5, 10, 15 and 20˚C/min	- Cotton: $E_a = 63.3$; $A = 9.2\times10^{6}$; $k = 0.37$ - Corn: $E_a = 77.7$; $A = 2.4\times10^{8}$; $k = 0.43$	[7]

Lipid	Experimental protocol	Kinetics parameters (E_a, kJ/mol; A min^{-1}; k, min)	Ref.
linoleic safflower, high oleic sunflower, soybean and sunflower		- Canola: E_a = 88.4; A = 7.6x10^9; k = 0.51 - Sufflower: E_a = 75.2; A = 1.8x10^8; k = 0.44 - High oleic sufflower: E_a = 88.7; A = 3.1x10^9; k = 0.48 - High linoleic sufflower: E_a = 73.5; A = 1.1x10^8 k = 0.42 - High oleic sunflower: E_a = 86.5; A = 1.8x10^9; k = 0.47 - Soybean: E_a = 79.6; A = 4.1x10^8; k = 0.44 - Sunflower: E_a = 63.8; A = 1.1x10^7; k = 0.38	
Base oil lubricants	- 0.5 mg in a sealed pan - β = 5, 10 and 20˚C/min - Flow rate = 100 mL/min	- E_a = 13.8-83.9 kJ/mol; k = 0.11-0.75 min^{-1}.	[25]
Rapeseed, soybean, sunflower, lard and highly rancid oils	- 100-360˚C - 3-4 mg in an open pan - β = 5-20˚C/min - 166 mL/min - Antioxidants: BHT, BHA and PG	- Rapeseed: E_a = 64-73; A = 7.4x10^5-4.4x10^6 - Soybean: E_a = 62-64; A = 6.1x10^5-7.1x10^5 - Sunflower: E_a = 60-62; A = 4.1x10^5-6.6x10^5 - Lard: E_a = 92-93; A = 3.5x10^8-2.7x10^8	[65]
Fat extracted from baby formulas	- 3-4 mg - β = 4, 6, 7.5, 10, 12.5, 15˚C/min - Flow rate = 50 mL/min	- E_a = 80-106, A = 7.4x10^9-1.1x10^{-13}; $k_{120˚C}$ = 0.079-0.122	[19]
High oleic sunflower (HOS) and castor oil with blends of antioxidants	- 100 to 250˚C - 3-3.3 mg in an open pan - β = 10˚C/min - Flow rate = 10 mL/min - Antioxidants (0.5-2 wt %): TOC, PG, AA and MBP	- HOS: OOT = 191˚C - HOS + PG: OOT = 225-233˚C - HOS + AA: OOT = 206˚C - HOS + TOC: OOT = 190-214˚C - HOS + MBP: OOT = 226-242˚C - Castor: OOT = 197 ˚C - Castor + PG: OOT = 233-242˚C - Castor + AA: OOT = 208-212˚C - Castor + TOC: OOT = 195-203˚C - Castor + MBP: OOT = 212-234˚C	[66]
Linolenic acid (LNA) with different phenols	- 50-300˚C - 3-5 mg in an open pan - β = 5, 10, 20, 40, 40 and 80˚C/min - Flow rate = 250 mL/min	- LNA: E_a = 70.4; A = 5.6x10^9 min^{-1}; $k_{90˚C}$ = 0.792 - 0.26 BHT/LNA: E_a = 73±3.7; A = 8.9x10^9; $k_{90˚C}$ = 0.541 - 0.82 BHT/LNA: E_a = 79.6±8.4; A = 2.4 x 10^{10}; $k_{90˚C}$ = 0.17 - 1.51 BHT/LNA: E_a = 87±5.5; A = 2.0x10^{11}; $k_{90˚C}$ = 0.134 - 3.0 BHT/LNA: E_a = 91.9±4.5; A = 6.5x10^{11}; $k_{90˚C}$ = 0.091 - 4.0 BHT/LNA: E_a = 95.8±2.4; A = 2.1x10^{12}; $k_{90˚C}$ = 0.049	[67]
LNA with BHT, olivetol and DHZ	- 50-350˚C - 5 mg in an open pan - β = 2-20˚C/min - Flow rate = 250 mL/min	- LNA: E_a = 69.8±7.8; A = 2.2x10^9 - 1.2-11.6 BHT/LNA: E_a = 82-141 - 1.1-8.2 DHZ/LNA: E_a = 88-187 - 1.1-5.6 Olivetal/LNA: E_a = 84-129	[68]

AA – ascorbic acid; TOC – tocopherol; MBP – methylenebis (2,6 ditert-butylphenol); PG – propyl gallate; BHT- Butylated hydroxytoluene, DHZ-dehydrozingerone, AMF-anhydrous milk fat, CLA-conjugated linoleic acid, OOT – oxidation onset temperature.

Table 2. Summary of non-isothermal oxidation studies of lipids using differential scanning calorimetry

According to the Arrhenius principle, oil with a high E_a value oxidizes faster at high temperatures, while oil with a low E_a value oxidizes faster at low temperatures. Unfortunately, calculated values of E_a should not be used as a single parameter to rank the oxidative stability of lipid systems. This was exemplified in blends of soybean/anhydrous milk fat [63] that were non-isothermally oxidized. Interestingly, as the percentage of unsaturated fatty acids increased, the onset temperature of oxidation decreased and the only kinetic parameter that exhibited the same pattern was the constant rate of oxidation. The calculated E_a value is the cumulative effect of all the E_a values available in the system during oxidation, including intermediate compounds that have their own kinetic values. An equation representing the overall activation energy for autoxidation of lipids was earlier proposed [54, 56]. The overall effect included activation energies of initiation (E_i), propagation (E_p) and termination (E_t) based on the classical rate equation for autoxidation of hydrocarbons.

Unfortunately, equation (10) has been applied to limited fatty acids (C_{12}-C_{18}) [56, 57] and correlations between other kinetic parameters and the initiation and termination activation energies are needed.

$$E = E_p + \frac{1}{2} E_i - \frac{1}{2} E_t \quad (10)$$

The non-isothermal oxidation of different fatty acids and their esters [56, 57] showed that the calculated E_a values are similar among the different tested samples, indicating that the oxidation does not occur on free or esterified carboxyl groups of fatty acids. The non-isothermal oxidation of anhydrous milk fat with different ratios of unsaturated/saturated fatty acids showed that the start temperature of oxidation shifted to lower values as the ratio increased [9]. More importantly, the kinetics parameters (E_a, A and k) calculated also decreased. The onset temperature of oxidation not only is affected by the amount of saturated fatty acids but also by the presence and abundance of aromatic compounds and their alkyl substitutions [7, 25]. Kinetic parameters obtained from different oils were compared with structural parameters obtained with NMR spectroscopy. Moreover, an increase in the methylene carbons of the fatty acid chains increased the oxidative stability while conjugated structures were rapidly oxidized.

The addition of antioxidants to enhance oxidative stability of oils can be evaluated using the DSC in non-isothermal mode. A novel approach based on DSC data, named protective factor (PF) was provided in the literature [65]. In this investigation, the oxidation of methyl esters derived from rapeseed and waste frying oil was monitored at different concentrations of BHT and pyrogallol (PG). The oxidation onset temperature asymptotically increased with the addition of BHT and PG, without finding an optimum antioxidant concentration. In addition, the increase in the heating rate might change the reaction mechanisms in which antioxidants can capture free radicals, making difficult to compare their effectiveness. Therefore, the protective factor concept [65] was developed according to:

$$Protective\ factor\ (PF) = \frac{onset\ temperature\ of\ oil\ with\ antioxidant}{onset\ temperature\ of\ oil\ without\ antioxidant} \quad (11)$$

Values of PF lower than 1 means that the antioxidant has a pro-oxidant effect. On the other hand, PF values greater than one can be considered as a measurement of antioxidant effectiveness. Another important factor is the physical stability of the antioxidants. In a comparative study [65, 68], the addition of BHT, BHA and PG was evaluated in rapeseed, soybean, sunflower and lard oils. BHA and BHT were not effective antioxidants due to their volatility. These antioxidants escaped from the heated oil before they can react to neutralize free radicals. An optimum antioxidant concentration was found for BHT (8.4 mmol), BHA (2.8 mmol) and olivetol (4.5 mmol) in oxidized linolenic acid [63-65]. After the optimal concentration, a decrease in the antioxidant activity was observed. Interestingly, at high temperatures (<180°C), the antioxidants are no longer stable and their effectiveness significantly decreased. Similarly, α-tocopherol and *L*-ascorbic acid 6-palmitate were not effective even at high concentrations (up to 2% wt) during the oxidation of either high oleic sunflower oil or castor oil [66, 69].

6.3. Compensation theory

A proper kinetic interpretation should include the compensation effect, which is usually used to explain whether the variations on effective activation energy values have physical meaning or they are caused by either variations of process conditions or complexity of the reaction systems. In previous investigations [69, 70], the compensation effect is illustrated for various thermal degraded materials, such as polymers, cellulosic materials and $CaCO_3$. The compensation effect can be evaluated by plotting ln *k*, obtained by equation (7), against *1/T*. Figure 9 shows that an increase in the effective activation energy causes an increase in *ln k* (equation 7). Similarly, a decrease in the effective activation energy results in a lower value of *ln k*. The point of concurrence, where the different lines intercept, corresponds to $ln\ k_{iso}$ and $1/T_{iso}$ (k_{iso} is the isokinetic rate constant and T_{iso} is the isokinetic temperature), which indicates the existence of the compensation effect.

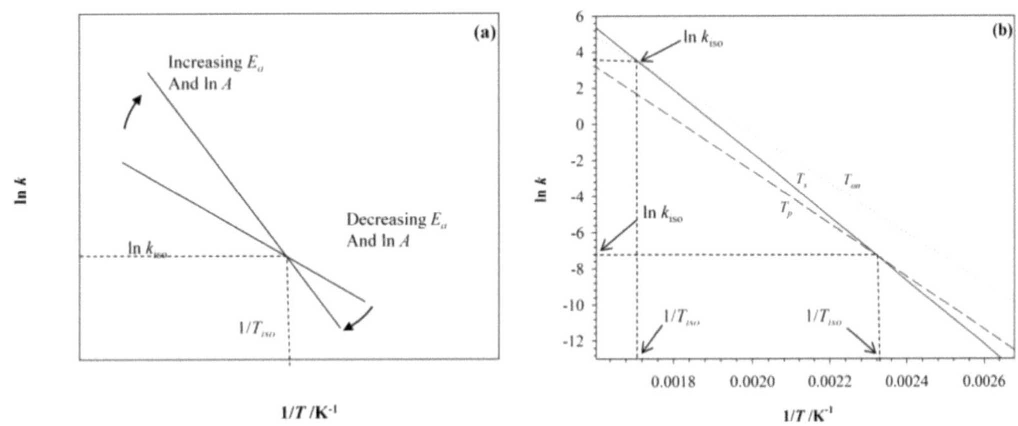

Figure 9. Arrhenius plot (*ln k vs 1/T*) for the non-isothermal oxidation: (a) compensation effect theory adapted from reference [69], and (b) anhydrous milk fat [9].

In studies of non-isothermal oxidation of anhydrous milk fat (AMF) [9], there is no concurrence at a single point. Therefore, the non-isothermal oxidation of AMF does not exhibit a compensation effect and thus the variations in kinetic parameters have no physical background. In fact, AMF is a very complex fat mainly composed of triacylglycerols that have a glycerol backbone to which three fatty acid moieties are esterified. These triacylglycerols are extremely diverse in chain lengths, position and number of unsaturations of their fatty acids [72]. Moreover, more than 400 fatty acids in milk fat were found [73]. Therefore, several reactions with different constant rates simultaneously occur and DSC only detects those reactions that have the greatest exothermal effect. This could explain why the variations in effective activation energy values have no physical meaning or there is no compensation effect.

The isokinetic temperature can also be defined as the temperature at which two rate constants of two different reactions are equal. For a set of kinetic parameters of two different reactions, the isokinetic temperature can be expressed as:

$$A_1 \cdot exp^{\left(\frac{-E_1}{R \cdot T_{iso}}\right)} = A_2 \cdot exp^{\left(\frac{-E_2}{R \cdot T_{iso}}\right)} \quad (12)$$

The isokinetic temperature of autoxidation of lecithin and linolenic acid calculated using equation (12) was 167°C [57]. This observation exemplifies the difficulty in determining the oxidative stability of multicomponent systems. For example, if the oxidation test is conducted at a temperature below T_{iso}, linolenic acid oxidizes faster than lecithin. But, if the same test is conducted at a temperature equal to T_{iso}, both lipids have the same oxidative stability. Contrary, lecithin oxidizes faster than linolenic acid above T_{iso}. Consequently, the estimation of the oxidative stability based only on the onset temperature is misleading. Therefore, interpretation of the shape of non-isothermal oxidation curves in combination with kinetic parameters can provide a better interpretation of the oxidative behavior in multicomponent systems.

7. Case studies of lipid oxidation after processing using new technologies

7.1. Kinetics of non-isothermal oxidation of anhydrous milk fat (AMF) rich in conjugated linoleic acid (CLA) after hydrolysis

AMF is the richest source of CLA composed of geometrical and positional isomers of linoleic acid. CLA has potential benefits, such as cancer prevention, atherosclerosis, weight control, and bone formation [74]. Additionally, CLA concentration in milk can be markedly enhanced through diet manipulation and nutritional management of dairy cattle [75]. In CLA-enriched AMF, CLA is distributed throughout different triacylglycerols together with other fatty acids, limiting its applicability as ingredient in different milk fat-based products. One known approach to produce free fatty acids (FFA) is through enzymatic hydrolysis. The enzymatic hydrolysis of AMF rich in CLA yielded around 88% of free fatty acids (FFA) [76, 77]. Unfortunately, FFAs are more susceptible to oxidation than those fatty acids attached to the triacylglycerol backbone.

Figure 10 shows the DSC curves of non-hydrolyzed and hydrolyzed AMF rich in CLA. Although the DSC curves were quite different between non-hydrolyzed AMF (blue and black lines) and hydrolyzed AMF (green and red lines), the maximum peak temperatures were in the same range. This observation supports the hypothesis that changes in the DSC signal below the temperature of the first peak can be attributed to oxidation and changes in DSC signal above T_P correspond to thermal decomposition and advanced oxidation products as shown in Figure 8. Figure 10 also shows that the oxidation of hydrolyzed AMF starts at low temperatures (~108°C), making difficult to obtain a consistent baseline. Therefore, the kinetic parameters were calculated using the onset temperature obtained as described in Section 5.1.

Table 3 shows the onset oxidation temperature and kinetic parameters of non-hydrolyzed and enzymatic hydrolyzed AMF rich in CLA. As expected, the E_a for the hydrolyzed fat is greater than that obtained for the non-hydrolyzed fat. Interestingly, the constant reaction

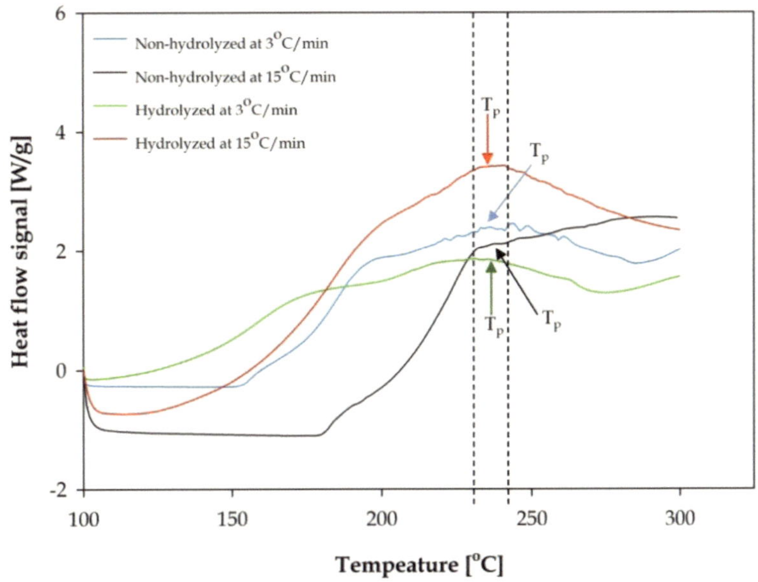

Figure 10. Non-isothermal oxidation of non-hydrolyzed and hydrolyzed anhydrous milk fat rich in CLA. T_p - maximum heat flow temperature.

Heating rate [°C/min]	Non-hydrolyzed AMF [°C]	Hydrolyzed AMF [°C]
3	173.47 ± 0.1	125.06 ± 1.4
6	184.71 ± 2.3	132.46 ± 0.3
9	192.13 ± 0.7	133.25 ± 1.2
12	197.85 ± 3.7	132.46 ± 0.3
15	198.66 ± 0.3	136.78 ± 0.1
E_a [kJ/mol]	82.42	175.78
A [min^{-1}]	8.7 x 10^7	2.3 x 10^{20}
$k_{130°C}$ [min]	0.002	0.003
$k_{200°C}$ [min]	0.075	10.28

Table 3. Onset temperature of oxidation for anhydrous milk fat (AMF)

rates (k) calculated at 130 °C were similar for the non-hydrolyzed and hydrolyzed samples. However, the constant rate of hydrolyzed fat was dramatically higher than that of non-hydrolyzed fat at 200°C.

The isokinetic temperatures of these two fats were calculated mathematically and graphically using equation (12) and Arrhenius plot, respectively. Figure 11 shows the compensation effect for the oxidation of non-hydrolyzed and hydrolyzed AMF rich in CLA. From this figure, the T_{iso} and k_{iso} were obtained (T_{iso} = 120°C and k_{iso} = 0.0011 min). Similarly, the T_{iso} and k_{iso} calculated mathematically with equation (12) were 118°C and 0.0010 min, respectively. However, T_{iso} for non-hydrolyzed AMF is unrealistic since the start temperature of oxidation is around 155°C. Therefore, oxidation of either non-hydrolyzed or hydrolyzed AMF rich in CLA does not exhibit a compensation effect.

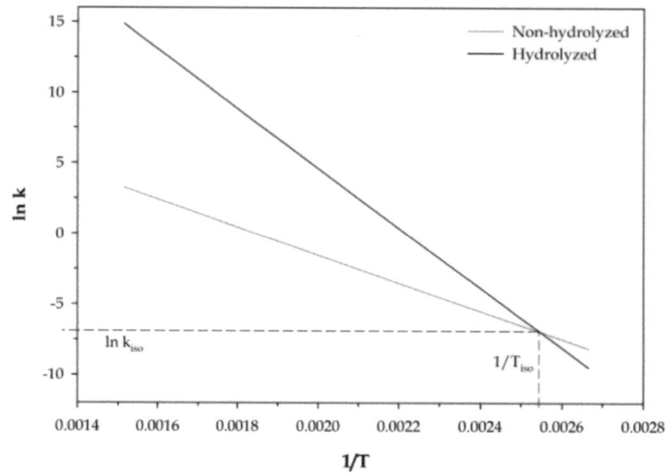

Figure 11. Arrhenius plot for the non-isothermal oxidation of non-hydrolyzed and hydrolyzed anhydrous milk fat rich in CLA.

7.2. Oxidative stability of AMF rich in CLA treated with pressure assisted thermal processing

CLA is not stable upon thermal processing and significant losses of its biological activity occurred through oxidation [78]. The application of high pressure (100-600 MPa) to a preheated sample can preserve the biological activity of functional compounds [79]. This is because the rise in temperature due to adiabatic heating is used to reach the target temperature, reducing the thermal damage due to the lack of temperature uniformity that occurs in traditional thermal processes [80, 81]. This technology is known as pressure-assisted thermal processing (PATP). Pressure alters interatomic distance, acting mainly on those weak interactions which bond energy is distance–dependent, such as van der Waals forces, electrostatic forces, hydrogen bonding and hydrophobic interactions of proteins. Based on the distance dependence, any pressurized sample would have its covalent bonds intact. This has been the central hypothesis in preserving the biological activity of functional compounds, such as ascorbic acid, folates, vitamins and anthocyanins [82-85].

The effects of PATP conditions on the antiradical ability of CLA in AMF were reported [86]. CLA can donate hydrogen to form a CLA-free radical that further reacts to inhibit hydroperoxides formation, depending on the final CLA retention. This suggests that the retained CLA after PATP treatment might enhance the oxidative stability of AMF. After PATP treatments, samples of AMF rich in CLA were oxidized at 6°C/min to calculate the start temperature of oxidation using the method described earlier in Section 5.1 of this chapter.

Figure 12 shows the influence of the retained CLA on the start temperature of oxidation (T_s). The stability of AMF is influenced in a non-linear mode by the CLA retention. Values of the start temperature of oxidation (~165-155°C) were obtained when the CLA retention was between 85 to 100%. On the other hand, the lowest value of the start temperature of oxidation (141°C) was obtained at 40% of CLA retention.

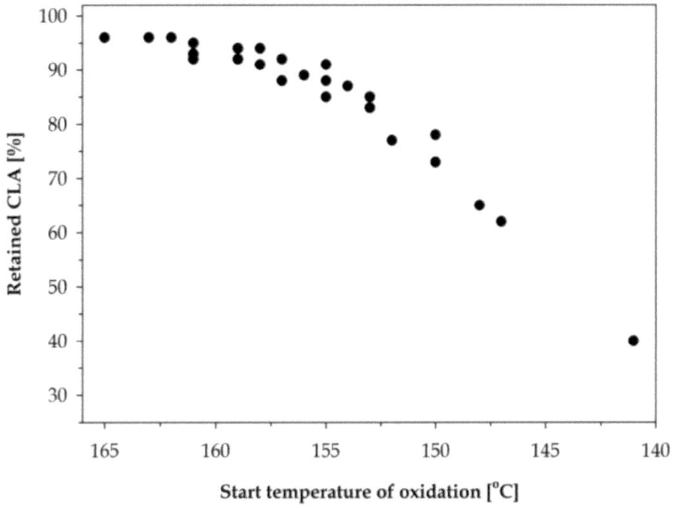

Figure 12. Influence of CLA retention on the start temperature (T_s) of oxidation in anhydrous milk fat rich in CLA.

A possible reason for the CLA-T_s relationship is that CLA can act as an antioxidant capturing those free radicals responsible for the lipid oxidation. This antioxidant behavior was clearly demonstrated in an earlier study by inducing lipid oxidation of fish oil by adding tert-butyl hydroperoxide. CLA effectively reduces lipid peroxidation as measured by chemiluminescence [86]. In addition, the antiradical or scavenging ability of CLA isomers measured by DPPH (2,2-diphenyl-1-picryl-hydrazyl-hydrate) technique depended on the CLA concentration [87].

8. Conclusions

The use of DSC to analyse lipid oxidation is a reliable, simple and convenient technique. It provides qualitative and quantitative information and offers unique advantages, such as the small amount of sample use, short test time, and good reproducibility. Additionally, the effectiveness of a particular antioxidant can be evaluated using DSC, measuring changes in the oxidation onset times or temperatures. The data obtained from DSC lipid oxidation have been extensively studied and correlated with other oxidation methods, such as the Rancimat method, PV, spectrophotometric and GC analysis of fats and oils from various sources.

A proper interpretation of DSC oxidation experiments should include sample composition, kinetic parameters (E_a, A, and k) in combination with the compensation theory. In multicomponent systems, the fat sometimes needs to be extracted from the matrix. Therefore, the lipid oxidative behaviour might be different from its original matrix. The results obtained from DSC oxidation depend upon the conditions used to prepare the sample and the heating protocol used. Factors, such as degree of saturation, amount of free fatty acids, chain length and the presence of natural antioxidants influence the oxidative stability and kinetic parameters. DSC can be coupled with other analytical techniques, such as GC, NMR, HPLC, etc, to provide a better description of the oxidative stability of fats and oils.

Author details

M.D.A. Saldaña and S.I. Martínez-Monteagudo
Department of Agricultural, Food and Nutritional Science, University of Alberta, Edmonton, AB, Canada

Acknowledgement

The authors thank to Alberta Livestock and Meat Agency Ltd. (ALMA) and to the Natural Sciences and Engineering Research Council of Canada (NSERC) for funding this project. Martinez-Monteagudo expresses his gratitude to Consejo Nacional de Ciencia y Tecnologia (CONACYT, Mexico) and Instituto de Inovacion y Transferencia Tecnologica (I^2T^2, Mexico) for the financial support (nr 187497).

9. References

[1] Kamal-Eldin A, Pokorn J (2005) Lipid oxidation products and methods used for their analysis. In Kamal-Eldin A, Pokorny J, editors. Analysis of lipid oxidation. AOCS Publishing, eBook ISBN: 978-1-4398-2239-5.

[2] O'Connor, TP, O'Brien NM (2006) Lipid oxidation. In Fox, PF, McSweeney P, editors. Advanced Dairy Chemistry, Volume 2: Lipids (3rd Edition). Springer-Verlag. pp. 557-600.

[3] Gray JI (1978) Measurement of lipid oxidation: A review. J. Am. Oil Chem. Soc. 55:539-546.

[4] Bradley DG, Min DB (1992) Singlet oxygen of foods. Crit. Rev. Food Sci. Nutr. 31:211-236.

[5] Shahidi F, Zhong Y (2010) Lipid oxidation and improving the oxidative stability. Chem. Soc. Rev. 39:4067-4079.

[6] Labuza TP (1971) Kinetics of lipid oxidation. Crit. Rev. Food Technol. 2:355-404.

[7] Adhvaryu A, Erhan SZ, Perez JM (2000) Oxidation kinetic studies of oils derived from unmodified and genetically modified vegetables using pressurized differential scanning calorimetry and nuclear magnetic resonance spectroscopy. Thermochim. Acta. 364: 97-97.

[8] Privett,OS, Blank ML (1962) Initial stages of autoxidation. J. Am. Oil Chem. Soc. 39: 465-468.

[9] Martínez-Monteagudo SI, Saldaña MDA, Kennelly JJ (2012) Kinetics of non-isothermal oxidation of anhydrous milk fat rich conjugated linoleic acid using differential scanning calorimetry. J. Therm. Anal. Calorim. 107: 973-981.

[10] Choe E, Min DB (2006) Chemistry and reactions of reactive oxygen species in foods. Crit. Rev. Food Sci. 46: 1-22.

[11] Brimberg UI, Kamal-Eldin A (2003) On the kinetics of the autoxidation of fats substrates with conjugated double bonds. Eur. J. Lipid Sci. Tech. 105: 17-22.

[12] Hamalainen TI, Sundberg S, Makkinen M, Hase T, Hopia A (2001) Hydroperoxide formation during autoxidation of conjugated linoleic acid methyl ester. Eur. J. Lipid Sci. Tech. 103: 588-593.

[13] Kenar JA, Gunstone FD, Knothe G (2007) Chapter 8: Chemical properties. In Harwood JL, Frank D, Gunstone FD, Dijkstra AJ, editors. The lipid handbook (3rd Edition). CRC Press. pp. 535-590.

[14] Lopez C, Lavigne F, Lesieur P, Bourgaux C, Ollivon M (2001) Thermal and structural behavior of milk fat. 1. Unstable species of anhydrous milk fat. J. Dairy Sci. 84: 756-766.

[15] Jensen RG (2002) The composition of bovine milk lipids: January 1995 to December 2000. J. Dairy Sci. 85: 295-300.

[16] Pokorny J (2005) Volumetric analysis of oxidized lipids. In Kamal-Eldin A, Pokorny J, editors. Analysis of lipid oxidation. AOCS Publishing.

[17] Simon P (2006) Induction periods theory and applications. J. Therm. Anal. Calorim. 84: 263-270.

[18] Fasina OO, Craig-Schmidt M, Colley Z, Hallman H (2008) Predicting melting characteristics of vegetable oils from fatty acid composition. LWT – Food Sci. Technol. 41: 1501-1505.

[19] Wirkowska M, Ostrowska-Ligeza E, Gorska A, Koczon P (2012) Thermal properties of fats extracted from powdered baby formulas. J. Therm. Anal. Calorim. 110:137-143

[20] Kousksou T, Jamil A, Zeraouli Y, Dumas JP (2006) DSC study and computer modelling of the melting process in ice slurry. Thermochim. Acta. 448: 123-129.

[21] Castello ML, Dweck J, Aranda DAG (2011) Kinetic study of thermal processing of glycerol by thermogravimetry. J. Therm. Anal. Calorim. 105:737-746

[22] Roy S, Riga AT, Alexander KS (2002) Experimental design aids the development of differential scanning calorimetry standard test procedure for pharmaceuticals. Thermochim. Acta. 392-393:399-404.

[23] ISO 11357-6:2008 (E) Plastics – differential scanning calorimetry (DSC) – Part 6: Determination of oxidation induction time (isothermal OIT) and oxidation induction temperature (dynamic OIT).

[24] Tsang W, Walker JA (1985) Characterization of RDF properties through high pressure differential scanning calorimetry. Resour Conserv. 9:355-365.

[25] Adhvaryu A, Perez JM, Singh ID (1999) Application of quantitative NMR spectroscopy to oxidation kinetics of base oils using pressurized differential scanning calorimetry technique. Energ Fuels. 13:493-498.

[26] Butrow AB, Seyler RJ (2003) Vapor pressure by DSC: extending ASTM E 1782 below 5 kPa. Thermochim. Acta. 402:145-152.

[27] Kousksou T, Jamil A, El-Omari K, Zeraouli Y, Le Guer Y (2011) Effect of heating rate and simple geometry on the apparent specific heat capacity: DSC applications. Themochim. Acta. 519:59-64.

[28] Litwinienko G, Daniluk A, Kasprzycka-Guttman T (2000) Study on autoxidation kinetics of fats by differential scanning calorimetry. 1. Saturated C_{12}-C_{18} fatty acids and their esters. Ind. Eng. Chem. Res. 39:1.12.

[29] Vyazovkin S, Wight CA (1997) Kinetics in solids. Annu. Rev. Phys. Chem. 48:125-149.

[30] Bouzidi L, Boodhoo M, Humphrey KL, Narine SS (2005) Use of the first and second derivatives to accurately determine key parameters of DSC thermographs in lipid crystallization studies. Thermochim. Acta. 439:94-102.

[31] Šimon P (2007) The single-step approximation attributes, strong and weak sides. J. Therm. Anal. Calorim. 88:709-715.

[32] Vyazovkin S, Wight CA (1999) Model-free and model-fitting approaches to kinetic analysis of isothermal and non-isothermal data. Thermochim. Acta. 341:53-68.

[33] Šimon P, Kolman L, Niklova I, Schmidt S (2000) Analysis of the induction period of oxidation of edible oils by differential scanning calorimetry. J. Am. Oil Chem. Soc. 77:639-642.

[34] Šimon P, Kolman L (2001) DSC study of oxidation induction periods. J. Therm. Anal. Calorim. 64:813-820.

[35] Schmid M, Ritter A, Affolter S (2006) Determination of oxidation induction time and temperature by DSC. J. Therm. Anal. Calorim. 2:367-371.

[36] Šimon P, Thomas PS, Okuliar J, Ray AS (2003) An incremental integral isoconversional method determination of activation parameters. J. Therm. Anal. Calorim. 72:867-874.

[37] Šimon P, Hynek D, Malikova M, Cibulkova Z (2008) Extrapolation of accelerated thermooxidative tests to lower temperatures applying non-Arrhenius temperature functions. J. Them. Anal. Calorim. 3:817-821.

[38] Šimon P (2009) Material stability predictions applying a new non-Arrhenian temperature function. J. Therm. Anal. Calorim. 2:391-396.

[39] Schmid M, Affolter S (2003) Interlaboratory tests on polymers by differential scanning calorimetry (DSC): determination and comparison of oxidation induction time (OIT) and oxidation induction temperature (OIT*). Polym. Test. 22:419-428.

[40] Cross CK (1970) Oil stability: A DSC alternative for the active oxygen method. J. Amer. Oil Chem. Soc. 47:229-230.

[41] Yamazaki M, Nagao A, Komomiya K (1980) High pressure differential thermal analysis of fatty acid methyl esters and triglycerides. J. Amer. Oil Chem. Soc. 57:59-60.

[42] Raemy A, Froelicher I, Loeliger J (1987) Oxidation of lipids studied by isothermal heat flux calorimetry. Thermochim. Acta. 114:159-164.

[43] Arain S, Sherazi STH, Bhanger MI, Talpur FN, Mahesar SA (2009) Oxidative stability assessment of *Bauhinia purpurea* seed oil in comparison to tow conventional vegetable oils by differential scanning calorimetry and Rancimat methods. Thermochim. Acta. 484:1-3.

[44] Pardauil JJR, Souza LKC, Molfetta F, Zamian JR, Filho GNR, da Costa CEF (2011) Determination of the oxidative stability by DSC of vegetable oils from the Amazonian area. Bioresource Technol. 102:5873-5877.

[45] Gromodzka J, Wardencki W, Pawlowicz R, Muszyn G (2010) Photoinduced and thermal oxidation of rapeseed and sunflower oils. Eur. J. Lipid Sci. Tech. 112:1229-1235.

[46] Velasco J, Adersen ML, Skibsted LH (2004) Evaluation of oxidative stability of vegetable oils by monitoring the tendency to radical formation. A comparison of electron spin resonance spectroscopy with the Rancimat method and differential scanning calorimetry. Food Chem. 85:623-632.

[47] Kowalski B (1989) Determination of oxidative stability of edible vegetable oils by pressure differential scanning calorimetry. Thermochim. Acta. 156:347-358.

[48] Kowalski B, Ratusz K, Miciula A, Krygier K (1997) Monitoring of rapeseed oil autoxidation with a pressure differential scanning calorimeter. Themochim. Acta. 307:177-121.

[49] Rudnik E, Szczucinska A, Gwardiak H, Szulc A, Winiarska A (2001) Comparative studies of oxidative stability of linseed oil. Thermochim. Acta. 370:135-140.

[50] Kowaslki B, Ratusz K, Kowalska D, Bekas W (2004) Determination of the oxidative stability of vegetable oils by differential scanning calorimetry and Rancimat measurements. Eur. J. Lipid Sci. Tech. 106:165-169.

[51] Tan CP, Che Man YB, Selamat J, Yusoff MSA (2001) Application of Arrhenius kinetics to evaluate oxidative stability in vegetable oils by isothermal differential scanning calorimetry. J. Am. Oil Chem. Soc. 78:1133-1138.

[52] Tan CP, Che Man YB, Selamat J, Yusoff MSA (2002) Comparative studies of oxidative stability of edible oils by differential scanning calorimetry and oxidative stability index methods. Food Chem. 76:385-389.

[53] Litwinienko G, Kasprzycka-Guttman, Jarosz-Jarszewska M (1995) Dynamic and isothermal DSC investigation of the kinetic of thermooxidative decomposition of some edible oils. J. Therm. Anal. 45:741-750.

[54] Litwinienki G, Daniluka A, Kasprzycka-guttman T (1999) A differential scanning calorimetry study on the oxidation of C_{12}-C_{18} saturated fatty acids and their esters. J. Am. Oil Chem. Soc. 76:655-657.

[55] Cifti ON, Kowalski B, Gogus F, Fadiloglu S (2009) Effect of the addition of a cocoa butter-like fat enzymatically produced from olive pomace oil on the oxidative stability of cocoa butter. J. Food Sci. 4:E184-E190.

[56] Litwinienko G (2001) Autoxidation of unsaturated fatty acids and their esters. J. Therm. Anal. Calorim. 65:639-646.

[57] Litwinienko G, Kasprzycka-Guttman T, Studzinski M (1997) Effects of selected phenol derivatives on the autoxidation of linolenic acid investigated by DSC non-isothermal methods. Thermochim. Acta. 307:97-106.

[58] Ulkowski M, Musialik M, Litwinienko G (2005) Use of differential scanning calorimetry to study lipid oxidation. 1. Oxidative stability of lecithin and linolenic acid. J. Agric. Food. Chem. 53:9073-9077.

[59] Litwinienko G, Kasprzycka-Guttman T (1998) A DSC study on thermoxidation kinetics of mustard oil. Thermochim. Acta. 319:185-191.

[60] Ostrowska-Ligeza E, Bekas W, Kowalska D, Lobacz M, Wroniak M, Kowalski B (2010) Kinetics of commercial olive oil oxidation: Dynamic differential scanning calorimetry and Rancimat studies. Eur. J. Lipid Sci. Tech. 112:268-274.

[61] Jimenez-Marquez J, Beltran-Maza G (2003) Aplicación de la calorimetría diferencial de barrido (CDB) en la caracterización del aceite de oliva virgen. Grasas Aceites. 54:403-409.

[62] Vittadini E, Lee JH, Frega NG, Min DB, Vodovotz Y (2003) DSC determination of thermally oxidized olive oil. J. Am. Oil Chem. Soc. 8:533-537.

[63] Thurgood J, Ward R, Martini S (2007) Oxidation kinetics of soybean oil/anhydrous milk fat blends: A differential scanning calorimetry study. Food Res. Int. 40:1030-1037.

[64] Litwinienko G, Kasprzycka-Guttman T (2000) Study on the autoxidation kinetics of fat components by differential scanning calorimetry. 2. Unsaturated fatty acids and their esters. Ind. Eng. Chem. Res. 39:13-17.

[65] Polavka J, Paligova J, Cvengroš J, Šimon P (2005) Oxidation stability of methyl esters studied by differential thermal analysis and Rancimat. J. Am. Oil Chem. Soc. 82:519-524.

[66] Quinchia LA, Delgado MA, Valencia C, Franco JM, Gallegos C (2011) Natural and synthetic antioxidant additives for improving the performance of new biolubricant formulations. J. Agr. Food Chem. 59:12917-12924.

[67] Kowalski B (1991) Thermal-oxidative decomposition of edible oils and fats. DSC stidies. Thermochim. Acta. 184:49-57.

[68] Musialik M, Litwinienko G (2007) DSC study of linolenic acid autoxidation inhibited by BHT, dehydrozingerone and olivetol. J. Therm. Anal. Calorim. 88:781-785.

[69] Agrawal RK (1989) The compensation effect – a fact or a fiction. J. Therm. Anal. 35:909-917.

[70] Agrawal RK (1986) On the compensation effect. J. Therm. Anal. 31:73-86.

[71] Lopez C, Lavigne F, Lesieur P, Bourgaux C, Ollivon M (2001) Thermal and structural behavior of milk fat. 1. Unstable species of anhydrous milk fat. J. Dairy Sci. 84:756-766.

[72] Jensen RG (2002) The composition of bovine milk lipids: January 1995 to December 2000. J. Dairy Sci. 85:295-300.

[73] Lock AL, Bauman DE (2004) Modifying milk fat composition of dairy cows to enhance fatty acids beneficial to human health. Lipids. 39:1197-1206.

[74] Bell JA, Griinari JM, Kennelly JJ (2006) Effect of safflower oil, flaxseed oil, monensin, and vitamin E on concentration of conjugated linoleic acid in bovine milk fat. J. Dairy Sci. 89:733-748.

[75] Martinez-Monteagudo SI, Khan M, Saldaña MDA, Temelli F (2011) Obtaining a milk fraction rich in conjugated linoleic acid. Abstract 45, In 24th Canadian Conference on Fats and Oils, September 26-27th, Edmonton, AB, Canada.

[76] Prado GHC, Khan M, Saldaña MDA, Temelli F (2012) Enzymatic hydrolysis of conjugated linoleic acid-enriched anhydrous milk fat in supercritical carbon dioxide. J. Supercrit. Fluids. 66:198-206.

[77] Campbell W, Drake MA, Larick DK (2003) The impact of fortification with conjugated linoleic acid (CLA) on the quality of fluid milk. J. Dairy Sci. 86:43-51.

[78] Mathys A, Reineke K, Heinz V, Knorr D (2009) High pressure thermal sterilization – development and application of temperature controlled spore inactivation studies. High Pressure Res. 29:3-7.

[79] Knoerzer K, Buckow R, Versteeg C (2010) Adiabatic compression heating coefficients for high-pressure processing – a study of some insulating polymer materials. J. Food Eng. 98:110-119.

[80] Wilson DR, Dabrowski L, Stringer S, Moezelaar R, Brocklehurst TF (2008) High pressure in combination with elevated temperature as a method for the sterilization of food. Trends Food Sci. Tech. 19:289-299.

[81] Oley I, Verlinde P, Hendrickx M, Van Loey A (2006) Temperature and pressure stability of L-ascorbic acid and/or [6s] 5-methyltetrahydrofolic acid: A kinetic study. Eur. Food Res. Tech. 223:71-77.

[82] Butz P, Serfert Y, Fernandez-Garcia A, Dieterich S, Lindauer R, Bognar A, Tauscher B (2004) Influence of high-pressure treatment at 25°C and 80°C on folates in orange juice and model media. J. Food Sci. 69:117-121.

[83] Matser AA, Krebbers B, van den Berg RW, Bartels PV (2004) Advantages of high pressure sterilisation on quality of food products. Trends Food Sci. Tech. 15:79-85.

[84] Verbeyst L, Oey I, Van der Plancken I, Hendrickx M, Van Loey A (2010) Kinetic study on the thermal and pressure degradation of anthocyanins in strawberries. Food Chem. 123:269-274.

[85] Martinez-Monteagudo SI, Saldaña MDA, Torres JA, Kennelly JJ. (2012). Effect of pressure-assisted thermal sterilization on conjugated linoleic acid (CLA) content in CLA-enriched milk, Innov Food Sci Emerg Technol, doi 10.1016/j.ifset.2012.07.004.

[86] Martinez-Monteagudo SI, Saldaña MDA (2011) Changes in conjugated linoleic acid (CLA) and antiradical ability of anhydrous milk fat treated with pressure assisted thermal sterilization. Abstract 43, In 24th Canadian Conference on Fats and Oils, September 26-27th, Edmonton, AB, Canada.

[87] Fagali N, Catala A (2008) Antioxidant activity of conjugated acid isomers, linoleic acid and its methyl ester determined by photoemission and DPPH techniques. Biophys. Chem. 137:56-62.